PRECALCULUS
MATHEMATICS

PRECALCULUS MATHEMATICS

Vivian Shaw Groza
Susanne Shelley
Sacramento City College

Holt, Rinehart and Winston, Inc.
New York   Chicago   San Francisco
Atlanta   Dallas   Montreal   Toronto
London   Sydney

Copyright © 1972 by Holt, Rinehart and Winston, Inc.
All Rights Reserved
Library of Congress Catalog Card Number: 76–158479
ISBN: 0-03-077670-8
AMS 1970 Subject Classification 00–01
Printed in the United States of America
2 3 4 5 038 1 2 3 4 5 6 7 8 9

PREFACE

This book is intended as a text for an advanced course in mathematics, preceding or concurrent with an introductory course in calculus. It is assumed that the student has achieved competence in the subjects of intermediate algebra and trigonometry.

Since the modern trend is to require an additional course beyond intermediate algebra and trigonometry as preparation for calculus, the objective of this book is to provide a strong foundation for calculus and to develop insights and skills pertaining to algebraic structures. Most intermediate algebra courses omit many algebraic topics that the student should know before studying calculus. Among these topics are adequate coverage of inequalities and absolute values, solution of polynomial equations of degree greater than 2, mathematical induction, the binomial theorem, partial fractions, matrices and determinants, and sequences and series. All of these topics are treated in detail in this text. Since many calculus courses presuppose a knowledge of analytic geometry, this text provides a background in plane and solid analytic geometry, and polar coordinates.

The text also includes topics from abstract algebra, such as the Peano Postulates and related theorems, a brief exposition of groups, rings, and integral domains, and the solution of cubic and quadratic equations which are often omitted in precalculus courses due to a lack of time. Sections dealing with these topics are marked "optional" and are designed to motivate and challenge the superior student.

The exercises are designed to promote competence in and understanding of all kinds of real and imaginary numbers and all varieties of algebraic expressions. Concepts or manipulations developed in one chapter are consistently reinforced in subsequent chapters. In view of the nature of the text, it is assumed that the student is able to apply various techniques learned in intermediate algebra and trigonometry before they are formally treated in this text.

Answers to most of the odd-numbered exercises can be found in the back of the book. The more difficult exercises are designated by a star and are included as motivation for the superior student.

The presentation is modern, with the exposition embedded within the framework of set theory and logic. Chapter 1 presents those topics that are sufficient for an understanding of the text and could be omitted if the students are sufficiently well-versed in these topics. The course outline on page vii suggests a selection of topics for a minimum course of 45 instructional days, for a refresher and enrichment course in algebra and trigonometry, for a semester course of 80 instructional days in precalculus mathematics with analytic geometry, or for a one-year course in freshman mathematics. The text may thus be adapted to a variety of curricula. It also meets the needs of mathematics majors and satisfies the CUPM recommendations for a precalculus course.

The authors would like to express their appreciation to J. Dorsett, St. Petersburg Junior College, St. Petersburg, Florida, Doris Stockton, University of Massachusetts, Amherst, Massachusetts, and Kenneth Humphreys, Sacramento City College, Sacramento, California, who offered many helpful suggestions with the preparation of this text.

V.S.G.
S.S.

September 1971
Sacramento, California

Suggested Course Outline

Precalculus Mathematics with Analytic Geometry	Review of Algebra and Trigonometry	Precalculus Mathematics and Analytical Geometry	Precalculus Mathematics and Analytical Geometry
Minimum course 45 days		1 semester–daily or 2 quarters	1 year 2 semesters or 3 quarters
Chapter 1 Chapter 5 Chapter 7 Chapter 8 Chapter 9 Sections 9.1–9.8 Chapter 11 Chapter 12 (Omit all optional sections.)	Chapters 1–7 Chapter 11 Chapter 12 (Omit all optional sections.)	All chapters except Chapter 6, omitting optional sections. (It is assumed that the student has competence in logarithmic and trigonometric functions.) If the class is very well prepared, some review sections could be omitted in favor of optional sections, particularly for math majors.	All chapters including optional sections.

CONTENTS

Preface   v

1 SOME BASIC CONCEPTS   **1**

    1.1   Sets   2
    1.2   Logic; TF-Statements, Open Statements, Quantifiers, Truth Sets   11
    1.3   Logic; Connectives and Truth Tables   17
    1.4   Logic; Tautologies, Negations of Open Statements   24
        Chapter Summary   30
        Review Exercises   31

2 THE NATURAL NUMBERS   **34**

    2.1   Relations, Functions, Operations   36
    2.2   The Peano Postulates (Optional)   42
    2.3   Theorems of the Peano System (Optional)   47
    2.4   Order   48
    2.5   Primes, Subtraction, Division   52
        Chapter Summary   59
        Review Exercises   61

3 THE INTEGERS   **63**

    3.1   The Set of Integers   64
    3.2   Groups, Rings, Integral Domains (Optional)   72
    3.3   Polynomials and Factoring   79
    3.4   Division Algorithm; Synthetic Division   83

	3.5	Factor and Remainder Theorems; Applications to Equations	88
		Chapter Summary	94
		Review Exercises	95

4 THE RATIONALS 98

4.1	The Set of Rationals, Q	99
4.2	Fields (Optional)	106
4.3	Rational Functions	108
4.4	Partial Fractions	116
4.5	Solution of Rational Equations	122
	Chapter Summary	126
	Review Exercises	128

5 THE REALS 130

5.1	The Set of Real Numbers	130
5.2	Order Theorems and Inequalities	140
5.3	Additional Order Theorems and Inequalities	146
5.4	Absolute Value	151
5.5	Integral Exponents	155
5.6	Radicals	162
5.7	Rational Exponents	166
5.8	Radical Equations and Inequalities	171
	Chapter Summary	176
	Review Exercises	180

6 EXPONENTIAL, LOGARITHMIC, AND TRIGONOMETRIC FUNCTIONS 182

6.1	Exponential and Logarithmic Functions	184
6.2	Logarithmic Functions (Continued)	193
6.3	Natural Trigonometric Functions	205
6.4	Natural Trigonometric Functions (Continued)	217
6.5	Inverse Trigonometric Functions	225
	Chapter Summary	231
	Review Exercises	234

7 THE COMPLEX NUMBERS 236

7.1	The Complex Number System	238
7.2	Polar Form	246
7.3	Powers and Roots; DeMoivre's Theorem	253
7.4	Graphic Representations (Optional)	258
	Chapter Summary	264
	Review Exercises	267

8 THEORY OF EQUATIONS 269

8.1	Quadratic Equations; Fundamental Theorem of Algebra; Roots and Factors	269
8.2	Rational Roots and Special Pairs of Roots	278

CONTENTS   xi

	8.3	Upper and Lower Bounds; Descartes' Rule of Signs	284
	8.4	Weierstrass Zero Theorem; Approximations	290
	8.5	Algebraic Solution of the Cubic Equation (Optional)	294
	8.6	Trigonometric Solution of the Cubic Equation (Optional)	302
	8.7	The Quartic Equation (Optional)	306
		Chapter Summary	309
		Review Exercises	311

9 PLANE ANALYTIC GEOMETRY   314

	9.1	Rectangular Coordinates; Distance; Midpoint	315
	9.2	Slope; Parallel and Perpendicular Lines	320
	9.3	Analytic Proofs	325
	9.4	The Straight Line	329
	9.5	Intercepts; Extent; Symmetry	333
	9.6	Circles and Standard Parabolas	339
	9.7	Ellipses and Hyperbolas	345
	9.8	Translation of the Axes	356
	9.9	Rotation of the Axes	360
	9.10	Polar Coordinates	367
	9.11	Parametric Equations	375
	9.12	Two-Dimensional Vectors	380
		Chapter Summary	388
		Review Exercises	390

10 SOLID ANALYTIC GEOMETRY   392

	10.1	Rectangular Coordinates; Distance; Midpoint	392
	10.2	Lines in Space	398
	10.3	Three-Dimensional Vectors	407
	10.4	Planes in Space	414
	10.5	Spheres; Cylinders; Cones	421
		Chapter Summary	426
		Review Exercises	427

11 SYSTEMS OF EQUATIONS   430

	11.1	Linear Systems	430
	11.2	Matrices	438
	11.3	Matrix Solution of Linear Systems	445
	11.4	Determinants	451
	11.5	Determinant Solution of Linear Systems	459
	11.6	Quadratic Systems	468
	11.7	The Resultant (Optional)	477
		Chapter Summary	481
		Review Exercises	483

12 SEQUENCES AND SERIES   485

	12.1	General Discussion	485
	12.2	Arithmetic Progressions	489
	12.3	Geometric Progressions	492

12.4	The Binomial Theorem	497
12.5	Generalized Binomial Expansion	501
	Chapter Summary	504
	Review Exercises	505

APPENDIX 507

Table I   Powers and Roots   509
Table II   Common Logarithms   510
Table III   Trigonometric Functions   512
Proof of the Partial Fraction Theorem   517

ANSWERS 521

Index   605

PRECALCULUS
MATHEMATICS

1
SOME BASIC CONCEPTS

A mathematical system is a logically ordered set of undefined concepts, definitions, assumptions, and theorems. Elementary algebra and Euclidean geometry are examples of mathematical systems. Very often, but not always, a mathematical system is created by abstracting specialized subject matter from the real world. Similar properties are generalized and individual differences or dissimilarities are omitted. After a system has been developed, the theorems may serve as predictions for the original specialization. Moreover, the mathematical system, since it is abstract, may be reinterpreted and applied in a variety of ways in the real world.

For example, Boolean algebra was created from the subject matter of logic. It has also been applied to set theory and, as a most surprising and practical development, it was seen to be a description of the theory of electrical circuits.

In a mathematical system, undefined concepts are stated first. Since it is impossible to define every word without becoming involved in a circular process, the simplest or most elementary terms are designated as the **undefined concepts**, or **primitives**. The selection of these terms is often motivated by the special subject matter, but in the mathematical system no specialized meaning is ascribed. This permits an objective development and a wide area for applications.

A **definition** is an agreement to use words, phrases, or symbols as substitutes for other words, phrases, or symbols. Definitions are usually made to achieve simplicity, to promote clarity, or to reduce the amount of writing necessary.

An order is maintained in the list of definitions by restricting the words used in the definition to be primitive terms or terms that have been previously defined.

The statements of a mathematical system are ordered by a deductive reasoning process that involves the addition of a third statement to two other statements. Thus, a logical theory must begin with at least two general statements which are assumed true. These statements that are assumed true are called **assumptions**, **axioms**, or **postulates**. Today these words are considered equivalent in mathematic meaning.

The statements that are obtained from the assumptions by the process of deductive reasoning are called **theorems**. A theorem is referred to as a **proved** statement.

1.1 SETS

The concept of a set is fundamental in mathematics. It is especially useful in developing the subject matter of elementary algebra.

1.1.1 The Set Concept

A set is a well-defined collection of objects called " elements " or " members " of the set.

A collection is **well-defined** if it is always possible to determine whether or not a particular object or element belongs to the set.

Set membership is indicated by the notation $a \in S$, which means that the element a is a member of the set S.[1]

For example, if S is the set of counting numbers and if $a = 5$, then $5 \in S$ means that 5 is a counting number.

1.1.2 Methods for Defining Particular Sets

Particular sets may be defined by two general methods: the listing method and the description method.

In the **listing method**, the set is defined by listing or stating the names of its members enclosed by braces.

For example,
$$A = \{\text{red, white, blue}\}$$
means "A is the set whose members are the colors: red, white, and blue."

---

[1] Sets are usually denoted by capital letters. However, lower-case letters such as r, f, and g are also used as the names of sets.

In the **description** method, the set is defined by describing the set or stating a property possessed by each member of the set and by those members only.

For example,

$A =$ the set of colors on the flag of the United States.

A formal variation of the description method is often referred to as the **rule** or **set-builder** method.

For example,

$S = \{x \mid x = 3n,$ where n is an integer$\}$

$T = \{(x, y) \mid 2x - 5y = 10,$ where x and y are real numbers$\}$.

The first example above is read "S is the set of all numbers, x, such that x equals the product of 3 and n, where n is an integer." (The vertical bar | means "such that.")

The second example is read "T is the set of all ordered pairs, (x, y), such that $2x$ minus $5y$ equals 10, where x and y are real numbers."

The triple dot notation, ..., is useful in listing very large sets. For example, $\{1, 2, 3, \ldots, 99, 100\}$ indicates the set consisting of the first hundred counting numbers. Use of the three dots instructs to continue according to the designated pattern.

The set $\{2, 4, 6, 8, 10, \ldots\}$ indicates the set of all even positive integers. The three dots at the end mean that the pattern is to be continued without ending.

1.1.3 Special Sets

The **empty set** is the set that has no members. It is denoted symbolically by $\emptyset$ or by $\{\ \}$.

For example, the set of human beings over 15 feet tall is the empty set. Also, $\emptyset = \{x \mid x \neq x\}$.

A **universal set**, U, is a set to which the elements of all other sets in a particular discussion must belong.

For example, let $U = \{1, 2, 3, 4, 5, 6, 7, 8, 9\}$. Now, if A is the set of multiples of 3, then $A = \{3, 6, 9\}$. The number 12 can not be a member of A because 12 does not belong to the universal set U that was selected. In a particular discussion, the universal set is always stated.

1.1.4 Set Relations

SUBSETS

Set A is a subset of set B if every element of A is an element of B. In symbols, $A \subset B$.[2]

---

[2] Some authors prefer to write $A \subseteq B$, reserving the notation $A \subset B$ to mean A is a proper subset of B.

For example,
$$\{3, 6, 9\} \subset \{1, 2, 3, 4, 5, 6, 7, 8, 9\}.$$

If A is a subset of B and does not consist of all of the elements of B, then A is called a **proper** subset of B. The set B is called an **improper** subset of itself. The empty set is considered to be a subset of every set. This does not violate the definition of subset because there are no elements in $\emptyset$ which are not in every other set.

EQUAL SETS

Two sets, A and B, are said to be equal, or identical, if and only if $A \subset B$ and $B \subset A$. In symbols, $A = B$.

For example, $\{3, 6, 9\} = \{9, 6, 3\}$. Note that the order in which the elements are written is not important; it is only necessary that the two sets have exactly the same objects in them.

Also, $\{3, 3, 6, 9\} = \{3, 6, 9\}$. An element does not have to be listed more than once. In general, the shorter form is preferred.

EQUIVALENT SETS

Two sets are said to be equivalent if and only if their elements can be placed in **one-to-one correspondence**; that is, each element of the first set can be paired with exactly one element of the second set and each element of the second set can be paired with exactly one element of the first set. In symbols, $A \leftrightarrow B$.

For example, $\{\text{red, white, blue}\} \leftrightarrow \{3, 6, 9\}$ because the elements can be "paired off" by a one-to-one correspondence such as

$$\begin{array}{ccc} \text{red}, & \text{white}, & \text{blue} \\ \updownarrow & \updownarrow & \updownarrow \\ 3, & 6, & 9 \end{array}$$

(Five other pairings are possible for this case, but only one needs to be shown to demonstrate the equivalence.)

1.1.5 Finite and Infinite Sets

An **infinite set** is a set that can be placed in one-to-one correspondence with at least one of its proper subsets.

For example, the set of all counting numbers, $\{1, 2, 3, \ldots\}$ is an infinite set because it can be placed in one-to-one correspondence with its proper subset, the set of even counting numbers $\{2, 4, 6, \ldots\}$

$$\begin{array}{ccccc} 1, & 2, & 3, & \ldots, & n, & \ldots \\ \updownarrow & \updownarrow & \updownarrow & & \updownarrow \\ 2, & 4, & 6, & \ldots, & 2n, & \ldots \end{array}$$

SETS

A **finite set** is a set that is not infinite; that is, it cannot be placed in one-to-one correspondence with *any* of its proper subsets.

A finite set, S, can be assigned a counting number, $n(S)$, to designate the number of elements in the set S. For example, if $S = \{a, b, c, d\}$, then $n(S) = 4$. The empty set is assigned the number 0; $n(\emptyset) = 0$.

It is also possible to assign numbers to infinite sets. For example, the Hebrew letter, $\aleph_0$, (read "aleph null") has been assigned as the name of the number of the set $\{1, 2, 3, \ldots\}$, the set of all counting numbers.

1.1.6 Venn Diagram

Geometric figures may be used to illustrate sets. One model commonly used is obtained by letting the points inside a rectangle represent the universal set U and the points inside circles within the rectangle represent subsets of U.

In Figure 1.1, called a **Venn diagram** for two sets, the area inside the rectangle is divided into four compartments, each indicated by one of the numerals 1, 2, 3, or 4.

When there is no special information concerning sets A and B, this figure provides a general way to describe all possible relationships. For example, by considering compartment 3 to be empty, the Venn diagram illustrates the special case where A and B have no common element (Figure 1.2).

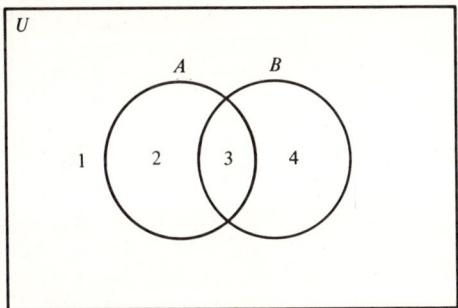

Figure 1.1 Venn Diagram for Two Sets

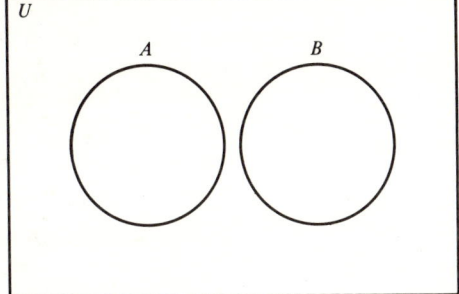

Figure 1.2 Sets with No Common Element

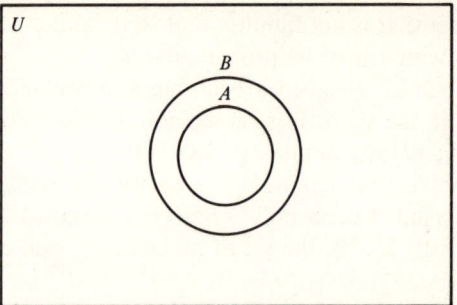

Figure 1.3 Set A Is a Subset of Set B

If compartment 2 is considered empty, then the Venn diagram illustrates the case where A is a subset of B (Figure 1.3).

Thus, by using the Venn diagram in Figure 1.1, and by reserving the right to consider one or more compartments empty, only one figure needs to be drawn to illustrate all possible cases.

A Venn diagram for three sets is illustrated in Figure 1.4. The area inside the rectangle is divided into 8 compartments, each indicated by one of the numerals 1, 2, 3, 4, 5, 6, 7, or 8.

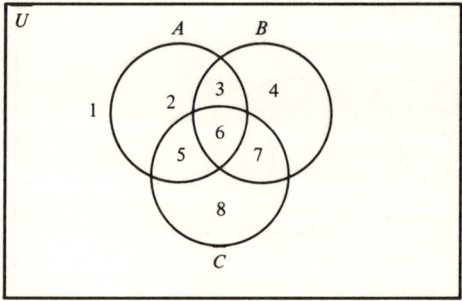

Figure 1.4 Venn Diagram for Three Sets

It should be noted that Venn diagrams are not used to prove theorems, but these diagrams serve as visual aids for understanding.

1.1.7 Set Operations

COMPLEMENT

The complement of a set A, $\bar{A}$, is the set of all elements in the universal set U that do *not* belong to A.[3]

---

[3] Other notations used for $\bar{A}$ are A', $\tilde{A}$, $\sim A$, or $-A$.

SETS   7

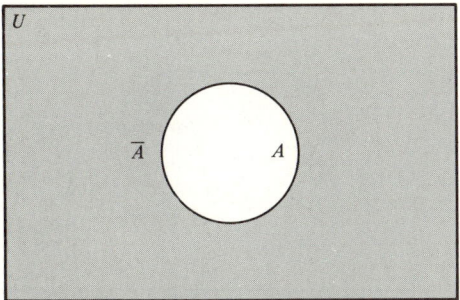

Figure 1.5 $\bar{A}$, Shaded

For example, if $U = \{1, 2, 3, 4, 5, 6, 7, 8, 9\}$ and $A = \{3, 6, 9\}$, then $\bar{A} = \{1, 2, 4, 5, 7, 8\}$.

The shaded area of the Venn diagram in Figure 1.5 illustrates $\bar{A}$.

RELATIVE COMPLEMENT

The relative complement of set A with respect to set B, $B - A$, is the set of elements in B which are not in A.

For example, if $A = \{3, 6, 9\}$ and $B = \{3, 4, 5, 6, 7\}$, then $B - A = \{4, 5, 7\}$. The shaded area of the Venn diagram in Figure 1.6 illustrates $B - A$.

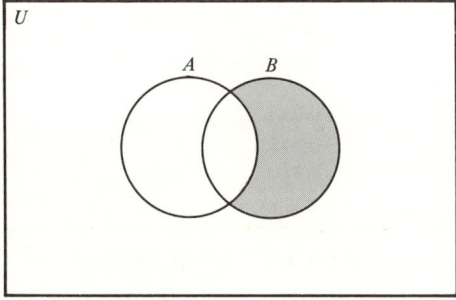

Figure 1.6 $B - A$, Shaded

$\bar{A}$, the complement of A, is also called the absolute complement of A and may be designated symbolically by $U - A$. The term absolute is used to distinguish $\bar{A}$ from $B - A$, the relative complement.

UNION

The union of two sets, $A \cup B$, is the set of all elements that are in A, or in B, or in both A and B.

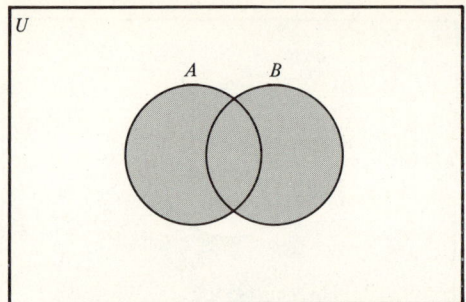

Figure 1.7 $A \cup B$, Shaded

For example, if $A = \{1, 2, 3, 4\}$ and $B = \{3, 4, 5\}$, then $A \cup B = \{1, 2, 3, 4, 5\}$. The shaded area of the Venn diagram in Figure 1.7 illustrates $A \cup B$.

INTERSECTION

The intersection of two sets, $A \cap B$, is the set of all elements that are in both A and B.

For example, if $A = \{1, 2, 3, 4\}$ and $B = \{3, 4, 5\}$, then $A \cap B = \{3, 4\}$. The shaded area of the Venn diagram in Figure 1.8 illustrates $A \cap B$.

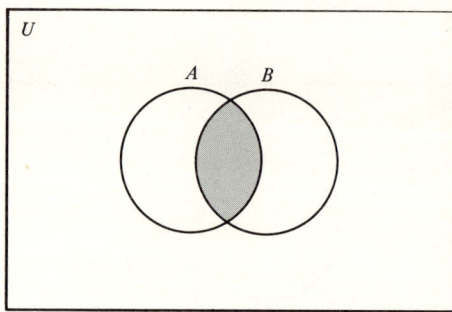

Figure 1.8 $A \cap B$, Shaded

EXERCISES 1.1

1–10. Use the listing method to describe each of the following sets, where

U = the set of digits = $\{0, 1, 2, 3, 4, 5, 6, 7, 8, 9\}$.

1. The set of even digits = $\{x \mid x = 2d, \text{ where } d \in U\}$
2. The set of odd digits = $\{x \mid x = 2d + 1, \text{ where } d \in U\}$
3. The set of digits less than 5 = $\{x \mid x < 5\}$
4. The set of digits greater than or equal to 6 = $\{x \mid x \geq 6\}$

SETS

5. $\{x \mid x = 3d \text{ where } d \in U\}$
6. $\{(x, y) \mid x \in U, y \in U \text{ and } x + 3y = 6\}$
7. $\{(x, y) \mid x \in U, y \in U \text{ and } x + 3y < 6\}$
8. $\{x \mid 3 < x < 7\} =$ the set of digits between 3 and 7
9. $\{(x, y) \mid 2x + y = 5, \text{ where } x \in U \text{ and } y \in U\}$
10. $\{(x, y) \mid 2x + y < 5, \text{ where } x \in U \text{ and } y \in U\}$

11–20. Use the rule or set-builder notation to describe each of the following sets, where
$$U = N = \text{the set of natural numbers} = \{1, 2, 3, 4, 5, \ldots\}.$$

11. $\{1, 2, 3, 4, 5\}$
12. $\{5, 6, 7, 8, 9\}$
13. $\{3, 6, 9, 12, \ldots\}$
14. $\{5, 10, 15, 20, \ldots\}$
15. $\{1, 4, 9, 16, 25, 36\}$
16. $\{12, 13, 14, 15, \ldots\}$
17. $\{(1, 2), (2, 4), (3, 6), (4, 8), \ldots\}$
18. $\{(3, 1), (6, 2), (9, 3), (12, 4), \ldots\}$
19. $\{2, 3, 5, 7, 11, 13, \ldots\} =$ the set of prime numbers
20. $\{4, 6, 8, 9, 10, 12, 14, 15, \ldots\} =$ the set of composite numbers

21–34. Let
$$U = \{-3, -2, -1, 0, 1, 2, 3\},$$
$$A = \{-2, -1, 0, 1\},$$
$$B = \{0, 1, 2\}.$$

Describe each of the following sets by the listing method.

21. $A \cup B$
22. $A \cap B$
23. $A - B$
24. $B - A$
25. $\bar{A}$
26. $\bar{B}$
27. $\bar{A} \cap \bar{B}$
28. $\bar{A} \cup \bar{B}$
29. $\overline{A \cup B}$
30. $\overline{A \cap B}$
31. $\bar{A} \cap B$
32. $A \cap \bar{B}$
33. $A \cup \bar{B}$
34. $\bar{A} \cup B$

35–40. Referring to the sets in Exercises 21–34, determine whether each of the following statements is true or false.

35. (a) $A - B = B - A$ (b) $A - B = A \cap \bar{B}$ (c) $B - A = B \cup \bar{A}$
36. (a) $\overline{A \cap B} = \bar{A} \cup \bar{B}$ (b) $\overline{A \cap B} = \bar{A} \cap \bar{B}$ (c) $\overline{A \cap B} = A \cup B$
37. (a) $\overline{A \cup B} = \bar{A} \cup \bar{B}$ (b) $\overline{A \cup B} = \bar{A} \cap \bar{B}$ (c) $\overline{A \cup B} = A \cap B$

38. (a) $B - A \subset B$ (b) $A \subset B - A$ (c) $B \subset B - A$
39. (a) $A \subset A \cup B$ (b) $A \subset A \cap B$ (c) $A \cap B \subset A$
40. (a) $A \cup B = B \cup A$ (b) $A \cap B = B \cap A$ (c) $A - B = B - A$

41–60. Describe each of the following by the set-builder method where

U = the set of integers = $\{\ldots, -3, -2, -1, 0, 1, 2, 3, \ldots\}$,

$A = \{x \mid x > -3\}$,
$B = \{x \mid x < 2\}$,
$C = \{x \mid -1 \leq x \leq 5\}$.

41. $A \cap B$
42. $B \cap C$
43. $A \cup B$
44. $A \cup C$
45. $B \cup C$
46. $A \cap C$
47. $\bar{A}$
48. $\bar{B}$
49. $\bar{C}$
50. $A - B$
51. $C - A$
52. $C - B$
53. $\bar{A} \cup B$
54. $\bar{A} \cap B$
55. $A \cap \bar{B}$
56. $A \cup \bar{B}$
57. $A \cap B \cap C$
58. $(A \cap B) \cup C$
59. $(A \cap B) \cup (\bar{A} \cap \bar{B})$
60. $(A \cap \bar{B}) \cup (\bar{A} \cap B)$

61–90. Let $U = R$ = the set of real numbers (the set of numbers that are in 1-1 correspondence with the points on a number line). Let Q = the set of rational numbers = $\{x/y \mid x$ and y are integers and $y \neq 0\}$. For each of the following: if the set is finite, list its elements; if the set is infinite, write "infinite" and list three elements in the set.

61. $\{x \mid -3 < x < 2$, where x is an integer$\}$
62. $\{x \mid -3 < x < 2$, where $x \in Q\}$
63. $\{(x, y) \mid 2x + y = 7\}$.
64. $\{(x, y) \mid 2x + y = 7$, where x and y are integers$\}$
65. $\left\{x \mid \dfrac{x}{6} \text{ is a natural number}\right\}$
66. $\left\{x \mid \dfrac{6}{x} \text{ is a natural number}\right\}$
67. $\{x \mid x - 3 = 0\} \cup \{x \mid 2x + 1 = 0\}$
68. $\{x \mid x^2 - 4 = 0\} \cup \{x \mid 3x - 2 = 0\}$
69. $\{x \mid x^2 - 4 = 0\} \cap \{x \mid x^3 + 8 = 0\}$
70. $\{(x, y) \mid x + y = 7\} \cap \{(x, y) \mid x - y = 3\}$
71. $\{(x, y) \mid 2x + y = 7\} \cap \{(x, y) \mid x + y = 5\}$
72. $\{(x, y) \mid xy = 6\} \cap \{(x, y) \mid y = x + 1\}$
73. $\{(x, y) \mid y = 2x\} \cap \{(x, y) \mid x^2 + y^2 = 20\}$
74. $\{(x, y) \mid x^2 - y^2 = (x - y)(x + y)\}$

75. $\{(x, y) \mid x^2 + y^2 = (x + y)^2\}$
76. $\{x \mid \sqrt{x + 6} = x\}$
77. $\{x \mid \sqrt{3x + 10} = x\}$
78. $\{x \mid \sqrt{3x - 2} = x\}$
79. $\{(x, y) \mid x^2 = 9\} \cap \{(x, y) \mid y^2 = 4\}$
80. $\{(x, y) \mid x^2 < 9\} \cap \{(x, y) \mid y^2 < 4\}$
81. $\{x \mid x - 2 > 0\} \cap \{x \mid x - 3 > 0\}$
82. $\{x \mid x - 2 > 0\} \cap \{x \mid x - 3 < 0\}$
83. $\{x \mid x - 2 < 0\} \cap \{x \mid x - 3 > 0\}$
84. $\{x \mid x - 2 < 0\} \cap \{x \mid x - 3 < 0\}$
85. $\{x \mid x - 2 \geq 0 \text{ and } x - 3 \geq 0\} \cup \{x \mid x - 2 < 0 \text{ and } x - 3 < 0\}$
86. $\{x \mid x - 2 > 0 \text{ and } x - 3 < 0\} \cup \{x \mid x - 2 < 0 \text{ and } x - 3 > 0\}$
87. $\{x \mid x - 2 \leq 3 \text{ and } x - 2 \geq 0\} \cup \{x \mid -(x - 2) \leq 3 \text{ and } x - 2 < 0\}$
88. $\{x \mid x - 2 > 3 \text{ and } x - 2 > 0\} \cup \{x \mid -(x - 2) > 3 \text{ and } x - 2 < 0\}$

☆89. (a) Show that the set described in Exercise 85 is the solution set of $(x - 2)(x - 3) \geq 0$.
  (b) Show that the set described in Exercise 86 is the solution set of $(x - 2)(x - 3) < 0$.
  (c) Make a general statement regarding the relationship between the solution sets of $p(x) \geq 0$ and $p(x) < 0$, where $p(x)$ is a polynomial in the variable x.

☆90. (a) Show that the set described in Exercise 87 is the solution set of $|x - 2| \leq 3$.
  (b) Show that the set described in Exercise 88 is the solution set of $|x - 2| > 3$.
  (c) Make a general statement regarding the relationship between the solution set of $|p(x)| > a > 0$ and the solution set of $|p(x)| \leq a$, where $p(x)$ is a polynomial in x.

1.2 LOGIC; TF-STATEMENTS, OPEN STATEMENTS, QUANTIFIERS, TRUTH SETS

1.2.1 TF-statements

Symbolic logic is concerned with statements that can be classified as true or false. Such statements are called *TF-statements*.

A **TF-statement** is a declarative sentence that can be classified as either "true" or "false" but not both.

EXAMPLES OF TF-STATEMENTS
$1 + 1 = 2$.	(true)
$2^3 = 6$.	(false)
London is in England.	(true)
February has 30 days.	(false)

EXAMPLES THAT ARE NOT TF-STATEMENTS
$3 + 2$.	(not a sentence)
$x + 2 = 5$.	(cannot be classified as true or false until the value of x is given)
Countries of North America.	(not a sentence)
An animal is a mammal.	(cannot be classified as true or false until a specific animal is named)
Where is Tahiti?	(not a declarative statement)

Notation A TF-statement will be expressed symbolically by a lowercase letter such as *p*, *q*, or *r*.

1.2.2 Open Statements

A **variable** is the name of an unspecified element from a universal set having more than one member.

A **constant** is the name of an element from a set having exactly one member.

For example, in the statement "If x is an integer, then $2x = x + 2$," x is a variable and 2 is a constant.

The statement $2x = x + 2$ has the property that it is neither true nor false, but it becomes either true or false when the variable x is replaced by a constant.

If $x = 1$, then $2x = x + 2$ becomes $2(1) = 1 + 2$ or $2 = 3$, false.
If $x = 2$, then $2x = x + 2$ becomes $2(2) = 2 + 2$ or $4 = 4$, true.

A statement such as this is called an *open statement*.

An **open statement** is a declarative sentence that contains one or more variables and it is not a TF-statement, but it becomes a TF-statement when each of its variables is replaced by a constant.

Notation An open statement will be expressed symbolically by a lowercase letter followed by one or more variables such as *px* or *qx,y*.

Example 1 Let *px* be *x is divisible by 2*.
Then *p3* is *3 is divisible by 2*   (false)
and *p8* is *8 is divisible by 2*.   (true)

LOGIC; TF-STATEMENTS, OPEN STATEMENTS, QUANTIFIERS, TRUTH SETS

Example 2 Let qx,y be $2x - y = 10$.
Then $q3,4$ is $6 - 4 = 10$ (false)
and $q8,6$ is $16 - 6 = 10$.   (true)

1.2.3 Quantifiers

Words such as "all," "some," and "none" are called *quantifiers* because they indicate quantity; that is, they state how many.

The expression "for all x" is called the **universal quantifier**.

"For all x" may be replaced by any one of the following expressions which are considered to have the same meaning:

> For each x
> For every x
> All x are (such that)
> Each x is (such that)
> Every x is (such that).

The expression "for some x" is called the **existential quantifier**. Expressions considered to have the same meaning as "for some x" are as follows:

> There is an x such that
> There exists an x such that
> Some x are (such that)
> There is at least one x such that.

When a quantifier is used, it is understood that the replacement values for each variable must belong to a specified universal set, U.

An open statement may become a TF-statement when one or more appropriate quantifiers are prefixed to the open statement. Such TF-statements are called *closed* statements.

Example 1 For each open statement given, determine whether the resulting statement is true or false, first after the universal quantifier is prefixed, and second after the existential quantifier is prefixed. ($U = R$, the set of real numbers.)

(a) $x + 3 = 3 + x$.
(b) $x + 3 = 10$.

Solution

(a) For all x, $x + 3 = 3 + x$.   (true)
    For some x, $x + 3 = 3 + x$.   (true)
(b) For all x, $x + 3 = 10$.   (false, for $x = 2$, $2 + 3 \neq 10$)
    For some x, $x + 3 = 10$.   (true, for $x = 7$, $7 + 3 = 10$)

If more than one variable occurs in an open statement, then a quantifier must be supplied for each variable in order to obtain a TF-statement.

Example 2 Supply appropriate quantifiers so that each of the resulting statements is true. ($U = R \times R$; that is, $x \in R$ and $y \in R$.)

(a) $xy = yx$
(b) $xy = x$
(c) $(x - 2)^2 + (y - 3)^2 = 0$

Solution
(a) For all x and for all y, $xy = yx$.
(b) For all x, there is some y so that $xy = x$.   ($y = 1$)
    There is some x so that for all y, $xy = x$.   ($x = 0$)
(c) For some x and for some y, $(x - 2)^2 + (y - 3)^2 = 0$.   ($x = 2, y = 3$)

As a result of the above discussion, it follows that there are two ways to obtain a TF-statement from an open statement:

(1) replace each variable by a constant, or
(2) prefix the universal or existential quantifier for each variable.

1.2.4 Truth Sets

The *truth set* of an open statement is the set of elements from a specified universal set for which the open statement is true.
    In symbols, let

U = the universal set,
px = the open statement,
P = the truth set of px,
$P = \{x \mid x \in U \text{ and } px \text{ is true}\}$.

For convenience, when there is no ambiguity as to the universal set, this symbolic form is shortened to
$$P = \{x \mid px\}.$$

Example 1 For $U = N$, the set of counting numbers, describe the truth set for the open statement: x is less than 10 by (a) the set-builder method and (b) the listing method.

Solution
(a) $P = \{x \mid x < 10\}$
(b) $P = \{1, 2, 3, 4, 5, 6, 7, 8, 9\}$

LOGIC; TF-STATEMENTS, OPEN STATEMENTS, QUANTIFIERS, TRUTH SETS 15

Example 2 If U is $N \times N$ and if $p_{x,y}$ is $x + 2y = 10$, describe the truth set of $p_{x,y}$ by both the set-builder method and the listing method.

Solution

(a) $P = \{(x, y) \mid x + 2y = 10\}$
(b) $P = \{(2, 4), (4, 3), (6, 2), (8, 1)\}$

EXERCISES 1.2

1–15. Determine whether each of the following statements is a TF-statement or an open statement or neither of these two.

1. $3 + 2 = 5$
2. $3 + x = 5$
3. $-x$ is a negative number.
4. -3 is a natural number.
5. For some real number x, $x + 3 = 5$.
6. For all real numbers x, $x - 3 = 3 - x$.
7. $2^5 = 10$
8. $3(2 + 5)$
9. $3(x + 5)$
10. $(x + y)(x - y) = x^2 - y^2$
11. $x^2 - y^2 = 9$
12. For all integers x and for all integers y, $x - y = -(y - x)$.
13. For all integers x and for all integers y, $(x + y)^2 = x^2 + y^2$.
14. If $\log_2 32 = 5$, then $2^5 = 32$.
15. $10^y = x$ and $\log_{10} x = y$.

16–30. For each of the following, determine if the resulting statement is true or false (a) after the universal quantifier is prefixed and (b) after the existential quantifier is prefixed.

16. $x - 3 = 3 - x$; $U =$ the set of real numbers.
17. $2(x - 3) = 2x - 6$; $U =$ the set of real numbers.
18. $5 - x$ is an integer; $U =$ the set of integers.
19. $5 - x$ is a natural number; $U =$ the set of natural numbers.
20. $\frac{x}{3}$ is an integer; $U =$ the set of integers.

21. $\dfrac{x}{3}$ is a rational number; $U =$ the set of rational numbers.

22. $\dfrac{x}{3} = \dfrac{3}{x}$; $U =$ the set of real numbers.

23. $x = x + 1$; $U =$ the set of real numbers.
24. $x^2 + 1 = 0$; $U =$ the set of real numbers.
25. $(x + 3)(x - 3) = x^2 - 9$; $U =$ the set of real numbers.
26. $(x + 2)(x - 2)\left(\dfrac{3}{x - 2}\right) = 3(x + 2)$; $U =$ the set of real numbers.
27. $x^2\left(\dfrac{6}{x^2} + \dfrac{3}{x}\right) = 6 + 3x$; $U =$ the set of real numbers.
28. $x\left(\dfrac{1}{x}\right) = 1$; $U =$ the set of real numbers.
29. $|x - 2| > 0$; $U =$ the set of real numbers.
30. $|x - 2| < -1$; $U =$ the set of real numbers.

31–40. To each of the following, prefix one or more quantifiers so that the resulting statement is true. If more than one way is possible, state all possibilities. Let $U =$ the set of real numbers.

31. $x + y = y + x$
32. $xy = yx$
33. $x + y = x$
34. $xy = x$
35. $x - y = y - x$
36. $(x + y)^2 = x^2 + y^2$
37. $x^2 = y$
38. $x^3 = y$
39. $x^2 + y^2 = 0$
40. $(x - 2)^2 + (y - 3)^2 = 0$

41–50. Describe the truth set of each of the following by (a) the set-builder method and (b) the listing method.

41. $U =$ the set of natural numbers; x is greater than 5.
42. $U =$ the set of integers; x is less than -3.
43. x and y are natural numbers; $3x + y = 12$.
44. x and y are natural numbers; $x^2 + y^2 = 25$.
45. $U =$ the set of real numbers; $(x + 3)(3x - 1) = 0$.
46. $U =$ the set of real numbers; $(2x + 5)(x - 4) = 0$.
47. $U =$ the set of integers; $x(2x + 1) = 0$.
48. $U =$ the set of integers; $x(x - 2)(2x + 1) = 0$.
49. $U =$ the set of real numbers; $\sqrt{3^2} = x$.
50. $U =$ the set of real numbers; $\sqrt{(-3)^2} = x$.

1.3 LOGIC; CONNECTIVES AND TRUTH TABLES

Certain words that are used to determine the structure or form of a statement are called *connectives*. The most common connectives are *not, and, or, implies,* and *if and only if.*

The truth value of a statement obtained by using one or more of the connectives depends on the truth value of each component statement. A table indicating this dependence is called a *truth table*. The truth table is considered to be the definition of the connective.

By using the concept of truth sets, a relationship between logic and set theory can be exhibited. A set diagram accompanies each of the following truth tables as an aid in clarifying the meaning of the connective. For the diagrams that follow, $P = \{x \mid px\}$ and $Q = \{x \mid qx\}$.

NEGATION, NOT p[4]

The word "not" is used to change the truth value of a statement. If p is true, then *not p* is false. If p is false, then *not p* is true.

Truth Table

p	NOT p
T	F
F	T

Set Diagram

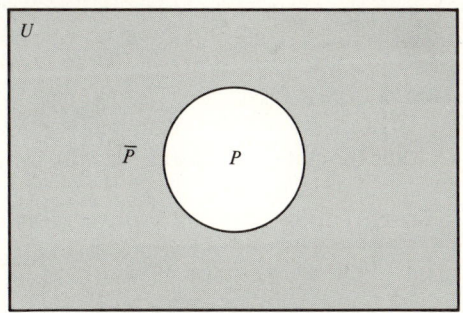

Figure 1.9 $\bar{P} = \{x \mid \text{not } px\}$

Example 1 If p is $3^2 = 9$, then *not p* is $3^2 \neq 9$.
If p is $\pi = 3$, then *not p* is $\pi \neq 3$.

Example 2 If $U = N$, the set of counting numbers and if $P = \{x \mid x < 5\}$, then $\bar{P} = \{x \mid x \not< 5\} = \{x \mid x \geq 5\}$.

CONJUNCTION, p AND q[5]

The word "and" is used with the meaning that the statement (p and q) is true only when both p and q are true.

---

[4] Some common symbolic forms for "not p" are: $\bar{p}, p', -p, \sim p, \tilde{p}$.
[5] A common symbolic form for "p and q" is: $p \wedge q$.

Truth Table

p	q	p AND q
T	T	T
T	F	F
F	T	F
F	F	F

Set Diagram

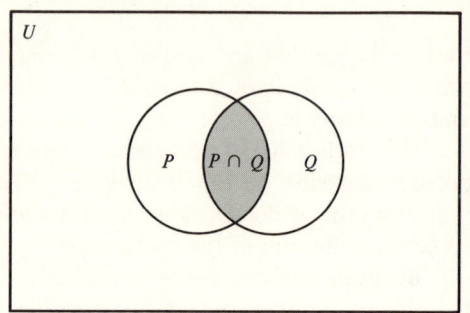

Figure 1.10 $P \cap Q = \{x \mid px \text{ and } qx\}$

Example 1 $(1 < \sqrt{2})$ and $(\sqrt{2} < 2)$. (true)

Example 2 $3^2 = 9$ and $3^3 = 9$. (false)

Example 3 $3 < 5$ and $-3 < -5$. (false)

Example 4 Let $P = \{(x, y) \mid x + y = 7\}$ and let $Q = \{(x, y) \mid x - y = 3\}$. Then $P \cap Q = \{(x, y) \mid x + y = 7 \text{ and } x - y = 3\} = \{(5, 2)\}$.

DISJUNCTION, p OR q[6]

In most branches of mathematics, the word "or" is used with the meaning that the statement (p or q) is false only when both p and q are false.

Truth Table

p	q	p OR q
T	T	T
T	F	T
F	T	T
F	F	F

Set Diagram

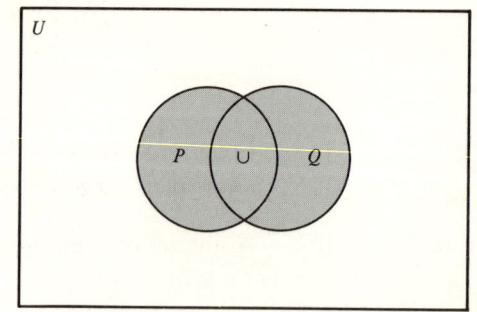

Figure 1.11 $P \cup Q = \{x \mid px \text{ or } qx\}$

Example 1 $3^2 = 9$ or $(-3)^2 = 9$. (true)

Example 2 $3^2 = 9$ or $3^3 = 9$. (true)

---

[6] A common symbolic form for "p or q" is: $p \lor q$.

LOGIC; CONNECTIVES AND TRUTH TABLES

Example 3 $\sqrt{-1} = -1$ or $(-1)^2 = -1.$ (false)

Example 4 Let $P = \{x \mid x - 2 = 0\}$ and let $Q = \{x \mid x - 3 = 0\}$. Then $P \cup Q = \{x \mid x - 2 = 0 \text{ or } x - 3 = 0\} = \{2, 3\}$.

IMPLICATION, p IMPLIES q[7]

In mathematics, the statement "p implies q" is considered to have the same meaning as the statement "if p, then q" and the same meaning as the statement "not p or q." The truth table for "not p or q" can be derived from the truth tables for "not" and "or." As a result, the implication is false only when the first component statement p is true and the second component statement q is false.

Truth Table

p	NOT p	q	NOT p OR q
T	F	T	T
T	F	F	F
F	T	T	T
F	T	F	T

Truth Table

p	q	p IMPLIES q (IF p, THEN q)
T	T	T
T	F	F
F	T	T
F	F	T

Set Diagram

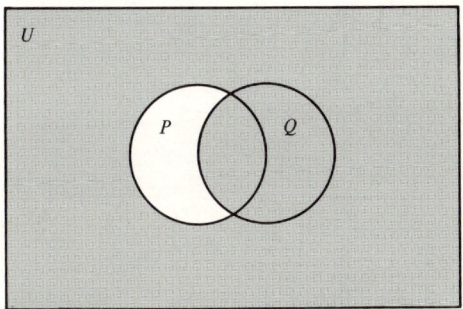

Figure 1.12 $\overline{P} \cup Q = \{x \mid \text{not } px \text{ or } qx\}$
$= \{x \mid \text{if } px, \text{ then } qx\}$

Example 1 $3 + (-3) = 0$ implies $(-3) + 3 = 0.$ (true)

Example 2 If $-3 < 2$, then $3 < -2.$ (false)

Example 3 If $2^3 = 6$, then $2^2 = 4.$ (true)

---

[7] A common symbolic form for "p implies q" is: $p \rightarrow q$.

Example 4 $\pi = 3$ implies $\pi^2 = 9$. (true)

Example 5 Let $U =$ the set of integers $= \{\ldots, -3, -2, -1, 0, 1, 2, 3, \ldots\}$.
Let $P = \{x \mid x^2 > 0\}$ where px is $x^2 > 0$.
Let $Q = \{x \mid x > 0\}$ where qx is $x > 0$.

Then $\bar{P} \cup Q = \{x \mid \text{If } x^2 > 0, \text{ then } x > 0\} = \{0, 1, 2, 3, \ldots\}$. (See Figure 1.13.) Since $\bar{P} \cup Q$ consists of some elements of U but not all, "for some x; if $x^2 > 0$, then $x > 0$" is true, while "for all x; if $x^2 > 0$, then $x > 0$" is false.

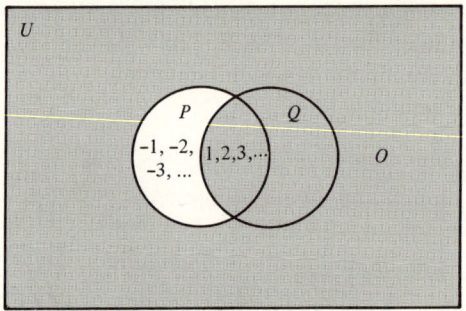

Figure 1.13 $\bar{P} \cup Q = \{0, 1, 2, 3, \ldots\}$

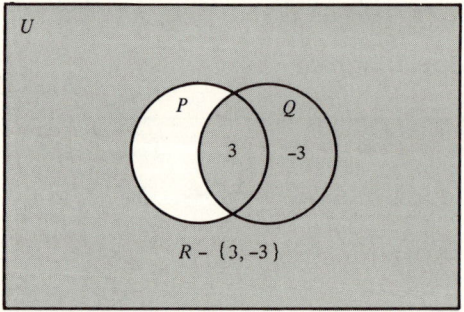

Figure 1.14 $\bar{P} \cup Q = R - \{3\} \cup \{3, -3\} = R$

Example 6 Let $U = R$, the set of real numbers.
Let $P = \{x \mid x = 3\}$ and let $Q = \{x \mid x^2 = 9\}$.

Then $\bar{P} \cup Q = R$. (See Figure 1.14.) Therefore, "for all x; $x = 3$ implies $x^2 = 9$" is true.

BICONDITIONAL, *p* IF AND ONLY IF *q*[8]

The statement "*p if and only if q*" is considered true only when both *p* and *q* are true or when both *p* and *q* are false. Mathematicians also express "*p if and only if q*" as "*p* is a necessary and sufficient condition for *q*."

[8] A common symbolic form for "*p* if and only if *q*" is: $p \leftrightarrow q$.

LOGIC; CONNECTIVES AND TRUTH TABLES

Truth Table

p	q	p IF AND ONLY IF q
T	T	T
T	F	F
F	T	F
F	F	T

Set Diagram

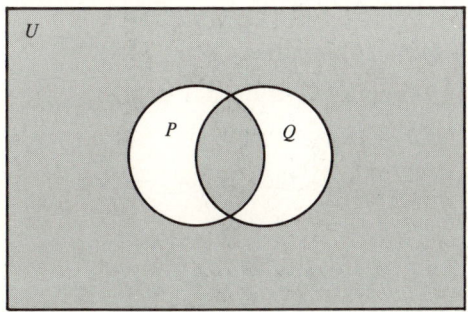

Figure 1.15 Truth Set of "px if and only if qx," shaded
$(P \cap Q) \cup \overline{(P \cup Q)} = \{x | px \text{ if and only if } qx\}$

Example 1 $\log_3 81 = 4$ if and only if $3^4 = 81$. (true)

Example 2 $-2 < 5$ if and only if $2 < -5$. (false)

When a biconditional statement is true, the two component statements are said to be **equivalent**. An important property of equivalent statements is that one may be substituted for the other in a given statement without changing the truth value of the statement.

Two open statements are equivalent if and only if they have the same truth set.

EXERCISES 1.3

1–25. Determine the truth value of each of the following composite statements. (Use your knowledge of algebra to determine the truth values of the component statements.)

1. Not $(5 + 2\sqrt{3} = 7\sqrt{3})$
2. Not $(-4 > -3)$
3. Not $(3^4 = 81)$
4. Not $(\log_2 1 = 0)$
5. $2 < 3$ and $-3 < -2$
6. $\log_2 1 = 0$ and $2^0 = 1$
7. $(\sqrt{3})^2 = 3$ and $(\sqrt{-3})^2 = 3$
8. $\sqrt{9} = 3$ and $\sqrt{9} = -3$

9. $3 < 5$ or $3 = 5$
10. $-2 \leq -4$ or $-4 < -2$
11. $\dfrac{3+5}{3} = 5$ or $\dfrac{2+2\sqrt{5}}{2} = \sqrt{5}$
12. $(2^{-1} + 5^{-1})^{-1} = 2 + 5$ or $\sqrt{9 + 16} = 3 + 4$
13. If $\sqrt{9} = 3$, then $3^2 = 9$.
14. If $(-3)^2 = 9$, then $\sqrt{9} = -3$.
15. If $\sqrt{9} = -3$, then $(-3)^2 = 9$.
16. $(2 > 3$ and $3 > 7)$ implies $(2 > 7)$.
17. $(4^2 = 16$ and $(-4)^2 = 16)$ implies $\sqrt{16} = \pm 4$.
18. $(2 \cdot 3 = 6$ or $6 < 9)$ implies $(2 < 9$ and $3 < 9)$.
19. $\log_3 9 = 2$ implies $9 = 3^2$.
20. If $(|2| = 2)$ and $(|-2| = 2)$, then $|2| = |-2|$.
21. If $-2 < -3$, then $-3 < -5$.
22. $\sqrt{16} = 4$ if and only if $4^2 = 16$.
23. $\sqrt{-16} = -4$ if and only if $(-4)^2 = -16$.
24. 210 is divisible by 12 if and only if 210 is divisible by 2 and 3.
25. The sum of any two even numbers is even if and only if the sum of any two odd numbers is odd.

26–50. Determine the truth set of each of the following, where

$$U = \text{the set of real numbers.}$$

26. x is not a positive number.
27. $-x$ is not a negative number.
28. $x \not\geq 2$
29. $x^2 \neq x$
30. $x^3 \neq x$
31. $(x-3)(x-5) = 0$ and $(x-2)(x-5) = 0$
32. $x > -2$ and $x < 2$
33. $2 < x < 7$ and $4 < x < 9$
34. $x + y = 10$ and $x - y = 4$
35. $y = x^2$ and $y = 3x + 4$
36. $(x-3)(x-5) = 0$ or $(x-2)(x-5) = 0$
37. $x > -2$ or $x < 2$

LOGIC; CONNECTIVES AND TRUTH TABLES

38. $2 < x < 7$ or $4 < x < 9$
39. $(x^2 + y^2 = 20$ and $y = 2x)$ or $(x^2 + y^2 = 20$ and $x = 3y)$
40. $(y = x^2 - 6$ and $y = x)$ or $(y = x^2 - 6$ and $y = -x)$
41. If $x = 5$, then $x^2 = 25$.
42. If $x^2 = 25$, then $x = 5$.
43. $x^2 = 9$ implies $x = 3$.
44. $5x - 5 = 2x + 7$ implies $3x = 12$.
45. If $\dfrac{1}{(x-5)^2} + \dfrac{1}{x-5} = \dfrac{x^2 - 24}{(x-5)^2}$, then $x^2 - x - 20 = 0$.
46. $x + 6 = x^2$ implies $\sqrt{x+6} = x$.
47. $(-x)^2 = x^2$ implies $\sqrt{-x^2} = -x$.
48. $(-x)^2 = x^2$ implies $\sqrt{(-x)^2} = x$.
49. If $(x-2)(x-3) = 6$, then $(x - 2 = 2$ or $x - 3 = 3)$.
50. If $x^2 + 1 \geq 0$, then $x^2 + 1 = 0$.

51-62. For each given pair of open statements (a) find the truth set of each open statement and (b) determine if the open statements are equivalent.

Let $U =$ the set of real numbers.

51. $\dfrac{1}{(x-5)^2} + \dfrac{1}{x-5} = \dfrac{x^2 - 24}{(x-5)^2}$; $x^2 - x - 20 = 0$
52. $\dfrac{1}{x^2} + \dfrac{x+2}{x} = \dfrac{x+1}{x^2}$; $x^2 + x = 0$
53. $\dfrac{2}{x} + \dfrac{15}{x^2} = 1$; $x^2 - 2x - 15 = 0$
54. $\dfrac{2}{x} + \dfrac{15}{x^2} = 1$; $2x + 15 = 1$
55. $\sqrt{x+6} = x$; $x + 6 = x^2$
56. $\sqrt{x+4} = x - 2$; $x + 4 = (x-2)^2$
57. $\sqrt{6x - 8} = x$; $6x - 8 = x^2$
58. $\sqrt[3]{2x^2 + x - 2} = x$; $2x^2 + x - 2 = x^3$
59. $2x - 4y = 6$; $3x - 6y = 9$
60. $3x + y = 2$; $6x + 2y = 8$
61. $x^2 + y^2 = 16$ and $x^2 - y^2 = 2$; $x^2 + y^2 = 16$ and $x^2 = 9$
62. $y = 5 - x^2$ and $x = 2y$; $y = 5 - 4y^2$ and $x = 2y$

1.4 LOGIC; TAUTOLOGIES, NEGATIONS OF OPEN STATEMENTS

1.4.1 Tautologies

A **tautology** is a composite statement that is true for every case of its component statements.

The truth of a tautology depends on the structure of the statement and is independent of the subject matter. A statement may be shown to be a tautology by constructing its truth table. Some important tautologies are presented below. The construction of their truth tables is left as an exercise for the student.

THE TRIVIAL TAUTOLOGY
   p if and only if p.

Example 1 $2 + 3 = 5$ if and only if $2 + 3 = 5$.

Example 2 $2^3 = 6$ if and only if $2^3 = 6$.

THE DOUBLE NEGATIVE
   Not (not p) if and only if p.

Example 1 -3 is not nonnegative if and only if -3 is negative.

Example 2 $\sqrt{2}$ is not irrational if and only if $\sqrt{2}$ is rational.

THE EXCLUDED MIDDLE
   p or not p.

Example 1 $(-2)(-3)$ is positive or $(-2)(-3)$ is not positive.

Example 2   12 is a prime number or 12 is not a prime number.

THE CONTRADICTION TAUTOLOGY
   Not (p and not p).

Example 1 $\sqrt{9}$ cannot be positive and nonpositive.

Example 2 $\sqrt{2}$ is not rational and irrational.

It follows from the contradiction tautology that the statement "p and not p" is always false. A composite statement that is always false for every case of its component statements is called a **contradiction**.

De Morgan's Tautologies
  I. Not (p or q) if and only if (not p and not q).
  II. Not (p and q) if and only if (not p or not q).

De Morgan's tautologies indicate that a negation may be distributed over a disjunction or a conjunction by changing the *or* to *and*, or the *and* to *or*, respectively.

Example 1 (It is not so that ($\sqrt{-1}$ is positive or $\sqrt{-1}$ is negative)) if and only if ($\sqrt{-1}$ is not positive and $\sqrt{-1}$ is not negative.)

Example 2 (For all x, it is not so that $x = 2$ or $x = 3$) if and only if (for all x, $x \neq 2$ and $x \neq 3$).

Example 3 (For all x and for all y, it is not so that $x > 0$ and $y > 0$) if and only if (for all x and for all y, $x \not> 0$ or $y \not> 0$).

The Contrapositive Tautology
  (p implies q) if and only if (not q implies not p).

Example 1 ($2^3 = 8$ implies $\log_2 8 = 3$) if and only if ($\log_2 8 \neq 3$ implies $2^3 \neq 8$).

Example 2 ($[\sqrt[3]{8} - 1][\sqrt[3]{8} + 1] = 8 - 1$ implies $[2 - 1][2 + 1] = 7$) if and only if ($[2 - 1][2 + 1] \neq 7$ implies $[\sqrt[3]{8} - 1][\sqrt[3]{8} + 1] \neq 8 - 1$).

Example 3 (For all x and for all y, $xy = 0$ implies $x = 0$ or $y = 0$) if and only if (for all x and for all y, $x \neq 0$ and $y \neq 0$ implies $xy \neq 0$).

Related to an implication and its contrapositive are two other statements called the *converse* and the *inverse*. These statements are often erroneously assumed to be true when the original implication is considered true. However, the converse and the inverse are *not* equivalent to the original implication if either (but not both) of its components is false.

The converse (or the inverse) of an open implication may be shown not equivalent to the original implication by providing a **counterexample** (a case for which one of the statements is true and for which the other statement is false).

Example 1

STATEMENT	If $-3 > 0$, then $(-3)^2 > 0$.	(true)
CONVERSE	If $(-3)^2 > 0$, then $-3 > 0$.	(false)
INVERSE	If $-3 \leq 0$, then $(-3)^2 \leq 0$.	(false)

Example 2

OPEN STATEMENT	If $x = 2$, then $x^2 = 4$.	(true for all x)
CONVERSE	If $x^2 = 4$, then $x = 2$.	(false for $x = -2$)
INVERSE	If $x \neq 2$, then $x^2 \neq 4$.	(false for $x = -2$)

1.4.2 Negations of Open Statements

The basic negation forms of open statements are derived from the following agreements:

AGREEMENTS
    Not (all cases are true)   means Some case is false.
    Not (some case is true)   means All cases are false.
    Not (all cases are false)   means Some case is true.
    Not (some case is false)   means All cases are true.

The basic negation forms listed below are consistent with these agreements. These equivalences may be considered as axioms in the theory of quantifiers.

BASIC NEGATION FORMS
    Not (for all x, px)   if and only if For some x, not px.
    Not (for some x, px)   if and only if For all x, not px.
    Not (for all x, not px)   if and only if For some x, px.
    Not (for some x, not px)   if and only if For all x, px.

Example 1 (a) State its truth value, (b) write its negation, and (c) state the truth value of the negation for each of the following. $U = R$, the set of real numbers.

(1) For all x, $x + 0 = x$.
(2) For all x, $x^3 > x^2$.
(3) For some x, $(x - 2)(x - 3) = 0$.
(4) For some x, $x^2 + 1 = 0$.

Solution

(1) a. True.
    b. For some x, $x + 0 \neq x$.
    c. False.
(2) a. False (for $x = \frac{1}{2}$, $\frac{1}{8} > \frac{1}{4}$ is false).
    b. For some x, $x^3 \not> x^2$.
    c. True (for $x = \frac{1}{2}$).
(3) a. True (for $x = 2$ and for $x = 3$).
    b. For all x, $(x - 2)(x - 3) \neq 0$.
    c. False (for $x = 2$ and for $x = 3$).

LOGIC; TAUTOLOGIES, NEGATIONS OF OPEN STATEMENTS 27

(4) a. False ($x^2 + 1 = 0$ means $x^2 = -1$ and there is no real number whose square is negative).
  b. For all x, $x^2 + 1 \neq 0$.
  c. True.

Example 2 Write the negations of each of the following:

(a) For all x, px or qx.
(b) For some x, px and not qx.
(c) For all x, if px, then qx.

Solution

(a) For some x, not (px or qx).
  For some x, not px and not qx (simplified form).
(b) For all x, not (px and not qx).
  For all x, not px or qx (simplified form).
(c) For some x, not (if px, then qx).
  For some x, not (not px or qx).
  For some x, px and not qx (simplified form).

The negation of a quantified statement involving two or more variables may be obtained by changing each quantifier (from universal to existential or from existential to universal) and by negating the open statement.

Example 3 Write the negation of

There is some x so that for all y, px,y.

Solution For all x, there is some y so that not px,y.

Example 4 Write the negation of

For all x, there is some y so that if $x > 0$, then $x = y^2$.

Solution There is some x so that for all y, not (if $x > 0$, then $x = y^2$).
  There is some x so that for all y, not ($x \not> 0$ or $x = y^2$).
  There is some x so that for all y, ($x > 0$ and $x \neq y^2$).

The expression "for no x" (or its alternate form, "there is no x such that") is also a quantifier.[9] It is used with the understanding that "for no x, px" is equivalent to "for all x, not px."

---

[9] The symbolic forms Ax and $\forall x$ are used for "for all x" and the symbolic forms $\exists x$ and Ex are used for "for some x." It is not necessary to have a symbolic form for "for no x" since "for no x, px" can be replaced by "for all x, not px."

Example 5 Express each of the following in an equivalent form in which the quantifier "for no x" is used. ($U =$ the set of real numbers.)

(a) For all x, $x + 1 \neq x + 2$.
(b) For all x, $x^2 \geq 0$.

Solution

(a) There is no x such that $x + 1 = x + 2$.
(b) For no x, $x^2 \ngeq 0$. (Or, for no x, $x^2 < 0$.)

EXERCISES 1.4

1–10. (a) Determine the truth value of each of the following

by letting fx be: $\dfrac{x}{2}$ is an integer,

by letting gx be: $\dfrac{x}{3}$ is an integer, and

by selecting (1) $p = f6$ and $q = g6$
(2) $p = f4$ and $q = g4$
(3) $p = f9$ and $q = g9$
(4) $p = f5$ and $q = g5$.

(b) Determine which of the following statements are tautologies.

1. (Not p implies q) if and only if (p or q).
2. (p implies not q) if and only if (not p or q).
3. (Not p implies q) if and only if (q or not p).
4. Not (p or not q) if and only if (not p and q).
5. Not (not p and q) if and only if (p or not q).
6. If (p implies not q) and q, then not p.
7. If (p implies q) and q, then p.
8. If (p implies q) and p, then q.
9. If (p implies q) and not q, then not p.
10. If (p implies q) and not p, then not q.

11–16. For each of the following statements, write its converse, its inverse, its contrapositive, and cite a counterexample for any of these three statements that is not equivalent to the original statement.

11. If $x + 2 = 5$, then $x = 3$.
12. If $-3x < 6$, then $x > -2$.
13. If $\sqrt{x + 2} = x$, then $x + 2 = x^2$.

LOGIC; TAUTOLOGIES, NEGATIONS OF OPEN STATEMENTS

14. If $x = 2$, then $x^4 = 16$.
15. If $xy = 0$ and $x \neq 0$, then $y = 0$.
16. If $xy > 0$, then $x > 0$ and $y > 0$.

17–22. Write the negation and the contrapositive of each of the following statements. (Simplify, by using the double negation tautology, or De Morgan's tautologies, or by replacing an implication by an equivalent disjunction, as necessary.)

17. If $ab = 0$, then $a = 0$ or $b = 0$.
18. If $(a > 0$ and $b > 0)$ or $(a < 0$ and $b < 0)$, then $ab > 0$.
19. If $a > 0$ and $b > 0$, then $a + b > 0$ and $ab > 0$.
20. If $a > b$ and $n > 0$, then $a^n > b^n$.
21. If $0 < b < 1$ and $n > 0$, then $b^n < b$.
22. If $a^2 + b^2 = 0$, then $a = 0$ and $b = 0$.

23–34. Write the negation of each of the following statements.

23. Every natural number is positive.
24. Some parallelograms are squares.
25. Nothing is perfect.
26. Not every real number is rational.
27. All squares are rectangles.
28. Each integer is a rational number.
29. Some problems cannot be solved.
30. Some real numbers are rational.
31. There is at least one solution for the inequality $x^2 > 1$.
32. All real numbers satisfy the equation $x + 3 = 3 + x$.
33. No real number satisfies the equation $x^2 + 1 = 0$.
34. For all x, there is no y such that $y = x/0$.

35–38. Illustrate that the truth or falsity of a quantified statement depends on the choice of the universal set by selecting for each of the following: (a) a universal set for which the statement is true, and (b) a universal set for which the statement is false. Select the universal set from the set of digits, the set of natural numbers, the set of integers, the set of rational numbers, the set of real numbers.

35. For all x and for all y, $x + y \in U$.
36. For some x and for some y, $3y = 4x$ and $x^2 + y^2 = 1$.
37. For all x there is a y so that $x > 0$ and $y > 0$ and $y = \sqrt{x}$.
38. There is an x for all y so that $x \leq y$.

CHAPTER SUMMARY

Sets

$a \in S$	a is an element of S.
$T \subset S$	T is a subset of S (every element of T is an element of S).
U	The universal set.
$\emptyset$ or $\{\}$	The empty set.
$A = B$	A is a subset of B and B is a subset of A.
$A \leftrightarrow B$	The elements of A are in one-to-one correspondence with the elements of B.
$\bar{A}$	Complement of A, elements in U not in A.
$A \cup B$	Union, elements in A or in B.
$A \cap B$	Intersection, elements in A and in B.
$B - A$	Relative complement, elements in B not in A.

Logic, Connectives

not p	Negation.
p and q	Conjunction, true only when both p and q are true.
p or q	Disjunction, false only when both p and q are false.
p implies q	Implication, false only when p is true and q is false. (Other forms: If p, then q; not p or q.)
p if and only if q	Biconditional, true only when p and q have the same truth values.

Tautologies

not (not p) if and only if p	Double negative.
p or not p	Excluded middle.
not (p and not p)	Contradiction.
(p implies q) if and only if (not q implies not p)	Contrapositive.
not (p or q) if and only if (not p and not q)	De Morgan's
not (p and q) if and only if (not p or not q)	De Morgan's

REVIEW EXERCISES

QUANTIFIERS
  for all x,   Universal quantifier.
  for some x,   Existential quantifier.
  for some x, not px Negation of "for all x, px."
  for all x, not px Negation of "for some x, px."
  for no x, px Equivalent to "for all x, not px."

REVIEW EXERCISES

1–4. Use the *listing* method to describe each of the following sets:

1. $\{x \mid -4 < x \leq 2 \text{ where } x \text{ is an integer}\}$
2. $\left\{p \mid p = \sqrt{\dfrac{n}{4}} \text{ where } n \text{ is a natural number and } p \text{ is a rational number}\right\}$
3. $\{n \mid \sqrt{2-n} \text{ is a real number and } n \text{ is an integer}\}$
4. $\{y \mid (y^2 - y)(y^2 - 2) = 0 \text{ where } y \text{ is a real number}\}$

5–8. Use the *set-builder* notation to describe each of the following sets:

5. $\{\ldots, -4, -2, 0, 2, 4, \ldots\}$
6. $\{\frac{1}{5}, \frac{2}{5}, \frac{3}{5}, \frac{4}{5}, 1\}$
7. $\{(6, 216), (7, 343), (8, 512), (9, 729)\}$
8. $\{(1, -1), (2, -2), (3, -3), (4, -4), \ldots\}$

9–14. Let $U = \{1, 2, 3, 4, 5, 6, 7, 8, 9\}$, $A = \{4, 5, 6, 7, 8, 9\}$, $B = \{7, 8, 9\}$, and $C = \{1\}$. List the members of each of the following:

9. $(A \cap B) \cup (\bar{A} \cap \bar{B})$
10. $(A \cap \bar{B}) \cup (\bar{A} \cap B)$
11. $(A \cap B) \cup (\bar{A} \cap C)$
12. $(A \cap \bar{B}) \cup (\bar{A} \cap \bar{C})$
13. $(\bar{A} \cap \bar{C}) \cup B$
14. $(A \cup C) \cap \bar{B}$

15–20. Given the *true* statements p: $2 + 3 = 5$ and q: $5 - 3 = 2$, determine the truth value of each of the following statements:

15. $2 + 3 = 5$ only if $5 - 3 = 2$.[10]
16. If $2 + 3 \neq 5$, then $5 - 3 \neq 2$.
17. A necessary condition for $5 - 3 = 2$ is that $2 + 3 \neq 5$.[10]
18. It is not the case that $2 + 3 = 5$ or $5 - 3 \neq 2$.
19. $2 + 3 \neq 5$ or $5 - 3 = 2$.
20. It is not the case that $2 + 3 = 5$ if $5 - 3 \neq 2$.[10]

---

[10] Other forms used by mathematicians for "p implies q" are: q if p, p only if q, p is sufficient for q, and q is necessary for p.

21–24. Determine if the two given statements are equivalent for each of the following pairs of statements:

21. It is not true that p implies not q; p and (p implies q).
22. p and (p implies q); p and q.
23. (q and not q) or p; p.
24. (p and q) implies q; p or not q.

25–29. Write the negation of each of the following statements:

25. For all x, $x \cdot 1 = x$.
26. There is no x such that $\dfrac{1}{x} = 0$.
27. There is an x such that $x + 2 = 0$.
28. For every x, $x + 2 = 2 + x$.
29. There is no x that satisfies the equation $x + 1 = x$.

30–35. Supply quantifiers so that each of the following statements becomes true:

30. $xy = yx$ where $U = $ the set of integers.
31. $x + y = 1$ where $U = $ the set of natural numbers.
32. $x + y = x$ where $U = $ the set of real numbers.
33. $x + y = 0$ where $U = $ the set of real numbers.
34. $xy = 1$ where $U = $ the set of real numbers.
35. $xy = 1$ where $U = $ the set of integers.

36–50. Describe the truth set for each of the following by the listing method. The universal set is N, the set of natural numbers.

36. $x < 5$
37. $x > 2$ and $x < 8$
38. $x \leq 5$ and $x < 9$
39. $x \leq 5$ or $x < 9$
40. $x \geq 5$ and $x \leq 2$
41. $x \geq 5$ or $x \leq 2$
42. $x^2 - 3x + 2 = 0$ or $x^2 - 4x + 3 = 0$
43. $x^2 - 3x + 2 = 0$ and $x^2 - 4x + 3 = 0$
44. $(x - 3 > 0$ and $x - 5 < 0)$ or $(x - 3 < 0$ and $x - 5 > 0)$
45. $(x - 3 \geq 2$ and $x - 3 \leq 0)$ or $(3 - x \geq 2$ and $x - 3 > 0)$
46. $x + y = 5$

REVIEW EXERCISES

47. $y = x + 1$
48. $x + y = 5$ and $y = x + 1$
49. $y = x^2$ and $y = 8x - 15$
50. $(y = x$ and $x \geq 0)$ or $(y = -x$ and $x < 0)$

2
THE NATURAL NUMBERS

Today elementary algebra[1] is considered to be a mathematical system consisting of an undefined set of objects called numbers, undefined and defined operations and relations, axioms, and theorems obtained from the definitions and axioms by the methods of logic.

Arithmetic and elementary algebra have evolved from subject matter concerned with rules for solving problems about numbers. These number problems can be traced to those of the early Egyptians and Babylonians around 1700 B.C. Later the Greeks, from 900 B.C. to 300 A.D., solved even more difficult problems and supplied proofs for their results. Unfortunately, the Greeks were limited to the set of positive real numbers, the other numbers not having been invented at that time. Also they resorted to geometric methods of solution involving very little symbolism. The modern symbolism used today was not in existence at that time but gradually evolved as the number system was developed.

After the collapse of the Roman Empire around 400 A.D., and with

---

[1] Elementary algebra is considered to be the generalization of arithmetic. It includes the subjects traditionally called "elementary algebra," "intermediate algebra," and "college algebra."

Algebra, in general, refers to a mathematical system having a certain type of structure. Examples of other algebras are Boolean algebra, the algebra of logic, the algebra of groups, rings, and fields, and the algebra of quaternions.

western Europe submerged in the Dark Ages for a period of 1000 years, the mathematical knowledge was preserved by the Hindus in India. The Hindus also developed our modern decimal system of numeration and contributed to the study of number problems.

From India the mathematical knowledge was transmitted to the Arabs. The word "algebra" first appeared around 825 in the work *Hisab al-jabr Wal-muqabala* written by al-Khowarizmi. Literally, "algebra" meant "reduction" or "restoration." A more common meaning of the Arabic word *al-jabr* was introduced into Spain by the Moors. During the sixteenth century in Italy, "algebra" meant the art of bonesetting, and a barber who practiced bonesetting as a sideline was called an *algebrista*.

As a result of al-Khowarizmi's work, "algebra" came to mean the study of equations and the word was adopted by both Arabic and European scholars. The name "algebra" was established by the book which became the model of algebra texts, *Vollstaendige Anleitung zur Algebra* (Complete Introduction to Algebra) written by the Swiss mathematician Leonard Euler in 1770.

The symbolism of algebra had a gradual development beginning around 250 A.D. with the works of the Greek Diophantus and culminating in the work *La Géométrie* written by the French mathematician René Descartes in 1637.

The universal set of numbers for elementary algebra, the set of complex numbers, also had a gradual development. The counting numbers and the fractions of arithmetic were known to the ancient Egyptians and Babylonians. The Greeks were familiar also with irrational numbers. The negative numbers were known to the Chinese at a very early date and the Hindu Brahmagupta stated rules for addition, subtraction, multiplication, and division around 628. In 1637 these new negative quantities were firmly established as numbers by the work of Descartes.

Decimal fractions were first introduced in print by Christoff Rudolff in 1530. He exhibited calculations with them but did not explain them. The theory was first explained by Simon Stevin in 1585 although his notation was poor. Stevin wrote $24\textcircled{0}$ $3\textcircled{1}$ $7\textcircled{2}$ $5\textcircled{3}$ for our modern 24.375. Around 1700 an historian remarked that there were as many opinions on the subject of decimal fractions as there were people. The modern forms were finally stabilized around 1800.

The Hindus, Mahavira around 850 and Bhaskara around 1150, were the first to indicate an awareness of the concept of an imaginary number by writing that a negative number cannot have a square root because a negative cannot be a square.

Cardan, in his *Ars Magna* of 1545, was the first to use the square root of a negative number in a computation. In 1572 Bombelli, in his *Algebra*, introduced a consistent theory of imaginary numbers. Descartes in 1637, in his *La Géométrie*, classified numbers as "real" and "imaginary" and discussed complex numbers as the solutions of equations. The development of complex numbers continued with the Irish mathematician William Rowan Hamilton presenting in 1835 the modern rigorous treatment of these quantities as number pairs.

For many years the subject matter of algebra remained an unordered set of rules for solving number problems. What Euclid did for geometry in 300 B.C., ordering its subject matter according to the principles of logic, was not done for algebra until over 2000 years later. As the number system and the symbolism developed, the properties of the numbers became better understood. Finally, in the twentieth century, elementary algebra emerged as a mathematical system.

This text is concerned with the presentation of elementary algebra as a mathematical system. Although the universal set of numbers is to be considered as undefined, we have in mind the set of complex numbers and are, of course, motivated by this interpretation. However, it is important in the development of the theory that conclusions are derived by logically valid arguments and not by reference to known statements about these special numbers.

The universal set of complex numbers will be introduced gradually by first considering a special subset, the set of natural numbers, and later extending this set until the universal set is obtained. In this way the special properties of the various subsets can be examined in detail, thus providing a better understanding of the structure of the universal set of complex numbers.

2.1 RELATIONS, FUNCTIONS, OPERATIONS

Elementary algebra is concerned not only with a set of numbers but also with *operations* and *relations*. In order to investigate the properties of a set of numbers, it is necessary to examine the concepts of relation and operation.

2.1.1 Ordered Pairs

Definition of Ordered Pair, (x, y) An ordered pair, (x, y), is a set of two elements of which one is named the first component and the other the second component. In symbols, the components are enclosed by parentheses and are separated by a comma, the first being written to the left of the second.

For example (San Francisco, California) is an ordered pair, (Smith, John) is an ordered pair, and (5, 3) is an ordered pair.

RELATIONS, FUNCTIONS, OPERATIONS

The order of the components in an ordered pair is important; (5, 3) is *not* the same as (3, 5). Usually when the variables x and y are used to indicate the components of an ordered pair, x designates the first component and y designates the second component.

2.1.2 Cartesian Product

Definition of Cartesian Product, $A \times B$ If A and B are sets, then the Cartesian product of A and B, $A \times B$, is defined as follows:

$$A \times B = \{(a, b) \mid a \in A \text{ and } b \in B\}.$$

The definition states that the Cartesian product, $A \times B$, is the set of ordered pairs whose first components belong to set A and whose second components belong to set B. Since (a, b) and (b, a) are the same only when $a = b$, it follows that $A \times B$ and $B \times A$ are not equal sets in general.

Example List the elements of $A \times B$ and $B \times A$ if

$$A = \{n, t\} \quad \text{and} \quad B = \{1, 2, 3\}.$$

Solution $A \times B = \{(n, 1), (n, 2), (n, 3), (t, 1), (t, 2), (t, 3)\}$

$B \times A = \{(1, n), (2, n), (3, n), (1, t), (2, t), (3, t)\}$

2.1.3 Relations

Definition of Relation A relation r on A to B is a subset of $A \times B$.
A relation r on A is a subset of $A \times A$.

From the definition it follows that a relation r on A to B is a set of ordered pairs and $r \subset A \times B$.

If (x, y) is an element of the relation r, it is often convenient to express this by writing $x\,r\,y$.

Definition of $x\,r\,y$ $x\,r\,y$ if and only if $(x, y) \in r$.

Definition of Domain The domain, Dm, of a relation is the set of all first components of the relation. In symbols,

$$\text{Dm} = \{x \mid (x, y) \in r\}.$$

Definition of Range The range, Rn, of a relation is the set of all second components of the relation. In symbols,

$$Rn = \{y \mid (x, y) \in r\}.$$

Example If $A = \{1, 2, 3, 4, 5\}$, and $B = \{1, 2, 3, 4, 5, 6\}$, and if r is a subset of $A \times B$ such that $r = \{(x, y) \mid y = 2x\}$, then list the elements of the relation r and state its domain and range.

Solution $r = \{(1, 2), (2, 4), (3, 6)\}$
$Dm = \{1, 2, 3\}$
$Rn = \{2, 4, 6\}$

2.1.4 Equivalence Relations

Definition of Equivalence Relation An equivalence relation on A is a subset of $A \times A$ such that for all x, y, and z in A

(1) $x\,r\,x.$ Reflexive property.
(2) If $x\,r\,y$, then $y\,r\,x.$ Symmetric property.
(3) If $x\,r\,y$ and $y\,r\,z$, then $x\,r\,z.$ Transitive property.

Equality is an important example of an equivalence relation. Using the substitution property as the definition of the equal relation, it follows that equality is an equivalence relation.

Definition of Equality, $x = y$ (Substitution Property) $x = y$ if and only if x may replace y or y may replace x without changing the truth value of the statement in which the replacement is made.

Theorem *Equality is an equivalence relation.*

(1) Reflexive property: $x = x$. (Since x may replace x without changing the truth or falsity of a statement.)
(2) Symmetric property: If $x = y$, then $y = x$. (Since if x may replace y or y may replace x, then y may replace x or x may replace y.)
(3) Transitive property: If $x = y$ and $y = z$, then $x = z$. (Since z may replace y, $x = z$ follows by replacing y by z in $x = y$.)

It is often convenient to indicate in symbols that one number does not equal another, $x \neq y$, read "x does not equal y."

Definition of $x \neq y$ $x \neq y$ if and only if it is not the case that $x = y$.

RELATIONS, FUNCTIONS, OPERATIONS

2.1.5 Functions

Definition of Function A function is a relation such that each element in the domain is paired with exactly one element in the range. In symbols,

$$\text{if } (x, y) \in r \text{ and } (x, z) \in r, \text{ then } y = z,$$

and

$$\text{if } x \, r \, y \text{ and } x \, r \, z, \text{ then } y = z.$$

A function is usually designated symbolically by the letter f or the letter g or any lowercase letter. Since a function is a relation, it has a domain, the set of its first components, and a range, the set of its second components.

It is often convenient to define a particular function by stating a rule which its members must satisfy. The notation $f(x)$, read "f of x," is useful for this purpose in designating the value of the second component of the function corresponding to the first component x. Thus, $y = f(x)$ means $(x, y) \in f$ and $(x, f(x)) \in f$.

Example 1 Which of the following relations are functions?

$$a = \{(1, 2), (2, 2), (3, 2)\}$$
$$b = \{(2, 1), (2, 2), (2, 3)\}$$

Solution Only a is a function since exactly one second component is assigned to each first component. b is not a function since more than one second component is assigned to the first component 2.

Example 2 Let $A = \{1, 2, 3, 4, 5\}$.
Let $f = \{(x, f(x)) \mid f(x) = x - 1\}$, where $f \subset A \times A$.

Then

(1) find $f(2), f(5),$ and $f(1)$
(2) list the elements of f, and
(3) state the domain and range of f.

Solution

(1) $f(2) = 2 - 1 = 1, f(5) = 5 - 1 = 4, f(1) = 1 - 1 = 0$. Since 0 is not in A, $(1, 0)$ is not in f.
(2) $f = \{(2, 1), (3, 2), (4, 3), (5, 4)\}$.
(3) $\text{Dm} = \{2, 3, 4, 5\}$.
   $\text{Rn} = \{1, 2, 3, 4\}$.

2.1.6 Operations

Definition of Operation An operation **in** a set A is a function whose domain is a subset of $A \times A$ and whose range is a subset of A.

The definition states that the domain of an operation is a set of ordered pairs whose components are elements of A. If the domain is all of $A \times A$, then the operation is said to be **on** A. Since an operation is a function, exactly one element of A is paired with each ordered pair of the domain. In other words, if $((x, y), z_1)$ and $((x, y), z_2)$ are elements of the operation, then $z_1 = z_2$.

If x, y, and z are elements of A and if a rule $f(x, y)$ is specified, then the operation could be described by writing $f(x, y) = z$. It is often convenient to use the notation $x \circ y = z$, read "x operation y equals z."

Definition of $x \circ y$ $x \circ y = z$ if and only if $\circ = \{((x, y), z) | z = \circ(x, y)\}$.

For example, $x \circ y = x + y$ where $\circ = +$ and $x \circ y = x \cdot y$ where $\circ = \cdot$.

When the operation is *on* A and thus the domain of the operation is $A \times A$, then for all x in A and for all y in A, there is exactly one z in A so that $x \circ y = z$.

An operation as defined is also referred to as a *binary* operation since two numbers are combined to obtain another.

EXERCISES 2.1

1–4. Consider the sets $A = \{a, b, c\}$, $B = \{x, y\}$, $C = \emptyset$; list the elements in the following sets:

1. $A \times B$
2. $B \times A$
3. $A \times A$
4. $A \times C$

5–10. Let $A = \{1, 2, 3, 4\}$.

5. List the elements in $A \times A$.
6. Select the largest subset of $A \times A$ such that each element x in the domain bears the following relation to each element y in the range: $x < y$.
7. Select the largest subset of $A \times A$ such that if x is in the domain of $A \times A$ and y is in the range of $A \times A$, $x = y$.
8. Select the largest subset of $A \times A$ such that if x and y are elements of A, x is a factor of y or y is a factor of x.
9. Which of the following properties of relations apply to the relations defined in Exercises 6, 7, and 8: reflexive, symmetric, transitive?
10. Define one or more equivalence relations on A.

RELATIONS, FUNCTIONS, OPERATIONS   41

11–20. Classify the following relations defined on the stated universal sets as reflexive, symmetric, transitive, or equivalence.

	RELATION	UNIVERSAL SET
11.	Is a neighbor of	people
12.	Is younger than	men
13.	Is a brother of	people
14.	Has the same area as	polygons
15.	Is perpendicular to	lines in a plane
16.	Is a multiple of	natural numbers
17.	Is a subset of	sets
18.	Intersects	lines in space
19.	Lives less than a mile from	people
20.	Is a brother of	men

21–26. Let R, S, and T stand for reflexive, symmetric, and transitive, and state an example of a set and a relation defined on that set having the following properties:

21. R, S, not T
22. R, not S, T
23. R, not S, not T
24. Not R, S, T
25. Not R, not S, T
26. Not R, S, not T

27. Let the relation $\equiv$ be defined on the set $\{0, 1, 2, 3, \ldots\}$ such that $x \equiv y$ if and only if x and y have the same remainder when divided by 5. Is the relation reflexive? Symmetric? Transitive? Justify.

28. Given set $S = \{1, 2, 3, 4\}$ and set $T = \{5, 6, 7, 8\}$. Define the relation $\sim$ such that $x \in S$, $y \in S$ and $x \sim y$ if and only if $xy + \in T$. Is $\sim$ an equivalence relation? Justify.

29–34. Which of the following relations are functions?

29. $R = \{(1, 2), (1, 3), (1, 4), (1, 5)\}$
30. $S = \{(2, 1), (3, 1), (4, 1), (5, 1)\}$
31. $T = \{(x, y) | x^2 + y = 5\}$
32. $A = \{(x, y) | y = x + 2\}$
33. $B = \{(x, y) | x = y^2 + 2\}$
34. $C = \{(x, y) | x^2 + y^2 = 4\}$

35–40. Let $A = \{-4, -2, 0, 2, 4\}$; let $f = \{(x, f(x)) | f(x) = x + 2,$ where $f \subset A \times A\}$.

35. Evaluate $f(-4)$, $f(0)$, and $f(2)$.
36. List the elements of f.

37. State the domain and range of f.
38. Let $f^{-1} = \{(f(x), x) | f(x) = x + 2,$ where $f \subset A \times A\}$. List the elements of f^{-1}.
39. Is f^{-1} a function? Justify.
40. What is the relationship between the domain and range of f and the domain and range of f^{-1}?

2.2 THE PEANO POSTULATES (Optional)

2.2.1 Introduction

An algebraic system consists of a nonempty set S, one or more relations on S, and one or more operations on S. In this section, S shall be selected as the set of natural numbers, N, which are also called the **counting numbers**.

$$N = \{1, 2, 3, 4, 5, \ldots\}$$

The objective is to exhibit an abstract mathematical system that can be interpreted as describing the natural numbers. The Italian mathematician Giuseppe Peano (1858–1932) proposed a system based on the fact that any natural number can be reached by adding 1 to the number that comes before it. If n is a natural number, then $n + 1$, the number that follows n in the counting process, is called the **successor** of n, denoted $s(n)$; s is called the **successor function**.

2.2.2 The System of Peano

UNDEFINED TERMS
  A set N, an element 1, a relation s on N.

Postulates

P1 $1 \in N$. (1 is a natural number.)

P2 For all x in N, there is exactly one element $s(x)$ in N. [Each natural number has exactly one successor. (Thus, s is a function.)]

P3 There is no x in N so that $s(x) = 1$. (1 is not the successor of any natural number.)

P4 For all x and y in N, if $s(x) = s(y)$, then $x = y$. (The set of natural numbers is in one-to-one correspondence with the set of their successors.)

P5 If S is a subset of N and if
  (1) $1 \in S$ and
  (2) $s(x) \in S$, whenever $x \in S$,
  then $S = N$
(The finite induction postulate.)

THE PEANO POSTULATES (OPTIONAL)

Definition of Addition

$$x + 1 = s(x)$$
$$x + s(y) = s(x + y)$$

Definition of Multiplication

$$x \cdot 1 = x$$
$$x \cdot s(y) = x \cdot y + x = xy + x$$

These two definitions are said to be recursive; that is, the expressions recur or are used repeatedly to obtain the functional value desired. For example, referring to the interpretation of N as the set of natural numbers, names can be provided for the elements $s(n)$ as follows: $s(1) = 2$, $s(2) = 3$, $s(3) = 4$, and so on.

Beginning with

$$x + 1 = s(x),$$

then

$$1 + 1 = s(1) = 2$$
$$2 + 1 = s(2) = 3$$
$$3 + 1 = s(3) = 4.$$

Beginning with

$$x + s(x) = s(x + y),$$

then

$$x + 2 = x + s(1) = s(x + 1) = (x + 1) + 1$$
$$x + 3 = x + s(2) = s(x + 2) = (x + 2) + 1$$
$$x + 4 = x + s(3) = s(x + 3) = (x + 3) + 1.$$

The recursive definition provides a general way to describe all these sums. It would actually be impossible to list all the special cases; the first few of which are listed above.

Some special cases of the multiplication operation are listed below.

Beginning with

$$x \cdot 1 = x,$$

then

$$1 \cdot 1 = 1$$
$$2 \cdot 1 = 2$$
$$3 \cdot 1 = 3.$$

Beginning with

$$x \cdot s(y) = x \cdot y + x,$$

then

$$x \cdot 2 = x \cdot s(1) = x \cdot 1 + x = x + x$$
$$x \cdot 3 = x \cdot s(2) = x \cdot 2 + x = (x + x) + x$$
$$x \cdot 4 = x \cdot s(3) = x \cdot 3 + x = [(x + x) + x] + x.$$

It may be seen from these examples that this definition preserves the arithmetical meaning of multiplication as a repeated sum; that is,

$$x \cdot y = \underbrace{x + x + \ldots + x}_{y \text{ terms}}$$

Although it is not explicitly stated, the equal relation is considered to be valid for this system; that is, the elements of N satisfy the following properties:

SUBSTITUTION
If $x = y$, then x may replace y or y may replace x.

REFLEXIVE
$x = x$.

SYMMETRIC
If $x = y$, then $y = x$.

TRANSITIVE
If $x = y$ and $y = z$, then $x = z$.

It is also assumed that the content of set theory and logic are valid for any algebraic system that is not especially concerned with set theory or logic itself.

2.2.3 Mathematical Induction

Peano's fifth postulate, the finite induction postulate, describes an especially important property of the set of natural numbers. It provides a very powerful method of proof called mathematical induction. If it can be shown that a certain open statement is true for $x = 1$ and also for $x = n + 1$ whenever it is true for $x = n$, then the finite induction postulate states that the set of natural numbers for which the open statement is true consists of *all* the natural numbers.

It is important to observe that a proof by mathematical induction consists of *two* parts: Part (1), the proof for $x = 1$ and Part (2), the proof for $x = n + 1$ assuming the statement is true for $x = n$.

THE PEANO POSTULATES (OPTIONAL)

Example 1 Prove by mathematical induction that for all natural numbers x,
$$1 + 2 + 3 + \ldots + x = \frac{x(x+1)}{2}.$$

Solution Let S be the set of natural numbers for which the statement is true.

PART (1) If $x = 1$, then
$$\frac{x(x+1)}{2} = \frac{1(1+1)}{2} = 1 \text{ and } 1 = 1.$$

Thus, $1 \in S$.

PART (2) Assume for $x = n$ that
$$1 + 2 + 3 + \ldots + n = \frac{n(n+1)}{2}.$$

Adding $n + 1$ to both sides of this equation,
$$1 + 2 + 3 + \ldots + n + (n+1) = \frac{n(n+1)}{2} + (n+1)$$
$$= (n+1)\left(\frac{n+2}{2}\right)$$
$$= \frac{(n+1)([n+1]+1)}{2}.$$

Thus, for $x = n + 1$,
$$1 + 2 + 3 + \ldots + x = \frac{x(x+1)}{2}.$$

Thus, if $n \in S$, then $n + 1 \in S$.

Combining Parts (1) and (2), $S = N$ and the statement is true for all natural numbers.

Example 2 Prove by mathematical induction that for all natural numbers x,
$$a^x - b^x = (a - b)(a^{x-1} + a^{x-2}b + \ldots + ab^{x-2} + b^{x-1}).$$

Solution Let S be the set of natural numbers for which the statement is true.

PART (1) If $x = 1$, $a^1 - b^1 = a - b$.
If $x = 2$, $a^2 - b^2 = (a - b)(a + b)$.
If $x = 3$, $a^3 - b^3 = (a - b)(a^2 + ab + b^2)$.
If $x = 4$, $a^4 - b^4 = (a - b)(a^3 + a^2b + ab^2 + b^3)$.

Thus, $1 \in S$.

(*Note:* The cases $x = 2$, $x = 3$, and $x = 4$ are not necessary for the induction proof but are included to aid understanding.)

PART (2) Assume for $x = n$ that
$$a^n - b^n = (a - b)(a^{n-1} + a^{n-2}b + \ldots + ab^{n-2} + b^{n-1}).$$

Now,
$$\begin{aligned}
a^{n+1} - b^{n+1} &= a^{n+1} - a^n b + a^n b - b^{n+1} \\
&= a^n(a - b) + b(a^n - b^n) \\
&= a^n(a - b) + b(a - b)(a^{n-1} + a^{n-2}b + \ldots + b^{n-1}) \\
&= (a - b)(a^n + a^{n-1}b + a^{n-2}b^2 + \ldots + b^n).
\end{aligned}$$

Thus, if $n \in S$, then $n + 1 \in S$.

Combining Parts (1) and (2), $S = N$.

EXERCISES 2.2 (Optional)

1–10. Prove by mathematical induction

1. $1 + 3 + 5 + \ldots + (2x - 1) = x^2$
2. $2 + 4 + 6 + \ldots + 2x = x(x + 1)$
3. $1^2 + 2^2 + 3^2 + \ldots + x^2 = \dfrac{x(x + 1)(2x + 1)}{6}$
4. $1^3 + 2^3 + 3^3 + \ldots + x^3 = \dfrac{x^2(x + 1)^2}{4}$
5. $1 + x + x^2 + \ldots + x^{n-1} = \dfrac{x^n - 1}{x - 1}$
6. $1 + 2 + 4 + 8 + \ldots + 2^{n-1} = 2^n - 1$
7. $\dfrac{1}{1 \cdot 3} + \dfrac{1}{3 \cdot 5} + \dfrac{1}{5 \cdot 7} + \ldots + \dfrac{1}{(2x - 1)(2x + 1)} = \dfrac{x}{2x + 1}$
8. $1 \cdot 3 + 2 \cdot 4 + 3 \cdot 5 + \ldots + x(x + 2) = \dfrac{x(x + 1)(2x + 7)}{6}$
9. $2^2 + 4^2 + 6^2 + \ldots + (2x)^2 = \tfrac{2}{3}x(x + 1)(2x + 1)$
10. $3 + 6 + 9 + \ldots + 3x = \dfrac{3x(x + 1)}{2}$
11. Show that if
$$1 + 2 + 3 + \ldots + x = \dfrac{(2x + 1)^2}{8}$$
is true for some natural number n, then it is true for $x = n + 1$. State whether this equation is valid for all natural numbers and give a reason for your answer.

THEOREMS OF THE PEANO SYSTEM (OPTIONAL)   47

12. Show that if
$$2 + 4 + 6 + \ldots + 2x = x(x+1) + 2$$
is true for some natural number n, then it is true for $x = n + 1$. Prove or disprove that this equation is valid for all natural numbers.

13. Show that $2^{x-1} = x$ is valid for $x = 1$ and $x = 2$. Prove or disprove that this equation is valid for all natural numbers.

14. Show that
$$1 + 2 + 3 + \ldots + x = x^2 - x + 1$$
is valid for $x = 1$ and $x = 2$. Prove or disprove that this equation is valid for all natural numbers.

15–18. For each of the following, show that it is possible to verify *only* Part (1) or Part (2) of the proof by mathematical induction.

15. $3^x = 6x^2 - 12x + 9$
16. $3 + 6 + 9 + \ldots + 3x = \frac{1}{2}(3x^2 + 3x + 2)$
17. $1 + 3 + 5 + \ldots + (2x - 1) = x^2 + 4$
18. $1 + \dfrac{1}{2} + \dfrac{1}{3} + \ldots + \dfrac{1}{x} = \dfrac{4 + 9x - x^2}{12}$

19. Prove by mathematical induction that $a^{2x} - b^{2x}$ is divisible by $a + b$ for all natural numbers x. (*Hint:* $a^{2x+2} - b^{2x+2} = a^{2x}(a^2 - b^2) + b^2(a^{2x} - b^{2x})$.)

20. Prove by mathematical induction that $a^{2x-1} + b^{2x-1}$ is divisible by $a + b$ for all natural numbers x.

2.3 THEOREMS OF THE PEANO SYSTEM (Optional)

Beginning with the Peano postulates, it is possible to derive the theorems listed below.

Theorems   Let x, y, and z be any numbers in N.

(1)	Closure, Addition	There is exactly one number $x + y$ in N.
(2)	Associative, Addition	$(x + y) + z = x + (y + z)$.
(3)	Commutative, Addition	$x + y = y + x$.
(4)	Distributive	$x(y + z) = xy + xz$ and $(y + z)x = yx + zx$.
(5)	Closure, Multiplication	There is exactly one number xy in N.
(6)	Associative, Multiplication	$(xy)z = x(yz)$.

(7) Commutative, Multiplication $\quad xy = yx$.

(8) Identity, Multiplication $\quad$ There is exactly one element, 1, in N so that $x \cdot 1 = 1 \cdot x = x$.

(9) $x + z = y + z$ if and only if $x = y$

(10) $xz = yz$ if and only if $x = y$.

The importance of the Peano postulates lies in the fact that these familiar number theorems can be logically proved beginning only with Peano's five assumptions.

EXERCISES 2.3 (Optional)

1–9. Let x, y, and z be any numbers in N.

1. Show that the set N contains no identity element for addition; that is, there exists no y such that for all x, $x + y = x$.
2. Prove: For every x, y, and z, $(y + z)x = yx + zx$. (*Note:* Since this is part of Theorem 4 in this section, only Theorems 1, 2, and 3 may be used in this proof.)
3. Prove: For every x, y, and z, $(xy)z = x(yz)$. (Do not use Theorems 7–10.)
4. Prove: For every x and y, $xy = yx$. (Do not use Theorems 8–10.)
5. Prove: There is exactly one element 1 in N so that $x \cdot 1 = 1 \cdot x = x$ for all x. (Do not use Theorems 9 and 10.)
6. Prove: For all x, y, z, if $x = y$, then $x + z = y + z$.
7. Prove: For all x, y, z, if $x + z = y + z$, then $x = y$.
8. Prove: For all x, y, z, if $x = y$, then $xz = yz$.
9. Prove: For all x, y, z, if $xz = yz$, then $x = y$.
10. Prove: For all x and y, xy is unique.

2.4 ORDER

A relation for which the transitive property holds is called an *order relation*.

Definition of Order Relation $\quad$ A relation r is an order relation if and only if for all (a, b) and (b, c) in r

$$a\ r\ b \text{ and } b\ r\ c \text{ implies } a\ r\ c.$$

From the definition it may be seen that equality is an order relation since if $a = b$ and $b = c$, then $a = c$. The concepts "a is less than b" and "a is greater

ORDER

than b" also describe order relations. Consider the following arrangement of the natural numbers.

$$1, s(1), s(s(1)), \ldots$$

Using the definition $s(1) = 1 + 1$ and $s(n) = n + 1$, this can be written

$$1, 1 + 1, 1 + 1 + 1, \ldots$$

Assigning the names $2 = 1 + 1$, $3 = 1 + 1 + 1$, and so on, this becomes

$$1, 2, 3, \ldots$$

It can be observed that a number a appears to the left of another number b in this arrangement if and only if there exists a natural number c so that $a + c = b$. In this case, a is said to be less than b, or, in symbols, $a < b$.

Definition of $a < b$ Let a and b be natural numbers. Then $a < b$ if and only if there exists a natural number c so that $a + c = b$.

If b is less than a, then a is said to be greater than b. In symbols, $a > b$.

Definition of $a > b$ $a > b$ if and only if $b < a$.

Theorem The relations $<$ and $>$ are order relations; that is, they are transitive. In symbols,

(1) If $a < b$ and $b < c$, then $a < c$, and
(2) If $a > b$ and $b > c$, then $a > c$.

Proof

(1) If $a < b$ and $b < c$, then there exist natural numbers n and m so that

$a + n = b$ and $b + m = c$	Definition of $<$
$(a + n) + m = c$	Substitution, $a + n$ for b
$a + (n + m) = c$	Associative, Addition
$n + m$ is in N.	Closure, Addition

Thus, $a < c$. Definition of $<$

(2) If $a > b$ and $b > c$, then

$b < a$ and $c < b$	Definition of $>$
$c < a$	Theorem, Part (1)
$a > c$.	Definition of $>$

It is convenient to have a compact notation to indicate that $a < b$ and $b < c$.

Definition of $a < b < c$ $a < b < c$ if and only if $a < b$ and $b < c$.

The $<$ and $>$ relations are distinguished from the $=$ relation by the fact that they are neither reflexive nor symmetric. For example, it is *not* the case that $3 < 3$ or $3 > 3$. Also, if $3 < 5$, then it is *not* the case that $5 < 3$. From the meaning of $=$, $<$, and $>$ it should be clear that exactly one of these relations can exist for any pair of natural numbers. This statement, although it can be derived from the postulates, will be accepted without proof and thus classified as an axiom.

The Trichotomy Axiom If a and b are any natural numbers, then exactly one of the following relations is valid:

$$a = b \quad \text{or} \quad a < b \quad \text{or} \quad a > b.$$

It is useful to have a symbolism that indicates that one relation *or* another is valid for a pair of numbers.

Definition of $\leq$ and $\geq$ $a \leq b$ if and only if $a = b$ or $a < b$. $a \geq b$ if and only if $a = b$ or $a > b$.

The relations $\leq$ and $\geq$ are called *inequalities* or *weak inequalities* in contrast to the inequalities, $<$ and $>$, called *strong inequalities*.

Definition of the Well-Ordering Principle An ordered set[2] is said to be well-ordered by the order relation r if and only if every nonempty subset S contains a least (or greatest) element; that is, there exists a number a in S so that for every x in S, $a \, r \, x$ or $a = x$.

The set of natural numbers N is a well-ordered set with respect to the relation $<$ and also with respect to the relation $\leq$.

Theorem The set of natural numbers N is a well-ordered set with respect to the relation $<$ or the relation $\leq$. In other words, if S is a nonempty subset of N, then there exists a number a in S so that for every x in S, $a \leq x$. (Proof is left to student.)

EXERCISES 2.4

1. Show that for all natural numbers a and b, the relation $\leq$ is transitive and reflexive but not symmetric.
2. Show that for all natural numbers a and b, the relation $\geq$ is transitive and reflexive but not symmetric.
3. Show that the set N is well-ordered with respect to $<$.

[2] An ordered set is a set on which an order relation is defined.

4. Show that the set N is well-ordered with respect to $\leq$.

5–8. Let A, B, and C be subsets of N such that

$$A = \{x \mid x < 8\}, \quad B = \{x \mid 5 < x\}, \quad C = \{x \mid x < 3\}.$$

Describe each of the following sets, using the order symbols:

5. $A \cup B$
6. $A \cap B$
7. $(A \cup C) \cap B$
8. $A \cap (B \cup C)$

9. Which of the following relations are order relations?

	RELATION	SET
(a)	younger than	people
(b)	is a parent of	males
(c)	is a child of	people
(d)	is parallel to	lines in a plane
(e)	is perpendicular to	lines in a plane
(f)	is an ancestor of	people
(g)	is shorter or equal in height	people
(h)	is a subset of	sets
(i)	is a proper subset of	sets
(j)	is less than	all numbers between 0 and 1
(k)	is greater in length than	chords of a circle
(l)	is lesser in length than	chords of a circle
(m)	is employed by	people
(n)	is greater than	natural numbers
(o)	implies	TF-statements
(p)	is a descendant of	people
(q)	is 20 or more than 20 miles away from	towns on a straight road
(r)	is less than a semicircle in distance from	3 points on a circle dividing the circle into 3 = arcs

10. Which of the sets in Exercise 9 are well-ordered with respect to the order relation defined on them?

11. Show that any nonempty finite subset of N has a greatest and a least element.

12. What can be said about infinite subsets of N concerning a least or greatest element?

13. Prove that the number c is unique in the definition $a < b$ if and only if $a = b + c$, where a, b, and c are natural numbers.

2.5 PRIMES, SUBTRACTION, DIVISION

2.5.1 Terminology

The study of the properties of the natural numbers is an important branch of mathematics known as *number theory*. It originated in the school of the Greek Pythagoras around 540 B.C. The distinction between odd and even numbers was known to the Pythagoreans. In Book VII of *Elements* Euclid has definitions for the concepts "unit," "number," "odd," "even," "prime," and "composite," as well as many theorems concerned with these concepts. The prime numbers can be thought of as the basic building units from which all the natural numbers are made. It is thought that a sufficient knowledge of the properties of prime numbers could provide a key for understanding the structural properties of all the numbers.

Definition of Multiple, Factor, Divisible, Divisor If a and b are natural numbers and if there exists a natural number q so that

$$a = b \cdot q,$$

then

a is a multiple of b, and
b is a factor of a, and
a is divisible by b, and
b is a divisor of a.

From the definition of multiplication it follows that for all a, $a = a \cdot 1$ and thus a is a divisor of a and 1 is a divisor of a. There are cases when it is desirable to exclude the divisor a so the expression *proper divisor* is introduced.

Definition of Proper Divisor If a and b are natural numbers, then b is a proper divisor of a (or proper factor of a) if and only if b is a divisor of a and $b \neq a$.

Definition of Prime A natural number p is a prime if and only if $p \neq 1$ and p has no proper divisors different from 1.

From the definition of a prime it follows that if p is a prime and if $p = nq$, then $n = 1$ or $n = p$.

Definition of Composite A natural number n is a composite if and only if $n \neq 1$ and n is *not* a prime.

Definition of Even A natural number n is even if and only if there exists a natural number k so that $n = 2k$.

PRIMES, SUBTRACTION, DIVISION

Definition of Odd A natural number is odd if and only if it is *not* even.

The set of even numbers = $\{2, 4, 6, 8, \ldots\} = \{x \,|\, x = 2k$ and $k \in N\}$.

The set of odd numbers = $\{1, 3, 5, 7, \ldots\} = \{x \,|\, x = 2k - 1$ and $k \in N\}$.

The set of the first 10 primes = $\{2, 3, 5, 7, 11, 13, 17, 19, 23, 29\}$.

2.5.2 The Fundamental Theorem of Arithmetic

An important property of prime numbers is that every composite number larger than 1 can be expressed as a product of primes in exactly one way, disregarding the order in which the factors are written. This statement is called *the fundamental theorem of arithmetic*.

Theorem (The Fundamental Theorem of Arithmetic) Every natural number larger than 1 is either prime or can be expressed as a product of primes in exactly one way, disregarding the order in which the factors are written.

For example,
$$360 = 2 \cdot 2 \cdot 2 \cdot 3 \cdot 3 \cdot 5$$
$$= 2^3 \cdot 3^2 \cdot 5.$$

Note that 360 can also be expressed as $72 \cdot 5$. However, this is not a prime factorization since the factor 72 is not a prime. While there are several factorizations possible for 360, there is only one prime factorization where each of the factors is a prime.

The fundamental theorem of arithmetic can be proved by mathematical induction beginning with a product having two prime factors and extending the result to a product having any number of factors.

Actually finding the prime factors of a number is in general a laborious process, especially if the number is very large. This is one task which is properly relegated to the electronic computer.

For smaller numbers, the prime factors are obtained by a trial-and-error process, trying all the primes less than the square root of the number.

Example 1 Find the prime factors of 868.

Solution

(1) Try 2: $868 = 2 \cdot 434$.
(2) Try 2: $868 = 2 \cdot 2 \cdot 217$.
(3) Try 3: 3 does not divide 217.
(4) Try 5: 5 does not divide 217.
(5) Try 7: $868 = 2 \cdot 2 \cdot 7 \cdot 31$.
(6) Since 31 is prime, the factorization is complete.

Thus,
$$868 = 2 \cdot 2 \cdot 7 \cdot 31.$$

2.5.3 Euclid's Theorem on Primes

Euclid proved in Book IX of *Elements* the statement "There are infinitely many prime numbers." His proof is described below.

Theorem   There are infinitely many prime numbers.

Proof (*Indirect method of proof*)   Assume there is a largest prime, P.
Consider the number $n = (1 \cdot 2 \cdot 3 \cdots P) + 1$.

Since n is larger than P, it must be composite. Then there is a number smaller than n that is a divisor of n. If this number is not prime, then it has a prime divisor. Let q be the smallest prime divisor of n.

Since q is a prime and since P is the largest prime, then q must be one of the numbers $2, 3, \ldots, P$. However, each of these numbers is *not* a divisor of n because each leaves a remainder of 1.

Thus, a contradiction has been reached, and therefore there is no largest prime.

2.5.4 Subtraction and Division

Definition of Subtraction   Let a and b be natural numbers. Then $a - b = d$ if and only if $a = b + d$ for exactly one natural number d.

Since $a = b + d$ if and only if $a > b$, it is noted that subtraction is defined only when $a > b$. For example, $5 - 2 = 3$ since $5 = 2 + 3$ but $2 - 5$ and $5 - 5$ are undefined. Thus, the set of natural numbers is *not* closed with respect to the subtraction operation.

Definition of Division   Let a and b be natural numbers. Then $\dfrac{a}{b} = q$ if and only if $a = bq$ for exactly one natural number q.

For example $\frac{6}{2} = 3$ since $6 = 2 \cdot 3$. On the other hand, the expressions $\frac{2}{6}$ and $\frac{1}{2}$ do not designate natural numbers. Thus, the set of natural numbers is *not* closed with respect to the division operation. On the other hand, it is possible to state a relationship between any two natural numbers that describes the division operation as it is used in arithmetic. This statement is called *the division algorithm*.

Theorem (The Division Algorithm)   If a and b are any two natural numbers with $a > b$, then there exists exactly one natural number q, called the *quotient*, and exactly one natural number r, called the *remainder*, so that either

$$a = bq$$

or

$$a = bq + r \quad \text{and} \quad r < b.$$

PRIMES, SUBTRACTION, DIVISION

For example, $68 = 7 \cdot 9 + 5$.
In the division of 68 by 7,

$$\begin{array}{r} 9 \\ 7\overline{)68} \\ 63 \\ \hline 5 \end{array}$$

9 is the quotient and 5 is the remainder.

2.5.5 The Euclidean Algorithm; GCD; LCM

Definition of GCD The **greatest common divisor**, GCD, of two natural numbers is the largest natural number that is a divisor of each of the two numbers.

For example, the GCD of 12 and 18 is 6.

$$12 = 2 \cdot 2 \cdot 3 \quad \text{and} \quad 18 = 2 \cdot 3 \cdot 3$$
$$12 = 6 \cdot 2 \quad \text{and} \quad 18 = 6 \cdot 3$$

The GCD is useful in simplifying fractions. For example,

$$\frac{12}{18} = \frac{6 \cdot 2}{6 \cdot 3} = \frac{2}{3}.$$

From the above example, it may be seen that one way to find the greatest common divisor of two numbers is to express each number as the product of prime factors. This is always possible by the fundamental theorem of arithmetic. The GCD can then be determined by inspection.

Another method for finding the GCD is attributed to Euclid and is called the *Euclidean algorithm*. It is based upon the division algorithm.

Theorem (The Euclidean Algorithm) (to find the GCD of a and b, $a > b$)

(1) Divide a by b to find q_1 and r_1 so that

$$a = bq_1 + r_1 \quad \text{where } r_1 < b.$$

(2) Divide b by r_1; similarly,

$$b = r_1 q_2 + r_2 \quad \text{where } r_2 < r_1.$$

(3) Divide r_1 by r_2; similarly,

$$r_1 = r_2 q_3 + r_3 \quad \text{where } r_3 < r_2.$$

Continue until there is no remainder,

$$r_{n-1} = r_n q_{n+1}.$$

Then, the last nonzero remainder r_n is the GCD of a and b.

The proof of the Euclidean algorithm is based on the division algorithm, $a = bq$ or $a = bq + r$ where $r < b$. The GCD of a and b is also the GCD of b and r. If d is the GCD of a and b, then $a = dk$ and $b = dl$, and $dk = (dl)q + r$. Thus, $r = dk - d(lq) = d(k - lq)$, and d is a divisor of r. It must also be the GCD of b and r since a larger divisor would also divide a and then d would not be the GCD of a and b.

Example 2 Find the GCD of 300 and 672, using the Euclidean algorithm.

Solution The work may be arranged in either of the two formats indicated below:

$$\begin{array}{r} 2 \\ 300\overline{)672} \\ \underline{600} \\ 72 \end{array} \quad \begin{array}{r} 4 \\ \overline{)300} \\ \underline{288} \\ 12 \end{array} \quad \begin{array}{r} 6 \\ \overline{)72} \\ \underline{72} \\ \end{array}$$

$$672 = 300(2) + 72$$
$$300 = 72(4) + 12$$
$$72 = 12(6)$$

Thus, 12 is the GCD of 300 and 672.

Definition of LCM The **least common multiple**, LCM, of two natural numbers a and b is the smallest natural number that is divisible by both a and b.

The definition states that if $m = ar$ and $m = bs$ and if for all $n < m$, a does not divide n or b does not divide n, then m is the LCM of a and b.

For example, the LCM of 3 and 4 is 12 and the LCM of 15 and 20 is 60.

The LCM of two numbers may be found by using the fundamental theorem of arithmetic; that is, by expressing each of the numbers as a product of prime factors. The product of the highest powers of all the different primes that occur in either prime factorization is the LCM.

Example 3 Find the LCM of 90 and 120.

Solution
$$90 = 2 \cdot 3 \cdot 3 \cdot 5 = 2 \cdot 3^2 \cdot 5$$
$$120 = 2 \cdot 2 \cdot 2 \cdot 3 \cdot 5 = 2^3 \cdot 3 \cdot 5$$

The LCM of 90 and 120 $= 2^3 \cdot 3^2 \cdot 5 = 360$.

The LCM of two numbers may also be found by dividing their product by their GCD. If d is the GCD of a and b, then $a = dr$ and $b = ds$, and

$$\text{LCM of } a \text{ and } b = rds = \frac{drds}{d} = \frac{ab}{\text{GCD of } a \text{ and } b}.$$

PRIMES, SUBTRACTION, DIVISION

Example 4 Find the LCM of 3456 and 10080.

Solution First find the GCD.

$$
\begin{array}{r}
2 \\
3456 \overline{)10080} \\
6912
\end{array}
\quad
\begin{array}{r}
1 \\
3168 \overline{)3456} \\
3168
\end{array}
\quad
\begin{array}{r}
11 \\
288 \overline{)3168} \\
288 \\
\hline
288 \\
288
\end{array}
$$

The GCD of 3456 and 10080 = 288

$$\text{The LCM of 3456 and 10080} = \frac{3456 \cdot 10080}{288}$$

$$= \frac{288 \cdot 12 \cdot 10080}{288}$$

$$= 120{,}960$$

The least common multiple is used in the addition of fractions. The least common denominator of two fractions is the LCM of their denominators.

Example 5 Add $\frac{2}{15}$ to $\frac{3}{20}$.

Solution The LCM of 15 and 20 = 60

$$\frac{2}{15} = \frac{8}{60} \quad \text{and} \quad \frac{3}{20} = \frac{9}{60}$$

Thus,

$$\frac{2}{15} + \frac{3}{20} = \frac{8}{60} + \frac{9}{60} = \frac{17}{60}.$$

2.5.6 Unproved Conjectures (Optional)

Twin primes are a pair of prime numbers whose difference is 2. For example, (3, 5), (5, 7), (11, 13), (17, 19) are pairs of twin primes. An unproved conjecture states that there are infinitely many twin primes. Although this statement resembles Euclid's theorem in its content and in its simplicity, no one has been able to prove it.

Another unproved statement is the famous Goldbach conjecture made in 1742 in a letter to Euler. Goldbach suggested that every even natural number, except 2, could be expressed as the sum of two primes. For example, $6 = 3 + 3$, $8 = 5 + 3$, and so on.

One of the most challenging problems in number theory is to find a formula that would produce prime numbers only, that is, a prime number generating function. The primes seem to be scattered among the natural numbers in some kind of pattern. Even though it seems reasonable to assume the existence of a pattern, no one has been able to give a precise description of it. A step in this direction is provided by the expression $n^2 - n + 41$ discovered by Euler. This expression produces prime numbers for each value of n from 1 to 40. When

$$n = 41, \quad n^2 - n + 41 = 41^2 - 41 + 41 = (41)^2$$

which is not prime.

EXERCISES 2.5

1–4. List the elements of the following subsets of the natural numbers, N. (Let $x \in N$.)

1. $A = \{x \mid 10 - x \in N\}$
2. $B = \left\{x \mid \dfrac{12}{x} \in N\right\}$
3. $C = \{x \mid x - 10 \in N\}$
4. $D = \left\{x \mid \dfrac{x}{12} \in N\right\}$

5–8. List the prime factors of each of the following:

5. 72
6. 780
7. 4235
8. 3289

9–10. Some prime numbers can be written in the form $n^2 + 1$, where $n \in N$. For example, $5 = 2^2 + 1$.

9. Find three more primes of the form $n^2 + 1$.
10. Are all numbers of the form $n^2 + 1$ prime? Justify your answer.
11. Find a prime number of the form $n^2 - 1$, where $n \in N$.
12. How many prime numbers of the form $n^2 - 1$, $n \in N$, exist? Make a conjecture.
13. A conjecture states that $f(n) = n^2 + n + 41$ is a formula for generating prime numbers. Disprove this conjecture by a counterexample.
14. A conjecture states that $f(n) = n^2 - 79n + 1601$ is a formula for generating prime numbers. Disprove this conjecture by a counterexample.

15–20. Find the GCD for the sets of numbers in each of the following exercises:

15. 630, 1540
16. 48, 126
17. 2340, 4140
18. 1925, 630
19. 27, 45, 120
20. 42, 56, 280

21–26. Find the LCM for the sets of numbers in each of the following exercises:

21. 18, 84
22. 24, 90
23. 96, 84
24. 90, 135
25. 8, 18, 27
26. 12, 20, 32

27–28. A number is said to be *perfect* if it is the sum of its proper divisors. For example, the proper divisors of 6 are 1, 2, and 3, and $1 + 2 + 3 = 6$.

27. Find the next largest perfect number after 6.
28. Show that 496 is a perfect number.

29–30. Euclid states in his *Elements* that if $2^n - 1$ is a prime number ($n \in N$), then $2^{n-1}(2^n - 1)$ is a perfect number.

29. Use Euclid's formula to show that 496 is a perfect number. What is the corresponding value for n?
30. For what value of n does Euclid's formula for perfect numbers yield the perfect number 8128?

CHAPTER SUMMARY

Cartesian product of two sets $A \times B = \{(a, b) | a \in A \text{ and } b \in B\}$.

Relation A relation on A to B is a subset of $A \times B$. A relation on A is a subset of $A \times A$.

Definition of $x \, r \, y$ $x \, r \, y$ if and only if $(x, y) \in r$, a relation.

Definition of Domain $\text{Dm} = \{x | (x, y) \in r\}$.

Definition of Range $\text{Rn} = \{y | (x, y) \in r\}$.

Function A function f is a relation such that if $(x, y) \in f$ and $(x, z) \in f$, then $y = z$.

Operation An operation on a set A is a function whose domain is $A \times A$ and whose range is a subset of A.

Definition of Equality $x = y$ (Substitution Property) $x = y$ if and only if x may replace y or y may replace x without changing the truth value of the statement in which the replacement is made.

Theorem (Equality is an Equivalence Relation)

(1) Reflexive $x = x.$
(2) Symmetric If $x = y$, then $y = x.$
(3) Transitive If $x = y$ and $y = z$, then $x = z.$

The Peano Postulates (for a set N, an element 1, a relation s on N)

(1) $1 \in N.$
(2) There is exactly one $s(x)$ in N for all x in $N.$
(3) There is no x in N so that $s(x) = 1.$
(4) If $s(x) = s(y)$, then $x = y.$
(5) Principle of mathematical induction: If S is a subset of N and if $1 \in S$ and if $s(x) \in S$ whenever $x \in S$, then $S = N.$

Theorems for Natural Numbers

(1) Closure, Addition There is exactly one $x + y$ in $N.$
(2) Closure, Multiplication There is exactly one xy in $N.$
(3) Associative, Addition $(x + y) + z = x + (y + z).$
(4) Associative, Multiplication $(xy)z = x(yz).$
(5) Commutative, Addition $x + y = y + x.$
(6) Commutative, Multiplication $xy = yx.$
(7) Distributive $x(y + z) = xy + xz.$
(8) Identity, Multiplication There is exactly one element, 1, in N so that $x \cdot 1 = 1 \cdot x = x.$

Definition of $a < b$ $a < b$ if and only if $c \in N$ and $a + c = b.$

Definition of $a > b$ $a > b$ if and only if $b < a.$

The Trichotomy Axiom Exactly one of the following is valid:

$$a = b \quad \text{or} \quad a < b \quad \text{or} \quad a > b.$$

Fundamental Theorem of Arithmetic Each natural number greater than 1 is either prime or can be expressed as a product of prime factors in exactly one way, disregarding the order in which the factors are written.

Division Algorithm If $a > b$, then there is exactly one $q \in N$ and exactly one $r \in N$ so that either $a = bq$ or $a = bq + r$ where $r < b.$

REVIEW EXERCISES

1–10. Let $A = \{1, 2, 3, 4\}$, $B = \{1, 2, 3\}$.

1. List the elements of $A \times A$.
2. List the elements of $A \times B$.
3. Consider the set C, where $C \subset (A \times A)$ such that
$$C = \{(1, 1), (2, 2), (3, 3), (4, 4)\}.$$
Does C define a function?
4. Describe set C in Exercise 3 by the rule method or set-builder notation.
5. Does set C in Exercise 3 define an equivalence relation? Justify.
6. Let the relation * be defined on the set of people living in California. Let a person p be related to a person q such that p is a son or daughter of q. Is the relation * thus defined reflexive, symmetric, or transitive? Justify.

7–10. Prove by mathematical induction.

7. $1 + 3 + 3^2 + \ldots + 3^{n-1} = \dfrac{3^n - 1}{2}$

8. $1 \cdot 3 + 2 \cdot 3^2 + 3 \cdot 3^3 + \ldots + n \cdot 3^n = \dfrac{(2n - 1)3^{n+1} + 3}{4}$

9. $1 \cdot 2 + 2 \cdot 3 + 3 \cdot 4 + \ldots + n(n + 1) = \dfrac{n(n + 1)(n + 2)}{3}$

10. $\dfrac{1}{2 \cdot 5} + \dfrac{1}{5 \cdot 8} + \dfrac{1}{8 \cdot 11} + \ldots + \dfrac{1}{(3n - 1)(3n + 2)} = \dfrac{n}{6n + 4}$

11–15. Let $W = \{0, 1, 2, 3, \ldots\}$. Let $\oplus$ be a binary operation defined on W as follows: Let x and y be in W; then $x \oplus y = x^y$, where x and y are not both zero.

11. Simplify $4 \oplus 3$.
12. Is W closed with respect to $\oplus$?
13. Is $\oplus$ commutative?
14. Is $\oplus$ associative?
15. Is there an identity element in W with respect to $\oplus$?

16–20. Let A, B, and C be subsets of N, the set of natural numbers, such that
$$A = \{x \mid x < 10\}$$
$$B = \{x \mid 3 < x \leq 12\}$$
$$C = \{x \mid 6 < x\}.$$

Describe each of the following sets as a single set using the order symbols.

16. $A \cup B$
17. $A \cap B$
18. $(A \cup B) \cap C$
19. $A \cap (B \cup C)$
20. $(A \cap B) \cup C$

21–24. List the elements of each of the following subsets of N. (Let $x \in N$.)

21. $A = \{x \mid 10 - 2x \in N\}$
22. $B = \{x \mid 2x - 10 \in N\}$
23. $C = \left\{x \left| \dfrac{10}{2x} \in N \right.\right\}$
24. $D = \left\{x \left| \dfrac{2x}{10} \in N \right.\right\}$

25–28. List the prime factors of each of the following:

25. 198
26. 295
27. 288
28. 8910

29. Find the GCD of 86 and 124.
30. Use the result of Exercise 29 to find the LCM of 86 and 124.
31. List three pairs of twin primes other than those listed in the chapter.
32. How many pairs of twin primes are there less than 100?

33–35. Fermat conjectured that $f(n) = 2^{2^n} + 1$ is prime for all natural numbers n.

33. Find the numbers generated by $n = 1, 2, 3, 4$. $[f(1), f(2), f(3), f(4)]$
34. Are all the numbers in Exercise 33 prime?
35. The formula has been proved to fail for $n = 5$. Show that $f(5)$, the number generated by 5, is not prime.
36. Two numbers are said to be *amicable* if each is the sum of the proper divisors of the other. Show that 284 and 220 are amicable numbers.

3
THE INTEGERS

Man created the natural numbers during the prehistorical stage of his existence, primarily for the purpose of keeping records. Fractions were known to the ancient Egyptians and Babylonians before 2000 B.C. In contrast, zero and the negative numbers struggled for recognition for many years. An awareness of the idea of zero can be traced to the Babylonians who, around 300 B.C., sometimes used a symbol to indicate an empty place in their positional system of numeration. The Hindus in India, however, are credited with establishing the concept of the number zero, sometime between 400 A.D. and 800 A.D.

While the idea of a negative quantity can be traced to the Chinese before 200 B.C. and to the Greek Diophantus around 275 A.D. who wrote in his *Arithmetica* that the equation $4x + 20 = 4$ is absurd (its solution being $x = -4$), the Hindu Brahmagupta (*c.* 628) is the first to actually write of "negative and affirmative quantities" and also the first to state the rules of signs. In 1225 Fibonacci interpreted a negative number as a financial loss instead of a gain. By the middle of the sixteenth century the operations with negative numbers were well understood even though their interpretation for practical applications was not. Cardan in 1545 referred to the positive numbers as true numbers (numeri veri) and to the negative numbers as fictitious numbers (numeri ficti). Descartes, in his *La Géométrie* of 1637, speaks of true roots and false roots (racines fausses). Finally, through the work of Descartes and others, the negative number was fully understood both theoretically and practically.

Theoretically, zero and the negative numbers need to be included in the number system so that it will be closed with respect to subtraction. This means that the equation $x + a = b$ will always have a solution, $x = b - a$.

Practically, negative numbers are interpreted as the measurements of quantities *opposite* in nature to other quantities, with zero assigned as a fixed reference point. For example, 30°F or +30°F means 30° above zero while −30°F means 30° below zero.

3.1 THE SET OF INTEGERS

If the set of natural numbers N is extended by uniting it with the number zero, designated symbolically by 0, and the negatives of the natural numbers, designated symbolically by $-n$ if n is a natural number, then the new set obtained is called the set of integers, I. Since N is a subset of I, $N \subset I$, N is said to be *embedded* in I.

3.1.1 Definitions

Definition of the Integers, I

$$I = \{x \mid x = n \quad \text{or} \quad x = 0 \quad \text{or} \quad x = -n \quad \text{where} \quad n \in N\}$$

Maintaining the order of the natural numbers, the integers may be indicated as follows:

$$I = \{\ldots, -3, -2, -1, 0, 1, 2, 3, \ldots\}.$$

The natural numbers are also called the **positive** integers and their negatives are called the **negative** integers. If a and b are two natural numbers, then $a < b$ means a precedes b or a is to the left of b in the order above. Similarly, $a > b$ means a follows b or a is to the right of b in the order above. It is consistent with this meaning to interpret $x > 0$ as x is to the right of 0 or x is positive and $x < 0$ as x is to the left of 0 or x is negative.

Definition of $x > 0$, $x \in I$ $x > 0$ if and only if x is a positive integer.

Definition of $x < 0$, $x \in I$ $x < 0$ if and only if x is a negative integer.

$\ldots, -3, -2, -1$	0	$1, 2, 3, \ldots$
negative integers	, zero ,	positive integers
$x < 0$	$x = 0$	$x > 0$

The definitions for the addition and multiplication of integers are motivated by a desire to retain as many properties of the natural numbers as possible. If two integers are natural numbers, then their sum and product should be the same as

THE SET OF INTEGERS

defined previously. If either of the two integers is not a natural number, then their sum and product should be defined so that the commutative, associative, and distributive properties will still be valid, if this is possible. Definitions which accomplish these objectives are stated below.

Definition of Addition of Integers (Let n and m be any natural numbers.)

(1) $n + m$ is the number defined previously for natural numbers.
(2) $n + 0 = n; 0 + n = n; 0 + 0 = 0$.
(3) $-n + 0 = -n; 0 + (-n) = -n$.
(4) $(-n) + (-m) = -(n + m)$.
(5) $n + (-m) = (-m) + n = \begin{cases} n - m & \text{if } n > m \\ -(m - n) & \text{if } n < m \\ 0 & \text{if } n = m. \end{cases}$

Definition of Subtraction of Integers For all integers x and y, $x - y = x + (-y)$.

Definition of Multiplication of Integers (Let n and m be any natural numbers.)

(1) nm is the number defined previously for natural numbers.
(2) $n \cdot 0 = 0; 0 \cdot n = 0; 0 \cdot 0 = 0$.
(3) $(-n) \cdot 0 = 0; 0 \cdot (-n) = 0$.
(4) $n(-m) = (-n)m = -(nm) = -nm$.
(5) $(-n)(-m) = nm$.

For natural numbers, $n < m$ if and only if there exists a natural number x so that $n + x = m$. For any two integers, a and b, there always exists an integer x so that $a + x = b$; namely, $x = b - a$. However, by observing the order in the arrangement, $\ldots, -3, -2, -1, 0, 1, 2, 3, \ldots$, and by retaining the meaning that $a < b$ if and only if a precedes b in this arrangement, the following definition preserves the order for the natural numbers:

Definition of $a < b$ (Let a and b be any integers.) $a < b$ if and only if there exists a *natural number* c so that $a + c = b$.

Definition of $a > b$ (Let a and b be any integers.) $a > b$ if and only if $b < a$.

3.1.2 A System for the Integers

There are two essentially different ways to develop an abstract mathematical system that can be interpreted as describing the set of integers. One way is to extend Peano's system based on the definitions of the previous section. Another way is to begin with Theorems (1)–(8) of Peano's system, modify these statements appropriately, and then accept the resulting statements as axioms. This is the system described below.

UNDEFINED CONCEPTS
    A set I,
    An operation $+$ in I (called addition),
    An operation $\cdot$ in I (called multiplication),
    A relation $>$ on I.

Notation $xy = x \cdot y$.

Axioms

(1)	Closure, Addition	For all x and y in I, there is exactly one $x + y$ in I.
(2)	Closure, Multiplication	For all x and y in I, there is exactly one xy in I.
(3)	Commutative, Addition	For all x and y in I, $x + y = y + x$.
(4)	Commutative, Multiplication	For all x and y in I, $xy = yx$.
(5)	Associative, Addition	For all x, y, and z in I, $(x + y) + z = x + (y + z)$.
(6)	Associative, Multiplication	For all x, y, and z in I, $(xy)z = x(yz)$.
(7)	Identity, Addition	There is exactly one element 0 in I so that for all x in I, $x + 0 = x$ and $0 + x = x$.
(8)	Identity, Multiplication	There is exactly one element 1 in I so that for all x in I, $x \cdot 1 = x$ and $1 \cdot x = x$.
(9)	Inverse, Addition	For each x in I, there is exactly one number $-x$ (called the **opposite** or **additive inverse** of x) in I so that $x + (-x) = 0$ and $(-x) + x = 0$.
(10)	Distributive	For all x, y, and z in I, $x(y + z) = xy + xz$.
(11)	Trichotomy	For all x in I, exactly one of the following is valid. $x > 0$ or $x = 0$ or $-x > 0$.
(12)	Positive Closure, Addition	If $x > 0$ and $y > 0$, then $x + y > 0$.
(13)	Positive Closure, Multiplication	If $x > 0$ and $y > 0$, then $xy > 0$.

(14)	Finite Induction	Let $P = \{x \mid x > 0 \text{ and } x \in I\}$ If S is a subset of P so that
		   (1) $1 \in S$, and
		   (2) $x + 1 \in S$ if $x \in S$,
		 then $S = P$.

THE SET OF INTEGERS

Definition $x + y + z = (x + y) + z.$

Definition $xyz = (xy)z.$

Definition of Subtraction $x - y = x + (-y).$

Definition of Positive x is positive if and only if $x > 0.$

Definition of Negative x is negative if and only if $-x > 0.$

Definition of $x > y$ $x > y$ if and only if $x - y > 0.$

Definition of $x < y$ $x < y$ if and only if $y > x.$

The closure axiom states that addition and multiplication are operations *on* I. By definition of an operation ∘ in a set S, there is exactly one element $x \circ y$ in S for each (x, y) in the domain of the operation. According to the closure axiom, the domain of both $+$ and $\cdot$ is $I \times I$.

It may be observed that there is no inverse axiom for multiplication. This is related to the fact that the set of integers is not closed with respect to division. For example, $\frac{1}{2}$ is not an integer.

The trichotomy axiom states that the set of integers may be divided into three mutually disjoint subsets (the intersection of any two is empty); namely, the set of negative integers ($x < 0$), the set consisting of zero only ($x = 0$), and the set of positive integers ($x > 0$).

The order axioms and definitions preserve the order relation for the set of natural numbers, which may be identified as the set of positive integers. Theorems previously established for the set of natural numbers may be shown to be valid for the set of positive integers.

Some of the important theorems of this system are presented below. For Theorems 1–11, x, y, and z are integers.

Theorem 1 Addition Theorem If $x = y$, then $x + z = y + z.$

> *Proof* $x + z = x + z$ Reflexive theorem of equality
> $x = y$ Given
> $x + z = y + z$ Substitution.

Theorem 2 Multiplication Theorem If $x = y$, then $xz = yz.$

> *Proof* $xz = xz$ Reflexive theorem of equality
> $x = y$ Given
> $xz = yz$ Substitution.

Theorem 3 Addition Cancellation Theorem If $x + z = y + z$, then $x = y$.

Proof $x + z = y + z$	Given
$-z \in I$	Inverse axiom, Addition
$(x + z) + (-z) = (y + z) + (-z)$	Addition theorem
$x + [z + (-z)] = y + [z + (-z)]$	Associative axiom, Addition
$x + 0 = y + 0$	Inverse axiom, Addition
$x = y$	Identity axiom, Addition.

Theorem 4 Zero Factor Theorem For all x in I, $x \cdot 0 = 0$.

Proof $0 + 0 = 0$	Identity axiom, Addition
$x(0 + 0) = x \cdot 0$	Multiplication theorem
$(x \cdot 0) + (x \cdot 0) = x \cdot 0$	Distributive axiom
$(x \cdot 0) + (x \cdot 0) = (x \cdot 0) + 0$	Identity axiom, Addition
$x \cdot 0 = 0$	Addition cancellation theorem.

Theorem 5 Opposite of an Opposite Theorem $-(-x) = x$.

Proof $-x + [-(-x)] = 0$	Inverse axiom, Addition
$x + (-x + [-(-x)]) = x + 0$	Addition theorem
$x + (-x + [-(-x)]) = x$	Identity axiom
$[x + (-x)] + [-(-x)] = x$	Associative axiom, Addition
$0 + [-(-x)] = x$	Inverse axiom, Addition
$-(-x) = x$	Identity axiom, Addition.

Theorem 6 Opposite of a Sum Theorem $-(x + y) = (-x) + (-y)$.

Proof $-(x + y) + (x + y) = 0$	Inverse axiom, Addition
$[-(x + y) + (x + y)] + [(-x) + (-y)]$ $= 0 + [(-x) + (-y)]$	Addition theorem
$[-(x + y) + (x + y)] + [(-x) + (-y)]$ $= (-x) + (-y)$	Identity axiom, Addition
$-(x + y) + [(x + (-x)) + (y + (-y))]$ $= (-x) + (-y)$	Associative axiom, Addition; Commutative axiom, Addition
$-(x + y) + (0 + 0) = (-x) + (-y)$	Inverse axiom, Addition
$-(x + y) = (-x) + (-y)$	Identity axiom, Addition.

Theorem 7 Opposite of a Difference $-(x - y) = y - x$.

Proof $-(x - y) = -(x + (-y))$	Definition of subtraction
$= (-x) + (-(-y))$	Opposite of a sum theorem
$= -x + y$	Opposite of an opposite theorem
$= y + (-x)$	Commutative axiom, Addition
$= y - x$	Definition of subtraction.

THE SET OF INTEGERS

Theorem 8 Opposite of a Product $-xy = (-x)y$.

Proof
$(-x)y + [xy + (-xy)] = (-x)y$	Identity and inverse axioms, Addition
$[(-x)y + xy] + (-xy) = (-x)y$	Associative axiom, Addition
$(-x + x)y + (-xy) = (-x)y$	Distributive axiom
$0 \cdot y + (-xy) = (-x)y$	Inverse axiom, Addition
$0 + (-xy) = (-x)y$	Zero factor theorem
$-xy = (-x)y$	Identity axiom, Addition.

Corollary 1 $(-x)y = -xy$.

Corollary 2 $x(-y) = -xy$.

Theorem 9 Product of Two Opposites $(-x)(-y) = xy$.

Proof
$-y + y = 0$	Inverse axiom, Addition
$-x(-y + y) = -x \cdot 0$	Multiplication theorem
$-x(-y) + (-x)y = -x \cdot 0$	Distributive axiom
$-x(-y) + (-x)y = 0$	Zero factor theorem
$-x(-y) + (-xy) = 0$	Opposite of a product theorem
$[-x(-y) + (-xy)] + xy = xy$	Addition theorem
$(-x)(-y) + [-xy + xy] = xy$	Associative axiom, Addition
$(-x)(-y) + 0 = xy$	Inverse axiom, Addition
$(-x)(-y) = xy$	Identity axiom, Addition

Theorem 10 Zero Product Theorem If $xy = 0$, then $x = 0$ or $y = 0$.

This theorem can be divided into the following two cases.

CASE 1 If $x = 0$ and $xy = 0$, then $x = 0$ or $y = 0$.
CASE 2 If $x \neq 0$ and $xy = 0$, then $x = 0$ or $y = 0$.

Case 1 is valid because of the definition of the connective "or." Case 2 is proved as follows:

Proof
$xy = 0$	Given
$xy + x = 0 + x$	Addition theorem
$xy + x = x$	Identity axiom, Addition
$xy + x \cdot 1 = x$	Identity axiom, multiplication
$x(y + 1) = x$	Distributive axiom
$x \cdot 1 = x$	Identity axiom, Multiplication
$y + 1 = 1$	Identity axiom, Multiplication (There is *exactly one* element in I so that for all x in I, $x \cdot 1 = x$)
$y + 1 = 0 + 1$	Identity axiom, Addition
$y = 0$	Addition cancellation theorem.

Theorem 11 Multiplication Cancellation Theorem If $z \neq 0$ and $xz = yz$, then $x = y$.

Proof	
$xz = yz$	Given
$xz + (-yz) = yz + (-yz)$	Addition theorem
$xz + (-yz) = 0$	Inverse axiom, Addition
$xz + (-y)z = 0$	Opposite of a product theorem
$[x + (-y)]z = 0$	Distributive axiom and commutative axiom
$z \neq 0$	Given
$x + (-y) = 0$	Zero product theorem
$[x + (-y)] + y = 0 + y$	Addition theorem
$[x + (-y)] + y = y$	Identity axiom, Addition
$x + (-y + y) = y$	Associative axiom, Addition
$x + 0 = y$	Inverse axiom, Addition
$x = y$	Identity axiom, Addition.

It may be noted that the definitions for addition and multiplication, stated in the preceding section, are now theorems, since they can be derived from the axioms of the system. One man's axiom may be another man's theorem, depending on how he chooses to develop a mathematical system.

The multiplication cancellation theorem implies that some equations involving the concept of division can be solved over the set of integers. For example if $2x = 6$, then $2x = 2 \cdot 3$ and $x = 3$. However, the equation $2x = 1$ can *not* be solved over the set of integers. Although $2x = 2 \cdot \frac{1}{2}$ and $x = \frac{1}{2}$, the fraction $\frac{1}{2}$ is not in the set of integers and thus the multiplication cancellation theorem for integers does not apply.

On the other hand, the addition cancellation theorem, together with the inverse axiom for addition, imply that all equations of the form $x + a = b$ are solvable over the integers; that is, $x = b + (-a)$. For example, if $x + 5 = 3$, then $x = 3 + (-5) = -2$.

EXERCISES 3.1

1–10. Justify each of the following statements by an axiom, definition, or theorem of the system for the integers. (Assume x, y, and z are integers.)

1. $2(x + 3) = 2x + 6$
2. $-3(2x) = -6x$
3. $3x(x + 2) = (x + 2)3x$
4. $3x(y) = 3(xy)$
5. $x + 0 = x$
6. $z \cdot 1 = 1 \cdot z = z$
7. xy is an integer
8. $(x + y) + z = x + (y + z)$
9. If $x + y = x + z$, then $y = z$
10. $x + (-x) = 0$

THE SET OF INTEGERS

11–18. Using only definitions and axioms, prove each of the following. (Assume all variables are integers.)

11. $x + 0 = 0 + x = x$
12. $x \cdot 1 = 1 \cdot x = x$
13. $(x + y)z = xz + yz$
14. $x(y - z) = xy - xz$
15. $x(y + z + t) = xy + xz + xt$
16. $(x + a)(x + a) = x^2 + 2ax + a^2$
17. $(x + a)(x + b) = x^2 + (a + b)x + ab$

18–24. Using definitions, axioms, or theorems, prove each of the following. (Assume all variables are integers.)

18. $(x + a)(x - a) = x^2 - a^2$
19. $x - (y + z) = (x - y) - z$
20. $x - (y - z) = (x - y) + z$
21. $-x(y + z) = -xy - xz$
22. $-x(y - z) = -xy + xz$
23. $(-x)(-y)(-z) = -xyz$
24. $x^4 + 4y^4 = (x^4 + 4x^2y^2 + 4y^4) - 4x^2y^2 = (x^2 + 2xy + 2y^2)(x^2 - 2xy + 2y^2)$

25–30. Using the definitions $2 = 1 + 1$, $3 = 2 + 1$, $4 = 3 + 1$, $5 = 4 + 1$, and $6 = 5 + 1$, prove:

25. $2 + 2 = 4$
26. $3 + 2 = 5$
27. $2 \cdot 2 = 4$
28. $3 \cdot 2 = 6$
29. $1 < 2$
30. $-2 < -1$

31–36. Solve over I or prove unsolvable over I. Justify each step involved.

31. $2x + 3 = 5$
32. $x(x - 2) = 0$
33. $3x + 2 = 6$
34. $2 - x = 6$
35. $3 - 2x = 6$
36. $3x + 2 = 2x + 3$

37. Show that $-$ is an operation on I; that is, show that the domain is $I \times I$, and that there is exactly one $x - y$ for each x and y in I.
38. Show that $<$ is an order relation; that is, prove that $<$ is transitive.
39. What is the domain of the operation defined as follows:

$$\frac{x}{y} = z \leftrightarrow \text{there is exactly one } z \text{ in } I \text{ such that } x = yz.$$

40. Is the set of integers well ordered with respect to $<$? (Does every nonempty subset contain a smallest element?)

3.2 GROUPS, RINGS, INTEGRAL DOMAINS (Optional)

Certain mathematical systems such as the number systems of elementary algebra may be classified with respect to their structural properties. The study of abstract systems concerned with these properties is called *modern algebra*.

3.2.1 Groups

A group is one of the simplest mathematical systems since it has only one operation.

Definition of Group A group is a system consisting of a set S and an operation $\circ$ such that

(1)	Closure	For all x and y in S, there is exactly one $x \circ y$ in S.
(2)	Associative	For all x, y, and z in S, $(x \circ y) \circ z = x \circ (y \circ z)$.
(3)	Identity	There exists an element e in S so that for all x in S, $x \circ e = x$.
(4)	Inverse	For each x in S, there exists an element x' in S so that $x \circ x' = e$.

If a system consisting of a set S and an operation $\circ$ is a group, it is conventional to say "S is a group with respect to $\circ$" or simply "S is a group" when there is no doubt as to what operation is meant.

If the commutative property is also valid for the group, then the group is called a **commutative group**.

Definition of Commutative Group A group S is a commutative group if and only if for all elements x and y in S, $x \circ y = y \circ x$.

A familiar example of a commutative group is the system consisting of the set of integers and the operation $+$. An example of a group that is not commutative is the system of a set of matrices and matrix multiplication, to be discussed in chapter 11. The set of all rotations in 3-space is another example of a noncommutative group.

Theorems that follow directly from the axioms are the following:

Theorem 1 $e \circ x = x \circ e = x$. (The right identity is a left identity.)

Theorem 2 e is unique. (There is exactly one identity element.)

GROUPS, RINGS, INTEGRAL DOMAINS (OPTIONAL)

Theorem 3 $x' \circ x = x \circ x' = e$. (The right inverse is also a left inverse.)

Theorem 4 x' is unique. (Each element has exactly one inverse.)

Theorem 5 If $x \circ y = x \circ z$, then $y = z$. (Left cancellation.)

Theorem 6 If $y \circ x = z \circ x$, then $y = z$. (Right cancellation.)

Theorem 7 $(x')' = x$. (The inverse of an inverse is the original element.)

3.2.2 Rings

A ring is one of the simplest algebraic systems having two operations.

Definition of Ring A ring is a system consisting of a set S, an operation $+$, and an operation $\cdot$ so that for all x, y, and z in S

(1) Closure, $+$ There is exactly one $x + y$ in S.
(2) Associative, $+$ $(x + y) + z = x + (y + z)$.
(3) Identity, $+$ e is in S so that for all x in S, $x + e = x$.
(4) Inverse, $+$ For each x in S, there is an element $-x$ in S so that $x + (-x) = e$.
(5) Commutative, $+$ $x + y = y + x$.
(6) Closure, $\cdot$ There is exactly one $x \cdot y$ in S.
    Convention $xy = x \cdot y$.
(7) Associative, $\cdot$ $(xy)z = x(yz)$.
(8) Distributive $x(y + z) = xy + xz$ (left),
     $(y + z)x = yx + zx$ (right).

Note that the system $S, +$ is a commutative group. If the system $S, \cdot$ is commutative, then the ring is called a **commutative ring**.

The set of even integers $\{\ldots, -4, -2, 0, 2, 4, \ldots\}$ is an example of a commutative ring without a multiplication ($\cdot$) identity and without multiplication inverses. The set of all integers is an example of a commutative ring with a multiplication ($\cdot$) identity but without multiplication inverses. A set of matrices furnishes an example of a noncommutative ring with a multiplication ($\cdot$) identity and without multiplication inverses for all its elements (some of its elements do have multiplication inverses).

3.2.3 Integral Domains

Certain commutative rings have been singled out for study because they describe certain number systems of elementary algebra. It may happen in a ring that $xy = 0$ (the additive identity) but $x \neq 0$ and $y \neq 0$. If x or y has this property, then it is called a **zero divisor**.

Definition of Zero Divisor x and y are zero divisors if and only if $x \neq 0$ and $y \neq 0$ and $xy = 0$.

Definition of Integral Domain An integral domain is a commutative ring in which there are no zero divisors.

An integral domain does not necessarily have an identity for the $\cdot$ operation. If it does, then it is called an integral domain with unit element. The set of integers is an example of an integral domain *with* unit element. The set of even integers is an example of an integral domain *without* unit element.

One important motivation for the study of these abstract systems is to determine, without being influenced by the special subject matter, how the properties of the number systems are related to one another.

The word "identity" has been selected as a general term to describe a number, e, with the special property that $x \circ e = x$. For ordinary addition, $e = 0$ and $x + 0 = x$ for all x. For ordinary multiplication, $e = 1$ and $x \cdot 1 = x$ for all x. Thus, the identity element may be different for different operations.

Similarly, the word "inverse" has been selected as a general term to describe a number x' so that $x \circ x' = e$, the identity. Again this inverse element may be different for different operations. For example, if $x = 2$, then -2 is the inverse for ordinary addition since $2 + (-2) = 0$, the addition identity, and $\frac{1}{2}$ is the inverse for ordinary multiplication since $2 \cdot \frac{1}{2} = 1$, the multiplication identity.

As another example, consider the system consisting of the set $A = \{2, 4, 6, 8\}$ and an operation * meaning $a * b$ is the remainder after dividing ab by 10. The identity element for this operation is 6 since for all x in A, $x * 6 = x$. ($2 * 6 = 2$, $4 * 6 = 4$, $6 * 6 = 6$, and $8 * 6 = 8$.)

For the inverses of this system,

$$2 * 2' = 6 \quad \text{and thus} \quad 2' = 8.$$
$$4 * 4' = 6 \quad \text{and thus} \quad 4' = 4.$$
$$6 * 6' = 6 \quad \text{and thus} \quad 6' = 6.$$
$$8 * 8' = 6 \quad \text{and thus} \quad 8' = 2.$$

EXERCISES 3.2 (Optional)

1. Show that the system $\{-1, 1\}$, with the operation multiplication, is a group. (Verify each of the axioms.)
2. Show that the tables below define a group. (Verify the axioms.)

*	e	a	b	c
e	e	a	b	c
a	a	b	c	e
b	b	c	e	a
c	c	e	a	b

$\circ$	e	a	b	c
e	e	a	b	c
a	a	e	c	b
b	b	c	e	a
c	c	b	a	e

GROUPS, RINGS, INTEGRAL DOMAINS (OPTIONAL)

Is it possible to rename the elements of these groups and to rearrange their order in the headings so that the resulting tables will be identical? Explain.

3. Which of the following tables describe a group? Justify either by verifying the axioms or by stating a counterexample to contradict one of the axioms.

*	A	B
A	A	A
B	B	A

#	0	1	2
0	0	0	0
1	0	2	0
2	0	0	1

$	0	1	2
0	0	1	2
1	1	2	0
2	2	0	1

¢	A	B	C	D
A	A	B	C	D
B	B	D	B	D
C	C	B	A	D
D	D	D	D	D

@	0	1	2
0	0	0	0
1	0	1	2
2	0	2	1

4–7. Which of the following sets and operations defined on these sets describe a group? Justify as in Exercise 3.

4. $S = \left\{ \dfrac{p}{q} \middle| p \text{ and } q \text{ are natural numbers} \right\}$; multiplication.

5. $S = \{0, 1, 2, 3, \ldots\}$; addition.

6. $S = $ Set of Integers; subtraction.

7. $S = \{x\}$; operation is defined by $x \circ x = x$.

8. Consider the rectangle $ABCD$:

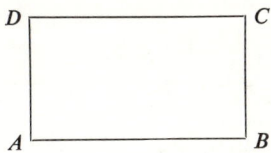

Let the following movements of the rectangle be defined:

$H = 180°$ rotation about the horizontal axis.

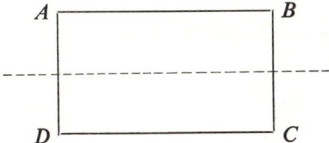

V = 180° rotation about the vertical axis.

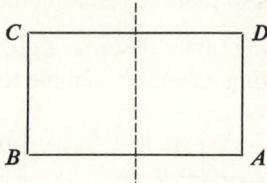

R = 180° rotation in clockwise direction.

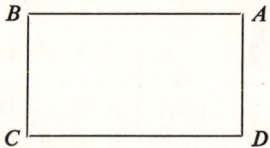

I = 360° rotation in clockwise direction.

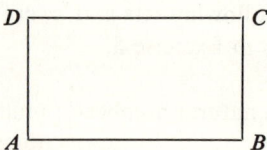

Let "op" define an operation on the set $\{H, V, R, I\}$, which means to perform the movements. For example, I op H means: rotate the rectangle 360° in clockwise direction,

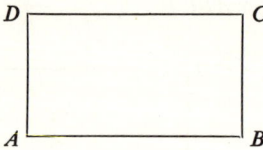

and then rotate 180° about the horizontal axis.

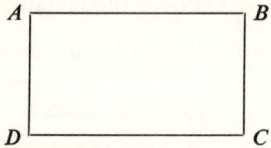

Thus I op H results in H. V op R means to rotate the rectangle 180° about its vertical axis

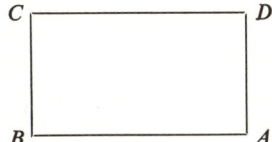

and then rotate this rectangle 180° in clockwise direction

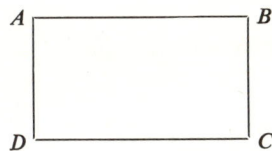

which results in position H. Thus V op $R = H$.

(a) Verify that

$$R \text{ op } V = H$$
$$V \text{ op } H = R$$
$$I \text{ op } I = I.$$

(b) Construct a table for the operation op.

op	H	V	R	I
H				
V				
R				
I				

(c) Does the system $\{H, V, R, I\}$, with the operation op, define a group? Justify.

9. Show that the set of even integers forms a ring without the multiplication identity. (Verify the axioms and prove that there is no multiplication identity.)

10–18. Consider the system $S = \{(a, b) | a \text{ and } b \text{ are integers}\}$, with addition and multiplication defined as follows:

$$(a, b) + (c, d) = (a + c, b + d)$$
$$(a, b) \cdot (c, d) = (ac, bd).$$

10. Show that S is closed for $+$ and $\cdot$.
11. Show that S is commutative for $+$ and $\cdot$.
12. Show that S is associative for $+$ and $\cdot$.
13. Show that $(0, 0)$ is the addition identity.

14. Find the addition inverse (x, y) such that $(a, b) + (x, y) = (0, 0)$.
15. Show that S is distributive for $\cdot$ over $+$. Is it distributive for $+$ over $\cdot$?
16. Find elements (x, y) and (r, s) such that $(x, y) \cdot (r, s) = (0, 0)$, but $(x, y) \neq (0, 0)$ and $(r, s) \neq (0, 0)$.
17. Is $(1, 1)$ a multiplication identity? Why?
18. Does each (a, b) except $(0, 0)$ have a multiplication inverse $[(a, b) \cdot (x, y) = (1, 1)]$? Why?

19–25. Consider the system $S = \{(n, m) | n$ and m are natural numbers$\}$ defined as follows:

$(n, m) \equiv (p, q)$ if and only if $n + q = m + p$.

$(n, m) + (p, q) \equiv (n + p, m + q)$.

$(n, m) \cdot (p, q) \equiv (np + mq, nq + mp)$.

(n, m) is "positive" if and only if for some natural number x, $m + x = n$.

19. Show that $\equiv$ is reflexive, symmetric, and transitive.
20. Show that S is closed with respect to $+$ and $\cdot$, and that $+$ and $\cdot$ are commutative and associative, and that $\cdot$ distributes over $+$.
21. Show that $(1, 1)$ is the addition identity.
22. Show that $(2, 1)$ is the multiplication identity.
23. Show that (m, n) is the addition inverse of (n, m).
24. Show that there are no zero divisors and thus the system is an integral domain.
25. Show that this system also decribes the integers by interpreting (n, m) as $n - m$, and thus $-n$ as $1 - (n + 1)$ and 0 as $1 - 1$. (Verify all the definitions for $+$, $\cdot$, and $<$ as given in Section 3.1.1.)

26–32. Which of the following are integral domains? Explain.

26. All integers divisible by 4.
27. All odd integers.
28. All positive integers.
29. $\{0\}$
30. $\{0, 1\}$ with

+	0	1
0	0	1
1	1	1

and

$\cdot$	0	1
0	0	0
1	0	1

31. $\{a + b\sqrt{2} | a$ and b are integers$\}$.
32. $S = \{0, 1, 2, 3\}$, where $a + b$ means the remainder when the ordinary sum is divided by 4, $a \cdot b$ means the remainder when the ordinary product is divided by 4.

POLYNOMIALS AND FACTORING

3.3 POLYNOMIALS AND FACTORING

Definition of x^n Let n be a natural number. Then
$$x^1 = x \quad \text{and} \quad x^n = x^{n-1} \cdot x.$$

As examples, $x^2 = x^1 x = xx$, $x^3 = x^2 x = (x^1 x)x = xxx$, and so on.

Note that the above definition agrees with the more informal description of x^n as $xx \ldots x$ (n factors of x).

Definition of Polynomial A polynomial in one variable x over the set S is an expression having the form
$$a_n x^n + a_{n-1} x^{n-1} + \ldots + a_1 x + a_0,$$
where n is a natural number and $a_0, a_1, \ldots, a_{n-1}, a_n$ are elements of S.

If $a_n \neq 0$, then n is the *degree* of the polynomial. The numbers $a_0, a_1, \ldots, a_{n-1}, a_n$ are called the *coefficients* of the polynomial, and a_n is called the *leading coefficient*. For convenience, any nonzero constant (any nonzero element of S) is called a **polynomial** of degree 0. The number 0 is called the **zero polynomial** and it is not assigned a degree.

For example,

$x^5 + 2x^3 - 3x$ is a polynomial of degree 5,
$3x^2 - x + 5$ is a polynomial of degree 2,
$6x + 7$ is a polynomial of degree 1,
3 is a polynomial of degree 0, and
0 is the zero polynomial of no degree.

If the set S is the set of integers, then all polynomials over S designate integers when x is replaced by an integer. For this reason, polynomials are also called **integral algebraic expressions**. However, the variable x is not always restricted to belong to the set of integers and a polynomial may represent numbers that are not integers.

An *integral polynomial* is a polynomial over I; that is, its coefficients are integers.

The set of polynomials in one variable x over the set of integers is an integral domain with the sum defined as
$$(a_n x^n + \ldots + a_1 x + a_0) + (b_n x^n + \ldots + b_1 x + b_0)$$
$$= (a_n + b_n)x^n + \ldots + (a_1 + b_1)x + (a_0 + b_0)$$
with 0's supplied as the coefficients of the higher degree terms if the polynomials do not have the same degree. The product is defined to be that obtained by applying the distributive, commutative, and associative axioms combined with the theorem $x^n x^m = x^{n+m}$.

Theorem $x^n x^m = x^{n+m}$ for all natural numbers n and m.

Note that
$$x^n x^m = \underbrace{(x \cdot x \ldots x)}_{n \text{ factors}} \underbrace{(x \cdot x \ldots x)}_{m \text{ factors}} = \underbrace{(x \cdot x \ldots x)}_{n + m \text{ factors}} = x^{n+m}.$$

A rigorous proof of this theorem is provided in Chapter 5.

Definition of Factor and Divisor   If $A(x)$ and $B(x)$ are polynomials and if there exists a polynomial $C(x)$ so that
$$A(x) = B(x)C(x),$$
then $B(x)$ is a factor of $A(x)$, and $B(x)$ is a divisor of $A(x)$.

Since $A(x) = A(x) \cdot 1 = -A(x) \cdot -1$, it follows that $1, -1, A(x)$, and $-A(x)$ are factors of $A(x)$, called the **trivial factors**.

If $A(x) = B(x) \cdot k$ where k is any element of S, then k is called a **constant factor** of $A(x)$. For example, 3 is a constant factor of $6x^2 - 12x + 9$ since $6x^2 - 12x + 9 = 3(2x^2 - 4x + 3)$.

There are cases when it is desirable to exclude the trivial factors so the expression **proper factor** is introduced.

Definition of Proper Factor $B(x)$ is a proper factor of $A(x)$ if and only if $B(x)$ is a nontrivial factor of $A(x)$; that is, $B(x) \neq 1$, $B(x) \neq -1$, $B(x) \neq A(x)$, and $B(x) \neq -A(x)$.

Definition of Prime Polynomial   A polynomial $P(x)$ is prime with respect to the set S if and only if the degree of $P(x) > 0$ and $P(x)$ has no proper factors whose coefficients are in S.

A prime polynomial over S is also called an **irreducible** polynomial over S. Some examples of polynomials that are prime over the set of integers are $x + 3$, $x^2 + 4$, and $2x^2 + x + 5$. On the other hand, $2x + 10$ is not prime since $2x + 10 = 2(x + 5)$, and $x^2 - 9$ is not prime since $x^2 - 9 = (x - 3)(x + 3)$.

There is a prime factorization theorem for polynomials similar to the fundamental theorem of arithmetic.

Prime Factorization Theorem   Every nonconstant polynomial over I, the set of integers, is either prime or can be expressed uniquely as the product of a constant factor and one or more prime polynomials over I, disregarding the order in which the factors are written.

To **factor a polynomial over** I means to express the polynomial as a product of a constant factor and prime polynomials over I.

POLYNOMIALS AND FACTORING

The following list of special products are, with the exception of the distributive axiom, theorems that can be derived from the axioms for the set of integers, I.

SPECIAL PRODUCTS

(1) Distributive Axiom $\qquad a(x + y) = ax + ay.$
(2) Difference of Squares $\qquad (x + a)(x - a) = x^2 - a^2.$
(3) Perfect Square $\qquad (x + a)^2 = x^2 + 2ax + a^2.$
(4) Sum of Cubes $\qquad (x + a)(x^2 - ax + a^2) = x^3 + a^3.$
(5) Difference of Cubes $\qquad (x - a)(x^2 + ax + a^2) = x^3 - a^3.$
(6) General Trinomial, I $\qquad (x + a)(x + b) = x^2 + (a + b)x + ab.$
(7) General Trinomial, II $\qquad (ax + b)(cx + d) = acx^2 + (ad + bc)x + bd.$
(8) General Distributive $\qquad (x + a)(y + b) = xy + ay + bx + ab.$

Recognition of these forms and their symmetric equivalents provides a rapid method for finding the prime factors of certain polynomials.

Example 1 Factor over I: $3x^4 - 48$.

Solution

$$3x^4 - 48 = 3(x^4 - 16) \qquad \text{Distributive axiom}$$
$$= 3(x^2 - 4)(x^2 + 4) \qquad \text{Difference of squares}$$
$$= 3(x - 2)(x + 2)(x^2 + 4) \qquad \text{Difference of squares.}$$

Example 2. Factor over I: $x^4 - x^3 + 8x - 8$.

Solution

$$x^4 - x^3 + 8x - 8 = (x^4 - x^3) + (8x - 8) \qquad$$
$$= x^3(x - 1) + 8(x - 1) \qquad \left\{ \begin{array}{l} \text{Recognition of general} \\ \text{distributive theorem} \end{array} \right.$$
$$= (x - 1)(x^3 + 8)$$
$$= (x - 1)(x + 2)(x^2 - 2x + 4) \qquad \text{Sum of cubes.}$$

Example 3 Factor over I: $2x^2 + x - 15$.

Solution $2x^2 + x - 15 = (2x + a)(x + b)$ where $ab = -15$ and $a + 2b = 1$. Trying the factors of -15: $(1, -15), (-1, 15), (3, -5), (-3, +5)$ it may finally be seen that $a = -5$ and $b = 3$. Thus, $2x^2 + x - 15 = (2x - 5)(x + 3)$.

Example 4 Factor over I: $x^4 + 6x^2 + 9 - 25x^2$.

Solution

$$\begin{aligned}
x^4 + 6x^2 + 9 - 25x^2 &= (x^2 + 3)^2 - 25x^2 & &\text{Perfect square} \\
&= (x^2 + 3 + 5x)(x^2 + 3 - 5x) & &\text{Difference of squares} \\
&= (x^2 + 5x + 3)(x^2 - 5x + 3) & &\text{Convention: arrange in} \\
& & &\text{descending powers of } x.
\end{aligned}$$

Example 5 Factor over I: $x^4 - 19x^2 + 9$.

Solution

$$\begin{aligned}
x^4 - 19x^2 + 9 &= (x^4 + 9) - 19x^2 & &\text{Recognizing form of} \\
& & &\text{Example 4} \\
&= (x^4 + 6x^2 + 9) - 19x^2 - 6x^2 & &\text{Completing the square} \\
&= (x^4 + 6x^2 + 9) - 25x^2 & & \\
&= (x^2 + 5x + 3)(x^2 - 5x + 3) & &\text{See Example 4.}
\end{aligned}$$

Note: The square in Step 2 could also have been completed as follows:

$$(x^4 - 6x^2 + 9) - 19x^2 + 6x^2 \quad \text{or} \quad (x^4 - 6x^2 + 9) - 13x^2.$$

However, 13 is not a perfect square and the difference of squares theorem cannot be used.

EXERCISES 3.3

1–24. Factor over I

1. $4a^3 + 12a^2 - a - 3$
2. $r^4 + 4t^4$
3. $r^2 + 4rs + 4s^2 - r - 2s - 6$
4. $(x^2 - 2x + 1)^2 - 3(x^2 - 2x + 1) + 2$
5. $t^6 - 64$
6. $a^4 + 3a^2b^2 + 4b^4$
7. $x^2 - 2x(y - z) + (y - z)^2$
8. $p^3(r - s)^3 + p^3$
9. $2p^3 - 3p^2 + 2p - 3$
10. $a^4 + a^2 + 1$
11. $1 - x - x^2 + x^3$
12. $5 + 5x^6$
13. $(x + 2)^2 - (x - 7)^2$
14. $x^3y - xy^3$
15. $(m + 1)(m + 2) + m + 1$
16. $4x^2(2x^2 - 3x + 2) - 100x(2x^2 - 3x + 2)$
17. $x^2 - 4z^2 + 9y^2 - 6xy$
18. $p^8 - 14p^4 + 25$
19. $x^6 - a^6$
20. $25x^2 - y^2 + 2yz - z^2$
21. $a^3 - ab^2 - a^2b + b^3$
22. $9x^3 - 9x^2 - 18x$
23. $x^4 + 2x^2y^2 + 9y^4$
24. $(y + z)^2 + (y + z) - 42$

25–28. Prove that each of the following polynomials is prime over I.

25. $x^2 + 4$
26. $x^2 + a^2$, $a \neq 0$
27. $x^2 + x + 1$
28. $x^2 \pm ax + a^2$, $a \neq 0$

29. Show that $x^4 + 4$ is *not* prime over I.

30. Factor $x^6 - y^6$ two ways:
    (a) as the difference of two squares
    (b) as the difference of two cubes.
    Justify each method of factorization.

31. Factor Example 2 in the text by associating the terms in a different way than the one shown.

32. Factor $ax + bx + ay + by$ in two different ways by associating the terms in two different ways.

33. Prove by induction that $x^n - a^n$ has a factor $x - a$ where $a \in I$, $n \in N$.

34. Prove by induction that $x^{2n+1} + a^{2n+1}$ has a factor $x + a$ where $a \in I$, $n \in N$.

35. Prove by induction that $x^{2n} - a^{2n}$ has factors $x + a$ and $x - a$ where $a \in I$, $n \in N$.

36. Prove that $x^4 + (2a - b^2)x^2 + a^2$ can be factored over I for all a and b in I.

3.4 DIVISION ALGORITHM; SYNTHETIC DIVISION

3.4.1 The Division Algorithms

Although neither the set of integers I nor the set of polynomials over I are closed with respect to the division operation, it is still possible to state a division algorithm similar to the one for the set of natural numbers ($a = bq + r$ where $a > b$ and $0 \leq r < b$) with the conditions $a > b$ and $0 \leq r < b$ replaced by the conditions the degree of $A(x) \geq$ degree of $B(x)$ and $R = 0$ or degree of $R(x) <$ degree of $B(x)$.

Theorem (The Division Algorithm for Polynomials over I) If $A(x)$ and $B(x)$ are any two polynomials over I with degree n of $A(x) \geq$ degree m of $B(x) > 0$ and the leading coefficient of $B(x)$ is 1, then there is exactly one integral polynomial $Q(x)$, called the **quotient**, and exactly one integral polynomial $R(x)$, called the **remainder** so that

$$A(x) = B(x)Q(x) + R(x),$$

where $R = 0$ or degree of $R(x) <$ degree of $B(x)$.

For example,
$$2x^5 - 3x^3 - x^2 - 3 = (x^2 - 2)(2x^3 + x - 1) + 2x - 5,$$
$$5x^3 + 15x^2 - 4x - 4 = (x + 3)(5x^2 - 4) + 8, \quad \text{and}$$
$$x^3 + 8 = (x + 2)(x^2 - 2x + 4).$$

The division algorithm may be proved by induction on n, the degree of $A(x)$. The proof involves subtracting multiples of $B(x)$ from $A(x)$ until the polynomial remaining has a degree less than that of $B(x)$. This is actually the division process illustrated for the special case below.

$$
\begin{array}{r}
2x^3 + x - 1 \\
x^2 - 2 \overline{\smash{)}2x^5 - 3x^3 - x^2 - 3} \\
\underline{2x^5 - 4x^3 } \\
x^3 - x^2 - 3 \\
\underline{x^3 - 2x } \\
-x^2 + 2x - 3 \\
\underline{-x^2 + 2} \\
2x - 5
\end{array}
$$

Thus,
$$\frac{2x^5 - 3x^3 - x^2 - 3}{x^2 - 2} = 2x^3 + x - 1 + \frac{2x - 5}{x^2 - 2}$$

with the degree of $2x - 5$ less than the degree of $x^2 - 2$.

Similarly if $B(x) \neq 0$, then $A(x) = B(x)Q(x) + R(x)$ can be expressed as

$$\frac{A(x)}{B(x)} = Q(x) + \frac{R(x)}{B(x)}.$$

3.4.2 Synthetic Division

For the division of a polynomial by a **linear divisor**, $x - r$, there is an abbreviated method called **synthetic division**. Consider the long division problem below. Note that a space has been provided for the missing x^2 term of the dividend.

$$
\begin{array}{r}
2x^3 - x^2 - 3x + 4 \\
x - 3 \overline{\smash{)}2x^4 - 7x^3 + 13x - 5} \\
\underline{2x^4 - 6x^3 } \\
-x^3 \\
\underline{-x^3 + 3x^2 } \\
-3x^2 + 13x \\
\underline{-3x^2 + 9x } \\
4x - 5 \\
\underline{4x - 12} \\
7
\end{array}
$$

DIVISION ALGORITHM; SYNTHETIC DIVISION

This work may be reduced by not writing a term that repeats the term above and by then raising the remaining terms as illustrated below.

$$
\begin{array}{r}
2x^3 - x^2 - 3x + 4 \\
x - 3 \overline{)2x^4 - 7x^3 + 13x - 5} \\
- 6x^3 \\
\hline
- x^3 \\
+ 3x^2 \\
\hline
- 3x^2 \\
+ 9x \\
\hline
4x \\
- 12 \\
\hline
7
\end{array}
\qquad
\begin{array}{r}
2x^3 - x^2 - 3x + 4 \\
x - 3 \overline{)2x^4 - 7x^3 + 13x - 5} \\
- 6x^3 + 3x^2 + 9x - 12 \\
\hline
- x^3 - 3x^2 + 4x + 7
\end{array}
$$

The work may be reduced even further by omitting the powers of x and writing only the coefficients, providing a 0 for any missing term.

$$
\begin{array}{r}
2 \; -1 \; -3 4 \\
1 - 3 \overline{)2 \; -7 0 13 \; - 5} \\
-6 3 9 \; -12 \\
\hline
-1 \; -3 4 7
\end{array}
$$

Since this division technique applies only for divisors of the form $x - r$, the coefficient of x in the divisor is always 1 and thus may be omitted. If the first coefficient of the quotient is written in the first position of the last line, then the coefficients of the quotient are repeated on this line and thus may be written on the last line only. If $-r$ is replaced by r and the products obtained by multiplying by r are added, instead of subtracting the products obtained by multiplying by $-r$, then the last line of the division remains unchanged. This follows from the fact that $a + (-b) = a - b$. It is convenient to place r on the last line since r multiplies the numbers in this line.

The quotient and remainder may be read from the last line of the synthetic division, with the last number recognized as the remainder and the other numbers excluding r recognized as the coefficients of the quotient polynomial.

$$
\begin{array}{r}
2 -7 0 13 \; -5 \\
6 -3 \; -9 12 \\
\hline
3 \overline{)2 -1 \; -3 4 7} \\
\downarrow \downarrow \\
\text{Quotient} \text{Remainder} \\
2x^3 - x^2 - 3x + 4
\end{array}
$$

The general description of synthetic division is rather complicated, but the format below illustrates the process for the division of the polynomial $P(x) = a_n x^n + a_{n-1} x^{n-1} + \ldots + a_1 x + a_0$ by $x - r$.

$$r \overline{)\begin{array}{ccccccc} a_n & a_{n-1} & a_{n-2} & \ldots & a_2 & a_1 & a_0 \\ & ra_n & rb_{n-2} & \ldots & rb_2 & rb_1 & rb_0 \\ \hline a_n & b_{n-2} & b_{n-3} & & b_1 & b_0 & R \end{array}}$$

where

$$b_{n-2} = a_{n-1} + ra_n$$
$$b_{n-3} = a_{n-2} + rb_{n-2}$$
$$\vdots$$
$$b_1 = a_2 + rb_2$$
$$b_0 = a_1 + rb_1$$
$$R = a_0 + rb_0$$

and $P(x) = (x - r)(a_n x^{n-1} + b_{n-2} x^{n-2} + \ldots + b_1 x + b_0) + R$.

Example 1 Using synthetic division, divide $3x^4 - 2x^2 + 4$ by $x - 2$.

Solution

$$2 \overline{)\begin{array}{ccccc} 3 & 0 & -2 & 0 & 4 \\ & 6 & 12 & 20 & 40 \\ \hline 3 & 6 & 10 & 20 & 44 \end{array}}$$

$$3x^4 - 2x^2 + 4 = (x - 2)(3x^3 + 6x^2 + 10x + 20) + 44$$

Example 2 Divide $7x^3 + 12x^2 - 5x + 4$ by $x + 2$.

Solution $x - r = x + 2 = x - (-2)$ and $r = -2$

$$-2 \overline{)\begin{array}{cccc} 7 & 12 & -5 & 4 \\ & -14 & 4 & 2 \\ \hline 7 & -2 & -1 & 6 \end{array}}$$

$$7x^3 + 12x^2 - 5x + 4 = (x + 2)(7x^2 - 2x - 1) + 6$$

Example 3 Divide $x^3 + 2x^2 - 2x + 3$ by $x + 3$.

Solution

$$-3 \overline{)\begin{array}{cccc} 1 & 2 & -2 & 3 \\ & -3 & 3 & -3 \\ \hline 1 & -1 & 1 & 0 \end{array}}$$

$$x^3 + 2x^2 - 2x + 3 = (x + 3)(x^2 - x + 1)$$

DIVISION ALGORITHM; SYNTHETIC DIVISION

EXERCISES 3.4

1–6. Express each of the following pairs of polynomials in the form $A(x) = B(x)Q(x) + R(x)$, where $R = 0$ or the degree of $R(x)$ is less than the degree of $B(x)$.

1. $A(x) = 2x^6 + 5x^4 - x^3 + 1$; $\quad B(x) = -x^2 + x + 1$
2. $A(x) = 3x^5 + 2x^2 - x - 2$; $\quad B(x) = x^3 + 1$
3. $A(x) = x^4 + 4x^3 + 5x^2 + 2x$; $\quad B(x) = x^2 + 3x + 2$
4. $A(x) = 1 - x^2 + x^4$; $\quad B(x) = 1 - x$
5. $A(x) = 4x^3 - 13x + 9$; $\quad B(x) = 2x - 3$
6. $A(x) = x^4 - 4x^2 - 16x - 16$; $\quad B(x) = x^2 + 2x + 4$

7–14. Perform the following by synthetic division:

7. $(3x^2 + 2x - 5) \div (x + 2)$
8. $(3x^3 + 2x^2 + 5x - 1) \div (x - 2)$
9. $(-x^4 + x^2 - 1) \div (x - 1)$
10. $(-x^4 + x^3 - x + 1) \div (x - 1)$
11. $(2x^4 - 3x^3 - 20x^2 - 6) \div (x - 4)$
12. $(2x^4 - 3x^3 - 20x^2 - 6) \div (x + 3)$
13. $(x^5 - 1) \div (x - 1)$
14. $(x^5 - 1) \div (x + 1)$

15–20. Use synthetic division to find the quotient and remainder for each of the following. Express the answer in the form $A(x) = B(x)Q(x) + R(x)$, where the degree of $R(x)$ is less than the degree of $B(x)$, or $R = 0$.

15. $A(x) = x^3 - 2x^2 + 3x - 4$; $\quad B(x) = x - 3$
16. $A(x) = x^3 - 2x^2 + 3x - 4$; $\quad B(x) = x + 3$
17. $A(x) = 3x^4 + 17x^3 + 30x - 16$; $\quad B(x) = x + 5$
18. $A(x) = x^4 - x^2 + 1$; $\quad B(x) = x - 1$
19. $A(x) = 3x^4 - 17x^3 + 30x^2 - 160$; $\quad B(x) = x - 4$
20. $A(x) = 4x^3 - 13x + 9$; $\quad B(x) = x - 3$

21–25. Determine $R(x)$, the remainder, for each of the following, and evaluate $P(r)$ for the given $P(x) \div (x - r)$.

21. $(2x^3 + 3x^2 - 18x - 4) \div (x - 2)$
22. $(x^3 - 2x^2 + x - 4) \div (x + 2)$
23. $(x^4 - 3x + 4) \div (x - 1)$
24. $(2x^6 + 12x^3 + x - 30) \div (x + 2)$

25. $(x^8 - x^5 - x^3 + 1) \div (x - 1)$
26. Let $P(x) = (x - r)Q(x) + R$ where $P(x)$ and $Q(x)$ are polynomials.
    (a) Find the relationship between $P(r)$ and R.
    (b) What can be said about the factorization of $P(x)$ if $R = 0$?
☆27. If a polynomial is divided by a quadratic divisor having the form $x^2 - r$, then a synthetic division process requires that the sums of two adjacent columns are written at the same time and the two products obtained by multiplying by r are each shifted two places to the right. The last two sums are the coefficients of the remainder which has the from $ax + b$.

For example, the synthetic division of $x^4 - 2x^3 + x^2 + 9x - 10$ by $x^2 - 2$ would be written as follows:

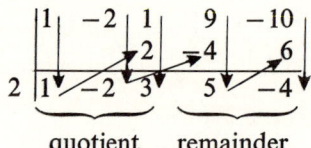

$$\underbrace{}_{\text{quotient}} \underbrace{}_{\text{remainder}}$$

Thus, $x^4 - 2x^3 + x^2 + 9x - 10 = (x^2 - 2)(x^2 - 2x + 3) + (5x - 4)$.

Using the synthetic division process described above, perform the following divisions:
(a) $(x^5 - 2x^3 + 2x^2 + x + 2) \div (x^2 - 3)$
(b) $(x^5 + 5x^4 + x^3 + 10x^2 - 7x + 2) \div (x^2 + 2)$
(c) $(x^4 + x^3 - x + 1) \div (x^2 + 1)$
(d) $(x^4 + 5x - 2) \div (x^2 - 5)$

☆28. Develop a synthetic division process for a divisor having the following form: $x^2 + ax + b$.

☆29. State the Euclidean algorithm for two polynomials over I and show that the GCD of the two polynomials is the last nonzero remainder in the algorithm.

☆30. By using the Euclidean algorithm for polynomials over I, find $d(x)$, the GCD of $x^5 + x^2 - 4x + 2$ and $x^4 - x^3 - 2x - 4$.

3.5 FACTOR AND REMAINDER THEOREMS; APPLICATIONS TO EQUATIONS

3.5.1 The Remainder Theorem

The Remainder Theorem Let $P(x)$ be an integral polynomial and let r be an integer.

$$\text{If } P(x) = (x - r)Q(x) + R, \text{ then } R = P(r).$$

Proof $P(r) = (r - r)Q(r) + R = 0 \cdot Q(r) + R = 0 + R = R$.

FACTOR AND REMAINDER THEOREMS; APPLICATIONS TO EQUATIONS

The remainder theorem states that if a polynomial $P(x)$ is divided by $x - r$, then the remainder is the value of the polynomial for $x = r$. Note in this case that either $R = 0$ or the degree of R must be 0 since the degree must be less than 1, the degree of $x - r$. Therefore R must be a constant.

Example 1 Find the remainder when $2x^{37} + x^5 - 3x^2 - 4$ is divided by $x - 1$.

Solution $R = P(1) = 2 \cdot 1^{37} + 1^5 - 3 \cdot 1^2 - 4 = 3 - 7 = -4$

Example 2 Find $P(5)$ if $P(x) = 4x^4 - 18x^3 - 11x^2 + 2x + 18$.

Solution Since $P(5) = R$, R may be found by synthetic division.

$$\begin{array}{r|rrrrr} & 4 & -18 & -11 & 2 & 18 \\ & & 20 & 10 & -5 & -15 \\ \hline 5) & 4 & 2 & -1 & -3 & 3 \end{array}$$

Thus, $P(5) = 3$.

3.5.2 The Factor Theorem

The Factor Theorem Let $P(x)$ be an integral polynomial and let r be an integer.

$$P(r) = 0 \text{ if and only if } x - r \text{ is a factor of } P(x).$$

Proof $P(x) = (x - r)Q(x) + R$ Division algorithm.
$P(x) = (x - r)Q(x) + P(r)$ Remainder theorem.

If $P(r) = 0$, then $x - r$ is a factor of $P(x)$.
If $x - r$ is a factor of $P(x)$, then $R = P(r) = 0$.

The factor theorem may be used to find linear factors of a polynomial.

Example 3 Of which is $x + 1$ a factor?
(a) $x^3 + 1$
(b) $x^4 + 1$

Solution
(a) If $P(x) = x^3 + 1$, then $P(-1) = (-1)^3 + 1 = -1 + 1 = 0$.
 $x + 1$ is a factor of $x^3 + 1$.
(b) If $P(x) = x^4 + 1$, then $P(-1) = (-1)^4 + 1 = 1 + 1 = 2 \neq 0$.
 $x + 1$ is not a factor of $x^4 + 1$.

Example 4 Show that $x + 3$ is a factor of $x^3 + 2x^2 - x + 6$.

Solution

$$\begin{array}{r} 1 2 -1 6 \\ -3 3 -6 \\ \hline -3 \overline{)1 -1 2 0} \end{array}$$

Thus, $x^3 + 2x^2 - x + 6 = (x + 3)(x^2 - x + 2)$ and $x + 3$ is a factor of $x^3 + 2x^2 - x + 6$.

Example 5 For what value of k is $x - 2$ a factor of $5x^3 - 8x^2 - 7x + k$?

Solution If $x - 2$ is a factor, then $R = 0$.

$$\begin{array}{r} 5 -8 -7 k \\ 10 4 -6 \\ \hline 2\overline{)5 2 -3 k - 6} \end{array}$$

If $R = 0$, then $k - 6 = 0$ and $k = 6$.

3.5.3 Solution of Polynomial Equations by Factoring

By applying the zero product theorem and the techniques of factoring, the solutions of some equations may be found by factoring.

Definition of Solution If r is a constant and if $P(x) = 0$ is an open statement, then r is a solution of $P(x) = 0$ if and only if $P(r) = 0$ is a true statement.

A solution of $P(x) = 0$ is also called a **root** of $P(x) = 0$, or a **zero** of $P(x)$.

Definition of Solution Set The solution set of $P(x) = 0$ over a set S is the set of all solutions of $P(x) = 0$ in S.

To **solve an equation** $P(x) = 0$ over a set S means to find all the solutions of $P(x) = 0$ in S.

The following theorem is useful in finding an integral root of a polynomial equation.

Theorem on Linear Integral Factors If r is an integer and if $x - r$ is a factor of the integral polynomial

$$P(x) = a_n x^n + a_{n-1} x^{n-1} + \ldots + a_1 x + a_0,$$

then r is a factor of a_0.

FACTOR AND REMAINDER THEOREMS; APPLICATIONS TO EQUATIONS

Proof Since the factor theorem states that $x - r$ is a factor of $P(x)$ if and only if $P(r) = 0$, thus

$$P(r) = a_n r^n + \ldots + a_1 r + a_0 = 0$$
$$r(a_n r^{n-1} + \ldots + a_1) + a_0 = 0$$
$$r(a_n r^{n-1} + \ldots + a_1) = -a_0.$$

Thus r is a factor of a_0.

Example 6 Solve $(x - 2)(x + 3)(x - 4) = 0$ over I.

  Solution $x - 2 = 0$ or $x + 3 = 0$ or $x - 4 = 0$ Zero product theorem

and

$x = 2$ or $x = -3$ or $x = 4$ Addition theorem

The solution set is $\{2, -3, 4\}$.

Example 7 Solve $x^3 - 5x^2 - x + 5 = 0$ over I.

  Solution

$$\begin{aligned} x^3 - 5x^2 - x + 5 &= (x^3 - 5x^2) - (x - 5) \\ &= x^2(x - 5) - (x - 5) \\ &= (x - 5)(x^2 - 1) \\ &= (x - 5)(x - 1)(x + 1) = 0 \end{aligned}$$

The solution set is $\{-1, 1, 5\}$.

Example 8 Solve $x^3 + x^2 - 16x + 20 = 0$ over I.

  Solution

(1) The possible integral roots are the integral factors of 20: $1, -1, 2, -2, 4, -4, 5, -5, 10, -10, 20, -20$.

(2) Apply the factor theorem.
$P(1) = 1 + 1 - 16 + 20 \neq 0$,
$P(-1) = -1 + 1 + 16 + 20 \neq 0$,
$P(2) = 8 + 4 - 32 + 20 = 0$ and $x - 2$ is a factor.

(3) Use synthetic division to find the quotient polynomial.

$$\begin{array}{r} 2 \overline{)\begin{array}{cccc} 1 & 1 & -16 & 20 \\ & 2 & 6 & -20 \end{array}} \\ \begin{array}{cccc} 1 & 3 & -10 & 0 \end{array} \end{array}$$

$x^3 + x^2 - 16x + 20 = (x - 2)(x^2 + 3x - 10)$

(4) Complete the factorization by repeating the above technique or by applying a special product theorem.

$$x^3 + x^2 - 16x + 20 = (x - 2)(x - 2)(x + 5)$$

(5) Solving, the roots are 2, 2, and -5 with 2 called a **repeated root**.
The solution set is $\{2, -5\}$.

Example 9 Solve $x^3 - 24x - 5 = 0$ over I.

Solution After applying the technique used in Example 8,
$$x^3 - 24x - 5 = (x - 5)(x^2 + 5x + 1) = 0$$
$$x - 5 = 0 \quad \text{or} \quad x^2 + 5x + 1 = 0$$
$$x = 5 \quad \text{and} \quad x^2 + 5x + 1 \text{ is prime over } I.$$
The solution set is $\{5\}$.

EXERCISES 3.5

1–8. Use the factor theorem to determine if the first polynomial is a factor of the second polynomial.

1. $x + 3$; $\quad x^3 - 4x^2 - 18x + 9$
2. $x - 3$; $\quad x^4 - 4x^3 - 7x^2 + 22x + 24$
3. $x - 1$; $\quad x^4 - 2x^2 - x + 7$
4. $x + 1$; $\quad 2x^3 - 3x^2 - 7x + 1$
5. $3x + 1$; $\quad 9x^3 + 6x^2 + 4x + 1$
6. $x - 1$; $\quad 9x^3 + x^2 - 7x - 3$
7. $x - 1$; $\quad 4x^3 - 2x^2 + x - 1$
8. $2x - 1$; $\quad 4x^4 - 2x^3 + 6x - 2$

9–12. Use synthetic division and the remainder theorem to compute each of the following:

9. If $P(x) = x^3 - 3x^2 + 7x - 4$, find $P(2)$, $P(-1)$.
10. If $P(x) = x^4 - 4x^3 + 7x^2 + 22x + 24$, find $P(3)$, $P(-2)$.
11. If $P(x) = 3x^4 - 5x^3 + 7x^2 - 6x + 9$, find $P(1)$, $P(-1)$, $P(3)$.
12. If $P(x) = 2x^5 - x^3 + x - 2$, find $P(1)$, $P(-1)$, $P(2)$.

13–16. Use the factor theorem to show that:

13. $x - 3$ is a factor of $x^4 - 6x^3 + 3x^2 + 17x + 3$.
14. $x + 2$ is a factor of $x^7 + 128$.
15. $x + a$ is a factor of $x^7 + a^7$.
16. $x - b$ is a factor of $x^9 - b^9$.

17–21. Use the remainder theorem to determine the remainder for each of the following problems. Compare the answers with the answers to Exercise 3.4, 21–25.

17. $(2x^3 + 3x^2 - 18x - 4) \div (x - 2)$
18. $(x^3 - 2x^2 + x - 4) \div (x + 2)$

FACTOR AND REMAINDER THEOREMS; APPLICATIONS TO EQUATIONS

19. $(x^4 - 3x + 4) \div (x - 1)$
20. $(2x^6 + 12x^3 + x - 30) \div (x + 2)$
21. $(x^8 - x^5 - x^3 + 1) \div (x - 1)$

22–32. Solve each of the following equations over I:

22. $(x^2 + 2x - 3)(x^2 + 2x - 8) = 0$
23. $x^3 - 4x^2 - 3x + 2 = 0$
24. $x^3 - 9x^2 + 26x - 24 = 0$
25. $(x + 2)^2 - (x - 7)^2 = 0$
26. $x^3 - 3x^2 - 9x + 27 = 0$
27. $x^4 - 16 = 0$
28. $x^4 - 2x^2 - 3x - 2 = 0$
29. $x^3 - x - 6 = 0$
30. $2x^5 - 3x^3 + 7x^2 - 8 = 0$
31. $x^3 - 6x^2 + 11x - 6 = 0$
32. $x^5 - 9x^3 + 3x^2 + 16x - 4 = 0$

33. For what value of k is $x + 3$ a factor of $x^3 - 3x^2 + 4x + k$?
34. For what value of k is $x - 5$ a factor of $x^3 - 4x^2 - 3x + k$?
35. For what value of k is $2x - 1$ a factor of $4x^3 - 4x^2 + 7x + k$?
36. For what value of k is $3x + 1$ a factor of $3x^3 - 5x^2 + 4x + k$?
37. For what value of k is $x - 4$ a factor of $x^3 - 3x^2 + kx - 12$?
38. For what value of k is $x + 2$ a factor of $x^4 - 4x^2 + kx + 6$?

☆39. (a) Prove that
$$P(x) = (x - a)(x - b)Q(x) + \frac{P(b) - P(a)}{b - a}x + \frac{bP(a) - aP(b)}{b - a}.$$

(b) Use the result of part (a) to find the remainder when $x^5 + 2x^3 - x + 5$ is divided by $(x - 1)(x - 2)$.

☆40. (a) Prove that
$$P(x) = P(a) + Q_1(a)(x - a) + Q_2(a)(x - a)^2 + \ldots + Q_n(a)(x - a)^n,$$

where
$$P(x) = (x - a)Q_1(x) + R_1$$
$$Q_1(x) = (x - a)Q_2(x) + R_2$$
$$\vdots$$
$$Q_{n-1}(x) = (x - a)Q_n(x) + R_n.$$

(b) By using repeated applications of synthetic division, determine the coefficients a_0, a_1, a_2, a_3, a_4 so that
$$x^4 - 2x^3 + x^2 - x + 1 = a_0 + a_1(x - 1) + a_2(x - 1)^2 + a_3(x - 1)^3 + a_4(x - 1)^4.$$

THE INTEGERS

CHAPTER SUMMARY

Axioms for the Set of Integers   For all x, y, and z in I,

(1) Closure, Addition — There is exactly one $x + y$ in I.
(2) Closure, Multiplication — There is exactly one xy in I.
(3) Commutative, Addition — $x + y = y + x$.
(4) Commutative, Multiplication — $xy = yx$.
(5) Associative, Addition — $(x + y) + z = x + (y + z)$.
(6) Associative, Multiplication — $(xy)z = x(yz)$.
(7) Identity, Addition — $x + 0 = x$, 0 in I and 0 is unique.
(8) Identity, Multiplication — $x \cdot 1 = x$, 1 in I and 1 is unique.
(9) Inverse, Addition — $x + (-x) = 0$, exactly one $-x$ in I for each x.
(10) Distributive — $x(y + z) = xy + xz$.
(11) Trichotomy — Exactly one of the following is valid: $x > 0$ or $x = 0$ or $-x > 0$.
(12) Positive Closure, Addition — If $x > 0$ and $y > 0$, then $x + y > 0$.
(13) Positive Closure, Multiplication — If $x > 0$ and $y > 0$, then $xy > 0$.
(14) Finite Induction is valid for the set $P = \{x \mid x > 0 \text{ and } x \text{ in } I\}$.

Definition $x + y + z = (x + y) + z$.

Definition $xyz = (xy)z$.

Definition of Positive x is positive if and only if $x > 0$.

Definition of Negative x is negative if and only if $-x > 0$.

Definition of Subtraction $x - y = x + (-y)$.

Definition of $x > y$ $x > y$ if and only if $x - y > 0$.

Definition of $x < y$ $x < y$ if and only if $y > x$.

Addition Theorem   If $x = y$, then $x + z = y + z$.

Multiplication Theorem   If $x = y$, then $xz = yz$.

Addition Cancellation Theorem   If $x + z = y + z$, then $x = y$.

CHAPTER SUMMARY

Zero Factor Theorem $x \cdot 0 = 0$.

Opposite of an Opposite Theorem $-(-x) = x$.

Opposite of a Sum Theorem $-(x + y) = (-x) + (-y)$.

Opposite of a Difference Theorem $-(x - y) = y - x$.

Opposite of a Product Theorem $-xy = (-x)y$.

Product of Two Opposites Theorem $(-x)(-y) = xy$.

Zero Product Theorem If $xy = 0$, then $x = 0$ or $y = 0$.

Multiplication Cancellation Theorem If $xz = yz$ and $z \neq 0$, then $x = y$.

Prime Factorization Theorem Every nonzero polynomial over I is either prime or can be expressed in exactly one way as the product of a constant factor and one or more prime integral polynomials, disregarding the order in which the factors are written.

The Division Algorithm for Integral Polynomials If the degree of $A(x) \geq$ degree of $B(x) > 0$ and if the leading coefficient of $B(x) = 1$, then for exactly one $Q(x)$ and exactly one $R(x)$ either $A(x) = B(x)Q(x)$ or $A(x) = B(x)Q(x) + R(x)$ and degree of $R(x) <$ degree of $B(x)$.

The Remainder Theorem If $P(x) = (x - r)Q(x) + R$, then $R = P(r)$.

The Factor Theorem $x - r$ is a factor of $P(x)$ if and only if $P(r) = 0$.

Theorem on Linear Integral Factors If $x - r$ is a factor of $P(x) = a_n x^n + \ldots + a_1 x + a_0$, then r is a factor of a_0.

REVIEW EXERCISES

1–7. Justify each step in the proof: If x, y, and z are integers, then
$$x + xy + z + zy = (x + z)(1 + y)$$
by using an axiom, definition, or theorem of the system for the integers.

1. $x + xy + z + zy = [x + (xy + z)] + zy$
2. $ = [x + (z + xy)] + zy$
3. $ = (x + z) + (xy + zy)$

4. $\qquad = (x+z) + (x+z)y$
5. $\qquad = (x+z) \cdot 1 + (x+z)y$
6. $\qquad = (x+z)(1+y)$
7. $x + xy + z + zy = (x+z)(1+y)$

8–10. Solve over I or prove unsolvable over I. Justify each step involved.

8. $x(x-2) = 3$
9. $\dfrac{x+5}{3} = 7$
10. $2x + 3 = 6$

11. Prove: The sum of two even integers is even.
12. Prove: The sum of two odd integers is even.
13. Under what operations is the set of integers closed? Justify.
14. Is the product of two odd integers odd or even? Prove your answer.

15–24. Factor completely over the integers.

15. $2ax - ay + 2x^2 - xy$
16. $12x^2 - 89xy + 60y^2$
17. $p^4 + p^2 + 25$
18. $x^3 + 3x^2y + 3xy^2 + y^3$
19. $x^2 - y^2 + 4yz - 4z^2$
20. $x^3 - 64y^3$
21. $xy^3 + 2y^2 - xy - 2$
22. $x^4 + 64y^4$
23. $4a^2 - x^2 + b^2 - y^2 - 4ab - 2xy$
24. $y^4 + 2y^2z^2 + 9z^4$

25–30. Given: Polynomials $A(x)$ and $B(x)$. Use synthetic division to write $A(x) = B(x)Q(x) + R(x)$, where $R = 0$ or the degree of $R(x)$ is less than the degree of $B(x)$.

25. $A(x) = 4x^3 - 2x^2 - x - 1;\quad B(x) = x - 1$
26. $A(x) = 2x^5 + 18x^2 + x - 2;\quad B(x) = x + 2$
27. $A(x) = x^4 + 2x^3 + x + 1;\quad B(x) = x + 1$
28. $A(x) = 3x^5 - 10x^4 + 3x^3 + x^2 - x - 4;\quad B(x) = x - 3$
29. $A(x) = 3x^3 + 2x^2 - 5x - 1;\quad B(x) = x - 1$
30. $A(x) = x^7 - 1;\quad B(x) = x - 1$

31. Use the factor theorem to find k so that $x + 2$ is a factor of $x^4 + 2x^3 + x - k$.
32. Use the factor theorem to find k so that 5 is a root of the equation $2x^2 - 3x + k = 0$.
33. If $P(x) = 3x^2 + 4x - k$, find k so that $P(3) = 6$.
34. If $P(x) = 2x^3 - 4x^2 + 3x - k$, find k so that $P(4) = 0$.

REVIEW EXERCISES

35–40. Solve each of the following equations over I.

35. $8x^3 + 10x^2 - x - 3 = 0$
36. $x^3 + 5x^2 + 36 = 0$
37. $x^3 - x^2 - 16x - 12 = 0$
38. $x^3 - 7x + 6 = 0$
39. $x^4 - 15x^2 + 10x + 24 = 0$
40. $2x^4 + 4x^3 + 6x^2 + 4x - 16 = 0$

4

THE RATIONALS

Throughout his existence man has had difficulty understanding fractions. For most purposes of primitive man, the natural numbers were sufficient. Measuring units could be divided into subunits and a natural number could still be used to describe the smaller measurement. For example, by dividing 1 foot into 12 equal parts called inches, a measurement of $\frac{1}{4}$ foot could be expressed as 3 inches.

As civilization became more complex, fractions were introduced to obtain greater accuracy in measurement. Unit fractions, fractions whose numerators are 1 such as $\frac{1}{2}$, $\frac{1}{3}$, $\frac{1}{4}$, $\frac{1}{5}$, and so on, were used around 1650 B.C. by the ancient Egyptians. The Egyptians seemed unable to grasp the concept of a pair of natural numbers representing one number such as our fraction $\frac{5}{6}$. The ancient Babylonians about this same time restricted their denominators to be 60 or powers of 60 and developed a system of positional fractions similar to our decimal fractions.

Before the time of Archimedes (287–212 B.C.) the Greeks disliked the idea of breaking unity into parts and worked with ratios of natural numbers instead. A measurement of $\frac{5}{6}$ would have been described as a measurement whose ratio to the whole was 5 to 6.

The Romans avoided fractions completely by the use of subunits. A twelfth part of the Roman unit was called *uncia*, from which is derived our modern "ounce" and "inch." A measurement of $\frac{5}{6}$ of a unit was regarded by the Romans as 10 uncias ($\frac{5}{6} = \frac{10}{12}$).

Our present method for writing fractions is probably of Hindu origin.

THE SET OF RATIONALS, Q

The Hindu fractions resembled ours, except that the bar was omitted. Brahmagupta (628 A.D.) and Bhaskara (1150 A.D.) both wrote $\genfrac{}{}{0pt}{}{2}{3}$ for our $\frac{2}{3}$. The bar was introduced by the Arabs.

The practical necessity of obtaining greater accuracy in measurement and the theoretical need to close the number system with respect to the division operation motivated the extension of the number system to include the positive fractions. The negative fractions and zero must also be included so that closure with respect to the subtraction operation is maintained. This new set of numbers is called the set of rational numbers, Q. The word "rational" is used to suggest the **ratio** concept of fractions and the letter "Q" is used to suggest the **quotient** concept. Since the set of rational numbers is closed with respect to the operations of addition, subtraction, multiplication, and division, these four operations are called the rational operations.

4.1 THE SET OF RATIONALS, Q

4.1.1 Definitions

Definition of the Set of Rationals, Q

$$Q = \left\{ \frac{a}{b} \,\Big|\, a \text{ and } b \text{ are integers and } b \neq 0 \right\}$$

The expression $\frac{a}{b}$, read "a over b," is considered to be an ordered pair of integers, with the first component a called the **numerator** and the second component b called the **denominator**. At this point it would not be valid to refer to $\frac{a}{b}$ as "a divided by b" since the division operation has not been defined for all pairs of integers a and b. However, if $a = bx$ where x is an integer, then it is intended that $\frac{a}{b}$ is another name for x, or $x = \frac{a}{b}$. For example, if $6 = 2x$, then $x = \frac{6}{2} = 3$.

Further definitions are necessary to clarify the meaning of $\frac{a}{b}$ and to justify that $\frac{a}{b}$ may be interpreted as a divided by b, especially for such cases as $2 = 5x$ and $x = \frac{2}{5}$.

Definition of $\frac{a}{1}$ If a is any integer, then $\frac{a}{1} = a$.

From the above definition it is seen that the set of integers is a proper subset of the set of rationals, $I \subset Q$, and Q is, indeed, an extension of the integers.

Definition of Equivalent Rational Numbers If a, b, c, and d are any integers such that $b \neq 0$ and $d \neq 0$, then

$$\frac{a}{b} \equiv \frac{c}{d} \text{ if and only if } ad = bc.$$

For example,

$$\frac{1}{2} \equiv \frac{3}{6} \quad \text{since} \quad 1 \cdot 6 = 2 \cdot 3 = 6$$

and

$$\frac{-4}{6} \equiv \frac{-10}{15} \quad \text{since} \quad -4 \cdot 15 = -10 \cdot 6 = -60.$$

The concept of equivalent rational numbers means that there are many different names for a rational number. The definition indicates when $\frac{a}{b}$ and $\frac{c}{d}$ are the names of the same number. In particular, $\frac{ak}{bk} \equiv \frac{a}{b}$ since $(ak)b = a(bk)$ by the associative and commutative properties for integers.

Fundamental Theorem for Rational Numbers If a, b, and k are any integers such that $b \neq 0$ and $k \neq 0$, then

$$\frac{ak}{bk} \equiv \frac{a}{b}.$$

Since, in general, an ordered pair (a, b) is equal to the ordered pair (c, d) if and only if $a = c$ and $b = d$, the ordered pairs in Q are restricted by the equivalence relation $\equiv$. Before $\equiv$ may be replaced by the equal sign and thus interpreted "is the name of the same number whose name is," it should be shown that $\equiv$ is an equivalence relation; that is, that the following properties are valid.

(1) Reflexive $\quad\quad \frac{a}{b} \equiv \frac{a}{b}.$

(2) Symmetric $\quad\quad$ If $\frac{a}{b} \equiv \frac{c}{d}$, then $\frac{c}{d} \equiv \frac{a}{b}$.

(3) Transitive $\quad\quad$ If $\frac{a}{b} \equiv \frac{c}{d}$ and $\frac{c}{d} \equiv \frac{e}{f}$, then $\frac{a}{b} \equiv \frac{e}{f}$.

THE SET OF RATIONALS, Q

The student may verify that these properties follow directly from the definition of $\equiv$ and properties of the integers. All statements derived from the reflexive, symmetric, and transitive properties of an equivalence relation are valid for all equivalence relations in general and for the equality relation in particular. Thus, $\frac{a}{b} \equiv \frac{c}{d}$ may be written as $\frac{a}{b} = \frac{c}{d}$.

Definition of Simplified Form If the GCD of the integers a and b is 1 or -1 and if b is positive and if $b \neq 1$, then $\frac{a}{b}$ is said to be in *simplified form* or *irreducible*. If $b = 1$, then a is the simplified form of $\frac{a}{1}$.

Note that since $\frac{a}{-b} = \frac{-a}{b}$ because $ab = (-a)(-b)$, it is always possible to rename a rational number so that it has a positive denominator.

To *simplify* a rational number means to write it in simplified form.

Definition of Addition Let a, b, c, and d be integers.

(1) $a + b$ is the same as defined previously for integers.

(2) $\frac{a}{b} + \frac{c}{d} = \frac{ad + bc}{bd}$ if $b \neq 0$ and $d \neq 0$.

Definition of Subtraction Let a, b, c, and d be integers.

(1) $a - b = a + (-b)$.

(2) $\frac{a}{b} - \frac{c}{d} = \frac{ad - bc}{bd}$ if $b \neq 0$ and $d \neq 0$.

Definition of Multiplication Let a, b, c, and d be integers.

(1) ab is the same as defined previously for integers.

(2) $\frac{a}{b} \cdot \frac{c}{d} = \frac{ac}{bd}$ if $b \neq 0$ and $d \neq 0$.

Definition of Division Let a, b, c, and d be integers.

$$\frac{\frac{a}{b}}{\frac{c}{d}} = \frac{ad}{bc} \quad \text{if } b \neq 0,\ c \neq 0,\ \text{and } d \neq 0$$

Using these definitions for the set of rationals, it may be verified that the closure, commutative, associative, distributive, identity, and inverse properties for addition and multiplication are valid. In contrast to the set of integers, every nonzero element of Q has a multiplication inverse and thus Q is closed with respect to the division operation. Also, it may be shown that $bx = a$ and $b \neq 0$ if and only if $x = \dfrac{a}{b}$ and thus $\dfrac{a}{b}$ is the quotient when a is divided by b.

The order relations may be defined for the set of rationals as follows.

Definition of Positive $\dfrac{a}{b}$ is positive if and only if ab is positive.

Definition of < $\dfrac{a}{b} < \dfrac{c}{d}$ if and only if there exists a positive rational number p so that $\dfrac{a}{b} + p = \dfrac{c}{d}$.

Definition of > $\dfrac{a}{b} > \dfrac{c}{d}$ if and only if $\dfrac{c}{d} < \dfrac{a}{b}$.

It may be verified that these definitions preserve the order properties for the set of integers.

The next section presents an abstract mathematical system that may be interpreted as the set of rational numbers defined in this section.

4.1.2 A System for the Rationals

UNDEFINED CONCEPTS

> A set Q
> An operation $+$, called addition
> An operation $\cdot$, called multiplication
> A relation $>$

Notation $xy = x \cdot y$.

Field Axioms For all x, y, and z in Q,

(1)	Closure, Addition	There is exactly one $x + y$ in Q.
(2)	Closure, Multiplication	There is exactly one xy in Q.
(3)	Commutative, Addition	$x + y = y + x$.
(4)	Commutative, Multiplication	$xy = yx$.

THE SET OF RATIONALS, Q

(5) Associative, Addition $(x + y) + z = x + (y + z)$.
(6) Associative, Multiplication $(xy)z = x(yz)$.
(7) Distributive $x(y + z) = xy + xz$.
(8) Identity, Addition There is exactly one element 0 in Q so that for all x in Q, $x + 0 = x$, and $0 + x = x$.
(9) Identity, Multiplication There is exactly one element 1 in Q so that for all x in Q, $x \cdot 1 = x$, and $1 \cdot x = x$.
(10) Inverse, Addition For each x in Q, there is exactly one element $-x$ in Q, called the **opposite** of x, so that $x + (-x) = 0$ and $(-x) + x = 0$.
(11) Inverse, Multiplication For each $x \neq 0$ in Q, there is exactly one element $\frac{1}{x}$ in Q, called the **reciprocal** of x, so that $x \cdot \frac{1}{x} = 1$ and $\frac{1}{x} \cdot x = 1$.

Order Axioms

(12) Trichotomy For all x in Q, exactly one of the following is valid: $x > 0$ or $x = 0$ or $-x > 0$.
(13) Closure of Positives, Addition If $x > 0$ and $y > 0$, then $x + y > 0$.
(14) Closure of Positives, Multiplication If $x > 0$ and $y > 0$, then $xy > 0$.

Definition of Subtraction $x - y = x + (-y)$.

Definition of Division $\frac{x}{y} = x \cdot \frac{1}{y}$ if $y \neq 0$.

Definition of Positive x is positive if and only if $x > 0$.

Definition of Negative x is negative if and only if $-x > 0$.

Definition of $x > y$ $x > y$ if and only if $x - y > 0$.

Definition of $x < y$ $x < y$ if and only if $y > x$.

The theorems for the integers, based on the first ten axioms remain valid for the set of rationals and are proved similarly. However, since there is now an inverse axiom for multiplication, there are some new theorems involving the concept of reciprocal. The proofs of these theorems are omitted and left as an exercise for the student.

Theorem (Reciprocal of a Reciprocal) If $x \neq 0$, then $\dfrac{1}{\frac{1}{x}} = x$.

Theorem (Product of Reciprocals) If $x \neq 0$ and $y \neq 0$, then $\dfrac{1}{x} \cdot \dfrac{1}{y} = \dfrac{1}{xy}$.

Theorem (Fundamental Theorem of Quotients) If $yz \neq 0$, then $\dfrac{xz}{yz} = \dfrac{x}{y}$.

Theorem (Sum of Quotients) If $yw \neq 0$, then $\dfrac{x}{y} + \dfrac{z}{w} = \dfrac{xw + yz}{yw}$.

Theorem (Product of Quotients) If $yw \neq 0$, then $\dfrac{x}{y} \cdot \dfrac{z}{w} = \dfrac{xz}{yw}$.

Theorem (Difference of Quotients) If $yw \neq 0$, then $\dfrac{x}{y} - \dfrac{z}{w} = \dfrac{xw - yz}{yw}$.

Theorem (Quotient of Quotients) If $yzw \neq 0$, then $\dfrac{\frac{x}{y}}{\frac{z}{w}} = \dfrac{x}{y} \cdot \dfrac{w}{z} = \dfrac{xw}{yz}$.

By comparing these theorems with the definitions for the rationals in the preceding section, it may be seen that this system describes the set of rational numbers. The order properties will be discussed in the next chapter.

EXERCISES 4.1

1–10. Use the commutative, associative, and distributive axioms to perform the calculations below in the simplest way possible. State which of these axioms was used.

1. $\dfrac{1}{17} + \left(\dfrac{2}{3} + \dfrac{16}{17}\right)$

2. $36\left(\dfrac{4}{9} - \dfrac{5}{12}\right)$

3. $45\left(\dfrac{22}{37}\right) + 45\left(\dfrac{15}{37}\right)$

4. $\dfrac{23}{25}\left(\dfrac{3}{4} \cdot \dfrac{21}{23}\right)$

THE SET OF RATIONALS, Q   105

5. $\dfrac{125}{3}\left(-\dfrac{19}{23}\right)\left(\dfrac{12}{5}\right)\left(-\dfrac{23}{38}\right)$ 8. $\dfrac{87}{92}\left(-\dfrac{20}{41}\right)-\dfrac{87}{92}\left(\dfrac{21}{41}\right)$

6. $\dfrac{-7}{31}-\left(\dfrac{13}{20}+\dfrac{24}{31}\right)$ 9. $\dfrac{2}{3}\left(\dfrac{5}{7}\cdot\dfrac{2}{11}\right)+\dfrac{1}{11}\left(\dfrac{2}{3}\cdot\dfrac{67}{7}\right)$

7. $-450\left(\dfrac{-7}{90}+\dfrac{11}{150}\right)$ 10. $\dfrac{3}{5}\left(\dfrac{4}{7}\cdot\dfrac{1}{4}\right)-\dfrac{1}{2}\left(\dfrac{3}{5}\cdot\dfrac{16}{7}\right)$

11–14. Express each of the following as a single rational number in simplified form. Justify each step using axioms, theorems, or definitions.

11. $\left(\dfrac{3}{8}+\dfrac{7}{16}+\dfrac{5}{8}\right)$ 13. $\dfrac{23}{125}\left(\dfrac{17}{23}\cdot\dfrac{3}{8}\right)$

12. $60\left(\dfrac{-11}{12}+\dfrac{4}{15}\right)$ 14. $\dfrac{1}{-\frac{2}{3}}+\dfrac{\frac{1}{5}}{-3}$

15–26. Simplify the following. Assume that no variables result in zero denominators.

15. $\dfrac{1}{3}+\dfrac{2}{5}$ 22. $\dfrac{5+\frac{5}{6}}{5-\frac{5}{6}}$

16. $\dfrac{2a-b}{2a}-\dfrac{a-2b}{7b}$ 23. $\dfrac{y+\frac{1}{4}}{y-\frac{1}{4}}$

17. $\dfrac{5}{a-2}-\dfrac{3}{2a-5}$ 24. $\dfrac{1+\dfrac{x}{y}}{\dfrac{y^2-x^2}{y^2}}$

18. $\dfrac{6}{x^2}+\dfrac{1}{3x}-\dfrac{3}{4x}$

19. $5-\dfrac{b}{b-2}$ 25. $3-\dfrac{1}{1-\dfrac{2}{3-\frac{3}{2}}}$

20. $\dfrac{a+b}{a}+\dfrac{a-b}{-b}$

21. $\dfrac{5}{2-\frac{1}{3}}$ 26. $\dfrac{\frac{1}{2}-\frac{1}{3}}{1+\frac{1}{2}\left(\frac{1}{3}\right)}$

27. Show that the relation ≡ is reflexive.
28. Show that the relation ≡ is symmetric.
29. Show that the relation ≡ is transitive.

30–32. Let $\dfrac{a}{b}$ and $\dfrac{c}{d}$ be rational numbers. Using definitions, axioms, and theorems for the system of rational numbers, prove each of the following:

30. If $\dfrac{a}{b} = \dfrac{c}{d}$, then $\dfrac{a}{c} = \dfrac{b}{d}$.

31. If $\dfrac{a}{b} = \dfrac{c}{d}$, then $\dfrac{b}{a} = \dfrac{d}{c}$.

32. If $\dfrac{a}{b} = \dfrac{c}{d}$, then $\dfrac{a+b}{b} = \dfrac{c+d}{d}$.

33. If m is a rational number such that $m \neq 0$ and $m \neq 1$, simplify

$$1 - \cfrac{1}{1 - \cfrac{1}{1-m}}.$$

Justify each step.

34. Prove the difference of quotients theorem: If $yw \neq 0$, then

$$\left(\dfrac{x}{y}\right) - \left(\dfrac{z}{w}\right) = \dfrac{xw - yz}{yw}.$$

35. Prove the quotient of quotients theorem: If $zwy \neq 0$, then

$$\dfrac{x}{y} \div \dfrac{z}{w} = \dfrac{x}{y} \cdot \dfrac{w}{z} = \dfrac{xw}{yz}.$$

36. Ancient Egyptians wrote all fractions, except $\frac{2}{3}$, as the sum of unit fractions. Thus $\frac{2}{7} = \frac{1}{4} + \frac{1}{28}$. Express 2/97 as a sum of 2 unequal unit fractions.

37. Verify the commutative property for addition of rational numbers, as defined in Section 4.1.1.

38. Verify the associative property for multiplication of rational numbers, as defined in Section 4.1.1.

39. If a and b are integers such that $a > 0$, $b > 0$, and $a < b$, how is $1/a$ related to $1/b$?

40. Prove: If a, b, and c are positive integers, then $(a/c) < (b/c) \leftrightarrow a < b$.

4.2 FIELDS (Optional)

Definition of Field A field is a mathematical system consisting of a set F, an operation $+$, and an operation $\cdot$, such that

(1) $F, +$ is a commutative group,
(2) $F - 0, \cdot$ is a commutative group, with 0 the addition identity,
(3) For every x, y, and z in F, $x \cdot (y + z) = (x \cdot y) + (x \cdot z)$.

FIELDS (OPTIONAL)

The definition states that the closure, commutative, associative, identity, and inverse axioms are valid for both $\cdot$ and $+$ and that $\cdot$ is distributive over $+$. For this reason, these axioms are called the **field axioms.** Since these axioms are valid for the set of rational numbers, the set of rational numbers is a field.

One important property of a field is that it has no zero divisors. If $ab = 0$ and $a \neq 0$, then $\frac{1}{a}(ab) = \frac{1}{a} \cdot 0 = 0$ and thus $b = 0$. Thus if $ab = 0$, then $a = 0$ or $b = 0$ and there are no zero divisors.

Division is always possible in a field; that is, for all a and b in F, $b \neq 0$, there exists an x so that $bx = a$. If $bx = a$, then $\frac{1}{b}(bx) = a \cdot \frac{1}{b}$ and $x = a \cdot \frac{1}{b}$. Therefore, the term "quotient" may be used and designated symbolically by $\frac{a}{b} = a \cdot \frac{1}{b}$.

It may be shown that the theorems on opposites and on reciprocals are valid for every field. Thus if a system is shown to be a field, these theorems are automatically valid for the system and do not need to be proved for the special interpretation. This is an advantage in considering an abstract system.

EXERCISES 4.2 (Optional)

1–3. Show that each of the following is a field:

1. $\{a + b\sqrt{2} \mid a \in Q \text{ and } b \in Q\}$
2. $\{a + bi \mid a \in Q, b \in Q, i^2 = -1\}$
3. $\{a + b\sqrt{3} \mid a \in Q \text{ and } b \in Q\}$
4. Show that the system $\{0, 1\}$, $+$, $\cdot$, defined by the tables below is a field.

+	0	1
0	0	1
1	1	0

$\cdot$	0	1
0	0	0
1	0	1

5. Write out $+$ and $\cdot$ tables for a field of three elements, $\{0, 1, a\}$.

6–9. Prove or disprove that each of the following is a field:

6. The set of integers.
7. The set of even integers.
8. The system $\{0, 1, a, b\}$, $+$, $\cdot$, defined in the following tables:

+	0	1	a	b
0	0	1	a	b
1	1	0	b	a
a	a	b	0	1
b	b	a	1	0

$\cdot$	0	1	a	b
0	0	0	0	0
1	0	1	a	b
a	0	a	b	1
b	0	b	1	a

9. $\{ax + b \mid a, b \in Q, x$ is a variable$\}$
10. Prove: Let F be a field and a, b, and $c \in F$; If $c \neq 0$ and if $ac = bc$, then $a = b$.
11. Let $S = \{(a, b) \mid a$ and b are integers, $b \neq 0\}$. Let $\equiv$, $+$, and $\cdot$ be defined as follows:

$$(a, b) \equiv (c, d) \leftrightarrow ad = bc$$
$$(a, b) + (c, d) \equiv (ad + bc, bd) \qquad [bd \neq 0]$$
$$(a, b) \cdot (c, d) \equiv (ac, bd).$$

Show that S is a field. Can this system be interpreted as the set of rational numbers with $(a, 1)$ identified as the integer a? Explain.

12–14. Define a relation *congruent modulo k* ($\equiv_k$) on the set of integers as follows: For each a, $b \in I$, and for $k \in N$, $a \equiv_k b \leftrightarrow k$ is a factor of $(a - b)$.

12. Prove that $\equiv_k$ is an equivalence relation.
13. The set of integers modulo 3 can be expressed as $I/3 = \{0, 1, 2\}$. Thus $7 \equiv_3 1$, $85 \equiv_3 1$, $14 \equiv_3 2$, $21 \equiv_3 0$, and so forth, where $1 \in I/3$, $2 \in I/3$, $0 \in I/3$. Define $+$ and $\cdot$ on $I/3$ as follows:

+	0	1	2		·	0	1	2
0	0	1	2		0	0	0	0
1	1	2	0		1	0	1	2
2	2	0	1		2	0	2	1

Is $I/3$ a field? Justify.

14. Are the integers modulo 5, $I/5$, a field? Write out tables for $\cdot$ and $+$. Justify your decision.
15. Are the integers modulo 6 a field? Write out tables for $+$ and $\cdot$ and justify your decision.
16. Are the integers modulo n a field, where n is any natural number? If so, justify your conclusion. If not, determine the set of natural numbers for which I/n is a field.

4.3 RATIONAL FUNCTIONS

If $A(x)$ and $B(x)$ are integral polynomials or polynomials with rational coefficients, then their quotient $\dfrac{A(x)}{B(x)}$ designates a rational number when x is replaced by any integer or rational number for which $B(x) \neq 0$. For this reason expressions of the form $\dfrac{A(x)}{B(x)}$ are called **rational functions.** They are functions

RATIONAL FUNCTIONS

since for each r in Q such that $B(r) \neq 0$, there is exactly one rational number $\frac{A(r)}{B(r)}$. This results from the fact that the set of rational numbers is closed with respect to the addition, subtraction, multiplication, and division operations. It may be shown that the set of quotients of polynomials over Q is a field; that is, the closure, commutative, associative, distributive, identity, and inverse axioms for addition and multiplication are valid. From this it follows that the theorems derived from these axioms are also valid. This illustrates an advantage in considering the abstract system called a field. Once a system is identified as a field, all the theorems for fields are automatically valid for the special interpretation and do not need to be proved again.

Theorem (The Set of Rational Functions over Q Is a Field) Let

$$\mathbf{Q}(x) = \left\{ \frac{A(x)}{B(x)} \;\middle|\; A(x) \text{ and } B(x) \text{ are polynomials over } Q \text{ and } B(x) \neq 0 \right\}$$

with equivalent quotients, addition, and multiplication defined as follows (addition and multiplication of polynomials over Q are defined similarly as for polynomials over I, the set of integers):

Definition of Equivalent Quotients

$$\frac{A(x)}{B(x)} = \frac{C(x)}{D(x)} \text{ if and only if } A(x)D(x) = B(x)C(x), \qquad \frac{A(x)}{1} = A(x).$$

Definition of Addition $\quad \dfrac{A(x)}{B(x)} + \dfrac{C(x)}{D(x)} = \dfrac{A(x)D(x) + B(x)C(x)}{B(x)D(x)}.$

Definition of Multiplication $\quad \dfrac{A(x)}{B(x)} \cdot \dfrac{C(x)}{D(x)} = \dfrac{A(x)C(x)}{B(x)D(x)}.$

Then $\mathbf{Q}(x)$ is a field.

The proof involves showing that the closure, commutative, associative, distributive, identity, and inverse properties for addition and multiplication are valid. The verification is left for the student.

In proving that the set of rational functions over Q is a field, the following statements are derived.

ADDITION IDENTITY

The zero polynomial $0 = \dfrac{0}{1} = 0 \cdot x^n + \ldots + 0 \cdot x + 0$ is the addition identity, n is any natural number.

Multiplication Identity

The unit polynomial $1 = \dfrac{1}{1} = 0 \cdot x^n + \ldots + 0 \cdot x + 1$ is the multiplication identity, n is any natural number.

Addition Inverse

The addition inverse of $\dfrac{A(x)}{B(x)}$, $-\dfrac{A(x)}{B(x)} = \dfrac{-A(x)}{B(x)} = \dfrac{-1 \cdot A(x)}{B(x)}$.

Multiplication Inverse

The multiplication inverse of $\dfrac{A(x)}{B(x)}$, $\dfrac{1}{\frac{A(x)}{B(x)}} = \dfrac{B(x)}{A(x)}$, where $A(x) \neq 0$ and $B(x) \neq 0$.

The subtraction and division operations are defined as follows:

Definition of Subtraction $\quad \dfrac{A(x)}{B(x)} - \dfrac{C(x)}{D(x)} = \dfrac{A(x)}{B(x)} + \left(-\dfrac{C(x)}{D(x)}\right).$

Definition of Division $\quad \dfrac{A(x)}{B(x)} \div \dfrac{C(x)}{D(x)} = \dfrac{A(x)}{B(x)} \cdot \dfrac{D(x)}{C(x)}.$

All of the theorems derived from the field axioms are valid for the field of quotients of polynomials over Q. In particular, the following theorem is called to attention because of its importance.

Fundamental Theorem of Quotients If $B(x) \neq 0$ and $C(x) \neq 0$, then

$$\dfrac{A(x)C(x)}{B(x)C(x)} = \dfrac{A(x)}{B(x)}.$$

For polynomials whose coefficients are rational numbers, every nonzero rational number is a constant factor of every polynomial since

$$P(x) = a \cdot \dfrac{1}{a} \cdot P(x), \quad \text{where } a \in Q.$$

For example,

$$2x^2 - 3x + 4 = 2\left(x^2 - \dfrac{3}{2}x + 2\right)$$

and

$$2x^2 - 3x + 4 = 5\left(\dfrac{2}{5}x^2 - \dfrac{3}{5}x + \dfrac{4}{5}\right) = \dfrac{5}{3}\left(\dfrac{6}{5}x^2 - \dfrac{9}{5}x + \dfrac{12}{5}\right).$$

RATIONAL FUNCTIONS

It is possible to express every polynomial over Q as a product of a constant factor and an *integral* polynomial that has no constant integral factor.

For example,

$$\frac{1}{3}x^2 + \frac{1}{2}x - \frac{1}{4} = \frac{1}{12} \cdot 12\left(\frac{1}{3}x^2 + \frac{1}{2}x - \frac{1}{4}\right) = \frac{1}{12}(4x^2 + 6x - 3)$$

and

$$\frac{5}{3}x^2 - \frac{5}{2}x - \frac{15}{4} = \frac{5}{12}(4x^2 - 6x - 9).$$

Definition of Simplified Form If $B(x) \neq 0$, then $\frac{A(x)}{B(x)}$ is in simplified form if and only if $A(x)$ and $B(x)$ are integral polynomials, $B(x) \neq 1$, and the GCD of $A(x)$ and $B(x)$ is 1 or -1; that is, $A(x)$ and $B(x)$ have no common integral constant factor or common prime polynomial factor.

If $B(x) = 1$, then $A(x)$ is the simplified form of $\frac{A(x)}{1}$.

Definition of Identical Polynomials $A(x)$ and $B(x)$ are identical polynomials over Q if and only if their corresponding coefficients are equal.

In symbols, $A(x) = a_n x^n + a_{n-1} x^{n-1} + \ldots + a_1 x + a_0$ is identical to $B(x) = b_n x^n + b_{n-1} x^{n-1} + \ldots + b_1 x + b_0$ if and only if $a_n = b_n$, $a_{n-1} = b_{n-1}, \ldots,$ $a_1 = b_1$, and $a_0 = b_0$.

Theorem on Equating Coefficients of Polynomials The corresponding coefficients of $A(x)$ and $B(x)$ are equal if and only if for all x in Q, $A(x) = B(x)$.

Proof

(1) If $a_n = b_n$, $a_{n-1} = b_{n-1}, \ldots, a_1 = b_1$, and $a_0 = b_0$, then $A(x) - B(x) = 0$, the zero polynomial, which is 0 for all x. Thus, for all x, $A(x) = B(x)$.

(2) If for all x, $A(x) = B(x)$, then for all x, $A(x) - B(x) = 0$, the zero polynomial, and for all x

$$(a_n - b_n)x^n + \ldots + (a_1 - b_1)x + (a_0 - b_0) = 0.$$

For $x = 0$, $0 + 0 + \ldots + 0 + (a_0 - b_0) = 0$ and thus $a_0 = b_0$. Then

$$A(x) - B(x) = x \cdot [(a_n - b_n)x^{n-1} + \ldots + (a_2 - b_2)x + (a_1 - b_1)]$$
$$= 0 \quad \text{for all } x.$$

Since $x \neq 0$ for all x,

$$(a_n - b_n)x^{n-1} + \ldots + (a_2 - b_2)x + (a_1 - b_1) = 0 \quad \text{for all } x.$$

For $x = 0$, $a_1 - b_1 = 0$ and $a_1 = b_1$. Continuing this process,

$$a_n = b_n, a_{n-1} = b_{n-1}, \ldots, a_1 = b_1, a_0 = b_0.$$

Theorem on Equating Coefficients of Numerators of Quotients If $\dfrac{A(x)}{C(x)} = \dfrac{B(x)}{C(x)}$ for all x except those x for which $C(x) = 0$, then $A(x) = B(x)$ for all x. ($A(x)$, $B(x)$, and $C(x)$ are polynomials over Q.)

Proof $\dfrac{A(x) - B(x)}{C(x)} = 0$ for all x such that $C(x) \neq 0$. Then

$$(A(x) - B(x)) \cdot \dfrac{1}{C(x)} = 0 \quad \text{and} \quad \dfrac{1}{C(x)} \neq 0.$$

Therefore, $A(x) - B(x) = 0$, the zero polynomial, which is 0 for all x. Thus, for all x, $A(x) = B(x)$ and their corresponding coefficients are equal.

Example 1 Find p and q if for all x except $x = 2$ and $x = -1$,

$$\dfrac{x+7}{x^2 - x - 2} = \dfrac{p(x+1) + q(x-2)}{x^2 - x - 2}.$$

Solution For all x, $p(x+1) + q(x-2) = (p+q)x + (p-2q) = x + 7$. Equating coefficients,

$$\begin{aligned} p + q &= 1 & \text{and} \quad p - 2q &= 7 \\ p &= 1 - q & \text{and} \quad 1 - q - 2q &= 7 \\ & & 1 - 3q &= 7 \\ & & -3q &= 6 \\ & & q &= -2. \end{aligned}$$

$$p = 1 - (-2) = 3.$$

Thus, $p = 3$ and $q = -2$.

Example 2 Simplify $\dfrac{x^3 + 8}{x^4 + 4x^2 + 16} \div \dfrac{x^3 + 2x^2 - x - 2}{x^3 + x^2 + 2x - 4}$ if x is restricted to avoid zero denominators.

Solution First factor each numerator and denominator:

$x^3 + 8$	$= (x+2)(x^2 - 2x + 4)$	Sum of cubes.
$x^4 + 4x^2 + 16$	$= (x^4 + 8x^2 + 16) - 4x^2$	Completing the square.
	$= (x^2 + 4)^2 - (2x)^2$	Perfect square.
	$= (x^2 + 2x + 4)(x^2 - 2x + 4)$	Difference of squares.

RATIONAL FUNCTIONS

$$x^3 + 2x^2 - x - 2 = x^2(x+2) - (x+2) \qquad \text{Distributive axiom.}$$
$$= (x+2)(x^2 - 1) \qquad \text{Distributive axiom.}$$
$$= (x+2)(x+1)(x-1) \qquad \text{Difference of squares.}$$
$$x^3 + x^2 + 2x - 4 = (x-1) \cdot Q(x) \qquad \text{Factor theorem, } P(1) = 0.$$
$$= (x-1)(x^2 + 2x + 4) \qquad \text{Synthetic division.}$$

Now,

$$\frac{x^3 + 8}{x^4 + 4x^2 + 16} \div \frac{x^3 + 2x^2 - x - 2}{x^3 + x^2 + 2x - 4}$$

$$= \frac{(x+2)(x^2 - 2x + 4)}{(x^2 + 2x + 4)(x^2 - 2x + 4)} \cdot \frac{(x-1)(x^2 + 2x + 4)}{(x-1)(x+1)(x+2)}$$

$$= \frac{x+2}{x^2 + 2x + 4} \cdot \frac{x^2 + 2x + 4}{(x+1)(x+2)} \qquad \text{Fundamental theorem of quotients.}$$

$$= \frac{(x+2)(x^2 + 2x + 4)}{(x+1)(x+2)(x^2 + 2x + 4)} \qquad \text{Definition of multiplication of quotients.}$$

$$= \frac{1}{x+1} \qquad \text{Fundamental theorem of quotients.}$$

Example 3 Express as one simplified quotient:

$$\frac{x+2}{x+5} + \frac{x-4}{x-5} + \frac{10}{25 - x^2} \quad \text{for } x \neq 5 \text{ and } x \neq -5.$$

Solution

$$\frac{(x+2)(x-5)}{(x+5)(x-5)} + \frac{(x-4)(x+5)}{(x-5)(x+5)} + \frac{-10}{x^2 - 25} \qquad \text{Fundamental theorem.}$$

$$= \frac{(x^2 - 3x - 10) + (x^2 + x - 20) - 10}{x^2 - 25} \qquad \text{Addition of quotients.}$$

$$= \frac{2x^2 - 2x - 40}{x^2 - 25} \qquad \text{Commutative, associative, distributive axioms.}$$

$$= \frac{2(x+4)(x-5)}{(x+5)(x-5)} \qquad \text{Theorems, special products.}$$

$$= \frac{2(x+4)}{x+5} \qquad \text{Fundamental theorem of quotients.}$$

Example 4 Simplify

$$\frac{\dfrac{1}{x} - \dfrac{1}{3}}{\dfrac{1}{x^2} - \dfrac{1}{9}} \quad \text{for } x \neq 0 \text{ and } x \neq 3 \text{ and } x \neq -3.$$

Solution

$$\frac{9x^2\left(\dfrac{1}{x} - \dfrac{1}{3}\right)}{9x^2\left(\dfrac{1}{x^2} - \dfrac{1}{9}\right)} \qquad \text{Fundamental theorem of quotients.}$$

$$= \frac{9x - 3x^2}{9 - x^2} \qquad \text{Distributive axiom.}$$

$$= \frac{3x(3 - x)}{(3 + x)(3 - x)} \qquad \text{Special products.}$$

$$= \frac{3x}{x + 3} \qquad \text{Fundamental theorem of quotients.}$$

EXERCISES 4.3

1–4. Express each of the following polynomials as a product of a constant factor and an integral polynomial that has no constant integral factor.

1. $\frac{1}{5}x^3 + \frac{2}{3}x - 2$
2. $6x^2 + \frac{1}{6}x + \frac{1}{2}$
3. $\frac{2}{3}y^5 + \frac{1}{4}y^3 - 2y + \frac{1}{7}$
4. $\frac{1}{2}a^2 - \frac{2}{3}a - \frac{3}{7}$

5–8. For each of the following, find p and q for all x such that the denominators do not equal zero.

5. $\dfrac{x + 3}{x^2 + 3x + 2} = \dfrac{p(x + 2) + q(x + 1)}{x^2 + 3x + 2}$

6. $\dfrac{2x - 1}{x^2 - 4x + 5} = \dfrac{p(x - 5) + q(x + 1)}{x^2 - 4x + 5}$

7. $\dfrac{5}{2x^2 - 3x + 54} = \dfrac{p(2x - 9) + q(x + 6)}{2x^2 - 3x + 54}$

8. $\dfrac{3x}{6x^2 + x - 3} = \dfrac{p(3x + 2) + q(2x - 1)}{6x^2 + x - 3}$

RATIONAL FUNCTIONS

9–28. Simplify and state what restrictions need to be placed on x or y.

9. $\dfrac{x^2 - 7x + 12}{x^2 - 2x - 3}$

10. $\dfrac{16x^3 - 54}{4x^2 + 6x + 9}$

11. $\dfrac{x^2 - 6x + 9}{x^2 - 7x + 12} \div \dfrac{x^4 - 81}{x^3 - 4x^2 + 9x - 36}$

12. $\left(\dfrac{x^3 - 4x^2 - 5x}{x^2 - 2x + 1} \cdot \dfrac{x^2 + x - 2}{x^4 + 8x} \right) \div \dfrac{x - 4}{x^2 - 2x + 4}$

13. $\dfrac{2x}{x^3 + 8} - \dfrac{x}{x^2 - 2x + 4}$

14. $\dfrac{4x^2}{4x - 1} + \dfrac{5x^2 - 1}{12x^2 + 5x - 2} - \dfrac{4x - 1}{3x + 2}$

15. $\dfrac{\dfrac{3}{x} - \dfrac{2}{3}}{\dfrac{9}{x^2} - \dfrac{4}{9}}$

19. $\dfrac{x^3 + x^2 - 4x - 4}{3x^5 - 3x^4 - 6x^3}$

16. $\dfrac{9x^2 - 4}{\dfrac{x - 1}{1 - 2x} - 1}$

20. $\dfrac{6x^2 - x - 12}{27x^3 + 64} \div \dfrac{4x^2 - 9}{9x^2 - 12x + 16}$

17. $\dfrac{\dfrac{3}{5}x^2 - 6x + \dfrac{63}{5}}{\dfrac{4x^2}{3} + \dfrac{8x}{3} - 20}$

21. $\dfrac{20}{y^2 - 25} - \dfrac{3}{5 + y} + \dfrac{2}{5 - y}$

18. $\dfrac{x^2 - \dfrac{1}{x}}{x + 1 + \dfrac{1}{x}}$

22. $\dfrac{x^2 - 2x}{x^4 + 4} - \dfrac{1}{x^2 + 2x + 2}$

23. $\dfrac{ax - ay - bx + by}{x^2 - y^2} \div \dfrac{a^2 - 2ab + b^2}{ax + ay - bx - by}$

24. $\dfrac{x^4 - x^2 - 12}{x^3 - 2x^2 + 3x - 6} \div \dfrac{3x^2 + 4x - 4}{9x^2 - 6x}$

25. $\dfrac{y^3 + 27}{y^4 - 6y^3 + 9y^2 - 81}$

26. $3 - 1 - \dfrac{\dfrac{1}{2}}{3 - \dfrac{3}{x + 1}}$

27. $\dfrac{x^2 - \dfrac{64}{x^4}}{1 + \dfrac{4}{x^2} + \dfrac{16}{x^4}}$

28. $1 - \dfrac{x+1}{x - 1 - \dfrac{1}{x + 1 + \dfrac{1}{x-1}}}$

☆29. Prove that **Q**(x), the set of rational functions over Q, is a field.

☆30. Show that quotients of polynomials over the set {0, 1}, with the operations + and · defined in the given tables, satisfy the field axioms.

+	0	1		·	0	1
0	0	1		0	0	0
1	1	0		1	0	1

☆31. Prove that if $A(x)$ and $B(x)$ are polynomials over Q such that the degree of $B(x)$ is less than or equal to the degree of $A(x)$, then there is exactly one polynomial $Q(x)$, and exactly one polynomial $R(x)$ such that $A(x) = B(x)Q(x) + R(x)$ where $R(x) = 0$ or the degree of $R(x)$ is less than the degree of $B(x)$.

☆32. Show that an integral polynomial can be factored over the integers if and only if it can be factored over the rationals.

4.4 PARTIAL FRACTIONS

It has been seen that a sum of quotients may always be expressed as a single quotient. For example,

$$\frac{1}{2} + \frac{1}{3} = \frac{5}{6} \quad \text{and} \quad \frac{1}{x+2} + \frac{1}{x+3} = \frac{2x+5}{x^2 + 5x + 6}.$$

There are problems in integral calculus that require a single quotient to be expressed as a sum of quotients. The process is called *resolving a quotient into a sum of partial fractions*. Finding the terms of this sum is a more difficult task than combining the terms, just as factoring a polynomial is more difficult than multiplying the factors. A sample problem follows.

Example Find rational numbers a and b so that

$$\frac{2x - 3}{(x - 5)(x + 4)} = \frac{a}{x - 5} + \frac{b}{x + 4}.$$

Solution For all x except $x = 5$ and $x = -4$,

$$\frac{a}{x - 5} + \frac{b}{x + 4} = \frac{a(x + 4) + b(x - 5)}{(x - 5)(x + 4)} = \frac{2x - 3}{(x - 5)(x + 4)}.$$

PARTIAL FRACTIONS

For all x,
$$a(x+4) + b(x-5) = 2x - 3.$$

For $x = 5$,
$$a(5+4) + b(5-5) = 10 - 3$$
$$9a = 7 \quad \text{and} \quad a = \frac{7}{9}.$$

For $x = -4$,
$$a(-4+4) + b(-4-5) = -8 - 3$$
$$-9b = -11 \quad \text{and} \quad b = \frac{11}{9}.$$

Thus
$$\frac{2x-3}{(x-5)(x+4)} = \frac{7/9}{x-5} + \frac{11/9}{x+4}.$$

Theorem (The Partial Fractions Theorem) If $N(x)$ and $D(x)$ are polynomials with the degree of $N(x)$ less than the degree of $D(x)$, then $\frac{N(x)}{D(x)}$ can be resolved into a sum of partial fractions as follows:

(1) If $D(x)$ has a linear factor $ax + b$, then one term of the sum is the partial fraction $\frac{p}{ax+b}$, where p is a constant.

(2) If $D(x)$ has n repeated linear factors, $(ax+b)^n$, then n terms of the sum are the n partial fractions as follows:

$$\frac{p_1}{ax+b} + \frac{p_2}{(ax+b)^2} + \ldots + \frac{p_n}{(ax+b)^n}$$

with $p_1, p_2, \ldots, p_n$ constants.

(3) If $D(x)$ has a quadratic factor, $ax^2 + bx + c$, then one term of the sum is the partial fraction $\frac{px+q}{ax^2+bx+c}$, where p and q are constants.

(4) If $D(x)$ has n repeated quadratic factors, $(ax^2 + bx + c)^n$, n terms of the sum are the n partial fractions as follows:

$$\frac{p_1 x + q_1}{ax^2+bx+c} + \frac{p_2 x + q_2}{(ax^2+bx+c)^2} + \ldots + \frac{p_n x + q_n}{(ax^2+bx+c)^n},$$

where the p_i's and q_i's are constants.

The proof of this theorem is given in the Appendix.

Example 1 Resolve $\dfrac{x+17}{2x^2+5x-3}$ into a sum of partial fractions.

Solution

$$2x^2 + 5x - 3 = (2x-1)(x+3)$$

$$\frac{x+17}{(2x-1)(x+3)} = \frac{p}{2x-1} + \frac{q}{x+3} = \frac{p(x+3) + q(2x-1)}{(2x-1)(x+3)}$$

For all x,
$$p(x+3) + q(2x-1) = x + 17.$$

For $x = \dfrac{1}{2}$,

$$p\left(\frac{1}{2} + 3\right) + q \cdot 0 = \frac{1}{2} + 17$$

$$\frac{7}{2}p = \frac{35}{2} \quad \text{and} \quad p = 5.$$

For $x = -3$,
$$p \cdot 0 + q(-6-1) = -3 + 17$$
$$-7q = 14 \quad \text{and} \quad q = -2.$$

Thus,
$$\frac{x+17}{2x^2+5x-3} = \frac{5}{2x-1} + \frac{-2}{x+3}.$$

Example 2 Resolve $\dfrac{3x-1}{x^2+4x+4}$ into a sum of partial fractions.

Solution
$$x^2 + 4x + 4 = (x+2)^2$$

and
$$\frac{3x-1}{(x+2)^2} = \frac{p}{x+2} + \frac{q}{(x+2)^2} = \frac{p(x+2) + q}{(x+2)^2}$$

For all x,
$$p(x+2) + q = 3x - 1.$$

For $x = -2$,
$$p(-2+2) + q = -6 - 1$$
$$q = -7.$$

For $x = 0$,
$$p(0 + 2) + q = 0 - 1$$
$$2p + (-7) = -1$$
$$2p = 6 \quad \text{and} \quad p = 3.$$

Thus,
$$\frac{3x - 1}{x^2 + 4x + 4} = \frac{3}{x + 2} + \frac{-7}{(x + 2)^2}.$$

Example 3 Resolve $\dfrac{x^3 - 19x - 17}{x^2 - 4x - 5}$ into partial fractions.

Solution Since the degree of the numerator is not less than the degree of the denominator, use the division algorithm to obtain
$$x^3 - 19x - 17 = (x^2 - 4x - 5)(x + 4) + 2x + 3$$
and
$$\frac{x^3 - 19x - 17}{x^2 - 4x - 5} = x + 4 + \frac{2x + 3}{x^2 - 4x - 5}.$$

Now
$$\frac{2x + 3}{x^2 - 4x - 5} = \frac{p}{x - 5} + \frac{q}{x + 1} = \frac{p(x + 1) + q(x - 5)}{(x - 5)(x + 1)}.$$

For all x,
$$p(x + 1) + q(x - 5) = 2x + 3.$$

For $x = 5$,
$$p(5 + 1) + q(5 - 5) = 10 + 3$$
$$6p = 13 \quad \text{and} \quad p = \frac{13}{6}$$

For $x = -1$, $p(-1 + 1) + q(-1 - 5) = -2 + 3$
$$-6q = 1 \quad \text{and} \quad q = -\frac{1}{6}.$$

Thus,
$$\frac{x^3 - 19x - 17}{x^2 - 4x - 5} = x + 4 + \frac{13/6}{x - 5} + \frac{-1/6}{x + 1}.$$

Example 4 Resolve $\dfrac{4x^2 + x + 1}{(x^2 + 1)(x - 1)}$ into partial fractions.

Solution

$$\frac{4x^2 + x + 1}{(x^2 + 1)(x - 1)} = \frac{ax + b}{x^2 + 1} + \frac{c}{x - 1} = \frac{(ax + b)(x - 1) + c(x^2 + 1)}{(x^2 + 1)(x - 1)}$$

For all x,
$$(ax + b)(x - 1) + c(x^2 + 1) = 4x^2 + x + 1.$$

For $x = 1$,[1]
$$0 + c(1 + 1) = 4 + 1 + 1$$
$$2c = 6 \quad \text{and} \quad c = 3.$$

For $x = 0$,
$$b(0 - 1) + c(0 + 1) = 0 + 0 + 1$$
$$-b + 3 = 1 \quad \text{and} \quad b = 2.$$

For $x = 2$,
$$2a + b + 5c = 16 + 2 + 1$$
$$2a = 19 - 2 - 15 = 2 \quad \text{and} \quad a = 1.$$

Thus,
$$\frac{4x^2 + x + 1}{(x^2 + 1)(x - 1)} = \frac{x + 2}{x^2 + 1} + \frac{3}{x - 1}.$$

Example 5 Resolve $\dfrac{2x^3 + 9x - 4}{x^4 + 6x^2 + 9}$ into partial fractions.

Solution

$$x^4 + 6x^2 + 9 = (x^2 + 3)^2$$

$$\frac{2x^3 + 9x - 4}{(x^2 + 3)^2} = \frac{ax + b}{x^2 + 3} + \frac{cx + d}{(x^2 + 3)^2}$$

For all x,
$$(ax + b)(x^2 + 3) + cx + d = 2x^3 + 9x - 4$$

and rearranging terms,
$$ax^3 + bx^2 + (3a + c)x + (3b + d) = 2x^3 + 9x - 4.$$

Equating coefficients,
$$a = 2, \quad b = 0, \quad 3a + c = 9, \quad \text{and} \quad 3b + d = -4.$$

---

[1] It is desirable to select values for x so that the computations are relatively easy.

PARTIAL FRACTIONS

Since $a = 2$,
$$3a + c = 6 + c = 9 \quad \text{and} \quad c = 3.$$
Since $b = 0$,
$$3b + d = d = -4.$$
Thus,
$$\frac{2x^3 + 9x - 4}{x^4 + 6x^2 + 9} = \frac{2x}{x^2 + 3} + \frac{3x - 4}{(x^2 + 3)^2}.$$

EXERCISES 4.4

1–10. Find rational numbers a, b, c, and d so that each of the following is a true statement:

1. $\dfrac{3x + 2}{(x + 5)(x - 4)} = \dfrac{a}{(x + 5)} + \dfrac{b}{(x - 4)}$

2. $\dfrac{2x}{(x - 2)(2x + 1)} = \dfrac{a}{x - 2} + \dfrac{b}{2x + 1}$

3. $\dfrac{3}{(2x + 1)(x - 1)} = \dfrac{a}{2x + 1} + \dfrac{b}{x - 1}$

4. $\dfrac{x - 2}{(2x - 3)^2} = \dfrac{a}{2x - 3} + \dfrac{b}{(2x - 3)^2}$

5. $\dfrac{2x + 1}{(5x + 1)^2} = \dfrac{a}{5x + 1} + \dfrac{b}{(5x + 1)^2}$

6. $\dfrac{7x - 10}{(3x - 4)^2} = \dfrac{a}{3x - 4} + \dfrac{b}{(3x - 4)^2}$

7. $\dfrac{5x^2 - 4x + 5}{(x^2 + 2)(x - 1)} = \dfrac{ax + b}{x^2 + 2} + \dfrac{c}{x - 1}$

8. $\dfrac{4x^2 + 3x - 1}{(x^2 + x + 1)(3x + 1)} = \dfrac{ax + b}{x^2 + x + 1} + \dfrac{c}{3x + 1}$

9. $\dfrac{5x^2 + x + 5}{(x^2 + 1)^2} = \dfrac{ax + b}{x^2 + 1} + \dfrac{cx + d}{(x^2 + 1)^2}$

10. $\dfrac{x^3 + x + 1}{(x^2 + 2)^2} = \dfrac{ax + b}{x^2 + 2} + \dfrac{cx + d}{(x^2 + 2)^2}$

11–34. Resolve each of the following into a sum of partial fractions:

11. $\dfrac{x}{x^2 + 3x + 2}$

12. $\dfrac{x - 22}{x^2 + x - 6}$

13. $\dfrac{2x + 3}{6x^2 + 7x + 2}$

14. $\dfrac{x - 8}{2x^2 + 3x - 2}$

15. $\dfrac{x}{(x+1)^2}$

16. $\dfrac{3x}{(x-2)^2}$

17. $\dfrac{x+2}{x^2-1}$

18. $\dfrac{x-2}{4x^2-1}$

19. $\dfrac{x^3}{x^2-4}$

20. $\dfrac{2x^3-18x+6}{x^2-9}$

21. $\dfrac{6}{(x+2)^2(x+1)}$

22. $\dfrac{7x-6}{x^3-2x^2}$

23. $\dfrac{x^2+5x-9}{x^3+3x}$

24. $\dfrac{x^2}{x^3-x^2+x-1}$

25. $\dfrac{x^3+1}{x^4+2x^2+1}$

26. $\dfrac{x^2}{x^4+4x^2+4}$

27. $\dfrac{4x^2-28}{x^4+x^2-6}$

28. $\dfrac{x^3+5x}{x^4+5x^2+6}$

29. $\dfrac{x-4}{x^2-2x-3}$

30. $\dfrac{4x^2+5x+8}{(x^2-5)(x+2)}$

31. $\dfrac{13x}{(3x^2-2)(2x^2+3)}$

32. $\dfrac{3x^2+20}{x^2+4x}$

33. $\dfrac{x^2+4x+4}{x^2+2x-3}$

34. $\dfrac{13x^2+12x-4}{30x^3+5x^2-10x}$

35. Show that for any polynomial $f(x)$ of degree ≤ 1,
$$\frac{f(x)}{(x-c)(x-d)} = \frac{f(c)}{c-d}\cdot\frac{1}{x-c} + \frac{-f(d)}{c-d}\cdot\frac{1}{x-d}.$$

36. Show that for any polynomial $f(x)$ of degree ≤ 1,
$$\frac{f(x)}{(x-c)^2} = \frac{f(c+1)-f(c)}{x-c} + \frac{f(c)}{(x-c)^2}.$$

4.5 SOLUTIONS OF RATIONAL EQUATIONS

Definition of Equivalent Open Statements The open statements px and qx are equivalent over a set S if and only if their solution sets over S are identical.

For example, $x = 2$ and $x + 3 = 5$ are equivalent. Also, $(x-2)(x-3) = 0$ is equivalent to the composite statement $x = 2$ or $x = 3$.

Since the solution of the equation $x = r$ can be immediately recognized as r, it is desirable to develop techniques for transforming an open equation into an equivalent equation having the form $x = r$ or an equivalent combination of equations having the form $x = r$.

SOLUTIONS OF RATIONAL EQUATIONS

Addition Equivalence Theorem If $A(x)$, $B(x)$, and $C(x)$ are rational functions over Q, then $A(x) = B(x)$ is equivalent to $A(x) + C(x) = B(x) + C(x)$.

Proof If $A(r) = B(r)$, then $A(r) + C(r) = B(r) + C(r)$ by the addition theorem.

If $A(r) + C(r) = B(r) + C(r)$, then $A(r) = B(r)$ by the addition cancellation theorem.

Therefore, any solution of $A(x) = B(x)$ is also a solution of $A(x) + C(x) = B(x) + C(x)$ and conversely.

Multiplication Equivalence Theorem If $A(x)$, $B(x)$, and $C(x)$ are rational functions over Q, then $A(x) = B(x)$ is equivalent to $A(x)C(x) = B(x)C(x)$ and $C(x) \neq 0$.

Proof If $A(r) = B(r)$, then $A(r)C(r) = B(r)C(r)$ by the multiplication theorem.

If $A(r)C(r) = B(r)C(r)$ and $C(r) \neq 0$, then $A(r) = B(r)$ by the multiplication cancellation theorem.

Therefore, any solution of $A(x) = B(x)$ is also a solution of $A(x)C(x) = B(x)C(x)$ and $C(x) \neq 0$ and conversely.

Example 1 Solve $\dfrac{2x}{x-2} - \dfrac{x-1}{x-3} = \dfrac{-4}{x^2 - 5x + 6}$.

Solution By the multiplication equivalence theorem, this equation is equivalent to the combination of statements $x \neq 2$, $x \neq 3$, and

$$(x-2)(x-3)\left[\frac{2x}{x-2} - \frac{x-1}{x-3}\right] = (x-2)(x-3)\left[\frac{-4}{x^2 - 5x + 6}\right]$$

$$2x(x-3) - (x-2)(x-1) = -4$$
$$2x^2 - 6x - (x^2 - 3x + 2) = -4$$
$$x^2 - 3x - 2 = -4$$
$$x^2 - 3x + 2 = 0$$
$$(x-1)(x-2) = 0$$
$$x - 1 = 0 \quad \text{or} \quad x - 2 = 0$$

and
$$x = 1 \quad \text{or} \quad x = 2.$$

The original equation is equivalent to the statement $(x = 1 \text{ or } x = 2)$ and $(x \neq 2 \text{ and } x \neq 3)$ which in turn is equivalent to $x = 1$.

The solution set is $\{1\}$.

A solution of the transformed equation that is not a solution of the original equation is called an **extraneous solution** or an **extraneous root**. The number 2 in Example 1 is an extraneous root. This terminology is misleading, however, since the so-called extraneous "root" is *not* a root of the original equation.

Example 2 Solve $\dfrac{8x - 11}{9 - x^2} + \dfrac{x - 5}{x - 3} = 2x$.

Solution The equation to be solved is equivalent to $x \neq 3$, $x \neq -3$, and

$$(9 - x^2)\left(\dfrac{8x - 11}{9 - x^2}\right) + (9 - x^2)\left(\dfrac{x - 5}{x - 3}\right) = (9 - x^2)2x$$

$$8x - 11 - (x + 3)(x - 5) = 18x - 2x^3$$
$$8x - 11 - (x^2 - 2x - 15) = 18x - 2x^3$$
$$2x^3 - x^2 - 8x + 4 = 0$$
$$x^2(2x - 1) - 4(2x - 1) = 0$$
$$(x^2 - 4)(2x - 1) = 0$$
$$(x + 2)(x - 2)(2x - 1) = 0$$

Thus, $(x = -2$ or $x = 2$ or $x = \tfrac{1}{2})$ and $(x \neq 3$ and $x \neq -3)$. The solution set is $\{-2, 2, \tfrac{1}{2}\}$.

EXERCISES 4.5

Solve each of the following equations:

1. $\dfrac{x + 5}{2x} + \dfrac{5}{x - 1} = \dfrac{1}{2}$

2. $\dfrac{2x - 5}{x - 3} - \dfrac{x^2 + 6}{x^2 + x - 12} = \dfrac{x - 2}{x + 4}$

3. $\dfrac{2}{x + 3} = \dfrac{3}{3x + 7} + \dfrac{1}{x + 1}$

4. $\dfrac{1}{x} - \dfrac{1}{x + 3} - \dfrac{1}{x + 2} + \dfrac{1}{x + 5} = 0$

5. $\dfrac{3}{x^2 + x - 2} = \dfrac{1}{x - 1} - \dfrac{1}{x - 4}$

6. $\dfrac{28}{2x - 11} + \dfrac{2x}{x + 7} = 2$

7. $\dfrac{x - 2}{x - 5} + \dfrac{x - 1}{x + 4} = \dfrac{2x^2 - 6x + 5}{x^2 - x - 20}$

8. $\dfrac{2x - 1}{x - 6} - \dfrac{x}{x + 1} - 1 = 0$

9. $2 + \dfrac{x^2}{3x - 9} = \dfrac{2x}{3 + x} + \dfrac{3 + x}{3}$

10. $\dfrac{3x + 1}{x + 2} = \dfrac{3x - 2}{x + 1}$

11. $\dfrac{1}{x} + \dfrac{3}{x + 1} + \dfrac{3}{x^2 + x} = 0$

12. $\dfrac{3 - x}{1 - x} + \dfrac{x - 5}{7 - x} = \dfrac{16 - 4x}{x^2 - 8x + 7}$

13. $\dfrac{x + 2}{2x - 3} + \dfrac{3x - 1}{4x + 5} = \dfrac{10x^2 + 2x + 13}{8x^2 - 2x - 15}$

14. $\dfrac{x + 3}{x + 4} + \dfrac{x + 4}{x + 3} = \dfrac{2x^2 - 16x}{x^2 + 7x + 12}$

15. $\dfrac{\dfrac{x}{x - 1} - 1}{\dfrac{x}{x - 1} + 1} = \dfrac{\dfrac{x}{x + 1} - 1}{\dfrac{x}{x + 1} + 1}$

16. $\dfrac{1 + \dfrac{1 + \dfrac{1+x}{2}}{4}}{3} = 1$

17. $\dfrac{1}{1 + \dfrac{1}{1 + \dfrac{2}{1+x}}} = \dfrac{1}{3}$

18. $\dfrac{2}{2x^2 - 5x + 6} = \dfrac{3}{3x^2 - 10x + 4}$

19. $\dfrac{4x + 3}{3x - 2} - \dfrac{3x - 2}{2x + 5} + \dfrac{x^2 - 10x + 3}{6x^2 + 11x - 10} = 0$

20. $\dfrac{8}{x^2 - 1} = \dfrac{4}{x - 1} - \dfrac{4}{x + 2}$

21. $\dfrac{x}{x - 5} - \dfrac{3}{x + 2} - \dfrac{21}{x^2 - 3x - 10} = 0$

22. $\dfrac{4x}{x - 1} - \dfrac{5x}{2x + 1} = \dfrac{x^2 + 12x + 2}{2x^2 - x - 1}$

23. $\dfrac{x - 6}{x - 1} - \dfrac{3 - x}{x + 5} + \dfrac{1}{2} = \dfrac{30}{5 - 4x - x^2}$

24. $\dfrac{5x}{x + 3} + \dfrac{15}{x - 3} + \dfrac{30x}{x^2 - 9} = x + 3$

25. $\dfrac{5x + 3}{x^2 - 8x - 20} - \dfrac{2x + 1}{x^2 - 100} = \dfrac{24x - 8}{x^3 + 2x^2 - 100x - 200}$

26. $\dfrac{x - 2}{x + 4} - \dfrac{x + 2}{x - 4} = 2$

27. $\dfrac{8x^2 - 32}{x^2 - 2x + 4} - \dfrac{16}{x + 2} = \dfrac{-101}{x^3 + 8}$

28. $\dfrac{x + 2}{x^2 + x + 1} - \dfrac{x}{x - 1} = 2$

29. $\dfrac{1}{4x} + \dfrac{1}{4x^2} = x + 1$

30. $\dfrac{x + 3}{x^2 + x} - \dfrac{1}{x^3 + x^2} = \dfrac{3}{4}$

31. $\dfrac{3}{x^3+1} - \dfrac{6}{(x+1)^3} = \dfrac{14x^3-8}{(x^3+1)(x+1)^2}$

32. $\dfrac{1}{(x-1)^3} + \dfrac{1}{(x+1)^3} = \dfrac{10x^2-18}{(x^2-1)^3}$

CHAPTER SUMMARY

The Set of Rationals $Q = \left\{\dfrac{a}{b} \,\Big|\, a \text{ and } b \text{ are integers and } b \neq 0\right\}$, where $\dfrac{a}{b} = \dfrac{c}{d}$ if and only if $ad = bc$ and $\dfrac{a}{1} = a$.

Field Axioms for Q For all x, y, and z in Q

(1) Closure, Addition — There is exactly one $x + y$ in Q.

(2) Closure, Multiplication — There is exactly one xy in Q.

(3) Commutative, Addition — $x + y = y + x$.

(4) Commutative, Multiplication — $xy = yx$.

(5) Associative, Addition — $(x + y) + z = x + (y + z)$.

(6) Associative, Multiplication — $(xy)z = x(yz)$.

(7) Distributive — $x(y + z) = xy + xz$.

(8) Identity, Addition — 0 is in Q and for all x, $x + 0 = x$ and $0 + x = x$.

(9) Identity, Multiplication — 1 is in Q and for all x, $x \cdot 1 = x$ and $1 \cdot x = x$.

(10) Inverse, Addition — Exactly one $-x$ in Q so that $x + (-x) = 0$ and $(-x) + x = 0$.

(11) Inverse, Multiplication — Exactly one $\dfrac{1}{x}$ in Q if $x \neq 0$ so that $x \cdot \dfrac{1}{x} = 1$ and $\dfrac{1}{x} \cdot x = 1$.

Definition of Subtraction $x - y = x + (-y)$.

Definition of Division If $y \neq 0$, then $\dfrac{x}{y} = x\left(\dfrac{1}{y}\right)$.

CHAPTER SUMMARY

Special Theorems for the Set of Rationals

(1) Reciprocal of a reciprocal — If $x \neq 0$, then $\dfrac{1}{1/x} = x$.

(2) Product of reciprocals — If $xy \neq 0$, then $\dfrac{1}{x} \cdot \dfrac{1}{y} = \dfrac{1}{xy}$.

(3) Fundamental theorem of quotients — If $yz \neq 0$, then $\dfrac{xz}{yz} = \dfrac{x}{y}$.

(4) Sum of quotients — If $yw \neq 0$, then $\dfrac{x}{y} + \dfrac{z}{w} = \dfrac{xw + yz}{yw}$.

(5) Product of quotients — If $yw \neq 0$, then $\dfrac{x}{y} \cdot \dfrac{z}{w} = \dfrac{xz}{yw}$.

(6) Difference of quotients — If $yw \neq 0$, then $\dfrac{x}{y} - \dfrac{z}{w} = \dfrac{xw - yz}{yw}$.

(7) Quotient of quotients — If $yzw \neq 0$, then $\dfrac{x}{y} \div \dfrac{z}{w} = \dfrac{xw}{yz}$.

Theorem *The set of quotients of polynomials over Q is a field;* that is, the closure, commutative, associative, distributive, identity, and inverse axioms for addition and multiplication are valid.

Definition of Simplified Form of Quotients of Polynomials over Q If $B(x) \neq 0$ and $B(x) \neq 1$, then $\dfrac{A(x)}{B(x)}$ is in simplified form if and only if $A(x)$ and $B(x)$ are integral polynomials with no common integral constant factor and no common prime integral polynomial factor. If $B(x) = 1$, then $A(x)$ is the simplified form of $\dfrac{A(x)}{1}$.

Theorem on Equating Coefficients of Polynomials The corresponding coefficients of $A(x)$ and $B(x)$ are equal if and only if for all x in Q, $A(x) = B(x)$.

Theorem on Equating Coefficients of Numerators on Quotients If for all x such that $C(x) \neq 0$, $\dfrac{A(x)}{C(x)} = \dfrac{B(x)}{C(x)}$, then for all x, $A(x) = B(x)$.

The Partial Fractions Theorem If the denominator of a partial fraction is linear, $ax + b$ or $(ax + b)^n$, then the numerator is a constant. If the denominator of a partial fraction is quadratic, $ax^2 + bx + c$ or $(ax^2 + bx + c)^n$, then the numerator is of the form $px + q$.

Equivalence Theorems If $A(x)$, $B(x)$, and $C(x)$ are rational functions over Q, then $A(x) = B(x)$ is equivalent to

$$A(x) + C(x) = B(x) + C(x) \qquad \text{Addition Theorem}$$

and

$$A(x)C(x) = B(x)C(x) \text{ and } C(x) \neq 0 \qquad \text{Multiplication Theorem}$$

REVIEW EXERCISES

1–10. Simplify and state the restrictions on the variable.

1. $\dfrac{3 + \frac{2}{3}}{3 - \frac{2}{3}}$

2. $\dfrac{4x + 3}{9} - \dfrac{3x - 1}{12}$

3. $\dfrac{7 - 4x}{9x^2 - 144} + \dfrac{2x - 4}{4x^2 - 64}$

4. $\dfrac{x - 5}{3x^2 + 12x} - \dfrac{2x - 16}{4x^2 - 64}$

5. $\left(\dfrac{x}{x + 2}\right) \div \left(1 - \dfrac{x}{x + 2}\right)$

6. $\dfrac{6x + 30}{6x + 18} \cdot \dfrac{5x + 15}{5x^2 - 125}$

7. $x - 3 - \dfrac{x^2 + 9}{x - 3}$

8. $\dfrac{x^2 - 5x}{x - 1} \div \left(\dfrac{x^2 - 25}{x^2 + x - 20} \div \dfrac{x^2 + x - 2}{x^2 - 2x - 8}\right)$

9. $\left(x + \dfrac{4x^2}{6x^2 - 2x - 4}\right) \cdot \left(1 - \dfrac{4x}{6x^2 + 2x - 4}\right)$

10. $\dfrac{\dfrac{x + 3}{x - 1} - \dfrac{x - 1}{x + 3}}{\dfrac{1}{(x + 3)(x - 1)}}$

11–18. Resolve each of the following into a sum of partial fractions.

11. $\dfrac{2x}{3x^2 + 5x - 2}$

12. $\dfrac{4}{x^2 - x - 30}$

13. $\dfrac{2x - 3}{2x^2 + 3x - 2}$

14. $\dfrac{x}{(2x + 1)^2}$

15. $\dfrac{x^3 + 2x + 1}{x^2 - 4}$

16. $\dfrac{3}{(x + 2)^2(x + 3)}$

17. $\dfrac{x^2 + 3x + 1}{(x^2 - 3)(x + 2)}$

18. $\dfrac{3x + 5}{2x^3 + 8x}$

REVIEW EXERCISES

19–30. Solve for x.

19. $\dfrac{3}{x+2} + \dfrac{4}{x-2} = \dfrac{5}{x^2-4}$

20. $\dfrac{2x+3}{x-1} = \dfrac{2x-5}{x+3}$

21. $\dfrac{x+5}{5x} - \dfrac{3}{2x-7} - \dfrac{1}{5} = 0$

22. $3 - \dfrac{2x+3}{3x-2} + \dfrac{x-1}{2-3x} = 0$

23. $3x + 1 - \dfrac{3x-1}{x+1} = 3(x+1)$

24. $\dfrac{3x^2-1}{x} = x - 6 - \dfrac{1}{x}$

25. $\dfrac{x^2-8x}{x^2-4x} + \dfrac{x^2-4x}{16-4x} + \dfrac{x}{x-4} = 0$

26. $\dfrac{1}{1 - \dfrac{1}{1 - \dfrac{1}{x}}} = 2$

27. $\dfrac{1}{(1-x)^3} + \dfrac{1}{1-x^3} = \dfrac{2}{(1-x^3)(1-x)^2}$

28. $\dfrac{2x^3 - 9}{x} = 18 - x$

29. $\dfrac{\dfrac{1}{x} - \dfrac{1}{4}}{\dfrac{1}{16} - \dfrac{1}{x^2}} = x$

30. $\dfrac{2}{3x^3 + x^2 - 3x - 1} + \dfrac{1}{x^3 - 3x^2 - x + 3} = 0$

5
THE REALS

5.1 THE SET OF REAL NUMBERS

5.1.1 Irrationality of $\sqrt{2}$

The length of the diagonal of a unit square presented a difficult problem to the Greek mathematician Pythagoras and his followers around 580–500 B.C. If x represents the length of the diagonal, then they knew that $x^2 = 1^2 + 1^2 = 2$. Therefore, they reasoned, there must be some "number" that can be assigned as the length of the diagonal and moreover, $x^2 = 2$. However, these Greeks also proved that there is no *rational* number x such that $x^2 = 2$. At this time the concept of number was limited to the set of positive rational numbers and thus the existence of a number that is not rational led to a major dilemma.

Theorem There is no rational number x so that $x^2 = 2$.

Proof (*Indirect method*) Assume there is a rational number, $\frac{p}{q}$, and $\frac{p^2}{q^2} = 2$. It may also be assumed that $\frac{p}{q}$ is in simplified form and thus p and q have no common prime factors.

Now, $p^2 = 2q^2$ and, by the unique factorization theorem, 2 must be a factor of p^2 and 2 must also be a factor of p since 2 is prime.

THE SET OF REAL NUMBERS

Let $p = 2n$. Then $p^2 = 4n^2 = 2q^2$ and $2n^2 = q^2$. Consequently, 2 is a factor of q^2 and also of q. But it was assumed that p and q have no common prime factors and thus 2 cannot be a factor of both p and q. A contradiction has been reached and thus the negation of the assumption is valid; there is no rational number x so that $x^2 = 2$.

5.1.2 Other Irrational Numbers

Another famous example of an irrational number known to the early Greeks involved the side of a cube whose volume is 2 cubic units; or $x^3 = 2$. In a similar way, it was shown that there is no rational number x so that $x^3 = 2$.

At a much later date, 1525 A.D., the symbolism $\sqrt{2}$ and $\sqrt[3]{2}$ (read "principal square root of 2" and "principal cube root of 2") was introduced to indicate the positive numbers such that $(\sqrt{2})^2 = 2$ and $(\sqrt[3]{2})^3 = 2$. Since $\sqrt{2}$ and $\sqrt[3]{2}$ are not rational numbers, they are called **irrational** numbers.

The number π, the ratio of the circumference of any circle to its diameter, is another example of a number that is not rational. Moreover, π is not a zero of any polynomial with rational coefficients, unlike $\sqrt{2}$ and $\sqrt[3]{2}$, which are zeros of $x^2 - 2$ and $x^3 - 2$, respectively.

Table 5.1 The Set of Real Numbers

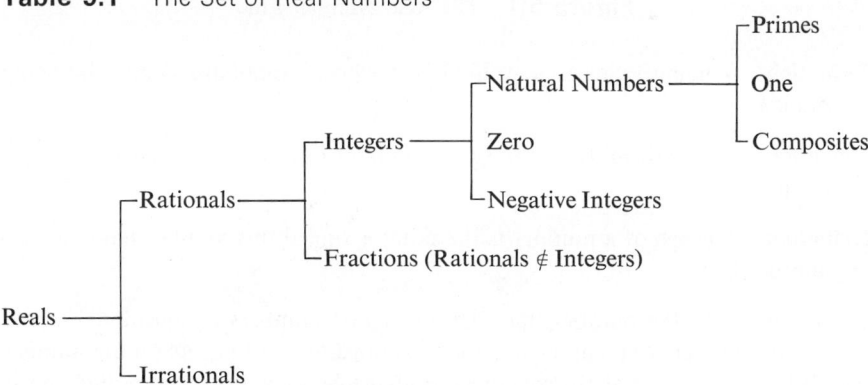

The set of real numbers, R, is the union of the set of rational numbers and the set of irrational numbers. At this point, however, the questions "What is an irrational number?" and consequently "What is a real number?" are still unanswered. One major difficulty involves finding a convenient symbolism to designate all irrational numbers. While a negative integer was represented by prefixing a minus sign to a natural number, $-n$, and a nonintegral fraction was represented as a quotient of 2 integers, $\dfrac{p}{q}$, no such simple symbolic representation is available for a general irrational real number. To provide insight regarding this difficulty, a geometric model, the real number line, is introduced.

5.1.3 The Real Number Line

A point on a straight line is selected as a fixed reference point, called the **origin**, and is assigned the number 0. A positive direction and the corresponding negative direction are selected and a unit of measurement is selected. Rational numbers are assigned to points whose distance from the origin is a rational number; positive numbers are assigned to points on the positive side of 0 and negative numbers are assigned to points on the negative side of 0. A point corresponding to $\sqrt{2}$ can be located on the number line by constructing a unit square and by using the diagonal of this square as the radius of a circle. Points corresponding to $\sqrt{3}$, $\sqrt{5}$, and so on can be constructed similarly. See Figure 5.1.

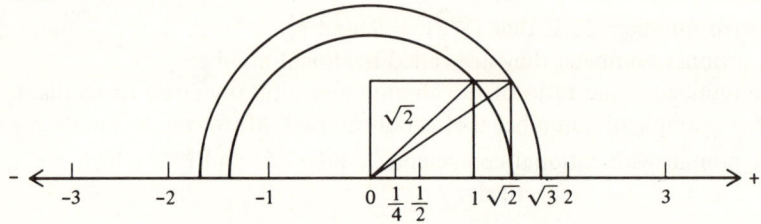

Figure 5.1 The Real Number Line

Definition A **number line** is a straight line whose points are named by using numbers.

Definition A **coordinate** of a point on a number line is the number that names the point.

Definition A **graph** of a number is the point assigned this number name on the number line.

Referring to the number line, the set of real numbers is considered as the totality of numbers that can be assigned as coordinates of points on the number line so that there is a one-to-one correspondence between the set of points on the line and the set of real numbers. This technique still does not provide a symbolism for designating the real numbers but it does suggest a possible approach.

5.1.4 Decimals

For the case of $\sqrt{2}$, it may be observed from the number line that there are many rational numbers on both sides of $\sqrt{2}$, and $\sqrt{2}$ may be approximated by rational numbers. For this purpose it is convenient to use decimal fractions, rational numbers whose denominators are powers of ten.

Using tenths, $\quad 1.4 < \sqrt{2} < 1.5.$

Using hundredths, $\quad 1.41 < \sqrt{2} < 1.42.$

THE SET OF REAL NUMBERS 133

Using thousandths, $\quad 1.414 < \sqrt{2} < 1.415.$

Using ten thousandths, $\quad 1.4142 < \sqrt{2} < 1.4143.$

An old Babylonian tablet (c. 1700 B.C.) in the collection at Yale University exhibits a very good approximation to $\sqrt{2}$. A transcription of the Babylonian base 60 numerals yields

$$\sqrt{2} = 1 + \frac{24}{60} + \frac{51}{3600} + \frac{10}{216{,}000}$$

or

$$\sqrt{2} = 1.414213 \text{ correct to the sixth decimal place.}$$

The process of decimal approximation suggests the use of an infinite decimal notation to describe the real numbers, as follows.

$R = \{c.d_1 d_2 d_3 \ldots | c \text{ is an integer and } d_i \text{ is a digit } (0, 1, 2, 3, 4, 5, 6, 7, 8, 9)\}.$

For example,

$\dfrac{49}{2} = 24.500 \ldots \quad$ (repeating zeros).

$\dfrac{49}{2} = 24.499 \ldots \quad$ (repeating nines).

$\dfrac{1}{3} = 0.333 \ldots \quad$ (repeating threes).

$\sqrt{2} = 1.414213 \ldots \quad$ (nonrepeating digits).

If the infinite decimal representation contains repeating zeros, then these zeros may be omitted. A repeating decimal of this kind is also called a **terminating decimal**. It may be shown that every repeating decimal can be expressed as the quotient of two integers and thus represents a rational number. Conversely, every rational number can be expressed as a repeating decimal.

Example 1 Show that $5.3\overline{21}$ represents a rational number. (The bar is drawn over the block of digits 21 to indicate that this block of digits repeats.)

Solution Let

$$x = 5.3\overline{21}$$
$$1000x = 5321.2121\ldots$$
$$10x = 53.2121\ldots$$
$$\overline{}$$
$$990x = 5268$$
$$x = \frac{5268}{990}$$

Example 2 Express $\frac{5}{12}$ as a repeating decimal.

Solution

$$\begin{array}{r} 0.4166\ldots \\ 12\overline{)5.0000\ldots} \\ 48 \\ \overline{20} \\ 12 \\ \overline{80} \\ 72 \\ \overline{80} \end{array}$$ and $\frac{5}{12} = 0.41\overline{6}$

Theorem The infinite decimal representation of a real number is repeating if and only if the number is rational.

Proof

PART 1 Let $x = cd_1d_2 \ldots d_k \overline{d_{k+1} \ldots d_{k+n}}$ where the block of digits $d_{k+1} \ldots d_{k+n}$ repeats.
Then,

$$10^{k+n}x = cd_1d_2 \ldots d_k d_{k+1} \ldots d_{k+n} \cdot d_{k+1} \ldots d_{k+n} \ldots$$
$$10^k x = cd_1d_2 \ldots d_k \cdot d_{k+1} \ldots d_{k+n} \ldots$$
$$\overline{(10^{k+n} - 10^k)x = cd_1d_2 \ldots d_{k+n} \cdot - cd_1d_2 \ldots d_k \cdot}$$
$$x = \frac{cd_1d_2 \ldots d_{k+n} \cdot - cd_1d_2 \ldots d_k}{10^{k+n} - 10^k}$$

PART 2 The proof that any rational number $\frac{a}{b}$ can be expressed as an infinite repeating decimal is based on the division algorithm; $a = bq + r$ where $0 \leq r < b$.

If $a = b$, then $\frac{a}{b} = 1.000\ldots$ (repeating zeros) $= 0.999\ldots$ (repeating nines).
If $a < b$, then $q = 0$. In all cases,

$$a = bq + r_0 \quad \text{where } 0 \leq r_0 < b.$$

First digit d_1 after decimal point,

$$10r_0 = bd_1 + r_1 \quad \text{where } 0 \leq r_1 < b.$$

Next digit d_2 after decimal point,

$$10r_1 = bd_2 + r_2 \quad \text{where } 0 \leq r_2 < b.$$

THE SET OF REAL NUMBERS

The other digits are obtained similarly. (This process is the familiar division calculation of arithmetic.) Since there are only b values for the remainders r_i; that is, $0, 1, 2, \ldots, b-1$, at some point in the division, one of these b values must occur a second time. The block of digits from this point until the next repeat constitutes the set of repeating digits.

As a consequence of the preceding theorem, the irrrational real numbers can be identified as those numbers that can be represented by infinite nonrepeating decimals.

5.1.5 Upper and Lower Bounds

While the decimal notation provides insight for understanding both the rational and irrational real numbers, there are certain difficulties that result from this representation. For example, some numbers can be expressed in more than one way, such as $2.999\ldots$ and $3.000\ldots$ which both represent the integer 3. Moreover, it is very difficult to define general rules for addition and multiplication and even more difficult to show that the field axioms are valid and, in particular, to show that the reciprocal $\dfrac{1}{x}$ exists for every nonzero real number x. A resolution of this problem is provided by the concept of upper and lower bounds.

Definition of Upper Bound Let set A be a nonempty subset of an ordered set B. Then u is an upper bound of A if and only if, for all a in A, $a \leq u$.

Definition of Least Upper Bound (lub) $\bar{u}$ is the least upper bound of A if and only if, for all a in A, $a \leq \bar{u}$ and if $a \leq u$, then $\bar{u} \leq u$.

In other words, the least upper bound, $\bar{u}$, is the smallest number that is larger than or equal to every number in set A.

For example, let $A = \left\{2\tfrac{1}{2}, 2\tfrac{3}{4}, 2\tfrac{7}{8}, 2\tfrac{15}{16}, \ldots, 2 + \dfrac{2^n - 1}{2^n}, \ldots\right\}$. Then upper bounds for A are 3, 4, 5, 3.5, and in fact any number equal to or larger than 3. Since 3 is the smallest number that is larger than every number in set A, 3 is the least upper bound of A.

Definition of Lower Bound Let set A be a nonempty subset of the ordered set B. Then l is a lower bound of set A if and only if, for all a in A, $l \leq a$.

Definition of Greatest Lower Bound (glb) $\bar{l}$ is the greatest lower bound of A if and if only, for all a in A, $\bar{l} \leq a$ and if $l \leq a$, then $l \leq \bar{l}$.

In other words, the greatest lower bound, l, is the largest number that is smaller than or equal to every number in set A.

For example, let $A = \{5.1, 5.01, 5.001, 5.0001, \ldots\}$. Then 5, 4, 3, and any number smaller than 5 are lower bounds for A. Of all of these, 5 is the largest and thus 5 is the greatest lower bound.

Now, suppose A is a set of rational approximations to $\sqrt{2}$,

$$A = \{1, 1.4, 1.41, 1.414, 1.4142, \ldots\}.$$

Then, the irrational number $\sqrt{2}$ may be defined as the least upper bound of this set. Another possible definition is the following one:

$\sqrt{2}$ = the least upper bound of $\{x \mid x^2 < 2$ and x is a rational number$\}$

Generalizing this procedure, a definition of the set of real numbers can now be stated.

Definition of the Set of Real Numbers The **set of real numbers**, R, is the set of all least upper bounds and greatest lower bounds of sets of rational numbers. Of these bounds, those that are not rational are called **irrational numbers**.

Using the definition above, it is possible to derive the important properties of the set of real numbers. Such a derivation is beyond the scope of this book. Instead, a system for the real numbers will be presented in the next section and the properties of this system will then be discussed.

In summary, the discussion so far has been concerned with answering the question, What is a real number? The following observations were made:

(1) The set of real numbers is the set of all least upper bounds and greatest lower bounds of sets of rational numbers.

(2) The set of real numbers can be represented as infinite decimals. The rationals are the repeating decimals and the irrationals are the nonrepeating decimals. If decimals containing infinitely many repeating nines (or equivalently, repeating zeros) are excluded, then the representation is unique.

(3) There is a one-to-one correspondence between the set of real numbers and the set of points on the number line.

5.1.6 A System for the Real Numbers

UNDEFINED CONCEPTS
    A set of numbers, R
    Two operations, $+$ and $\cdot$ (where $xy = x \cdot y$)
    A relation $>$

THE SET OF REAL NUMBERS

Field Axioms   For all x, y, and z in R

(1)	Closure, Addition	There is exactly one $x + y$ in R.
(2)	Closure, Multiplication	There is exactly one xy in R.
(3)	Commutative, Addition	$x + y = y + x$.
(4)	Commutative, Multiplication	$xy = yx$.
(5)	Associative, Addition	$(x + y) + z = x + (y + z)$.
(6)	Associative, Multiplication	$(xy)z = x(yz)$.
(7)	Distributive	$x(y + z) = xy + xz$.
(8)	Identity, Addition	There is exactly one element, 0, in R so that for all x in R, $x + 0 = x$ and $0 + x = x$.
(9)	Identity, Multiplication	There is exactly one element, 1, in R so that for all x in R, $x \cdot 1 = x$ and $1 \cdot x = x$.
(10)	Inverse, Addition	For each x in R, there is exactly one element, $-x$, in R (called the **opposite** of x), so that $x + (-x) = 0$ and $(-x) + x = 0$.
(11)	Inverse, Multiplication	For each x in R such that $x \neq 0$, there is exactly one element, $\frac{1}{x}$, in R (called the **reciprocal** of x) so that $x \cdot \frac{1}{x} = 1$ and $\frac{1}{x} \cdot x = 1$.

Order Axioms

(12)	Trichotomy	For all x in R, exactly one of the following is valid: $x > 0$ or $x = 0$ or $-x > 0$.
(13)	Closure of Positives, Addition	If $x > 0$ and $y > 0$, then $x + y > 0$.
(14)	Closure of Positives, Multiplication	If $x > 0$ and $y > 0$, then $xy > 0$.

Completeness Axiom

(15) Every nonempty subset of R that has an upper bound has exactly one least upper bound in R.
Every nonempty subset of R that has a lower bound has exactly one greatest lower bound in R.

Definitions

Subtraction	$x - y = x + (-y)$.
Division	$\dfrac{x}{y} = x \cdot \dfrac{1}{y}$.
Positive	x is positive if and only if $x > 0$.
Negative	x is negative if and only if $-x > 0$.
$x > y$	$x > y$ if and only if $x - y > 0$.
$x < y$	$x < y$ if and only if $y > x$.

Axioms (1)–(14) of this system are also valid for the rational numbers. Axiom (15), the completeness axiom, is not valid for the set of rationals. This is the axiom that unites the set of irrational numbers with the set of rationals.

The theorems derived from the field axioms for the set of rationals are valid for any field and, in particular, for the field of real numbers. These theorems are restated in the Summary at the end of this chapter.

A field for which the Order Axioms (12)–(14) are valid is called an **ordered field**. The set of rationals and the set of reals are both ordered fields. The set of real numbers is called a **complete ordered field** since the completeness axiom is valid. Theorems derived from the order axioms are valid for all ordered fields. The next section is devoted to the concept of order.

EXERCISES 5.1

1–6. Express each of the following in the form p/q where p and q are relatively prime natural numbers:

1. $0.\overline{325}$
2. $0.\overline{123}$
3. $2.4\overline{16}$
4. $0.\overline{142857}$
5. $3.2\overline{14}$
6. $0.1\overline{234}$

7–11. Express each of the following as repeating decimals:

7. $3\frac{2}{9}$
8. $\frac{5}{7}$
9. $5\frac{7}{11}$
10. $\frac{14}{35}$
11. $8\frac{4}{13}$
12. Prove that $\sqrt{3}$ is irrational.
13. Prove that $3 + \sqrt{2}$ is irrational.
14. Geometrically plot 1.7, 1.73, and 1.732 using 100 squares for 1.00 to 2.00. Compare with the geometric construction of $\sqrt{3}$.

THE SET OF REAL NUMBERS

15–18. Approximate each of the following to the nearest hundredth:

15. $\sqrt{132}$ 17. $\sqrt{78.2}$
16. $\sqrt{7.82}$ 18. $\sqrt{782}$

19–24. Find the lub of each of the following sets. In each case, state if the lub is an element of the set.

19. $A = \{x \mid -2 < x < 11, x \text{ is a real number}\}$
20. $B = \{x \mid x^2 < 17, x \text{ is an integer}\}$
21. $C = \{\frac{1}{2}, \frac{3}{4}, \frac{7}{8}, \frac{15}{16}, \ldots\}$
22. $D = \{x \mid 17 < x^2 \leq 20, x \text{ is a rational number}\}$
23. $E = \{x \mid x^2 \leq 3, x \text{ is a rational number}\}$
24. $F = \{x \mid x^2 \leq 4, x \text{ is a rational number}\}$

25–30. Find the glb of each of the following sets. In each case state if the glb is an element of the set.

25. $A = \{x \mid 0 < x < 1, x \text{ is a real number}\}$
26. $B = \{x \mid 17 < x^2, x \text{ is a positive real number}\}$
27. Set C as defined in Exercise 21.
28. Set D as defined in Exercise 22.
29. $E = \{x \mid 4 \leq x^2, x \text{ is a positive rational number}\}$
30. $F = \{x \mid 3 \leq x^2, x \text{ is a positive rational number}\}$
31. Which is a better approximation of π, $\sqrt{10}$ or 22/7?
32. Express as the quotient of two integers the error made in using 0.1667 as an approximation of $\frac{1}{6}$.
33. Using 4000 miles as the radius of the earth, what is the error made in computing the length of the equator by using:

    (a) $\pi = 22/7$ (b) $\pi = 3.14$ (c) $\pi = 3.1416$.

34. A man wants to fence a plot which is in the shape of an isosceles right triangle each of whose equal sides is s feet long. [Perimeter, $P = s(2 + \sqrt{2})$.] To find the perimeter correct to the nearest foot, how accurately must $\sqrt{2}$ be approximated if

    (a) $s = 10$ feet   (b) $s = 100$ feet.

35. If the plot to be fenced as in Exercise 34 is in the shape of a 30°–60°–90° triangle such that $P = s(1 + \frac{1}{2} + \sqrt{3}/2)$, how accurately must $\sqrt{3}$ be approximated in order to find the perimeter correct to the nearest foot if

    (a) $s = 10$ feet   (b) $s = 100$ feet.

☆ 36. If $S = \{x \mid x^2 < 3, x \text{ is a positive rational number}\}$, prove that the lub of S is not an element of S.

☆ 37–38. Let $a = 0.a_1 a_2 a_3 \ldots a_n \ldots$ and $b = 0.b_1 b_2 b_3 \ldots b_n \ldots$ be infinite decimals with no repeating nines.

37. Express $a + b$ and ab as a least upper bound of a set of rational numbers.

38. For $a \neq 0$, express $-a$ and $\dfrac{1}{a}$ as a least upper bound of a set of rational numbers.

5.2 ORDER THEOREMS AND INEQUALITIES

5.2.1 Order Theorems

Theorem 1 (General Trichotomy Theorem) For all x and y in R, exactly one of the following is valid: $x > y$ or $x = y$ or $x < y$.

Proof
$x > y$ if and only if $x - y > 0$.	Definition.
$x = y$ if and only if $x - y = 0$.	Inverse, identity axioms.
$x < y$ if and only if $y > x$.	Definition.
$y > x$ if and only if $y - x > 0$.	Definition.
$-(x - y) = y - x$.	Theorem, inverse of difference.
$x < y$ if and only if $-(x - y) > 0$.	Substitution.

Since exactly one of $x - y > 0$, $x - y = 0$, $-(x - y) > 0$ is valid by the trichotomy axiom, therefore exactly one of their equivalents is valid.

Theorem 2 (Transitivity Order Theorem) For all x, y, and z in R, if $x > y$ and $y > z$, then $x > z$.

Proof
$x - y > 0$ and $y - z > 0$.	Definition of $>$.
$(x - y) + (y - z) > 0$.	Positive closure axiom.
$(x - z) + (y - y) > 0$.	Associative, commutative axioms.
$(x - z) > 0$.	Identity, inverse axioms.
$x > z$.	Definition of $>$.

Corollary For all x, y, and z in R, if $x < y$ and $y < z$, then $x < z$. (Proof left for student.)

Examples If $7 > 5$ and $5 > 2$, then $7 > 2$.
If $-2 > -5$ and $-5 > -7$, then $-2 > -7$.

The reasons in the proofs of Theorems 3, 4, and 5 are left for the student.

ORDER THEOREMS AND INEQUALITIES

Theorem 3 (Addition Order Theorem) For all x, y, and z in R, if $x > y$, then $x + z > y + z$.

Proof $x - y > 0$.
$x - y = (x - y) + (z - z)$.
$x - y = (x + z) + (-y - z)$.
$x - y = (x + z) - (y + z)$.
$(x + z) - (y + z) > 0$.
$x + z > y + z$.

Corollary For all x, y, and z in R, if $x < y$, then $x + z < y + z$. (Proof left for student.)

Examples If $2 > -3$, then $2 + 3 > -3 + 3$, or $5 > 0$.
If $x - 4 > 6$, then $x > 6 + 4$, or $x > 10$.

Theorem 4 (Order of Addition Inverses) For all x in R, if $x > 0$, then $-x < 0$ and if $x < 0$, then $-x > 0$.

Proof If $x > 0$, then $x + (-x) > 0 + (-x)$
$0 > -x$
$-x < 0$.

If $x < 0$, then $x + (-x) < 0 + (-x)$
$0 < -x$
$-x > 0$.

Theorem 5 (Multiplication Order Theorem) For all x, y, and z in R,

(1) Multiplication by Positive: If $x > y$ and $z > 0$, then $xz > yz$.

(2) Multiplication by Negative: If $x > y$ and $z < 0$, then $xz < yz$.

Proof of (1) $x - y > 0$.
$(x - y)z > 0$.
$xz - yz > 0$.
$xz > yz$.

Proof of (2) $x - y > 0$.
$-z > 0$.
$(x - y)(-z) > 0$.
$-xz + yz > 0$.
$yz > xz$.
$xz < yz$.

Corollary

(1) If $x < y$ and $z > 0$, then $xz < yz$.
(2) If $x < y$ and $z < 0$, then $xz > yz$.

(Proof left to student.)

Examples If $7 > 3$, then $7(+2) > 3(+2)$ or $14 > 6$.
If $7 > 3$, then $7(-2) < 3(-2)$ or $-14 < -6$.
If $7 > -3$, then $7(-2) < (-3)(-2)$ or $-14 < +6$.
If $3x > 12$, then $x > 4$.
If $-3x > 12$, then $x < -4$.

5.2.2 Inequalities and the Number Line

An **inequality** is a statement involving an order relation. There are four basic forms for open statements that are inequalities:

$$x > a, \quad x < a, \quad x \geq a, \text{ and } x \leq a.$$

The statement $x \geq a$ means "x is greater than a or x equals a." The statement $x \leq a$ means "x is less than a or x equals a."

The number line is a useful graphic aid in the solution of inequalities. For example, the set $\{x \mid x > 3\}$ is graphed as shown in Figure 5.2. The circle above the numeral 3 indicates that 3 is *excluded* from the solution set. The solution set is indicated by the half-line starting at 3 (but not including 3) and all values greater than 3 as shown by the direction of the line.

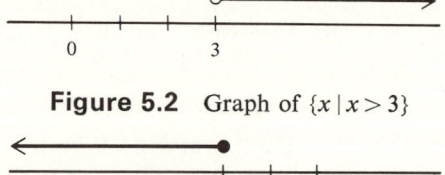

Figure 5.2 Graph of $\{x \mid x > 3\}$

Figure 5.3 Graph of $\{x \mid x \leq -2\}$

The set $\{x \mid x \leq -2\}$ is graphed as shown in Figure 5.3. This time a solid dot over the -2 coordinate indicates that -2 is *included* in the solution set, as well as all points to the left of -2, since x is *less than or equal to* -2.

It is often convenient to graph intersections and unions of inequalities. Since it follows from the transitivity order theorem that $a < b$ whenever $a < x$ and $x < b$, the statement $a < x$ and $x < b$ may be expressed as $a < x < b$, read "x is *between a* and *b*."

ORDER THEOREMS AND INEQUALITIES

Definition $a < x < b$ if and only if $a < x$ and $x < b$.

The equivalences below are useful for graphing unions and intersections:

$$\{x|px\} \cap \{x|qx\} = \{x|px \text{ and } qx\}.$$
$$\{x|px\} \cup \{x|qx\} = \{x|px \text{ or } qx\}.$$

The graph of $\{x|-2 < x\} \cap \{x|x < 1\} = \{x|-2 < x < 1\}$ is illustrated in Figure 5.4.

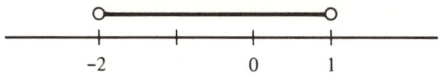

Figure 5.4 Graph of $\{x|-2 < x < 1\}$

The graph of $\{x|x \leq -2\} \cup \{x|x > 1\} = \{x|x \leq -2 \text{ or } x > 1\}$ is illustrated in Figure 5.5.

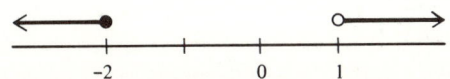

Figure 5.5 Graph of $\{x|x \leq -2 \text{ or } x > 1\}$

5.2.3 Solution of Inequalities

A *solution* of an inequality in one variable is a number from a specified set of numbers such that the inequality becomes a true statement when the variable is replaced by the name of this number.

The *solution set* of an inequality is the set of all solutions of the inequality. Two inequalities are *equivalent* if and only if they have the same solution set.

To *solve an inequality in one variable* x means to state the solution set of an equivalent inequality having the form $x > a$, $x < a$, $x \geq a$, or $x \leq a$, or to state the solution set of an equivalent union or intersection of sets whose corresponding inequalities have theses forms.

The theorems for inequalities provide the techniques for finding the equivalent inequalities. Solving an inequality is similar to solving an equation with the important exception that multiplication (or division) by a negative number changes the order of the inequality.

Equivalence Order Theorem 1 For all open rational expressions A, B, and C,

(1) $A > B$ is equivalent to $A + C > B + C$,
(2) $A > B$ is equivalent to $AC > BC$ if $C > 0$, and
(3) $A > B$ is equivalent to $AC < BC$ if $C < 0$.

Example 1 Solve $3x + 1 > x - 7$ and graph the solution set.

Solution $3x + 1 > x - 7$
$2x + 1 > -7$
$2x > -8$
$x > -4$

The solution set $= \{x \mid x > -4\}$. (See Figure 5.6.)

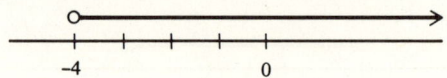

Figure 5.6

Example 2 Solve $2 - x \geq 3x + 1$.

Solution $2 - x \geq 3x + 1$
$2 - 4x \geq 1$
$-4x \geq -1$
$x \leq \frac{1}{4}$

The solution set $= \{x \mid x \leq \frac{1}{4}\}$. (See Figure 5.7.)

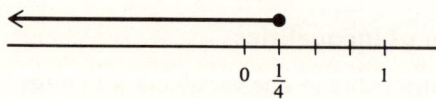

Figure 5.7

Example 3 Solve $\dfrac{x}{x+1} > 2$.

Solution Since the inequality has a variable in the denominator, it is necessary to consider two cases, $x + 1 > 0$ and $x + 1 < 0$. The solution set of the inequality is the union of the solution sets for the two cases.

CASE 1 $x + 1 > 0$ and thus $x > -1$.
Since $x + 1 > 0$, multiplying each side of the inequality by $x + 1$ does not change the order.

$$x > 2x + 2$$
$$-x > 2$$
$$x < -2$$

The solution set $= \{x \mid x > -1 \text{ and } x < -2\} = \emptyset$.

ORDER THEOREMS AND INEQUALITIES

CASE 2 $x + 1 < 0$ and thus $x < -1$.
Since $x + 1 < 0$, multiplying each side of the inequality by $x + 1$ *does* change the order.

$$x < 2x + 2$$
$$-x < 2$$
$$x > -2$$

The solution set $= \{x \mid x < -1 \text{ and } x > -2\} = \{x \mid -2 < x < -1\}$.

The union of the solution sets for Case 1 and Case 2 is
$$\{x \mid -2 < x < -1\} \cup \emptyset = \{x \mid -2 < x < -1\}.$$

EXERCISES 5.2

1. Prove: For all x, y, and z in R, if $x < y$ and $y < z$, then $x < z$.
2. Prove: For all x, y, and z in R, if $x < y$, then $x + z < y + z$.
3. Prove: For all x, y, and z in R, if $x < y$ and $z > 0$, then $xz < yz$.
4. Prove: For all x, y, and z in R, if $x < y$ and $z < 0$, then $xz > yz$.

5–30. Solve each of the following and graph the solution set.

5. $5x + 3 > 3x + 5$
6. $3 - 2x \geq x + 2$
7. $3(x + 2) < 4x$
8. $\dfrac{x}{x+2} > 1$
9. $\dfrac{x}{x+2} \leq 3$
10. $5 - 3x > 0$
11. $2 - 3x < \dfrac{2x}{3}$
12. $\dfrac{5 - 2x}{2} \geq 3$
13. $\dfrac{3}{x} < 0$
14. $\dfrac{2 + x}{x} < 5$
15. $\dfrac{x + 2}{3} \leq \dfrac{2 - x}{2}$
16. $3x + 5 < \dfrac{5x + 3}{2}$
17. $\dfrac{3}{x+1} > 2$
18. $\dfrac{2x}{x+2} \leq -1$
19. $\dfrac{1}{x-1} < 1$
20. $\{x \mid x > 3\} \cup \{x \mid x > -2\}$
21. $\{x \mid x > 3\} \cap \{x \mid x > -2\}$
22. $\{x \mid x < 3\} \cup \{x \mid x > -2\}$
23. $\{x \mid x < 3\} \cap \{x \mid x > -2\}$
24. $\{x \mid x < 3\} \cup \{x \mid x < -2\}$
25. $\{x \mid x < 3\} \cap \{x \mid x < -2\}$
26. $\{x \mid -3 < x < 2\} \cup \{x \mid x \geq 1\}$
27. $\{x \mid -3 < x < 2\} \cap \{x \mid x \geq 1\}$
28. $\{x \mid 2x + 3 > 0\} \cup \{x \mid 3x + 5 < 0\}$
29. $2(x + 3) > 3(x - 1) + 6$
30. $\dfrac{1}{x} + \dfrac{3}{5x} > \dfrac{8}{x+1}$

5.3 ADDITIONAL ORDER THEOREMS AND INEQUALITIES

Theorem 6 (Positive Product Theorem) For all x and y in R, $xy > 0$ if and only if either $(x > 0$ and $y > 0)$ or $(x < 0$ and $y < 0)$.

Theorem 7 (Negative Product Theorem) For all x and y in R, $xy < 0$ if and only if either $(x > 0$ and $y < 0)$ or $(x < 0$ and $y > 0)$.

The validity of these theorems follows immediately from the positive closure axiom for multiplication and the multiplication order theorem by noting that exactly one of the four cases listed is possible for any pair of nonzero real numbers.

(1) If $x > 0$ and $y > 0$, then $xy > 0$.
(2) If $x < 0$ and $y < 0$, then $xy > 0$.
(3) If $x > 0$ and $y < 0$, then $xy < 0$.
(4) If $x < 0$ and $y > 0$, then $xy < 0$.

An immediate consequence of the positive product and negative product theorems is the equivalence order theorem stated below which is useful for solving quadratic inequalities.

Equivalence Order Theorem 2

(1) The solution set of $A(x)B(x) > 0$ is equivalent to the union of the solution set of $(A(x) > 0$ and $B(x) > 0)$ and the solution set of $(A(x) < 0$ and $B(x) < 0)$.
(2) The solution set of $A(x)B(x) < 0$ is equivalent to the union of the solution set of $(A(x) > 0$ and $B(x) < 0)$ and the solution set of $(A(x) < 0$ and $B(x) > 0)$.

Example 1 Solve $(x + 2)(x - 3) > 0$.

Solution
CASE 1
$$x + 2 > 0 \quad \text{and} \quad x - 3 > 0$$
$$x > -2 \quad \text{and} \quad x > 3$$
$$\{x \mid x > -2 \text{ and } x > 3\} = \{x \mid x > 3\}$$
CASE 2
$$x + 2 < 0 \quad \text{and} \quad x - 3 < 0$$
$$x < -2 \quad \text{and} \quad x < 3$$
$$\{x \mid x < -2 \text{ and } x < 3\} = \{x \mid x < -2\}$$

The solution set is the union of the solution sets for Case 1 and Case 2, or

$$\{x \mid x > 3\} \cup \{x \mid x < -2\} = \{x \mid x > 3 \text{ or } x < -2\}.$$

(See Figure 5.8.)

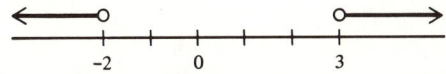

Figure 5.8

Example 2 Solve $(x + 2)(x - 3) \leq 0$.

Solution
CASE 1

$$x + 2 \geq 0 \quad \text{and} \quad x - 3 \leq 0$$
$$x \geq -2 \quad \text{and} \quad x \leq 3$$
$$\{x \mid x \geq -2 \text{ and } x \leq 3\} = \{x \mid -2 \leq x \leq 3\}$$

CASE 2

$$x + 2 \leq 0 \quad \text{and} \quad x - 3 \geq 0$$
$$x \leq -2 \quad \text{and} \quad x \geq 3$$
$$\{x \mid x \leq -2 \text{ and } x \geq 3\} = \emptyset$$

The solution set is the union of the solution sets for Cases 1 and 2, or

$$\{x \mid -2 \leq x \leq 3\} \cup \emptyset = \{x \mid -2 \leq x \leq 3\}.$$

(See Figure 5.9.)

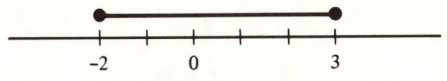

Figure 5.9

Note that the problem in Example 2 is the negation of the problem in Example 1, since $\overline{A(x) > 0} \leftrightarrow A(x) \leq 0$. Consequently, the solution set for Example 2 is the complement of the solution set for Example 1; that is,

$$\overline{\{x \mid x > 3\} \cup \{x \mid x < -2\}} = \overline{\{x \mid x > 3\}} \cap \overline{\{x \mid x < -2\}}$$
$$= \{x \mid x \leq 3\} \cap \{x \mid x \geq -2\}$$
$$= \{x \mid -2 \leq x \leq 3\}.$$

Compare Figures 5.8 and 5.9.

Example 3 Solve $x^2 - x \leq 6$.

Solution Using the addition theorem, $x^2 - x - 6 \leq 0$. Factoring, $(x+2)(x-3) \leq 0$. The solution is completed as shown in Example 2.

Theorem 8 **(Positivity of 1)** $1 > 0$.

Proof If $x > 0$, then $x \cdot 1 > 0$ since $x \cdot 1 = x$. If $1 = 0$, then $x = x \cdot 1 = x \cdot 0 = 0$. However, $x = 0$ and $x > 0$ is a contradiction by the trichotomy axiom. If $1 < 0$, then for $x > 0$, $x \cdot 1 < 0 \cdot 1$, by the multiplication order theorem. Then, $x < 0$ or $-x > 0$. However, $-x > 0$ and $x > 0$ is a contradiction by the trichotomy action. Therefore, $1 \neq 0$ and $1 \not< 0$ but $1 > 0$.

Theorem 9 (Order of Multiplication Inverses) If $x > 0$, then $\frac{1}{x} > 0$ and if $x < 0$, then $\frac{1}{x} < 0$.

Proof Since $x \cdot \frac{1}{x} = 1$ and since $1 > 0$, then $x \cdot \frac{1}{x} > 0$. By the positive product theorem, either $\left(x > 0 \text{ and } \frac{1}{x} > 0\right)$ or $\left(x < 0 \text{ and } \frac{1}{x} < 0\right)$.

Example 4 Solve $\dfrac{x-3}{x+2} \leq 5$.

Solution

$$\frac{x-3}{x+2} - 5 \leq 0$$

$$\frac{x - 3 - 5(x+2)}{x+2} \leq 0$$

$$\frac{-4x - 13}{x+2} \leq 0$$

CASE 1

$$-4x - 13 \geq 0 \quad \text{and} \quad x + 2 < 0$$
$$-4x \geq 13$$
$$x \leq \frac{-13}{4} \quad \text{and} \quad x < -2$$

ADDITIONAL ORDER THEOREMS AND INEQUALITIES

Thus,
$$x \le -\frac{13}{4}.$$

CASE 2
$$-4x - 13 \le 0 \quad \text{and} \quad x + 2 > 0$$
$$-4x \le 13$$
$$x \ge \frac{-13}{4} \quad \text{and} \quad x > -2$$

Thus,
$$x > -2$$

The solution set is $\left\{ x \mid x \le -\dfrac{13}{4} \text{ or } x > -2 \right\}$

The pair of inequalities $a > b$ and $c > d$ are called *inequalities of the same order* or *inequalities having the same sense* while the pair $a > b$ and $c < d$ are called *inequalities of opposite order* or *inequalities having opposite sense*.

Theorem 10 (Addition of Inequalities Having the Same Sense)
If $x > y$ and $z > w$, then $x + z > y + w$.

Proof		
$x - y > 0$ and $z - w > 0$.		Definition of $>$.
$(x - y) + (z - w)$	> 0.	Positive closure axiom, addition.
$(x + z) - (y + w)$	> 0.	Association and commutation.
$x + z$	$> y + w$.	Definition of $>$.

Corollary If $x < y$ and $z < w$, then $x + z < y + w$.
(Proof left for student.)

Theorem 11 (Multiplication of Inequalities Having the Same Sense)
If $x > y > 0$ and $z > w > 0$, then $xz > yw$.

The proof of Theorem 11 is similar to that of Theorem 10 and therefore the proof is left for the student.

Examples If $x - y > 10$ and $y > 4$, then $(x - y) + y > 10 + 4$ or $x > 14$.
If $x > 3$, then $x \cdot x > 3 \cdot 3$ or $x^2 > 9$.

EXERCISES 5.3

1–20. Solve each of the following and graph the solution set on a number line:

1. $(x-1)(x+5) \leq 0$
2. $(2x+3)(x-2) \geq 0$
3. $x^2 - x - 12 > 0$
4. $x^2 < 5x - 6$
5. $x^2 + 1 \leq 2x$
6. $\dfrac{x}{x+2} > 4x$
7. $\dfrac{x^2 - 4}{3x} \geq 1$
8. $(x+2)(x+3)(x-1) > 0$
9. $(x+1)(x-3)(x+2) < 0$
10. $x(x+1)(x+3) < 0$
11. $x^2 \geq 4$
12. $x^2 < 16$
13. $x^2 > 2x$
14. $x^2 - 3x - 4 < 0$
15. $\dfrac{x-2}{2} > \dfrac{4}{x}$
16. $\dfrac{1-x}{x-4} \geq 0$
17. $x(x+5)(x+2) > 0$
18. $\dfrac{x}{x-2} > 3$
19. $(x+2)(x+3)(x+4)(x-1)(x-2) > 0$
20. $x^2(x+3)(x-2)(x+10) \geq 0$

21–25. Solve each of the following:

21. $2 < \dfrac{1}{x} < 3$
22. $-1 \leq \dfrac{1}{x+2} < 2$
23. $-3 < \dfrac{2}{x} < 3$
24. $0 < x^2(x+2)$
25. $-2 < \dfrac{x}{x+1} \leq 1$

26. Prove: If $w, x, y,$ and z are real numbers such that $x > y > 0$ and $z > w > 0$, then $xz > yw$.

27. Prove: If $w, x, y,$ and z are real numbers such that $x < y$ and $z < w$, then $x + z < y + w$.

28. Prove: If x and y are real numbers such that $x \leq y$ and $y \leq x$, then $x = y$.

29. Prove: If x is a real number such that $0 < x < 1$, then $x^2 < x$.

30. Prove: If x is a real number such that $x > 1$, then $x^2 > x$.

31–40. By using Theorem 10 or Theorem 11, solve each of the following systems for x. The solution should not involve the constant k, a nonzero real number.

ABSOLUTE VALUE

Example 1 $2x + 5k > 11$
$\qquad\quad\;\; x - k > 2$

Solution $\;\; 2x + 5k > 11$
$\qquad\qquad 5x - 5k > 10$
$\qquad\qquad \overline{}$
$\qquad\qquad 7x \qquad\; > 21$
$\qquad\qquad\quad\; x > 3$

Example 2 $0 < kx < \dfrac{1}{4}$
$\qquad\qquad\; 5 < k\; < 6$

Solution $\;\; \dfrac{1}{6} < \dfrac{1}{k} \;\; < \dfrac{1}{5}$

$\qquad\qquad 0 < \dfrac{1}{k}(kx) < \dfrac{1}{5}\left(\dfrac{1}{4}\right)$

$\qquad\qquad 0 < x \quad\; < \dfrac{1}{20}$

31. $x + k < 3$ and $x - k < 5$
32. $x + k > 2$ and $x - k > -4$
33. $2x + k > 1$ and $x - 2k > 3$
34. $x + 3k < 2$ and $2x - k < 4$
35. $x + k < 180$ and $k > 60$
36. $2x + k < 180$ and $k > 120$
37. $0 < kx < 0.5$ and $2 < k < 3$
38. $0 < kx < 0.1$ and $\frac{1}{4} < k < \frac{1}{2}$
39. $0 < (k + 3)x < 1$ and $3 < 2k < 5$
40. $0 < (5 - k)x < 1$ and $1 < 2k < 3$

5.4 ABSOLUTE VALUE

5.4.1 Definition

On the real number line, any point a and its opposite, or additive inverse, $-a$, are the same distance from the origin, but on opposite sides of the origin. The algebraic sign of a number indicates on which side of the origin the corresponding point is located. The **distance** of the point from the origin, regardless of the side on which it is located, is called the **absolute value** of the number and is designated by the symbol $|a|$.

Definition (Absolute Value of a Real Number $|x|$)

$$|x| = x \text{ if } x \geq 0$$
$$|x| = -x \text{ if } x < 0$$

The definition indicates that the absolute value of a number is never negative. In symbols, $|x| \geq 0$. In particular,

$$|0| = 0$$
$$|5| = 5$$
$$|-5| = -(-5) = 5.$$

5.4.2 Solution of Absolute Value Equations

If $p(x)$ is a polynomial in the variable x and if a is a nonnegative real number, then the solution set of $|p(x)| = a$ is the union of the solution sets of $p(x) = a$ and $-p(x) = a$ since by definition, $|p(x)| = p(x)$ or $|p(x)| = -p(x)$. Thus, there are always two cases to be considered.

Example 1 Solve for x: $|x| = 4$.

Solution

CASE 1 $|x| \geq 0$, $|x| = x$, and $x = 4$
CASE 2 $|x| < 0$, $|x| = -x$ and $-x = 4$ or $x = -4$

The solution set is $\{4, -4\}$.

Example 2 Solve for x: $|x - 3| = 5$.

Solution

CASE 1
$$|x - 3| = x - 3$$
$$x - 3 = 5$$
$$x = 8$$

CASE 2
$$|x - 3| = -(x - 3)$$
$$-(x - 3) = 5$$
$$x - 3 = -5$$
$$x = -2$$

The solution set is $\{8-2,\}$.

5.4.3 Solution of Absolute Value Inequalities

If $p(x)$ is a polynomial in the variable x and a is a nonnegative real number, then the solution sets of the inequalities $|p(x)| < a$ and $|p(x)| > a$ may be obtained by using the following theorem:

ABSOLUTE VALUE

Theorem on Absolute Value Inequalities

(1) $|p(x)| < a$ if and only if $-a < p(x) < a$.
(2) $|p(x)| > a$ if and only if $p(x) > a$ or $p(x) < -a$.

Proof of (1)

CASE 1 $p(x) \geq 0$. Then $|p(x)| = p(x)$. The inequality becomes $p(x) < a$. The solution set $= \{x \mid 0 \leq p(x) < a\}$.
CASE 2 $p(x) < 0$. Then $|p(x)| = -p(x)$. The inequality becomes $-p(x) < a$ or $p(x) > -a$. The solution set is $\{x \mid -a < p(x) < 0\}$.

The solution set of $|p(x)| < a$ is the union of these two sets:

$$\{x \mid 0 \leq p(x) < a\} \cup \{x \mid -a < p(x) < 0\} = \{x \mid -a < p(x) < a\}.$$

Proof of (2)

CASE 1 $p(x) \geq 0$. Then $|p(x)| = p(x)$, and $p(x) > a$. Thus, the solution set is $\{x \mid p(x) \geq 0 \text{ and } p(x) > a\} = \{x \mid p(x) > a\}$.
CASE 2 $p(x) < 0$. Then $|p(x)| = -p(x)$. Thus, $-p(x) > a$ or $p(x) < -a$. The solution set is $\{x \mid p(x) < 0 \text{ and } p(x) < -a\} = \{x \mid p(x) < -a\}$.

The solution set of $|p(x)| > a$ is the union of these two sets:

$$\{x \mid p(x) > a \text{ or } p(x) < -a\}.$$

Example 3 Solve $|x - 3| < 2$.

Solution $-2 < x - 3 < 2$
$\phantom{-2 <} 1 < x < 5$ (adding 3)

The solution set $= \{x \mid 1 < x < 5\}$.

Example 4 Solve $|x + 2| > 4$.

Solution $x + 2 > 4$ or $x + 2 < -4$
$ x > 2$ or $\phantom{x + 2 <} x < -6$

The solution set is $\{x \mid x > 2 \text{ or } x < -6\}$.

Example 5 Solve $|x+2| \leq 4$, using the result of Example 4.

Solution Noting that $|x+2| \leq 4$ is equivalent to *not* $(|x+2| > 4)$, the desired solution set is

$$\overline{\{x \mid x > 2 \text{ or } x < -6\}} = \{x \mid \overline{x > 2} \text{ and } \overline{x < -6}\}$$
$$= \{x \mid x \leq 2 \text{ and } x \geq -6\}$$
$$= \{x \mid -6 \leq x \leq 2\}.$$

EXERCISES 5.4

1–20. Solve for x.

1. $|x - 2| = 3$
2. $|3x + 1| = 4$
3. $|x| \geq 3$
4. $|x| < 3$
5. $|3 - 2x| > 4$
6. $2 > |1 - x|$
7. $|x| + 2 \leq 3$
8. $|1 - 2x| \geq 3$
9. $2|x - 2| < 5$
10. $2 \geq \left|3 + \dfrac{x}{2}\right|$
11. $3 - |2 - x| < 6$
12. $\dfrac{2}{|2 - x|} < 3$
13. $|x| = |-3|$
14. $|x| = -3$
15. $|2x| = 2x$
16. $|x - 2| = 2 - x$
17. $\dfrac{3}{|x + 1|} < 4$
18. $|2x + \tfrac{1}{2}| - \tfrac{1}{4} = 0$
19. $|x - 2| = |2 - x|$
20. $\left|\dfrac{x + 2}{x}\right| < 3$

21. If $|x| < 2$, then between what values is $x^2 - 3x + 4$?

22. If $|x| < 3$, then between what values is $x^2 - 2x + 5$?

23. If $|2x - 5| < 0.1$, then solve for x.

24. If $|3x - 6| < 0.1$, then solve for $|x - 2|$.

25. If ε is any positive real number, find a condition that $|x - 1|$ must satisfy so that $|5x - 5| < \varepsilon$.

26. Find a sufficient condition for $|x - 2|$ so that $|3x - 6| < \varepsilon$ where ε is any positive real number.

27. Find δ (expressed in terms of ε) so that if $|x - 3| < \delta$ then $|5x - 15| < \varepsilon$.

INTEGRAL EXPONENTS

28. Find δ (expressed in terms of ε) so that if $|x+4| < \delta$ then
$$|(5-2x) - (5+8)| < \varepsilon.$$

29. Assuming that $|x| < 1.5$, solve for $|x-1|$ so that $|3x^2 - 4x + 1| < 0.5$.

30. Assuming that $2 < x < 4$, find a sufficient condition for $|x-3|$ so that $|5x^2 - 13x - 6| < 0.1$.

31. Assuming that $-1.5 < x < -0.5$, find δ so that if $|x+1| < \delta$ then $|2x^2 + 9x + 7| < \varepsilon$ (ε is any positive real number).

32. If ε is any positive real number and if $0 < x < 3$, then find δ so that if $|x-2| < \delta$, then $|2x^2 - x - 6| < \varepsilon$.

33. If ε is any positive real number and if $\dfrac{3}{2} < x < \dfrac{5}{2}$, then find δ so that if $|x-2| < \delta$, then $\left|\dfrac{1}{x} - \dfrac{1}{2}\right| < \varepsilon$.

34. If ε is any positive real number and if $-6 < x < -4$, then find δ so that if $|x+5| < \delta$, then $\left|\dfrac{1}{x} + \dfrac{1}{5}\right| < \varepsilon$.

35–42. Prove each of the following theorems. Assume a and b are real numbers.

35. $|ab| = |a| \cdot |b|$

36. $\left|\dfrac{a}{b}\right| = \dfrac{|a|}{|b|}, b \neq 0$

37. $|a + b| \leq |a| + |b|$

38. $|a - b| \geq |a| - |b|$

39. Prove: For every real number x, $|x^2| = |x|^2 = x^2$.

40. Does $\sqrt{x^2} = |x|$? Justify your answer.

41. If x is a real number other than zero, what is the value of $\dfrac{|x|}{x}$? Justify.

42. If $a < c < b$ and $a < d < b$, prove that $|c - d| < b - a$.

5.5 INTEGRAL EXPONENTS

Definition of x^n If x is any real number and n is any natural number, then
$$x^1 = x \quad \text{and} \quad x^n = x^{n-1} \cdot x.$$

For example, if $n = 5$, then

$$\begin{aligned}x^5 &= x^4 \cdot x \\ &= x^3 \cdot x \cdot x \\ &= x^2 \cdot x \cdot x \cdot x \\ &= x^1 \cdot x \cdot x \cdot x \cdot x \\ &= x \cdot x \cdot x \cdot x \cdot x.\end{aligned}$$

The definition of x^n is a recursive definition and, from the example, it may be seen that it has the familiar meaning of n factors of x. The advantage of the recursive definition is that it permits mathematically valid proofs of theorems concerning exponents.

The operation designated symbolically by x^n is called "exponentiation" or "raising to a power." The real number x is called the *base*, the natural number n is called the *exponent*, and x^n is called the *nth power of x* and read "x to the n." The special case x^2 is read "x squared" or "the square of x" and the special case x^3 is read "x cubed" or "the cube of x."

The Exponent Theorems For all real numbers x and y, and for all natural numbers m and n,

(1) $x^m x^n = x^{m+n}$. Product of powers, like bases.

(2) $\dfrac{x^m}{x^n} = \begin{cases} x^{m-n} & \text{if } m > n \\ 1 & \text{if } m = n \\ \dfrac{1}{x^{n-m}} & \text{if } m < n. \end{cases}$ Quotient of powers, like bases.

(3) $(x^m)^n = x^{mn}$. Raising a power to a power.

(4) $(xy)^n = x^n y^n$. Power of a product.

(5) $\left(\dfrac{x}{y}\right)^n = \dfrac{x^n}{y^n}$. Power of a quotient.

The proofs of Theorems 1 and 3 are shown below. The other three theorems are proved similarly and are left for the student to prove.

Proof of Theorem 1 (by induction on n) Let S be the subset of N for which $x^m x^n = x^{m+n}$.

PART 1 $\begin{aligned} x^{m+1} &= x^{(m+1)-1} \cdot x = x^m \cdot x & \text{Definition.} \\ &= x^m \cdot x^1. & \text{Definition.} \end{aligned}$

Thus, $x^m x^1 = x^{m+1}$ and 1 is in S.

INTEGRAL EXPONENTS

PART 2 Assume $x^m x^k = x^{m+k}$. Then

$(x^m x^k)x = x^{m+k} \cdot x.$	Multiplication theorem.
$x^m(x^k \cdot x) = x^{m+k} \cdot x.$	Associative axiom.
$x^m \cdot x^{k+1} = x^{(m+k)+1}.$	Part 1.
$x^m \cdot x^{k+1} = x^{m+(k+1)}.$	Associative axiom.

Thus, $k+1$ is in S whenever k is in S.

Combining Parts 1 and 2, $S = N$.

Proof of Theorem 3 (by induction on m) Let S be the subset of N for which $(x^n)^m = x^{nm}$.

PART 1
$(x^n)^1 = x^n$	Definition.
$ = x^{n \cdot 1}.$	Identity axiom.

Thus, 1 is in S.

PART 2 Assume $(x^n)^k = x^{nk}$. Then

$x^n(x^n)^k = x^n x^{nk}$	Multiplication theorem.
$ = x^{n+nk}$	Exponent theorem 1.
$ = x^{n(k+1)}.$	Distributive, commutative axioms.
$x^n(x^n)^k = (x^n)^{k+1}.$	Definition.
$(x^n)^{k+1} = x^{n(k+1)}.$	Substitution.

Thus, $k+1$ is in S whenever k is in S. Therefore, $S = N$.

Definition of x^0 If x is a real number and $x \neq 0$, then $x^0 = 1$.

Definition of x^{-n} If x is any nonzero real number and n is any natural number, then

$$x^{-n} = \frac{1}{x^n}.$$

Using the two preceding definitions it may be shown that the five exponent theorems are valid for all integral exponents. The proofs are left for the student.

Theorems on the Order of Natural Powers For x and y any positive real numbers, and for n and m any natural numbers,

(1) If $x < 1$, then $x^n < 1$ and $x^{n+1} < x$. $\qquad \left(\frac{1}{2}\right)^3 = \frac{1}{8} < \frac{1}{2}$.

(2) If $x > 1$, then $x^n > 1$ and $x^{n+1} > x$. $\qquad 2^3 = 8 > 2$.

(3) If $x > 1$ and $n > m$, then $x^n > x^m$. $\qquad 32 = 2^5 > 2^3 = 8$.

(4) If $x < 1$ and $n > m$, then $x^n < x^m$. $\qquad \frac{1}{32} = \left(\frac{1}{2}\right)^5 < \left(\frac{1}{2}\right)^3 = \frac{1}{8}$.

(5) If $x > y$, then $x^n > y^n$ and conversely. $\qquad 3 > 2$ and $3^4 > 2^4$.

$\qquad \frac{1}{2} > \frac{1}{3}$ and $\left(\frac{1}{2}\right)^4 > \left(\frac{1}{3}\right)^4$.

Proof of Theorem 1 (*by induction on* n) Let S be the subset of N for which $0 < x^{n+1} < x$ and $0 < x^n < 1$.

PART 1 $\quad x > 0$ and $x < 1$. $\qquad$ Given.
$\qquad\qquad\quad x^2 < x$. $\qquad\qquad\quad$ Multiplication order theorem 1.

Thus, 1 is in S.

PART 2 $\quad$ Assume $0 < x^{k+1} < x$.

$\qquad (x^{k+1})x < x(x)$. $\qquad$ Multiplication order theorem 1.
$\qquad x^{(k+1)+1} < x^2$. $\qquad\quad$ Exponent theorem 1.
$\qquad x^2 < x$. $\qquad\qquad\qquad$ Part 1.
$\qquad x^{(k+1)+1} < x$. $\qquad\quad\;$ Transitivity order theorem.

Thus, $k + 1$ is in S whenever k is. It follows, then, that $S = N$ or for all n, if $0 < x < 1$, then $0 < x^{n+1} < x$. By multiplying each term by $\frac{1}{x}$ which is positive, $0 < x^n < 1$.

Theorem 2 is proved similarly and therefore is left for the student.

Proof of Theorem 3 If $x > 1$ and $n > m$, then $x^n > x^m$.

Let $n = m + k$. (For two natural numbers n and m such that $n > m$ there always exists a natural number k so that $n = m + k$.) Then

$\qquad x^k > 1$ $\qquad\qquad$ Theorem 2.
$\qquad x^m x^k > x^m$ $\qquad\;$ Multiplication theorem 1.
$\qquad x^{m+k} > x^m$ $\qquad\;$ Exponent theorem 1.

or

$\qquad x^n > x^m$.

Theorem 4 is proved similarly and therefore is left for the student.

INTEGRAL EXPONENTS

Proof of Theorem 5 For $x > 0$ and $y > 0$, $x > y$ if and only if $x^n > y^n$.

Induction on n: Let S be the subset of N for which $x > y$ if and only if $x^n > y^n$.

PART 1 Since $x > y$ if and only if $x > y$, 1 is in S.

PART 2 Assume $x > y$ if and only if $x^k > y^k$.
(a) If $x > y$ and $x^k > y^k$, then $x^{k+1} > y^{k+1}$ by the theorem on multiplication of inequalities having the same order.
(b) If $x^{k+1} > y^{k+1}$, then $x^{k+1} - y^{k+1} > 0$ by definition. Now

$$x^{k+1} - y^{k+1} = (x - y)(x^k + x^{k-1}y + x^{k-2}y^2 + \ldots + xy^{k-1} + y^k).$$

If the product is positive, then both factors are positive or both factors are negative.
Since

$$x^k + x^{k-1}y + \ldots + xy^{k-1} + y^k > 0$$

by the positive closure axioms for addition and multiplication, therefore, the factor $x - y$ must be positive, or $x - y > 0$. Thus, $x > y$.

Example 1

$$5^{-2} = \frac{1}{5^2} = \frac{1}{25}$$

Example 2

$$\frac{1}{3^{-4}} = \left(\frac{1}{3}\right)^{-4} = 3^4 = 81$$

Example 3

$$\left(\frac{2}{3}\right)^{-3} = \left(\frac{3}{2}\right)^3 = \frac{3^3}{2^3} = \frac{27}{8}$$

Example 4

$$\left(-\frac{1}{5}\right)^{-2} = (-5)^2 = 25$$

Example 5

$$\begin{aligned}
(x^{-2} + y^{-2})^{-2} &= \left(\frac{1}{x^2} + \frac{1}{y^2}\right)^{-2} \\
&= \left(\frac{y^2 + x^2}{x^2 y^2}\right)^{-2} \\
&= \left(\frac{x^2 y^2}{x^2 + y^2}\right)^2 \\
&= \frac{x^4 y^4}{(x^2 + y^2)^2} = \frac{x^4 y^4}{x^4 + 2x^2 y^2 + y^4}
\end{aligned}$$

Example 6

$$(x^{-2}y^{-2})^{-2} = (x^{-2})^{-2}(y^{-2})^{-2}$$
$$= x^4 \cdot y^4 = x^4 y^4$$

Example 7

$$\frac{2^{-3} + 2^{-5}}{2^{-7}} = \frac{2^{-3} + 2^{-5}}{2^{-7}} \cdot \frac{2^7}{2^7} = \frac{2^4 + 2^2}{1} = 16 + 4 = 20$$

EXERCISES 5.5

1–26. Let m and n be integers. Simplify each of the following and leave the answers without zero or negative exponents. (Assume no variables represent zero denominators, and no zero exponents have zero as a base.)

1. $x^{m+1} x^{n-2}$
2. $\dfrac{x^m y^n}{(xy)^{m+n}}$
3. $\dfrac{(x^m)^{n+1}}{(x^{m+1})^n}$
4. $\dfrac{x^{-n} y^{n+1}}{x^{-1} y^{n-1}}$
5. $\dfrac{x^m(x^2 + x^3 - 1)}{x^{m+2}}$
6. $\dfrac{x^2(x^{m+2})}{x^{m+4}}$
7. $x^{-2} y^2$
8. $x^{-2} - y^{-2}$
9. $2x^{-m}$
10. $[x^{-m}(x^{-n})]^{-m}$
11. $x^{-n}(x^{n+2} + x^{n+1} + x^n)$
12. $(x^{-1} + y^{-1})^{-1}$
13. $(x^{-n} y^{-n})^{-n}$
14. $x^{-3} y^3 (x^3 y^{-1} + x^4 y^{-3})$
15. $\dfrac{1 - x^{-3}}{x^{-1} + x^{-2} + x^{-3}}$
16. $\dfrac{(xy)^{-1} + x^{-1} + y^{-1} + 1}{x^{-1} + (xy)^{-1}}$
17. $x^{1-n}(x^{n-1} + x^n + x^{n+1})$
18. $(x^{m-n})^n (x^n)^{n-m}$
19. $\dfrac{(x^n y)^3}{(xy^3)^n}$
20. $\dfrac{2^{-2} - x^{-2}}{2 - x}$
21. $2^n + 2^n$
22. $3^n + 3^n$
23. $3^n + 3^n + 3^n$
24. $2^{-n} + 2^{-n}$
25. $3^{-n} + 3^{-n} + 3^{-n}$
26. $2^{-n} - 2^{-n-1}$

27–33. Prove each of the following:

27. For all real numbers x, y ($x \neq 0$), and for all natural numbers m and n,

$$\frac{x^n}{x^m} = x^{n-m} \text{ if } n > m,$$

$$\frac{x^n}{x^m} = 1 \text{ if } x = m.$$

$$\frac{x^n}{x^m} = \frac{1}{x^{m-n}} \text{ if } n < m.$$

INTEGRAL EXPONENTS

28. For all real numbers x and y, and for all natural numbers n, $(xy)^n = x^n y^n$.
29. For all real numbers x and y ($y \neq 0$), and for all natural numbers n, $\left(\dfrac{x}{y}\right)^n = \dfrac{x^n}{y^n}$.
30. For all real numbers x and y, and for all integral values of n, $(xy)^n = x^n y^n$.
31. For all real numbers x and y ($x \neq 0$, $y \neq 0$), and for all integral values of n, $\left(\dfrac{x}{y}\right)^n = \dfrac{x^n}{y^n}$.
32. For all positive real numbers x and for any natural numbers m and n, if $x < 1$ and $n > m$, then $x^n < x^m$.
33. For all real numbers x such that $x > 1$, and for all natural numbers n, $x^n > 1$ and $x^{n+1} > x$.
34. Prove: For all real numbers x and y, $(x - y)^2 \geq 0$.
35. Prove: For all real numbers x and y,
$$x^2 + y^2 \geq 2xy.$$
36. Prove: If x and y are real numbers such that $x > y$, then
$$x^3 + 3xy^2 > y^3 + 3x^2 y.$$

37–46. Insert $<$ or $>$ between each of the following pairs of real numbers so that the resulting statements are true.

37. $(\tfrac{1}{2})^{15}$, $(\tfrac{1}{2})^{30}$
38. $(\tfrac{1}{2})^{15}$, $(\tfrac{1}{3})^{15}$
39. $(\sqrt{2})^{30}$, $(\sqrt{3})^{30}$
40. $(\sqrt{2})^{15}$, $(\sqrt{2})^{30}$
41. $\left(\dfrac{x}{x+1}\right)^5$, $\left(\dfrac{x}{x+1}\right)^3$, where $x > 0$
42. $\left(\dfrac{x+1}{x}\right)^5$, $\left(\dfrac{x+1}{x}\right)^3$, where $x > 0$
43. $\left(\dfrac{x}{y+1}\right)^7$, $\left(\dfrac{x}{y}\right)^7$, where $x > 0$ and $y > 0$
44. $\left(\dfrac{x+1}{y}\right)^7$, $\left(\dfrac{x}{y+1}\right)^7$, where $x > 0$ and $y > 0$
45. $\left(\dfrac{x+1}{y+1}\right)^8$, $\left(\dfrac{x+1}{y+1}\right)^5$, where $x > 0$, $y > 0$, and $y > x$
46. $\left(\dfrac{x+1}{y+1}\right)^8$, $\left(\dfrac{x+1}{y+1}\right)^5$, where $x > 0$, $y > 0$, and $y < x$

5.6 RADICALS

Not all real numbers are included in the range of the operation defined by $f(x, n) = x^n$ with domain $R \times N$. In particular, there is no real number x such that $x^2 = -1$, since the product of two positive real numbers is always positive and the product of two negative real numbers is always positive. On the other hand, all positive real numbers are in the range as the following theorem indicates.

Theorem (Existence of Positive Roots) If a is any positive real number and n is any natural number, then there is exactly one *positive* real number x so that $x^n = a$. This unique positive real number is called the principal nth root of a and is designated symbolically as $\sqrt[n]{a}$.

Although it is possible to present a rigorous proof of this theorem at this time, the proof is somewhat lengthy and involved, and thus will not be presented.

Definition of $\sqrt[n]{a}$ For any real number a and any natural number n,

(1) if $a \geq 0$, then $\sqrt[n]{a}$ is the unique positive real number for which $x^n = a$; that is,

$$\sqrt[n]{a} = \text{lub of } \{x \mid x \geq 0 \text{ and } x^n \leq a \text{ where } x \in R\},$$

(2) if $-a < 0$ and n is odd, then $\sqrt[n]{-a}$ is the unique negative real number, $-x$, for which $x^n = a$; that is,

$$\sqrt[n]{-a} = -\sqrt[n]{a} \text{ for } n \text{ odd and } a > 0.$$

Notations $\sqrt[1]{a} = a$ and $\sqrt[2]{a} = \sqrt{a}$.

The proofs of the following theorems are left for the student.

Theorem For any real number x,

(1) If n is odd, then $\sqrt[n]{x^n} = x$,
(2) If n is even, then $\sqrt[n]{x^n} = |x|$.

Theorem For all positive real numbers x and y and for all natural numbers n, $x^n = y^n$ if and only if $x = y$.

The number designated by the symbol $\sqrt[n]{a}$ is called the *principal nth root of a*. The symbol $\sqrt[n]{a}$ is called a *radical* with a called the *radicand* and n called the *index*.

RADICALS

It should be noted that $\sqrt[n]{a}$ is undefined if $a < 0$ and n is even since there is no real number that has a negative even power. On the other hand, if n is even and $a > 0$, then there are two real numbers for which $x^n = a$, a positive real number and its additive inverse. For example, if $x^2 = 9$, then $x = 3$ or $x = -3$. The symbol $\sqrt{9}$ designates only the positive number 3; that is, $\sqrt{9} = 3$. The other square root is designated symbolically by $-\sqrt{9}$; that is, $-\sqrt{9} = -3$. Note also that $\sqrt{(-3)^2} = |-3| = 3$.

Radical Theorems Let x and y be any positive real numbers and let m and n be any natural numbers.

(1) Product of roots $\qquad\qquad \sqrt[n]{xy} = \sqrt[n]{x}\,\sqrt[n]{y}.$

(2) Quotient of roots $\qquad\qquad \sqrt[n]{\dfrac{x}{y}} = \dfrac{\sqrt[n]{x}}{\sqrt[n]{y}}.$

(3) Repeated roots $\qquad\qquad \sqrt[m]{\sqrt[n]{x}} = \sqrt[mn]{x}.$

(4) Order of roots $\qquad\qquad \sqrt[n]{x} > \sqrt[n]{y}$ if and only if $x > y$.

The proofs of the radical theorems are based on the theorem that if $x > 0$ and $y > 0$, then $x^n = y^n$ if and only if $x = y$.

(1) $(\sqrt[n]{xy})^n = xy = (\sqrt[n]{x})^n (\sqrt[n]{y})^n = (\sqrt[n]{x}\,\sqrt[n]{y})^n.$

(2) $\left(\sqrt[n]{\dfrac{x}{y}}\right)^n = \dfrac{x}{y} = \dfrac{(\sqrt[n]{x})^n}{(\sqrt[n]{y})^n} = \left(\dfrac{\sqrt[n]{x}}{\sqrt[n]{y}}\right)^n.$

(3) $(\sqrt[m]{\sqrt[n]{x}})^{mn} = [(\sqrt[m]{\sqrt[n]{x}})^m]^n = (\sqrt[n]{x})^n = x.$

(4) Since $\sqrt[n]{x} > \sqrt[n]{y}$ if and only if $(\sqrt[n]{x})^n > (\sqrt[n]{y})^n$, thus $\sqrt[n]{x} > \sqrt[n]{y}$ if and only if $x > y$.

Expressions involving radicals with rational radicands are said to be *simplified* if

(1) the radicand is positive, integral, and contains no powers whose exponent is greater than or equal to the index;
(2) no radicals occur in denominators; and
(3) the index is as small as possible; that is, the radicand is not a power whose exponent is a factor of the index.

To *rationalize the denominator* of a quotient means to rename the quotient so that no radicals are in the denominator.

To *rationalize the numerator* of a quotient means to rename the quotient so that no radicals are in the numerator.

Example 1 Simplify $\sqrt{180}$.

Solution $\sqrt{180} = \sqrt{36 \cdot 5} = \sqrt{36}\sqrt{5} = 6\sqrt{5}$

Example 2 Simplify $\sqrt[3]{-24}$.

Solution $\sqrt[3]{-24} = -\sqrt[3]{24} = -\sqrt[3]{8}\sqrt[3]{3} = -2\sqrt[3]{3}$

Example 3 Simplify $\sqrt{\frac{50}{3}}$.

Solution $\sqrt{\frac{50}{3}} = \sqrt{\frac{50 \cdot 3}{3 \cdot 3}} = \frac{\sqrt{25 \cdot 6}}{\sqrt{9}} = \frac{5}{3}\sqrt{6}$

Example 4 Simplify $\frac{2}{\sqrt[3]{5}}$.

Solution $\frac{2}{\sqrt[3]{5}} = \frac{2\sqrt[3]{25}}{\sqrt[3]{5}\sqrt[3]{25}} = \frac{2\sqrt[3]{25}}{5}$

Example 5 Rationalize the denominator $\frac{2}{3 - \sqrt{2}}$.

Solution $\frac{2}{3 - \sqrt{2}} \cdot \frac{3 + \sqrt{2}}{3 + \sqrt{2}} = \frac{2(3 + \sqrt{2})}{9 - 2} = \frac{2(3 + \sqrt{2})}{7}$

Example 6 Rationalize the numerator $\frac{\sqrt{5} + \sqrt{3}}{2}$.

Solution $\frac{\sqrt{5} + \sqrt{3}}{2} \cdot \frac{\sqrt{5} - \sqrt{3}}{\sqrt{5} - \sqrt{3}} = \frac{5 - 3}{2(\sqrt{5} - \sqrt{3})} = \frac{1}{\sqrt{5} - \sqrt{3}}$

Example 7 Simplify $\frac{1}{\sqrt[3]{2} - 1}$.

Solution $\frac{1}{\sqrt[3]{2} - 1} \cdot \frac{\sqrt[3]{4} + \sqrt[3]{2} + 1}{\sqrt[3]{4} + \sqrt[3]{2} + 1} = \frac{\sqrt[3]{4} + \sqrt[3]{2} + 1}{(\sqrt[3]{2})^3 - 1^3} = \sqrt[3]{4} + \sqrt[3]{2} + 1$

Example 8 Simplify $\sqrt[6]{27x^6}$.

Solution $\sqrt[6]{27x^6} = \sqrt[6]{27}\sqrt[6]{x^6} = \sqrt[2 \cdot 3]{3^3}\sqrt[6]{x^6} = \sqrt{\sqrt[3]{3^3}}\sqrt[6]{x^6} = \sqrt{3}|x|$

RADICALS

EXERCISES 5.6

1–12. Simplify.

1. $\sqrt{328}$
2. $\sqrt[3]{-54}$
3. $\sqrt{4+9}$
4. $\sqrt{12}+\sqrt{27}$
5. $3\sqrt{54}+2\sqrt{24}$
6. $3\sqrt[3]{24}+\sqrt[3]{375}-\sqrt[3]{81}$
7. $\sqrt{32}+\sqrt{50}$.
8. $8\sqrt{72}+2\sqrt{20}+3\sqrt{5}$
9. $\sqrt{3}\sqrt{27}$
10. $\sqrt[3]{8+27}$
11. $\sqrt[3]{8}+\sqrt[3]{27}$
12. $\sqrt[3]{250}$

13–25. Rationalize the denominators of each of the following and simplify. Assume all variables to be positive real numbers.

13. $\sqrt{1/2}$
14. $\sqrt{1/2}+4\sqrt{2}+6\sqrt{72}$
15. $\sqrt[3]{3/5}$
16. $\dfrac{2}{\sqrt{x+y}}$
17. $\dfrac{2}{\sqrt{x}+y}$
18. $\dfrac{2}{\sqrt{x}+\sqrt{y}}$
19. $\dfrac{3}{\sqrt[3]{y}}$
20. $\dfrac{1}{\sqrt[3]{6y}}$
21. $\dfrac{5}{\sqrt{11}+1}$
22. $\dfrac{4}{\sqrt{7}+\sqrt{3}}$
23. $\dfrac{1}{\sqrt{2}+\sqrt{3}+\sqrt{5}}$
24. $\dfrac{2}{\sqrt{7}-\sqrt{5}-\sqrt{2}}$
25. $\sqrt{7/8}+\sqrt{5/6}-\sqrt{7/2}$

26–36. Simplify. Assume all variables to be positive real numbers.

26. $\dfrac{\sqrt{3}}{\sqrt[3]{3}}$
27. $\dfrac{2}{\sqrt[3]{x}+\sqrt[3]{y}}$
28. $\dfrac{x}{\sqrt[3]{xy^2}}$
29. $\dfrac{1}{\sqrt[3]{x+1}}$
30. $\dfrac{2}{4-\sqrt[3]{2}}$
31. $\sqrt[4]{400}+\sqrt[6]{125}$
32. $\sqrt[9]{64}-\sqrt[6]{4}$
33. $\dfrac{\sqrt[8]{16}}{\sqrt[6]{27}}$
34. $\dfrac{\sqrt[12]{625}}{\sqrt[9]{1000}}$
35. $\sqrt[10]{4x^2y^6}-\sqrt[20]{16x^4y^{12}}$
36. $\sqrt[8]{256x^4y^2}-\sqrt[12]{x^6y^3}$

37–42. Rationalize the numerators of each of the following. Assume $x > 0$.

37. $\dfrac{\sqrt{x} - 2}{x - 4}$

38. $\dfrac{\dfrac{1}{\sqrt{x}} - \dfrac{1}{3}}{x - 9}$

39. $\dfrac{\sqrt[3]{x} + 2}{x + 8}$

40. $\dfrac{\sqrt[3]{x} - 1}{x - 1}$

41. $\dfrac{\dfrac{1}{\sqrt[3]{x}} + \dfrac{1}{2}}{x + 8}$

42. $\dfrac{\dfrac{1}{\sqrt[3]{x}} - 1}{x - 1}$

43–52. Solve for x.

43. $\sqrt[3]{x} > 5$
44. $\sqrt{x + 2} < 3$
45. $\sqrt{3x - 1} < 2$
46. $\sqrt[3]{x - 5} > 4$
47. $\sqrt{x^2} > 4$
48. $\sqrt{x^2} < 9$
49. $\sqrt{x^2 + 2x + 1} < 9$
50. $\sqrt{x^2 - 4x + 4} > 25$
51. $\sqrt[3]{x^2 - 2x + 2} > \sqrt[3]{5}$
52. $0 < \sqrt[3]{x^2 - 6x} < \sqrt[3]{7}$

53. Prove: For all positive real numbers x and y, and for all natural numbers n, $x^n = y^n$ if and only if $x = y$.

54. Prove: For any real number x, and any natural number n,
    (1) If n is odd, then $\sqrt[n]{x^n} = x$,
    (2) If n is even, then $\sqrt[n]{x^n} = |x|$.

5.7 RATIONAL EXPONENTS

By defining powers with fractional exponents as radicals, the exponential function may be extended to include rational exponents.

Exponent Definitions Let b be any nonzero real number and let m and n be any natural numbers.

(1) $b^1 = b$ and $0^n = 0$.
(2) $b^n = b \cdot b^{n-1}$.
(3) $b^0 = 1$.
(4) $b^{-n} = \dfrac{1}{b^n}$.

RATIONAL EXPONENTS

(5) $b^{m/n} = \sqrt[n]{b^m}$. (Undefined for n even and b^m negative.)

(6) $b^{-m/n} = \dfrac{1}{\sqrt[n]{b^m}}$. (Undefined for n even and b^m negative.)

Exponent Theorems Let a and b be any nonzero real numbers. Let x and y be any rational numbers.

(1) $b^x b^y = b^{x+y}$. Products of powers, like bases.
(2) $(b^x)^y = b^{xy}$. Raising a power to a power.
(3) $\dfrac{b^x}{b^y} = b^{x-y}$ Quotient of powers, like bases.
(4) $(ab)^x = a^x b^x$. Power of a product.
(5) $\left(\dfrac{a}{b}\right)^x = \dfrac{a^x}{b^x}$. Power of a quotient.

By using the product of roots theorem, it may be shown that for b nonnegative,

$$\sqrt[n]{b^m} = (\sqrt[n]{b})^m.$$

Theorems on Order of Rational Powers Let a and b be positive real numbers and let x and y be positive rational numbers.

(1) $a^x > b^x$ if and only if $a > b$.
(2) $a^{-x} < b^{-x}$ if and only if $a > b$.
(3) For $b > 1$, $b^x > b^y > 0$ if and only if $x > y$.
(4) For $0 < b < 1$, $0 < b^x < b^y$ if and only if $x > y$.

Proofs of Exponent Theorems (Optional) Let $x = \dfrac{m}{n}$ and let $y = \dfrac{p}{q}$ where m and p are any integers and n and q are positive integers.

Theorem 1

$$\begin{aligned}
b^x b^y &= b^{m/n} b^{p/q} \\
&= b^{mq/nq} b^{np/nq} \\
&= \sqrt[nq]{b^{mq}} \sqrt[nq]{b^{np}} \\
&= \sqrt[nq]{b^{mq} b^{np}} \\
&= \sqrt[nq]{b^{mq+np}} \\
&= b^{(mq+np)/nq} \\
&= b^{m/n + p/q} \\
&= b^{x+y}
\end{aligned}$$

Theorem 2

$$(b^x)^y = (b^{m/n})^{p/q}$$
$$= \sqrt[q]{(\sqrt[n]{b^m})^p}$$
$$= \sqrt[q]{[(\sqrt[n]{b})^m]^p}$$
$$= \sqrt[q]{(\sqrt[n]{b})^{mp}}$$
$$= \sqrt[q]{\sqrt[n]{b^{mp}}}$$
$$= \sqrt[nq]{b^{mp}}$$
$$= b^{mp/nq} = b^{xy}$$

Theorems 3, 4, and 5 are proved similarly and the proofs are left for the student.

Proofs of the Order Theorems (*Optional*) Let $x = \dfrac{m}{n}$ and let $y = \dfrac{p}{q}$, where $m, n, p,$ and q are any positive integers.

Theorem 1

$a^n > b^n$	if and only if $a > b$.	Order of powers theorem.
$\sqrt[m]{a^n} > \sqrt[m]{b^n}$	if and only if $a > b$.	Order of roots theorem.
$a^{m/n} > b^{m/n}$	if and only if $a > b$	Definition.

or

$a^x > b^x$ if and only if $a > b$.

Theorem 2 $a^x > b^x$ if and only if $a > b$ Theorem 1.

Since a^x, b^x, a^{-x}, b^{-x}, and $a^{-x}b^{-x}$ are all positive,

$$\frac{a^x}{a^x b^x} > \frac{b^x}{a^x b^x} \qquad \text{if and only if } a > b.$$

$$\frac{1}{b^x} > \frac{1}{a^x} \quad \text{or} \quad a^{-x} < b^{-x} \qquad \text{if and only if } a > b.$$

Theorem 3 Now, $x > y$ or $\dfrac{m}{n} > \dfrac{p}{q}$ if and only if $mq > np$ by the ordering of the set of rational numbers. For $b > 1$,

$\sqrt[nq]{b} > \sqrt[nq]{1} = 1$	Order of roots theorem.
$(\sqrt[nq]{b})^{mq} > (\sqrt[nq]{b})^{np} > 0$ if and only if $mq > np$	Order of powers, Theorem 3.

RATIONAL EXPONENTS

or

$b^{mq/nq} > b^{np/nq} > 0$ if and only if $\dfrac{m}{n} > \dfrac{p}{q}$

or

$b^x > b^y > 0$ if and only if $x > y$.

Theorem 4   If $0 < b < 1$, then

$0 < \sqrt[nq]{b} < 1$ Order of roots theorem.

Then,

$0 < (\sqrt[nq]{b})^{mq} < (\sqrt[nq]{b})^{np}$ if and only if $mq > np$ Order of powers, Theorem 4.

Thus,

$0 < b^x < b^y$ if and only if $x > y$.

Example 1   Simplify $(64)^{-2/3}$.

     Solution $(64)^{-2/3} = (2^6)^{-2/3} = 2^{-4} = \dfrac{1}{16}$

Example 2   Simplify $(\tfrac{1}{16})^{-1/2} + 3(4)^{-2} + (x+5)^0$.

     Solution

$$(\tfrac{1}{16})^{-1/2} = (16)^{1/2} = \sqrt{16} = 4$$
$$3(4)^{-2} = 3(\tfrac{1}{4})^2 = \tfrac{3}{16}$$
$$(x+5)^0 = 1$$

Thus,

$$4 + \tfrac{3}{16} + 1 = 5\tfrac{3}{16}.$$

Example 3   Simplify $(5^3 \cdot 27^{-2} \cdot x^0)^{-1/6}$.

     Solution

$$(5^3 \cdot 27^{-2} \cdot x^0)^{-1/6} = 5^{-3/6} \cdot 27^{2/6} \cdot x^0$$
$$= 5^{-1/2} \cdot 27^{1/3} \cdot 1$$
$$= \dfrac{\sqrt[3]{27}}{\sqrt{5}} = \dfrac{3}{\sqrt{5}} = \dfrac{3\sqrt{5}}{5}$$

Example 4 Simplify $[(-6)^2]^{1/2}$.

Solution $[(-6)^2]^{1/2} = \sqrt{(-6)^2} = \sqrt{36} = 6$

It is important to note that the theorems do not apply whenever the base is negative and an exponent indicates an even root. Thus

$$[(-6)^2]^{1/2} \neq (-6)^1.$$

EXERCISES 5.7

1–15. Write each of the following in radical form and simplify if possible. Assume all variables to be positive real numbers.

1. $8^{2/3}$
2. $8^{-2/3}$
3. $\left(\dfrac{81}{625}\right)^{3/4}$
4. $\left(\dfrac{81}{625}\right)^{-3/4}$
5. $\left(\dfrac{1}{27}\right)^{2/3}$
6. $\left(-\dfrac{1}{27}\right)^{2/3}$
7. $\left(\dfrac{1}{27}\right)^{-2/3}$
8. $\left(-\dfrac{1}{27}\right)^{-2/3}$
9. $2x^{2/3}(3x)^{-1/3}$
10. $3x^{-2/3}$
11. $(-8)^{-1/3}(16)^{3/4}$
12. $(\tfrac{1}{8})^{2/3}(8)^{-2/3}$
13. $(2^{1/3})^2 \cdot 2^0$
14. $\left(\dfrac{5^{1/2}5^{1/3}}{5^{5/12}}\right)^6$
15. $\left(\dfrac{x^{1/2}x^{-4/3}}{x^{-2/3}}\right)^{-1/2}$

16–22. Simplify each of the following, using theorems to justify. Assume all variables to be positive real numbers.

16. $(x^{-1/2} + y^{-1/2})^2$
17. $(x^{-1/2} + y^{-1/2})(xy)^{-2}$
18. $(x^{-1/2}y^{-1/2})^{-2}$
19. $(x^{-1/2} + y^{-1/2})(xy)^{1/2}$
20. $(x^{-2} + y^{-2})(xy)^{-1/2}$
21. $(x^{-2} + y^{-2})(xy)^{-2}$
22. $(x^{-2}y^{-2})^{-1/2}$
23. Supply the reasons for the proof of Exponent Theorem 1 in this section.
24. Supply the reasons for the proof of Exponent Theorem 2 in this section.
25. Prove Exponent Theorem 3.
26. Prove Exponent Theorem 4.
27. Prove Exponent Theorem 5.

RADICAL EQUATIONS AND INEQUALITIES

28–35. Simplify and give a reason for each step. State any restrictions on the variable if necessary.

28. $\sqrt{3}\sqrt{9x^6}$

29. $\sqrt[3]{8x^4}\sqrt{4x^2}$

30. $\sqrt{x^2y^2 - x^2z^2}$

31. $\sqrt{\dfrac{x+y}{x-y}} - \sqrt{\dfrac{x-y}{x+y}}$

32. $\sqrt[5]{x\sqrt{x\sqrt[3]{x}}}$

33. $\sqrt[3]{2x^2y^4}\sqrt[3]{4x}\sqrt[3]{y^2}$

34. $\dfrac{\sqrt{3}\sqrt[3]{3}}{\sqrt[6]{3}}$

35. $\sqrt{2\sqrt{2\sqrt{5}}}$

36–40. If $1 < \sqrt{2} < 2$, what can you say about each of the following?

36. $3^{\sqrt{2}}$

37. $3^{-\sqrt{2}}$

38. $3^{1/\sqrt{2}}$

39. $3^{-1/\sqrt{2}}$

40. $x^{\sqrt{2}}$

41–50. Solve for x.

41. $x^{2/3} > \sqrt[3]{16}$

42. $x^{3/5} > 2^{-3/5}$

43. $0 < x^{3/4} < \left(\dfrac{2}{5}\right)^{-3/4}$

44. $0 < x^{3/2} < (0.04)^{-1.5}$

45. $0 < x^{-2/5} < 9^{-1/5}$

46. $0 < x^{-3/4} < \left(\dfrac{1}{125}\right)^{-1/4}$

47. $3^{x^2 - x} > 3^2$

48. $0 < 2^{x^2 - 5x + 1} < 2$

49. $0 < (0.5)^3 < (0.5)^{x^2 - 1}$

50. $0 < (0.2)^{x^2} < 0.2$

51–54. Select the larger of each of the following pairs:

51. $\sqrt{5},\ \sqrt[3]{11}$

52. $3\sqrt{5},\ 4\sqrt{3}$

53. $\sqrt{2} - 1,\ \tfrac{1}{2}(\sqrt{2} - \sqrt{2})$

54. $2 - \sqrt{3},\ \dfrac{\sqrt{2} + \sqrt{3}}{2}$

5.8 RADICAL EQUATIONS AND INEQUALITIES

A radical equation or inequality is one in which the variable appears in a radicand. For example, $\sqrt{2x - 1} = 5$ and $\sqrt{x^2 - 3x - 1} < 5x + 2$.

Radical equations and inequalities may be solved by applying the following equivalence theorems:

Equivalence Theorem 3 *If A and B are open expressions and if n is any natural number, then the solution set of $A = B$ is a subset of the solution set of $A^n = B^n$.*

Corollary The solution set of $\sqrt[n]{A} = \sqrt[n]{B}$ is a subset of the solution set of $A = B$.

If a is a solution of $A = B$ so that $A(a) = B(a)$, then $[A(a)]^n = [B(a)]^n$ since exponentiation is an operation and thus x^n is unique. On the other hand, if b is a solution of $A^n = B^n$ so that $[A(b)]^n = [B(b)]^n$ and if n is even, then for $A(b) = -B(b)$, $[A(b)]^n = [-B(b)]^n = [B(b)]^n$. Therefore, a solution of $A^n = B^n$ is not always a solution of $A = B$. In solving a radical equation, the check is an important part of the solution.

Equivalence Theorem 4 If A and B are open expressions and if n is any natural number, then the solution set of $A > B > 0$ is a *subset* of the solution set of $A^n > B^n$.

Example 1 Solve $x - 6 = \sqrt{2x - 9}$.

Solution The solution set is a subset of

$$(x - 6)^2 = (\sqrt{2x - 9})^2$$

or

$$x^2 - 12x + 36 = 2x - 9$$
$$x^2 - 14x + 45 = 0$$
$$(x - 5)(x - 9) = 0$$
$$x = 5 \quad \text{or} \quad x = 9.$$

CHECKING If $x = 5$, then $x - 6 = 5 - 6 = -1$ and

$$\sqrt{2x - 9} = \sqrt{10 - 9} = 1.$$

$-1 \neq 1$ and 5 is *not* a solution.
If $x = 9$, then $x - 6 = 9 - 6 = 3$ and

$$\sqrt{2x - 9} = \sqrt{18 - 9} = \sqrt{9} = 3.$$

$3 = 3$, and 9 is a solution.
The solution set $= \{9\}$.

Note that the solution set of $x - 6 = \sqrt{2x - 9}$ is $\{9\}$ and the solution set of $(x - 6)^2 = (\sqrt{2x - 9})^2$ is $\{5, 9\}$ with $\{9\}$ a proper subset of $\{5, 9\}$.

Example 2 Solve $\sqrt{x^2 - 3x - 1} = 3$.

RADICAL EQUATIONS AND INEQUALITIES

Solution

$$x^2 - 3x - 1 = 9$$
$$x^2 - 3x - 10 = 0$$
$$(x + 2)(x - 5) = 0$$
$$x = -2 \quad \text{or} \quad x = 5$$

CHECKING If $x = -2$, $\sqrt{x^2 - 3x - 1} = \sqrt{4 + 6 - 1} = \sqrt{9} = 3$, and -2 is a solution.
If $x = 5$, $\sqrt{x^2 - 3x - 1} = \sqrt{25 - 15 - 1} = \sqrt{9} = 3$, and 5 is a solution. The solution set is $\{-2, 5\}$.

In this example, the solution set of $\sqrt{x^2 - 3x - 1} = 3$ is the same as the solution set of $x^2 - 3x - 1 = 9$ with $\{-2, 5\}$ an improper subset of $\{-2, 5\}$.

Example 3 Solve $\sqrt{3x + 4} - \sqrt{2x - 5} = 2$.

Solution

$$\sqrt{3x + 4} = 2 + \sqrt{2x - 5}$$
$$3x + 4 = 4 + 4\sqrt{2x - 5} + 2x - 5$$
$$x + 5 = 4\sqrt{2x - 5}$$
$$x^2 + 10x + 25 = 16(2x - 5) = 32x - 80$$
$$x^2 - 22x + 105 = 0$$
$$(x - 7)(x - 15) = 0$$
$$x = 7 \quad \text{or} \quad x = 15$$

After checking, it is seen that the solution set $= \{7, 15\}$.

Example 4 Solve $x^4 = 16$ over R.

REMARK Since $16 = 2^4$, $x^4 = 2^4$.

Now, the solution set of $\sqrt[4]{x^4} = \sqrt[4]{2^4}$ or $x = 2$ is a subset of the solution set of $x^4 = 16$, and it is possible that there are other solutions of $x^4 = 16$ besides $x = 2$. Equating like roots does not in general yield all possible solutions and thus does not qualify as a satisfactory method of solution.

Solution

$$x^4 - 16 = 0$$
$$(x + 2)(x - 2)(x^2 + 4) = 0$$
$$x = -2 \quad \text{or} \quad x = 2 \quad \text{or} \quad x^2 = -4$$

Since $x^2 = -4$ has no real solutions, the solution set over R is $\{-2, 2\}$.

Example 5 Solve $\sqrt{x+20} < |x|$.

Solution The solution set is a subset of the solution set of

$$x + 20 < x^2$$
$$x^2 - x - 20 > 0$$
$$(x+4)(x-5) > 0.$$

CASE 1 $x + 4 > 0$ and $x - 5 > 0$, or $x > -4$ and $x > 5$.
Solution set = $\{x \mid x > 5\}$.

CASE 2 $x + 4 < 0$ and $x - 5 < 0$, or $x < -4$ and $x < 5$.
Solution set = $\{x \mid x < -4\}$.

CHECKING It is noted that $x + 20 \geq 0$ or $x \geq -20$. Intersecting this set with each of the solution sets for the two cases, the solution set desired is

$$\{x \mid x > 5 \text{ or } -20 \leq x < -4\}.$$

Comparison of Equations and Inequalities

		EQUATIONS	INEQUALITIES
1	Reflexivity	$x = x$	
	Trichotomy		If $x \neq y$, then $x < y$ or $x > y$
2	Symmetry	If $x = y$, then $y = x$	
	Nonsymmetry		If $x < y$, then $y > x$
3	Transitivity	If $x = y$ and $y = z$, then $x = z$	If $x < y$ and $y < z$, then $x < z$
4	Addition	If $x = y$, then $x + z = y + z$	If $x < y$, then $x + z < y + z$
5	Multiplication	If $x = y$, then $xz = yz$	If $x < y$ and $z > 0$, then $xz < yz$ If $x < y$ and $z < 0$, then $xz > yz$
6	Zero Product	$ab = 0$ if and only if $a = 0$ or $b = 0$	
	Positive Product		$ab > 0$ if and only if either $(a > 0$ and $b > 0)$ or $(a < 0$ and $b < 0)$
	Negative Product		$ab < 0$ if and only if either $(a > 0$ and $b < 0)$ or $(a < 0$ and $b > 0)$

RADICAL EQUATIONS AND INEQUALITIES

Examples

	EQUATIONS	INEQUALITIES
(1)	$5 = 5$	$5 < 7$ and $5 > 3$
(2)	If $5 = x$, then $x = 5$.	If $5 < x$, then $x > 5$.
(3)	If $x = 3 + 2$ and $3 + 2 = 5$, then $x = 5$.	If $x < \sqrt{8}$ and $\sqrt{8} < 3$, then $x < 3$.
(4)	If $x - 2 = 5$, then $x = 7$.	If $x - 2 < 5$, then $x < 7$.
(5)	If $\dfrac{x}{2} = 5$, then $x = 10$.	If $\dfrac{x}{2} < 5$, then $x < 10$.
	If $\dfrac{x}{-2} = 5$, then $x = -10$.	If $\dfrac{x}{-2} < 5$, then $x > -10$.
(6)	If $(x - 3)(x + 2) = 0$, then $x - 3 = 0$ or $x + 2 = 0$.	If $(x - 3)(x + 2) > 0$, then $x - 3 > 0$ and $x + 2 > 0$ or $x - 3 < 0$ and $x + 2 < 0$. If $(x - 3)(x + 2) < 0$, then $x - 3 > 0$ and $x + 2 < 0$ or $x - 3 < 0$ and $x + 2 > 0$.

EXERCISES 5.8

1–20. Solve each of the following over R:

1. $\sqrt{2x + 3} = 5$
2. $\sqrt{2x + 3} = 5$
3. $\sqrt{x + 8} + x + 2 = 0$
4. $\dfrac{1}{x - 1} + \dfrac{1}{\sqrt{x - 1}} - 2 = 0$
5. $\sqrt[3]{x + 3} = 3$
6. $\sqrt{(3x + x)^3} = 27$
7. $\sqrt[3]{2x + 6} = 0$
8. $\sqrt{x^2 - 4x + 1} = 1$
9. $\sqrt{x^2 - 6x} = 4$
10. $\sqrt{8x + 25} - \sqrt{2x + 5} = \sqrt{2x + 8}$
11. $\sqrt{3 - x} - \sqrt{2 + x} = 3$
12. $\sqrt{x^2 - \sqrt{2x - 1}} + x = 1$
13. $\sqrt{3x + \sqrt{x + 7}} - 3 = 0$
14. $2(2x + 5)^{1/2} = (8x - 1)^{1/2} + 1$
15. $\sqrt{4x^2 - 1} = 2x$
16. $\dfrac{3}{\sqrt{x - 3}} - x = -\sqrt{x - 3}$
17. $\sqrt{x + 2} + \sqrt{x} = 3$
18. $\sqrt{2x + 3} = 3x + \dfrac{14}{3}$
19. $x^{2/3} - 6x^{1/3} + 5 = 0$
20. $2\sqrt{x} - \sqrt{4x - 3} = \dfrac{1}{\sqrt{4x - 3}}$

21–33. Solve each of the following over R:

21. $\sqrt{3x + 10} < |x|$
22. $\sqrt{2x + 20} \geq |2x|$
23. $\sqrt{5x - 6} < x$
24. $\sqrt{x + 2} - \sqrt{x} < 0$
25. $\sqrt{x + 2} - \sqrt{x} > 0$
26. $\sqrt{x^2 - 3x - 4} < x + 2$
27. $\sqrt{x^2 + 2x - 3} > 2 - x$
28. $\dfrac{x}{\sqrt{x + 2}} \geq 0$
29. $\dfrac{x}{\sqrt{x + 2}} \leq 0$
30. $x^4 - 256 \leq 0$
31. $x^4 - 49 > 0$
32. $x^6 - 64 \geq 0$
33. $x^6 - 729 < 0$

34. Solve $x^4 - 49 \leq 0$ by finding the complement over R of the solution set for Exercise 31.
35. Solve $x^4 - 256 > 0$ by finding the complement over R of the solution set for Exercise 30.

CHAPTER SUMMARY

The *set of real numbers*, R, is the set of all least upper bounds and all greatest lower bounds of sets of rational numbers.

Axioms for the System of Real Numbers

(1)	Closure, Addition	There is exactly one $x + y$ in R.
(2)	Closure, Multiplication	There is exactly one xy in R.
(3)	Commutative, Addition	$x + y = y + x$.
(4)	Commutative, Multiplication	$xy = yx$.
(5)	Associative, Addition	$(x + y) + z = x + (y + z)$.
(6)	Associative, Multiplication	$(xy)z = x(yz)$.
(7)	Distributive	$x(y + z) = xy + xz$.
(8)	Identity, Addition	Exactly one 0 in R and $x + 0 = x$ and $0 + x = x$.
(9)	Identity, Multiplication	Exactly one 1 in R and $x \cdot 1 = x$ and $1 \cdot x = x$.
(10)	Inverse, Addition	Exactly one $-x$ in R and $x + (-x) = 0$ and $(-x) + x = 0$.
(11)	Inverse, Multiplication	Exactly one $\dfrac{1}{x}$ in R for $x \neq 0$ and $x\left(\dfrac{1}{x}\right) = 1$ and $\left(\dfrac{1}{x}\right)x = 1$.

CHAPTER SUMMARY

(12)	Trichotomy	Exactly one is valid: $x > 0$, $x = 0$, $-x > 0$.
(13)	Closure of positives, Addition	If $x > 0$ and $y > 0$, then $x + y > 0$.
(14)	Closure of positives, Multiplication	If $x > 0$ and $y > 0$, then $xy > 0$.
(15)	Completeness	Every nonempty subset of R having an upper bound has exactly one least upper bound. Every nonempty subset of R having a lower bound has exactly one greatest lower bound.

Basic Theorems for R

(1)	Addition theorem	If $x = y$, then $x + z = y + z$.
(2)	Multiplication theorem	If $x = y$, then $xz = yz$.
(3)	Addition cancellation theorem	If $x + z = y + z$, then $x = y$.
(4)	Multiplication cancellation	If $xz = yz$ and $z \neq 0$, then $x = y$.
(5)	Zero factor	$x \cdot 0 = 0$.
(6)	Zero product	If $xy = 0$, then $x = 0$ or $y = 0$.

Theorems on Opposites for R

(1)	Opposite of an opposite	$-(-x) = x$.
(2)	Opposite of a sum	$-(x + y) = (-x) + (-y)$.
(3)	Opposite of a difference	$-(x - y) = y - x$.
(4)	Opposite of a product	$-xy = (-x)y$.
(5)	Product of two opposites	$(-x)(-y) = xy$.

Theorems on Reciprocals for R

(1)	Reciprocal of a reciprocal	If $x \neq 0$, then $\dfrac{1}{\frac{1}{x}} = x$.
(2)	Product of reciprocals	If $xy \neq 0$, then $\dfrac{1}{x} \cdot \dfrac{1}{y} = \dfrac{1}{xy}$.
(3)	Fundamental theorem, Quotients	If $yz \neq 0$, then $\dfrac{xz}{yz} = \dfrac{x}{y}$.
(4)	Sum of quotients	If $yw \neq 0$, then $\dfrac{x}{y} + \dfrac{z}{w} = \dfrac{xw + yz}{yw}$.
(5)	Product of quotients	If $yw \neq 0$, then $\dfrac{x}{y} \cdot \dfrac{z}{w} = \dfrac{xz}{yw}$.

Order Theorems for R

(1) General trichotomy — Exactly one is valid: $x > y$, $x = y$, $x < y$.

(2) Transitivity — If $x > y$ and $y > z$, then $x > z$.

(3) Addition — If $x > y$, then $x + z > y + z$.

(4) Order of opposites — If $x > 0$, then $-x < 0$. If $x < 0$, then $-x > 0$.

(5) Multiplication by positive — If $x > y$ and $z > 0$, then $xz > yz$.

(6) Multiplication by negative — If $x > y$ and $z < 0$, then $xz < yz$.

(7) Positive product — $xy > 0$ if and only if ($x > 0$ and $y > 0$) or ($x < 0$ and $y < 0$).

(8) Negative product — $xy < 0$ if and only if ($x > 0$ and $y < 0$) or ($x < 0$ and $y > 0$).

(9) Positivity of 1 — $1 > 0$.

(10) Order of reciprocals — If $x > 0$, then $\dfrac{1}{x} > 0$. If $x < 0$, then $\dfrac{1}{x} < 0$.

(11) Addition, same sense — If $x > y$ and $z > w$, then $x + z > y + w$.

(12) Multiplication, same sense — If $x > y > 0$ and $z > w > 0$, then $xz > yw$.

Absolute Value

$$|x| = x \quad \text{if } x \geq 0.$$
$$|x| = -x \quad \text{if } x < 0.$$

Radicals

(1) Definition: If $a \geq 0$, then $\sqrt[n]{a} = $ lub of $\{x \mid x \geq 0 \text{ and } x^n < a\}$. If $-a < 0$, then $\sqrt[n]{-a} = -\sqrt[n]{a}$ for n odd.

(2) Theorem: If n is odd, then $\sqrt[n]{x^n} = x$. If n is even, then $\sqrt[n]{x^n} = |x|$.

(3) Theorem (product of roots): $\sqrt[n]{xy} = \sqrt[n]{x}\sqrt[n]{y}$.

(4) Theorem (quotient of roots): $\sqrt[n]{\dfrac{x}{y}} = \dfrac{\sqrt[n]{x}}{\sqrt[n]{y}}$.

(5) Theorem (repeated roots): $\sqrt[m]{\sqrt[n]{x}} = \sqrt[mn]{x}$.

(6) Theorem (order of roots): $\sqrt[n]{x} > \sqrt[n]{y}$ if and only if $x > y$.

CHAPTER SUMMARY

Exponent Definitions $b \in R$, $b \neq 0$, and $m, n \in N$.

(1) $b^1 = b$ and $0^n = 0$
(2) $b^n = b(b^{n-1})$
(3) $b^0 = 1$
(4) $b^{-n} = 1/b^n$
(5) $b^{m/n} = \sqrt[n]{b^m}$ (Undefined for n even and b^m negative.)
(6) $b^{-m/n} = 1/\sqrt[n]{b^m}$ (Undefined for n even and b^m negative.)

Exponent Theorems $a, b \in R$, $ab \neq 0$, $x, y \in Q$.

(1) $b^x b^y = b^{x+y}$
(2) $(b^x)^y = b^{xy}$
(3) $b^x/b^y = b^{x-y}$
(4) $(ab)^x = a^x b^x$
(5) $(a/b)^x = a^x/b^x$

Equivalence Theorems A, B, and C are open rational expressions, k is a positive real number, and n is a natural number.

(1) $A = B$ is equivalent to $A + C = B + C$.
(2) $A = B$ is equivalent to $AC = BC$ and $C \neq 0$.
(3) $A > B$ is equivalent to $A + C > B + C$.
(4) $A > B$ is equivalent to $AC > BC$ and $C > 0$.
(5) $A > B$ is equivalent to $AC < BC$ and $C < 0$.
(6) $AB > 0$ is equivalent to $(A > 0$ and $B > 0)$ or $(A < 0$ and $B < 0)$.
(7) $AB < 0$ is equivalent to $(A > 0$ and $B < 0)$ or $(A < 0$ and $B > 0)$.
(8) $|A| = k$ is equivalent to $A = k$ or $A = -k$.
(9) $|A| < k$ is equivalent to $-k < A < k$.
(10) $|A| > k$ is equivalent to $A > k$ or $A < -k$.
(11) Solution set of $A = B$ is a subset of solution set of $A^n = B^n$.
(12) Solution set of $A > B$ is a subset of solution set of $A^n > B^n$.
(13) Solution set of $\sqrt[n]{A} = \sqrt[n]{B}$ is a subset of solution set of $A = B$.
(14) Solution set of $\sqrt[n]{A} > \sqrt[n]{B}$ is a subset of solution set of $A > B$.

REVIEW EXERCISES

1–4. Express each of the following in the form p/q, where p and q are relatively prime natural numbers.

1. $0.2\overline{15}$
2. $1.\overline{234}$
3. $2.\overline{3412}$
4. $0.\overline{4321}$

5–8. Find the lub and the glb of each of the following sets.

5. $A = \{x \mid -1 < x < 3,\ x \in R\}$
6. $B = \{x \mid |x| \leq 4,\ x \in R\}$
7. $C = \{x \mid |x| > 2,\ x \in R\}$
8. $D = \{x \mid x = \dfrac{1}{n},\ n \text{ is a natural number}\}$

9–20. Solve each of the following and graph the solution set.

9. $\dfrac{2x}{x+1} \leq 3$
10. $|2x - 5| < 2$
11. $3 - 2x < \dfrac{3x}{2}$
12. $\{x \mid x > \tfrac{1}{2}\} \cap \{x \mid 2x - 1 < 2\}$
13. $\{x \mid x + 1 < 0\} \cup \{x \mid |x| - 2x > 0\}$
14. $|x - 1| > |x| - 1$
15. $\dfrac{1}{x-1} < \dfrac{1}{2}$
16. $(x+2)(x-3)(x+4) < 0$
17. $x(x+1)(x-2) \geq 0$
18. $\dfrac{1-x}{x-1} > 0$
19. $x^2 > 4$
20. $x^2 \leq 2x$

21–30. Solve each of the following over R.

21. $|x - 3| = 2$
22. $4 \geq 2 - \dfrac{x}{2}$
23. $|2x - 5| < 4$
24. $|3 - 2x| > 2$
25. $\left|\dfrac{x+1}{x}\right| < 2$
26. $\dfrac{3}{|x-1|} \leq 2$
27. $2 - |1 - x| < 3$
28. $|x+1| - |x| > 0$
29. $|3x| = 3x$
30. $|x - 4| = |4 - x|$

31–35. Simplify. (Assume all variables to be positive real numbers.)

31. $\sqrt{50} + 2\sqrt{18}$
32. $\dfrac{3}{\sqrt{x+y}}$
33. $\dfrac{3}{\sqrt{x} + \sqrt{y}}$
34. $\dfrac{1}{\sqrt[3]{9y^2}}$
35. $\dfrac{1}{\sqrt{x} + \sqrt{y} + \sqrt{z}}$

REVIEW EXERCISES

36–40. Write each of the following in radical form and simplify if possible.

36. $\left(\dfrac{2^{1/2}2^{1/3}}{2^{5/12}}\right)^{-6}$

37. $\left(-\dfrac{1}{64}\right)^{2/3}$

38. $-\left(\dfrac{1}{64}\right)^{-2/3}$

39. $\left(\dfrac{1}{64}\right)^{-2/3} 8^{2/3}$

40. $(16)^{-3/4}$

41–45. Solve each of the following over R.

41. $\sqrt{3x+1} = 5$

42. $\sqrt{3x+1} = 5$

43. $\dfrac{2}{\sqrt{x-2}} - \sqrt{x} = \sqrt{x-2}$

44. $\sqrt{2x+3} < |x|$

45. (a) $\sqrt{x^2+2x-3} < x+2$

    (b) $\sqrt{x^2+2x-3} \geq x+2$

46. Prove: For all positive real numbers x and y such that $x \neq y$, $\dfrac{x}{y} + \dfrac{y}{x} > 2$.

47. Prove: For all real numbers, x, y, and z such that $x \neq y \neq z$,
$$x^2 + y^2 + z^2 > xy + yz + xz.$$

48. Prove: For all positive real numbers x and y such that $x \neq y$, $\dfrac{x+y}{2} > \sqrt{xy}$.

49. Prove: If x is any positive real number other than 1, then $\dfrac{x^2+1}{x} > 2$.

50. Prove: For all positive rational numbers x and y such that $x > y$,
$$x^x y^y > x^y y^x.$$

6

EXPONENTIAL, LOGARITHMIC, AND TRIGONOMETRIC FUNCTIONS

Logarithms were devised as a method for rapid and accurate computations related to problems in astronomy, engineering, surveying, navigation, and other areas. One of the earliest contributions was the work of the Swabian, Michael Stifel. In his *Arithmetica Integra*, published in Nuremburg in 1544, he stated the four laws of exponents for rational numbers and referred to the "upper numbers" as "exponents." He also presented the following table which could be considered as a primitive table of logarithms.

x	-3	-2	-1	0	1	2	3	4	5	6
2^x	$\frac{1}{8}$	$\frac{1}{4}$	$\frac{1}{2}$	1	2	4	8	16	32	64

Since decimal fractions were not developed until after 1600, it would have been impossible for Stifel to make a table of logarithms suitable for practical calculations.

HISTORICAL NOTE

The Scotsman John Napier (1550–1617), who worked for about 20 years on the theory, is generally acknowledged as the founder of logarithms. Although he first used the word "artificial number," he finally adopted the term "logarithm" which in Greek literally means "ratio number."

Napier's work, *Mirifici Logarithorum Canonis Descriptio* (*A Description of the Marvelous Law of Logarithms*), published in Edinburgh in 1614, contained the first real table of logarithms. The response was very enthusiastic. In a later work Napier also explained how to calculate a table of logarithms by a method which was essentially based on adding areas under the hyperbolic curve, $y = 1/x$.

About the same time but independently of Napier, the Swiss instrument maker, Jobst Bürgi (1552–1632) calculated a logarithm table. His *Arithmetische und Geometrische Progresstabuln* was published in Prague in 1620. This was actually a list of antilogarithms, with the logarithms written in red and the antilogarithms in black. Thus Bürgi referred to the logarithm as "Die Rothe Zahl" ("the red number"). While Napier selected the base $b = 0.9999999 = 1 - 10^{-7}$ and used geometric methods, Bürgi selected $b = 1.0001$ and used algebraic methods. Both selected a base close to 1 so that the powers of b would be close together, and thus the antilogarithms could be listed in intervals of 0.0000001 or 0.0001, respectively.

Henry Briggs (1561–1631), a professor of geometry at Gresham College, London, as a result of a mutual agreement resulting from a conversation with Napier, developed a table of logarithms using the base 10. His *Arithmetica Logarithmica*, published in 1624, contained 14-place tables for the integers from 1 to 20,000 and from 90,000 to 100,000. The interval from 20,000 to 90,000 was completed by the Dutch bookseller and publisher Adriaen Vlacq (1600–1666) who published a complete 14-place table of logarithms in 1628.

It was Briggs who introduced the word "mantissa" (originally meaning an "addition" and later an "appendix"), and who also suggested the term "characteristic." In the early tables, the characteristic was printed and was not dropped until about the middle of the eighteenth century.

Tables more accurate than those of Briggs and Vlacq were not calculated until the years 1924–1949, when 20-place tables were made.

Later developments were concerned with establishing the theory of logarithms on a logically sound foundation. The Swiss mathematician Leonhard Euler (1707–1783) made important contributions in his *Introductio*. It was Euler who introduced the letter e to represent the base of the Naperian or natural logarithms. Euler wrote, "Ponamus autem brevitatis gratia pro numero hoc 2.71828 ... constanter litteram e" which is the Latin for "For the sake of brevity we shall let the literal constant e represent this number 2.71828 ..."

The theory of logarithms was finally established through the works of the French mathematician Augustin Cauchy (1789–1857), particularly in his *Cours d'Analyse* published in Paris in 1821.

6.1 EXPONENTIAL AND LOGARITHMIC FUNCTIONS

In Chapter 5, the following two statements were proved for b, a real number and for x and y, any rational numbers:

If $b > 1$ and $x > y$, then $b^x > b^y > 0$.

If $0 < b < 1$ and $x > y$, then $0 < b^x < b^y$.

In general, a function for which $f(x) > f(y)$ when $x > y$ is called a **monotonic increasing function**.

The set of ordered pairs $\{(x, b^x) | b > 1 \text{ and } x \in Q\}$ is a function since there is exactly one power b^x for each x in the domain and it is a monotonic increasing function since $b^x > b^y$ for $x > y$.

A function for which $f(x) < f(y)$ when $x > y$ is called a **monotonic decreasing function**. The set of ordered pairs $\{(x, b^x) | 0 < b < 1 \text{ and } x \in Q\}$ is a monotonic decreasing function.

For $b = 1$, $1^x = 1^y = 1$ and $f = \{(x, b^x) | b = 1 \text{ and } x \in Q\} = \{(x, 1) | x \in Q\}$ is a constant function.

For $b = 0$, $0^x = 0$ for x a positive rational number but 0^0 and $0^{-x} = 1/0^x$ are meaningless or undefined. The set $\{(x, b^x) | b = 0, x > 0 \text{ and } x \in Q\} = \{(x,0) | x > 0 \text{ and } x \in Q\}$ is also a constant function.

For $b < 0$, the function $\{(x, b^x) | b < 0 \text{ and } x \in Q\}$ is not constant and it is neither monotonic increasing nor monotonic decreasing. Moreover, b^x is not a real number for rational exponents indicating even root extractions of negative numbers. The functional values "jump" back and forth between positive and negative values as indicated in the table below for the special case $b = -2$.

x	-2	-1	0	$\tfrac{1}{3}$	$\tfrac{1}{2}$	$\tfrac{2}{3}$	1	2	3	4
$(-2)^x$	$\tfrac{1}{4}$	$-\tfrac{1}{2}$	1	$-\sqrt[3]{2}$	Not Real	$\sqrt[3]{4}$	-2	4	-8	16

Functions that are monotonic increasing or monotonic decreasing have many important practical and theoretical applications. These functions are especially useful in the development of calculus. It is desirable then to focus attention on powers of a positive real base different from 1. Up to now, for $0 < b < 1$ or $b > 1$, the definitions have provided for exactly one real number b^x for each rational number x. It is also possible to define b^x for each *real* number x

EXPONENTIAL AND LOGARITHMIC FUNCTIONS

so that the fundamental theorems on exponents remain valid and so that the monotonic increasing or decreasing properties are preserved. The idea is illustrated for the special case $2^{\sqrt{3}}$.

The irrational number $\sqrt{3}$ is the least upper bound of the set of successive rational approximations to $\sqrt{3}$,

$\sqrt{3}$ = least upper bound of $\{1.7, 1.73, 1.732, 1.7320, 1.73205, \ldots\}$ where each rational number in the approximating set is less than $\sqrt{3}$.

Now consider the set B, where

$$B = \{2^{1.7}, 2^{1.73}, 2^{1.732}, 2^{1.7320}, 2^{1.73205}, \ldots\}.$$

Since $b = 2 > 1$, then $2^x \leq 2^y$ for $x < y$, x and y rational. Consequently, since

$$1.7 < 1.73 < 1.732 < 1.7320 < 1.73205 < \ldots < \sqrt{3} < 2,$$

then

$$2^{1.7} < 2^{1.73} < 2^{1.732} < 2^{1.7320} < 2^{1.73205} < \ldots < 2^2 = 4.$$

Therefore, B is a nonempty set of real numbers with an upper bound 4. By the completeness axiom, B has exactly one least upper bound. This least upper bound is defined as the value of $2^{\sqrt{3}}$.

$$2^{\sqrt{3}} = \text{lub of } \{2^{1.7}, 2^{1.73}, 2^{1.732}, 2^{1.7320}, 2^{1.73205}, \ldots\}.$$

Definition of b^u Let c be a nonnegative integer, d_i be a digit. If for all natural numbers n, $u - c \cdot d_1 d_2 \ldots d_n \geq 0$, where

$$u = \text{lub of } \{c \cdot d_1, c \cdot d_1 d_2, \ldots, c \cdot d_1 d_2 \ldots d_n, \ldots\},$$

then for $b > 1$,

$$b^u = \text{lub of } \{b^{c \cdot d_1}, b^{c \cdot d_1 d_2}, \ldots, b^{c \cdot d_1 d_2 \ldots d_n}, \ldots\}$$

and

$$b^{-u} = \frac{1}{b^u}.$$

As a consequence of this definition, it may be shown that for each *real* number x, there is exactly one real number b^x for $0 < b < 1$ or for $b > 1$.

Definition of Exponential Function, Exp

$$\text{Exp} = \{(x, b^x) \mid x \in R, b \in R, \quad \text{and} \quad 0 < b < 1 \text{ or } b > 1\}$$

It can be shown that the following theorems remain valid for an exponential function:

Exponential Function Theorems

(1) $b^x b^y = b^{x+y}$

(2) $(b^x)^y = b^{xy}$

(3) $\dfrac{b^x}{b^y} = b^{x-y}$

(4) $b^x > 0$

(5) $b^x = b^y$ if and only if $x = y$

(6) For $b > 1$, $b^x > b^y$ if and only if $x > y$

(7) For $b > 1$, $b^x = 1$ if and only if $x = 0$

(8) For $b > 1$, $b^x < 1$ if and only if $x < 0$

(9) For $b > 1$, $b^x > 1$ if and only if $x > 0$

The graphs of the two types of exponential functions are illustrated in Figure 6.1.

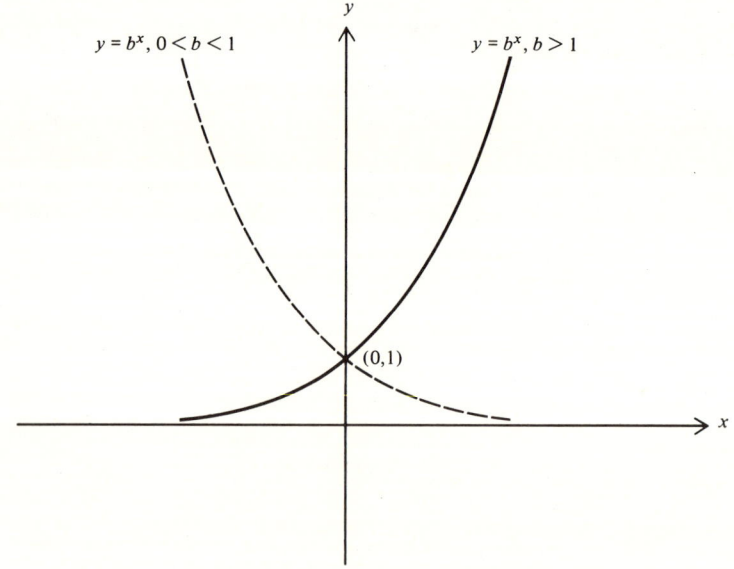

Figure 6.1 The Exponential Function

Definition of the Identity Function, I The identity function I is the function for which $I(x) = x$ for all x in R; that is,

$$I = \{(x, x) \mid x \in R\}.$$

EXPONENTIAL AND LOGARITHMIC FUNCTIONS

An operation, called *composition*, is defined on functions as follows:

$$f \circ g = \{(x, f(g(x)) \mid g(x) \in \text{Range of } g \cap \text{Domain of } f\}.$$

As a result of the two definitions above, it follows that for all functions $f, f \circ I = f$ and $I \circ f = f$. Note that $(f \circ I)(x) = f(I(x)) = f(x)$ and $(I \circ f)(x) = I(f(x)) = f(x)$.

For the composition operation, the function I has the same property as the identity 0 for the addition operation on real numbers and the identity 1 for the multiplication operation on real numbers.

Example 1 Find $(f \circ g)(x)$ and $(g \circ f)(x)$ if $f(x) = x^2$ and $g(x) = 2x - 1$.

Solution

$$(f \circ g)(x) = f(g(x)) = f(2x - 1) = (2x - 1)^2$$
$$(g \circ f)(x) = g(f(x)) = g(x^2) = 2x^2 - 1$$

Since $b^{x_1} = b^{x_2}$ if and only if $x_1 = x_2$, the exponential function has the special property that for each element in its range there is exactly one element in its domain. Any function with this property is called a **one-to-one** function.

Definition of One-to-One Function The function f is one-to-one if and only if for all x_1 and x_2 in the domain of f, $f(x_1) = f(x_2)$ if and only if $x_1 = x_2$.

The relation obtained by interchanging the components of the ordered pairs of a one-to-one function is also a function, called the **inverse function**. The domain of the original function is the range of the inverse function, and the range of the original function is the domain of the inverse function.

Definition of Inverse Function, f^{-1} Let

$$f = \{(x, y) \mid f \text{ is one-to-one}\}.$$

Then

$$f^{-1} = \{(y, x) \mid (x, y) \in f\}.$$

Definition of $f^{-1}(x)$

$$f^{-1}(x) = y \quad \text{if and only if} \quad x = f(y)$$

An alternate definition for an inverse function is the following which corresponds to the concept of additive inverse and multiplicative inverse as used in the system of real numbers.

Alternate Definition of Inverse Function, f^{-1} f^{-1} is the inverse function of f with respect to a nonempty set S if and only if for all x in S,

$$f \circ f^{-1} = I \quad \text{and} \quad f^{-1} \circ f = I.$$

In other words, $f(f^{-1}(x)) = x$ and $f^{-1}(f(x)) = x$ for all x in S.

Example 2 Is the function defined by $g(x) = \sqrt{x}$ for $x \geq 0$ the inverse of the function $f(x) = x^2$, where x is a real number?

Solution It is necessary to show that $f \circ g = I$ or $(f \circ g)(x) = x$ and $g \circ f = I$ or $(g \circ f)(x) = x$ for all x in the intersection of the domains of f and g; that is, for $x \geq 0$.

$$(f \circ g)(x) = f(\sqrt{x}) = (\sqrt{x})^2 = x \qquad \text{for } x \geq 0$$
$$(g \circ f)(x) = g(x^2) = \sqrt{x^2} = |x| = x \qquad \text{for } x \geq 0.$$

Note that
$$(g \circ f)(x) = |x| \neq x \qquad \text{if } x < 0.$$

Therefore, $g(x)$ is the inverse of $f(x) = x^2$ only if $x \geq 0$.

Example 3 Let $f = \{(x, y) | y = \sqrt{x - 4}\}$.

(1) Find $f(x)$.
(2) State the domain and range of f.
(3) Find f^{-1}.
(4) Find $f^{-1}(x)$.
(5) State the domain and range of f^{-1}.

Solution

(1) $f(x) = \sqrt{x - 4}$
(2) Dm of $f = \{x | x \geq 4\}$, Rn of $f = \{y | y \geq 0\}$
(3) $f^{-1} = \{(x, y) | x = \sqrt{y - 4}\}$
(4) Since $x = \sqrt{y - 4}$, then $x^2 = y - 4$ and $y = x^2 + 4$. $f^{-1}(x) = x^2 + 4$
(5) Dm of $f^{-1} = \{x | x \geq 0\}$, Rn of $f^{-1} = \{y | y \geq 4\}$

Inverse Function Theorem If f is one-to-one, then for all x in the domain of f and for all y in the range of f,

$$f^{-1}(f(x)) = x \qquad \text{and} \qquad f(f^{-1}(y)) = y.$$

Since $x = f^{-1}(y)$ if and only if $y = f(x)$, then $f(f^{-1}(y)) = f(x) = y$. Similarly, $f^{-1}(f(x)) = f^{-1}(y) = x$.

Example 4 Show that $f^{-1}(f(x)) = x$ for $f(x) = \sqrt{x - 4}$.

Solution $f^{-1}(f(x)) = [f(x)]^2 + 4$
$= (\sqrt{x - 4})^2 + 4$
$= (x - 4) + 4$
$= x$

EXPONENTIAL AND LOGARITHMIC FUNCTIONS

The domain of an exponential function is the set of all real numbers, but the range is the set of *positive* real numbers, since $b^x > 0$ for all real x. Since the exponential function is one-to-one, it has an inverse function called the **logarithm** function.

Definition of Logarithm Function, log

$$\log = \{(x, y) \mid x = b^y \text{ where } x > 0 \text{ and } b > 1 \text{ or } 0 < b < 1\}$$

The domain of log is the set of positive real numbers and the range of log is the set of all real numbers.

In the previous examples, the inverse function was found by interchanging x and y and then solving for y. In this case, $x = b^y$, there is no direct way to solve for y. To get around this difficulty, we *define* y to be the solution.

Definition of $\log_b x$ $\log_b x = y$ if and only if $x = b^y$ where $b > 1$ or $0 < b < 1$.

Inverse Logarithm Theorem   If x and y are any real numbers and $y > 0$, then $b^{\log_b y} = y$ and $\log_b b^x = x$.

For $f(x) = b^x = y$, then $f^{-1}(y) = \log_b y = x$.
$f^{-1}(f(x)) = f^{-1}(b^x) = \log_b b^x = x$, by the inverse function theorem.
$f(f^{-1}(y)) = f(\log_b y) = b^{\log_b y} = y$, by the inverse function theorem.

Corollary   For $x = 1$, $\log_b b^1 = \log_b b = 1$.

For $x = 0$, $\log_b b^0 = \log_b 1 = 0$.

The graph of the exponential function with base $b > 1$ and the graph of its inverse, the logarithm function with base $b > 1$, are illustrated in Figure 6.2.

Example 5   Write in logarithmic form

(1) $10^3 = 1000$

(2) $2^{-4} = \frac{1}{16}$

(3) $(125)^{-2/3} = \frac{1}{25}$.

Solution   Since $b^y = x$ if and only if $\log_b x = y$,

(1) $\qquad\qquad\qquad 10^3 = 1000 \quad \rightarrow \quad \log_{10} 1000 = 3$

(2) $\qquad\qquad\qquad 2^{-4} = \frac{1}{16} \quad \rightarrow \quad \log_2(\frac{1}{16}) = -4$

(3) $\qquad\qquad\qquad (125)^{-2/3} = \frac{1}{25} \quad \rightarrow \quad \log_{125}(\frac{1}{25}) = -\frac{2}{3}$.

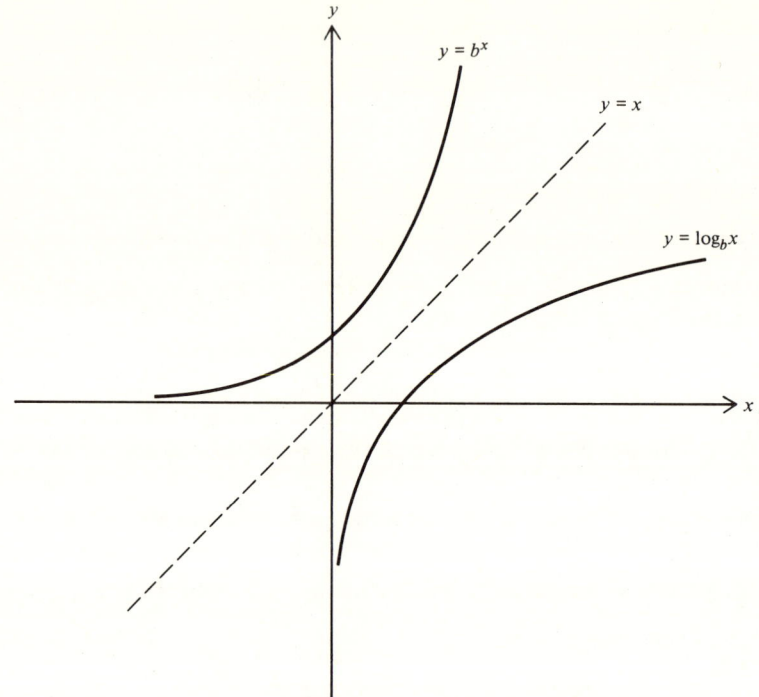

Figure 6.2 The Logarithmic and Exponential Functions, $b > 1$

Example 6 Write in exponential form

(1) $\log_{10} 100 = 2$
(2) $\log_2(\frac{1}{8}) = -3$
(3) $\log_{25} 5 = \frac{1}{2}$.

Solution $\log_b x = y$ if and only if $b^y = x$.

(1) $\qquad \log_{10} 100 = 2 \qquad \rightarrow \qquad 10^2 = 100$.
(2) $\qquad \log_2(\frac{1}{8}) = -3 \qquad \rightarrow \qquad 2^{-3} = \frac{1}{8}$.
(3) $\qquad \log_{25} 5 = (\frac{1}{2}) \qquad \rightarrow \qquad 25^{1/2} = 5$,

Logarithmic Operations Theorem If x, y, and b are real numbers such that $x > 0$, $y > 0$, and $b > 1$ or $0 < b < 1$, then

(1) $\log_b xy = \log_b x + \log_b y$

(2) $\log_b \dfrac{x}{y} = \log_b x - \log_b y$

(3) $\log_b x^y = y \log_b x$.

EXPONENTIAL AND LOGARITHMIC FUNCTIONS

Proofs

(1) $\log_b xy = \log_b b^{\log_b x} \cdot b^{\log_b y} = \log_b b^{\log_b x + \log_b y} = \log_b x + \log_b y$

(2) $\log_b \dfrac{x}{y} = \log_b \dfrac{b^{\log_b x}}{b^{\log_b y}} = \log_b b^{\log_b x - \log_b y} = \log_b x - \log_b y$

(3) $\log_b x^y = \log_b (b^{\log_b x})^y = \log_b b^{y \log_b x} = y \log_b x$

Example 7 Express $\log_b \tfrac{16}{7}$ in terms of the logarithms of prime integers.

Solution $\log_b \tfrac{16}{7} = \log_b 16 - \log_b 7$
  Logarithmic Operations Theorem (2)
  $= \log_b 2^4 - \log_b 7$
  $= 4 \log_b 2 - \log_b 7$
  Logarithmic Operations Theorem (3)

Example 8 Express $\tfrac{1}{3}(\log_b 5 + \log_b 7)$ as the logarithm of a single number.

Solution $\tfrac{1}{3}(\log_b 5 + \log_b 7) = \tfrac{1}{3} \log_b 35$
  Logarithmic Operations Theorem (1)
  $= \log_b 35^{1/3}$
  Logarithmic Operations Theorem (3)
  $= \log_b \sqrt[3]{35}$

EXERCISES 6.1

1–10. For each of the following relations, r, determine
(a) the inverse relation ($= \{(y, x) | (x, y) \in r\}$)
(b) the domain and range of the inverse relation, and
(c) whether the inverse relation is a function.

1. $\{(x, y) \mid 3x - 4y = 12\}$
2. $\{(x, y) \mid y = 9 - x^2\}$
3. $\{(x, y) \mid y = 3\}$
4. $\{(x, y) \mid xy = 1\}$
5. $\{(x, y) \mid 4x^2 + y^2 = 100\}$
6. $\{(x, y) \mid x^2 - y^2 = 9\}$
7. $\{(x, y) \mid y = \sqrt{x + 3}\}$
8. $\{(x, y) \mid y = x^3 - 8\}$
9. $\{(x, y) \mid y = \sqrt[3]{x + 1}\}$
10. $\{(x, y) \mid y = |x|\}$

11–14. For each of the following, (a) find $f^{-1}(x)$, (b) show that $f(f^{-1}(x)) = x$ and $f^{-1}(f(x)) = x$, and (c) graph $f(x)$ and $f^{-1}(x)$.

11. $f(x) = 2x - 6$
12. $f(x) = \sqrt{x + 3}$
13. $f(x) = \dfrac{6}{x}$
14. $(x) = \dfrac{x^3}{4}$

15–19. Solve each of the following if $f(x) = 4x - 8$:

15. $f(f^{-1}(x)) = 2$
16. $x = f^{-1}(f(-5))$
17. $f(x) = f^{-1}(x)$
18. $f^{-1}(x) = f(3)$
19. $f(x) + f^{-1}(x) = 0$

20–25. Solve each of the following if $f(x) = 2^x$:

20. $f^{-1}(x) = 3$
21. $f^{-1}(x) = 0$
22. $f^{-1}(x) = -\frac{1}{2}$
23. $f^{-1}(1) = x$
24. $f^{-1}(2) = x$
25. $x = f^{-1}(0.125)$

26–30. Solve each of the following if $f(x) = 2x + 4$:

26. $f^{-1}(f(x)) = 5$
27. $x = f(f^{-1}(0))$
28. $f^{-1}(x) = f(x)$
29. $f(x) = f^{-1}(\sqrt{3})$
30. $f(6) = f^{-1}(x)$

31–36. Solve each of the following if $f(x) = 4^{-x}$:

31. $x = f^{-1}(1)$
32. $x = f^{-1}(4)$
33. $x = f^{-1}(\frac{1}{16})$
34. $f^{-1}(x) = 3$
35. $f^{-1}(x) = -1$
36. $f^{-1}(x) = 0$

37–44. Write in logarithmic form each of the following:

37. $10^4 = 10{,}000$
38. $5^0 = 1$
39. $9^{1/2} = 3$
40. $4^{-3} = \frac{1}{64}$
41. $3^1 = 3$
42. $(64)^{2/3} = 16$
43. $9^{-3/2} = \frac{1}{27}$
44. $(\frac{1}{6})^{-2} = 36$

45–52. Write in exponential form each of the following:

45. $\log_3 9 = 2$
46. $\log_{10} 1 = 0$
47. $\log_{10} 10 = 1$
48. $\log_2 0.125 = -3$
49. $\log_{36} 6 = \frac{1}{2}$
50. $\log_{10} 100{,}000 = 5$
51. $\log_{10} 0.1 = -1$
52. $\log_8 0.25 = -\frac{2}{3}$

53–60. Evaluate each of the following:

53. $\log_{10} 1000$
54. $\log_{10} 0.01$
55. $\log_5 0.0016$
56. $\log_4 32$
57. $\log_9 1/27$
58. $\log_2 0.0625$
59. $\log_{10} 10^{-2.5}$
60. $10^{\log_{10} 3}$

LOGARITHMIC FUNCTIONS

61–66. Solve each of the following:

61. $\log_2 64 = x$
62. $\log_3 x = -4$
63. $\log_{10} x = 0$
64. $\log_2 0.25 = x$
65. $\log_x 0.04 = -2$
66. $\log_x 16 = \frac{2}{3}$

67–76. By using logarithmic operations theorems, express each of the following in terms of the logarithms of prime integers:

67. $\log_b 15$
68. $\log_b \frac{3}{5}$
69. $\log_b 3^5$
70. $\log_b \sqrt{3}$
71. $\log_b \sqrt{\frac{5}{3}}$
72. $\log_b \frac{1}{\sqrt[3]{5}}$
73. $\log_b \frac{81}{25}$
74. $\log_b 24 \sqrt[4]{3}$
75. $\log_b \frac{5\sqrt{5}}{7}$
76. $\log_b \sqrt[3]{\frac{625}{9}}$

77–86. By using logarithmic operations theorems, express each of the following as the logarithm of a single number:

77. $\log_b 8 + \log_b 9$
78. $\log_b 21 - \log_b 8$
79. $5 \log_b 3$
80. $\frac{1}{2} \log_b 7$
81. $2 \log_b 5 + 4 \log_b 3$
82. $\frac{1}{3} \log_b 7 - 2 \log_b 6$
83. $\frac{1}{5}(\log_b 3 + 2 \log_b 5)$
84. $3(\log_b 4 - 3 \log_b 6)$
85. $2 \log_b 3 + \frac{1}{4} \log_b 5 - 3 \log_b 7$
86. $\frac{1}{3}(2 \log_b 9 + 4 \log_b 4 - \log_b 85)$

6.2 LOGARITHMIC FUNCTIONS (Continued)

6.2.1 Common Logarithms

Since our numeral system is a base 10 positional system, the logarithmic function whose base is 10 is most useful for computations. The values of $\log_{10} x$ are called **common logarithms** or **logarithms to the base 10**. To reduce the amount of writing involved in a calculation using common logarithms, the numeral 10 designating the base is usually omitted.

CONVENTION

$\log x = \log_{10} x$

The following theorem is used for the evaluation of common logarithms.

Common Logarithm Theorem If k is any real number and x is any positive real number, then
$$\log x(10^k) = k + \log x.$$

Proof $\log x(10^k) = \log 10^k + \log x$ Logarithmic Operations Theorem (1)
$ = k + \log x.$ $\log_b b^k = k.$

Since every positive real number can be expressed as the product of a number between 1 and 10 and an integral power of 10, the common logarithm of any positive real number can be determined from the logarithms of the numbers between 1 and 10.

Definition of Characteristic and Mantissa If $\log x(10^k) = k + \log x$ where k is an integer and x is a real number such that $1 \leq x < 10$, then k is called the **characteristic** of $\log x(10^k)$ and $\log x$ is called the **mantissa** of $\log x(10^k)$.

Since $\log_{10} 1 = 0$ and $\log_{10} 10 = 1$, then for $1 \leq x < 10$, $0 \leq \log x < 1$. Thus it follows that the characteristic is the integral part of a common logarithm and the mantissa is the positive decimal part (or 0 for log 1). The characteristic is obtained by using the common logarithm theorem and the mantissa is obtained from a table of common logarithms such as Table II in the Appendix.

Example 1 Find log 2.34.

Solution $\log 2.34 = 0.3692$ located in the table in row 2.3 and in column 4.

Example 2 Find log 2340.

Solution $\log 2340 = \log 2.34(10^3)$
$ = \log 10^3 + \log 2.34$
$ = 3 + 0.3692$

Example 3 Find log 0.0234.

Solution $\log 0.0234 = \log 2.34(10^{-2})$
$ = -2 + 0.3692$

Definition of Antilogarithm The antilogarithm of y, antilog $y = x$ if and only if $y = \log x$.

The table of logarithms may also be used to find a number whose logarithm is given, that is, the antilogarithm of the given number.

Example 4 Find antilog $(-1 + 0.4518)$.

LOGARITHMIC FUNCTIONS

Solution Find x so that $\log x = -1 + 0.4518$.

$$\log 2.83 = 0.4518 \text{ (from the table).}$$
$$\log x = -1 + \log 2.83$$
$$= \log 2.83(10^{-1})$$
$$x = 2.83(10^{-1}) = 0.283$$

6.2.2 Linear Interpolation

A four-place common logarithm table, such as Table II in the Appendix, lists the mantissas of the logarithms of numbers having at most three significant digits. The logarithms of numbers having four significant digits may be approximated by the process of *linear interpolation*. The graph of the logarithmic function is approximated by the straight line joining two points on the curve whose ordinates are successive entries in the table. Figure 6.3 illustrates the procedure for determining log 2.346 by linear interpolation.

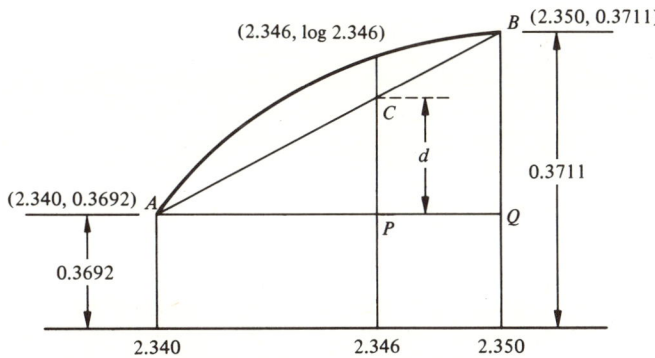

Figure 6.3 Linear Interpolation

The value of log 2.346 is approximated by $\log 2.340 + d = 0.3692 + d$. The fact that corresponding sides of similar triangles are proportional is used to determine d.

$$\frac{d}{BQ} = \frac{AP}{AQ}$$

$$\frac{d}{0.3711 - 0.3692} = \frac{2.346 - 2.340}{2.350 - 2.340}$$

$$\frac{d}{0.0019} = \frac{0.006}{0.010}$$

$$d = 0.6(0.0019) = 0.00114$$

Since the decimal approximations in the table are valid for only four decimal places, d is rounded off to 0.0011. Finally,

$$\log 2.346 = 0.3692 + 0.0011 = 0.3703.$$

The decimal points are usually omitted from the written work since this process can be performed mentally. After some practice, it is possible to perform all the linear interpolation calculation mentally.

Example 9 Find log 52.73.

Solution

$$\begin{array}{c|cc} & x & \log x \\ & 52.70 & 1.7218 \\ 10 \Big| 3 \Big\downarrow & 52.73 & \quad \Big\downarrow d \quad \Big| 8 \\ & 52.80 & 1.7226 \end{array}$$

$$\frac{d}{8} = \frac{3}{10} \quad \text{or} \quad d = 0.3(8) = 2.4$$

$$\log 52.73 = 1.7218 + 0.0002 = 1.7220$$

Example 10 Find antilog 3.1888.

Solution

$$\begin{array}{c|cc} & x & \log x \\ & 1540 & 3.1875 \\ 10 \Big| d \Big\downarrow & & \quad \Big\downarrow 13 \quad \Big| 28 \\ & 1550 & 3.1903 \end{array}$$

$$\frac{d}{10} = \frac{13}{28} \quad \text{or} \quad d = \frac{130}{28} = 5 \text{ approximately}$$

$$\text{antilog } 3.1888 = 1545$$

6.2.3 Computations

By using the theorems for logarithms, computations involving products, quotients, powers, and roots can be replaced by the simpler calculations involving sums, differences, products, and quotients.

LOGARITHMIC FUNCTIONS

Example 11 Compute $\sqrt{(0.725)(34.7)}$.

Solution Let $N = \sqrt{(0.725)(34.7)}$.

$\log N = \frac{1}{2}(\log 0.725 + \log 34.7)$

$$
\begin{aligned}
\log 0.725 &= -1 + 0.8603 \\
\log 34.7 &= 1 + 0.5403 \\
\hline
\text{sum} &= \overline{1}.4006 \\
\log N = \tfrac{1}{2} \text{ sum} &= 0.7003. \\
N &= 5.016 \text{ or } 5.02 \text{ rounded} \\
& \text{ to three significant digits.}
\end{aligned}
$$

Example 12 Compute $\sqrt[3]{\dfrac{1.380}{24.2}}$.

Solution Let $N = \sqrt[3]{\dfrac{1.380}{24.2}}$.

$\log N = \frac{1}{3}(\log 1.380 - \log 24.2)$

$$
\begin{aligned}
\log 1.380 &= 0.1399 = 2.1399 - 2 \\
\log 24.2 &= 1.3838 = 1.3838 \\
\hline
\text{difference} &= 0.7561 - 2 \\
&= 1.7561 - 3 \\
\log N = \tfrac{1}{3} \text{ difference} &= 0.5854 - 1 \\
N &= 0.3849
\end{aligned}
$$

(2 is added and subtracted so that the subtraction will yield a positive decimal.)
(1 is added and subtracted so that the division by 3 will yield an integer for the characteristic.)

$N = 0.385$ approximated to three significant digits

6.2.4 Other Bases

Common logarithms, or logarithms to the base 10 (also called Briggsian, after their founder Henry Briggs), are the most convenient for numerical computations. However, for more advanced mathematics, especially that involving calculus, **natural logarithms** (also called Naperian after their originator, John Napier), or logarithms to the base e, are more appropriate. The number e is irrational with an approximate value of 2.71828.

The following theorem is used to obtain the logarithm of a number to the base b from the logarithm of the number to the base a.

Theorem on Change of Logarithmic Base $\log_b x = \dfrac{\log_a x}{\log_a b}$

Proof $\log_b x = \log_b a^{\log_a x}$ since $x = a^{\log_a x}$

$\qquad\qquad = (\log_a x)\log_b a$ by Logarithmic Operations Theorem (3).

Now, letting $x = b$,

$$\log_b b = (\log_a b)\log_b a \quad \text{or} \quad 1 = (\log_a b)\log_b a.$$

Therefore,

$$\log_b a = \frac{1}{\log_a b}$$

and

$$\log_b x = \frac{\log_a x}{\log_a b}.$$

For the special case that $b = e$ and $a = 10$,

$$\log_e x = \frac{\log_{10} x}{\log_{10} 2.71828} = \frac{\log_{10} x}{0.4343} = 2.3026 \log_{10} x.$$

6.2.5 Logarithmic and Exponential Equations

An **exponential equation** is an equation in which the variable appears in an exponent. Exponential equations can be solved by using the property that the exponential function is one-to-one for $b > 0$ and $b \ne 1$; that is

$$b^x = b^y \quad \text{if and only if} \quad x = y$$

and

$$\log_b x = \log_b y \quad \text{if and only if} \quad x = y.$$

Example 13 Solve $2^x = 0.125$.

Solution $0.125 = (0.5)^3 = (\tfrac{1}{2})^3 = 2^{-3}$

Thus,

$$2^x = 2^{-3} \quad \text{and} \quad x = -3.$$

Example 14 Solve $2^x = 3$.

Solution Since 3 is not a rational power of 2, the logarithms of the two numbers are equated.

LOGARITHMIC FUNCTIONS

$$\log_{10} 2^x = \log_{10} 3$$
$$x \log 2 = \log 3$$
$$x = \frac{\log 3}{\log 2}$$

Approximately,
$$x = \frac{0.4771}{0.3010} = 1.585.$$

A **logarithmic equation** is an equation that contains logarithms. It is also solved by using the one-to-one property of the logarithmic and exponential functions.

Example 15 Solve $2 \log x - \log(x + 3) + \log 5 = \log 4$.

Solution
$$(\log x^2 + \log 5) - \log(x + 3) = \log 4$$
$$\log \frac{5x^2}{x + 3} = \log 4$$
$$\frac{5x^2}{x + 3} = 4$$
$$5x^2 - 4x - 12 = 0$$
$$(x - 2)(5x + 6) = 0$$
$$x = 2 \quad \text{or} \quad x = -\frac{6}{5}$$

CHECK $x = 2$. $2 \log 2 - \log 5 + \log 5 = 2 \log 2 = \log 4$. Thus, 2 is a solution.
$x = -\frac{6}{5}$. $2 \log x = 2 \log(-\frac{6}{5})$ which is undefined. Thus, $-\frac{6}{5}$ is *not* a solution.

Since the domain of the logarithmic function is the set of positive real numbers, the original equation required the restriction that $x > 0$ and $x + 3 > 0$. It is important, then, to check all proposed solutions of a logarithmic equation.

Example 16 Solve $x = \log_5 12$.

Solution Rewriting this equation in exponential form,
$$5^x = 12$$
$$x \log_{10} 5 = \log_{10} 12$$
$$x = \frac{\log 12}{\log 5}.$$

EXERCISES 6.2

1–12. Find each of the following logarithms:

1. log 21.3
2. log 4.76
3. log 0.123
4. log 405,000
5. log 0.0948
6. log 0.007
7. log 2.718
8. log 0.08354
9. log 179.6
10. log 5989
11. log 0.06421
12. log 31.47

13–20. Find each of the following antilogarithms:

13. antilog 1.8407
14. antilog 0.9750
15. antilog 6.9031
16. antilog $-1 + 0.0682$
17. antilog 4.5805
18. antilog 1.9999
19. antilog $-2 + 0.9517$
20. antilog 1.2180

21–30. (a) Write the logarithmic equation for the computation of each of the following; (b) Compute each of the following by using logarithms.

21. $\dfrac{(4.92)(0.0658)}{786}$

22. $\sqrt[3]{0.3583}$

23. $(0.00729)(2.06)^8$

24. $\sqrt[4]{\dfrac{936.4}{2875}}$

25. $\dfrac{450\sqrt[3]{75.2}}{(82.4)^2}$

26. $3.142(0.956)^3$

27. $\dfrac{\sqrt{0.843}}{(54.9)^2}$

28. $\sqrt[3]{\dfrac{24.5}{32.2}}$

29. $(1.025)^{20}(3995)$

30. $(-67.8)^3 \sqrt{\dfrac{(4.32)^2}{(0.402)^4}}$

31–40. Using $\log_{10} 2 = 0.3010$, evaluate each of the following without using the tables. Use the fact that $5 = \dfrac{10}{2}$.

31. log 5
32. log ½
33. log 0.4
34. $\log \sqrt[3]{25}$
35. $\log \sqrt{5}$
36. $\log(\log \sqrt[5]{10})$
37. log 2000
38. log 0.002
39. $\log \dfrac{25}{4}$
40. $\log 50\sqrt{2}$

LOGARITHMIC FUNCTIONS

41–45. Use the theorem on change of logarithmic base and the tables of common logarithms to compute each of the following:

41. $\log_2 14$
42. $\log_e 7$
43. $\log_5 14$
44. $\log_{27} 15$
45. $\log_e 2.35$

46–75. Solve for x.

46. $9^{x+2} = 1$
47. $(125)^{2-x} = 0.04$
48. $2^{x-1} = \sqrt[3]{16}$
49. $8^{2x+1} = 15$
50. $2(5^{3x}) = 5$
51. $(1/3)^x = 81$
52. $2y = y(1.05)^x$
53. $(1/8)^{x-1} = 4^{1-2x}$
54. $3^{x-2} = 243$
55. $4^x = 1/32$
56. $5^x = 2$
57. $4^{1-x} = 64$
58. $6^{x-3} = 4.5$
59. $(2.5)^{3x} = 6.25$
60. $(3^x)^2 = 1/27$
61. $7^{3x+5} = 1$
62. $2 \log(x+3) - \log(x+7) + 1 = \log 2$
63. $10^{-3 \log 2} = x$
64. $5^{2 \log_5 x} = 9$
65. $x = \log_2 26$
66. $\log(x+4) + \log(x-4) - \log 9 = 0$
67. $0.5 \log 3 + \log x = 2 \log 5 - \log 2$
68. $3 \log x + \log 3 - 2 \log 5 = -3 + \log 15$
69. $2 \log(x-1) - \log(x-4) = 4 \log 2$
70. $2 \log(x+5) - 3 \log 2 = 0$
71. $\log 64 - 3 \log x = 3 + \log 8$
72. $3^{2 \log_3 5} = x$.
73. $10^{-2 \log x} = 3$.
74. $x - 2 = \log_3 10$.
75. $x = \log_2 7$.
76. If a piece of paper 0.01 inch thick could be folded so that it is twice as thick but half as large, folded again in the same manner, and again until the process has been repeated 30 times, how thick would the resulting piece of paper be?
77. Radium decomposes according to $y = y_0 e^{-0.04t}$ where y_0 grams reduces to y grams in t centuries. What is the half-life of radium? (Find t when $y = \frac{1}{2} y_0$).

Trigonometry (literally "triangle measurement") arose from practical problems dealing with surveying, mensuration, and astronomy. So closely were astronomy and trigonometry connected that they were not considered to be two separate subjects until the thirteenth century. Beginning with tables of shadow lengths and tables of ratios of sides of right triangles (first appearing in the works of the ancient Egyptians and Babylonians) and later progressing to a systematic account of methods for finding the sides and angles of a triangle, trigonometry evolved until now it is regarded as a study of certain functions of real numbers and of complex numbers.

A Babylonian tablet, dated around 1900–1600 B.C. and referred to as Plimpton 322, exhibits a table of squares of ratios of sides of right triangles. Modern scholars have shown that these squares correspond to values of $\sec^2 \theta$ for θ ranging from 45° to 31° by degrees.

The Babylonians used a positional numeral system based on 60 with sexagesimal fractions that are similar to our decimal fractions except that the denominators were 60, 60 × 60, or higher powers of 60. These Babylonian sexagesimal fractions were used extensively in astronomical works. As these works entered Europe during the Renaissance, the sixtieths, the "first small parts" were translated into Latin as the *partes minutae primae* from which we have "minutes." Similarly, the 60 × 60 ths or "second small parts" were translated as "*partes minutae secundae*" from which we have "seconds."

The first concept of cotangent can be seen in the Egyptian papyrus written by Ahmes around 1650 B.C. There are four problems, relating to the measurement of pyramids, in which the word "seqt" is used to mean the ratio of a horizontal length to a vertical length.

On an Egyptian inscription from the thirteenth century B.C. there is a table relating shadow lengths with times of the day. This suggests the idea of a function; in particular, the tangent or cotangent function.

Although theorems on the ratios of sides of similar triangles were known to the ancient Egyptians and Babylonians, it is significant to note that apparently no concept of angle measure was known at this time.

The Greeks contributed the next major development, the chord function, which was a forerunner of the sine rather than the tangent. Although many of the laws and identities of trigonometry are found in geometric form in the works of Euclid (*c.* 300 B.C.) and other Greek writers, Hipparchus of Nicaea (*c.* 180–120 B.C.) is credited with compiling the first trigonometric table, and became known as the "father of trigonometry." While none of his works are now in existence, Hipparchus is said to have written a twelve-book treatise displaying the first systematic

HISTORICAL NOTE

use of the 360° circle and the first systematic study of relations between the lengths of chords of a circle and their central angles.

Around 150 A.D. Claudius Ptolemy wrote the *Syntaxis mathematica* (Mathematical Collection) which, though based on the books of Hipparchus, was so noted for its clarity and eloquence that it was called *magiste* (greatest) to distinguish it from lesser works by other writers. Later the Arabs prefixed their definite article *al* and from then on, Ptolemy's work became known as *Almagest* (the greatest). *Almagest* was the standard work on astronomy for over 1000 years until the time of Copernicus (1473–1543). It is of special interest in trigonometry because it contains not only a table of chords giving chord lengths for arcs from $\frac{1}{2}°$ to 180° in steps of $\frac{1}{2}°$ (roughly a table of sines from $\frac{1}{4}°$ to 90° in steps of $\frac{1}{4}°$) but also accounts of trigonometric identities that Ptolemy used in constructing his table. To see how these chord lengths are related to the sine function, let chord 2θ = the length of the chord of central angle 2θ. Then referring to Figure 6.4,

$$\sin\theta = \frac{AP}{OA} = \frac{AB}{2r} = \frac{\text{chord } 2\theta}{2r}$$

The radius of the circle was divided into 60 equal parts and the chord lengths were expressed sexagesimally in terms of parts of these. For example, chord $90° = r\sqrt{2} = 84^p\ 51'\ 10''$ for $r = 60$.

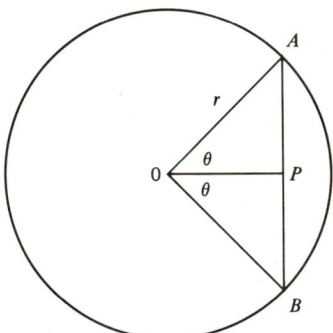

Figure 6.4

The birth of the sine function is credited to the Hindus by their use of the *half-chord*. The *Siddhantas* of the late fourth or early fifth century and the works of Aryabhata about 500 A.D. contain tables relating *half* a chord of a circle and *half* of its central angle. The Hindus called the sine *jya* meaning "half-chord." The Arabs changed this to *jyb* which was later read incorrectly as *jayb* (the Arabic for pocket or gulf) and translated into Latin as sinus from which we have our word *sine*.

The Arab writers on astronomy, especially Albategnus, the Ptolemy of Bagdad (c. 850–929), and Abu'l-Wefa (c. 940–998) improved the tables of the Hindus and, unlike the Hindus, they used a *unit circle* for the sine function. Abu'l-Wefa provided a new sine table for angles in steps of $\frac{1}{4}°$ and a table of tangents and he used all six of the common trigonometric functions. The Arabs, adopting the decimal numeral system of the Hindus, preferred to compute in decimals although they wrote their results in the traditional sexagesimal notation.

By the time the Arabic documents reached Europe, the subject matter of trigonometry was better organized with proofs given for the half-angle, double angle, and other theorems. Gradually decimals began to replace sexagesimals in the trigonometric tables.

The *De triangulis omnimodis* of Regiomontanus (1436–1476) is especially significant due to the systematic procedures he gave for the solution of triangles and due to the great influence this work had in establishing trigonometry as a subject independent of astronomy.

Georg Rheticus (1514–1576) in 1550 was the first to define the trigonometric functions using ratios of the sides of right triangles instead of the arc lengths or chord lengths of circles. From this point on, the functions began to be thought of in terms of pure numbers rather than lengths.

Rheticus spent twelve years with hired computers compiling ten-place trigonometric tables of all six functions and a fifteen-place sine table. These tables were perfected and edited in 1593 by Bartholomaus Pitiscus (1561–1613) whose important work was also the first on the subject to bear the word trigonometry in its title.

In 1583 Thomas Fincke contributed the name *secant* for its obvious meaning "cutting" and the name *tangent* probably because the vertical shadow length lies on the tangent line of the circle.

In 1620 Edmund Gunter introduced the term *cosine* for the sine of the complement of an angle by combining "complement" and "sine" to form *cosine*. He also introduced the term *cotangent* for similar reasons. The term *cosecant* appears first in the posthumous work of Rheticus, published in 1596.

The transition to analytic trigonometry was stimulated by Francois Viète (1540–1603) who applied algebra and symbolism to trigonometry and by Gilles Persone de Roberval (1602–1675) who exhibited in 1635 the first graph of half an arch of a sine curve.

Through the works of such men as John Wallis (1616–1703), Sir Isaac Newton (1642–1727), Johann Bernoulli (1667–1748), and Leonhard Euler (1707–1783), trigonometry became involved in new theoretical developments due to the invention of calculus. As examples, the infinite series $\sin x = x - \frac{x^3}{3} + \frac{x^5}{5!} - \frac{x^7}{7!} + \ldots$ was used as a better method for calculations and DeMoivre's theorem $(\cos \theta + i \sin \theta)^n = \cos n\theta + i \sin n\theta$ together with Euler's $e^{i\theta} = \cos \theta + i \sin \theta$ where $i = \sqrt{-1}$ were applied to problems in algebra and analysis.

The *radian* came into existence as the result of a meeting between the mathematician Thomas Muir and the physicist James T. Thomson. Desiring to adopt a new angular unit for the purpose of simplifying certain mathematical and physical formulas, they agreed upon "radian," probably a compound of "radial angle." The word "radian" first appeared in print in 1873 on an examination written by Thomson.

6.3 NATURAL TRIGONOMETRIC FUNCTIONS

6.3.1 Definition of Sine and Cosine

The natural trigonometric functions are also called the **circular functions** because of their relationship to the unit circle.

Definition of Unit Circle A unit circle is a circle whose radius is one unit in length.

With each real number may be associated a point on a unit circle by a function called the "wrapping" function, $\{(s, P(s))\}$. Letting the positive real number s designate the length of an arc and selecting a fixed point A on a unit circle, then by starting at A and wrapping this length counterclockwise around the circle, the terminal point of the arc determines a point $P(s)$ on the circle. For $s = 0$, $P(s)$ is selected as A. For $s < 0$, $P(s)$ is the point determined by wrapping the arc length $|s|$ clockwise around the circle.

Since the circumference of a circle $C = 2\pi r$, a unit circle has length $C = 2\pi$, approximately 6.2832. It follows that $P(s + 2\pi) = P(s)$, or the same point on the circle is determined by any two real numbers whose difference is 2π or any integral multiple of 2π. See Figure 6.5.

Choosing the origin of a rectangular coordinate system as the center of the unit circle, the equation of the unit circle is $x^2 + y^2 = 1$, an immediate consequence derived from the definition of a circle and the theorem of Pythagoras.

Definition of Sine

$$\text{sine} = \{(s, \sin s) | P(s) = (x, \sin s) \text{ and } x^2 + (\sin s)^2 = 1\}$$

Definition of Cosine

$$\text{cosine} = \{(s, \cos s) | P(s) = (\cos s, y) \text{ and } (\cos s)^2 + y^2 = 1\}$$

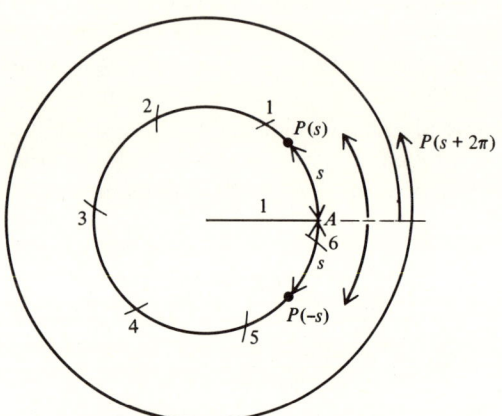

Figure 6.5 The Wrapping Function

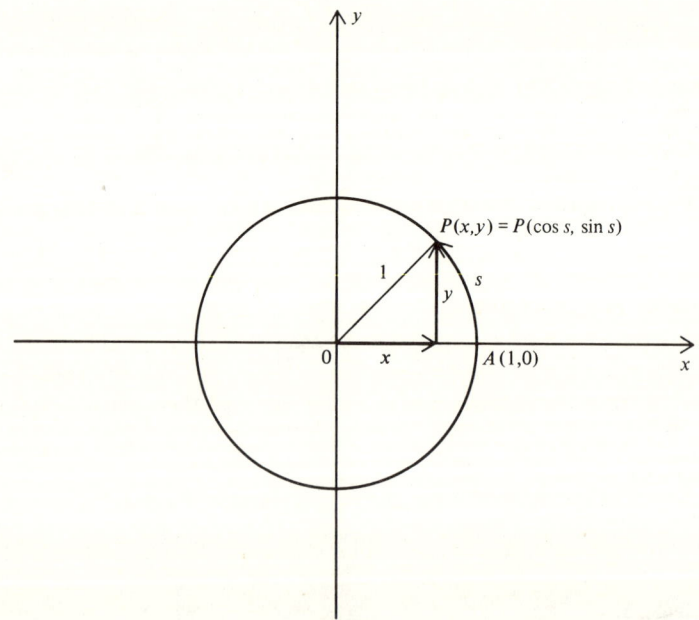

Figure 6.6 The Sine and Cosine Functions

NATURAL TRIGONOMETRIC FUNCTIONS

In other words, cos s and sin s are the x and y coordinates, respectively, of a point, $P(s)$, on the unit circle determined by the wrapping function with $A(1, 0)$ selected as the fixed reference point. The domain of both sine and cosine functions is the set of all real numbers. The range of both functions = $\{y \mid -1 \leq y \leq 1\}$.

6.3.2 Elementary Theorems

The two theorems stated below follow immediately from the definitions.

Theorem (First Pythagorean Identity) For all real numbers s,

$$(\sin s)^2 + (\cos s)^2 = 1.$$

Theorem (Functions of $s + 2k\pi$) For k any integer and s any real number,

$$\sin(s + 2k\pi) = \sin s$$

$$\cos(s + 2k\pi) = \cos s.$$

Theorem (Negative Values) If s is any real number, then

$$\sin(-s) = -\sin s \quad \text{and} \quad \cos(-s) = \cos s.$$

Since

$$P(s) = (x, y) = (\cos s, \sin s),$$

then

$$P(-s) = (x, -y) = (\cos(-s), \sin(-s)) = (\cos s, -\sin s).$$

Theorem (Related Values) If s is any real number, then

$$\sin(\pi - s) = \sin s, \qquad \cos(\pi - s) = -\cos s,$$

$$\sin(\pi + s) = -\sin s, \qquad \cos(\pi + s) = -\cos s,$$

$$\sin(2\pi - s) = -\sin s, \qquad \cos(2\pi - s) = \cos s.$$

From the geometry of the circle (see Figure 6.7), it follows that if

$$(x, y) = (\cos s, \sin s) \text{ are the coordinates of } P(s),$$

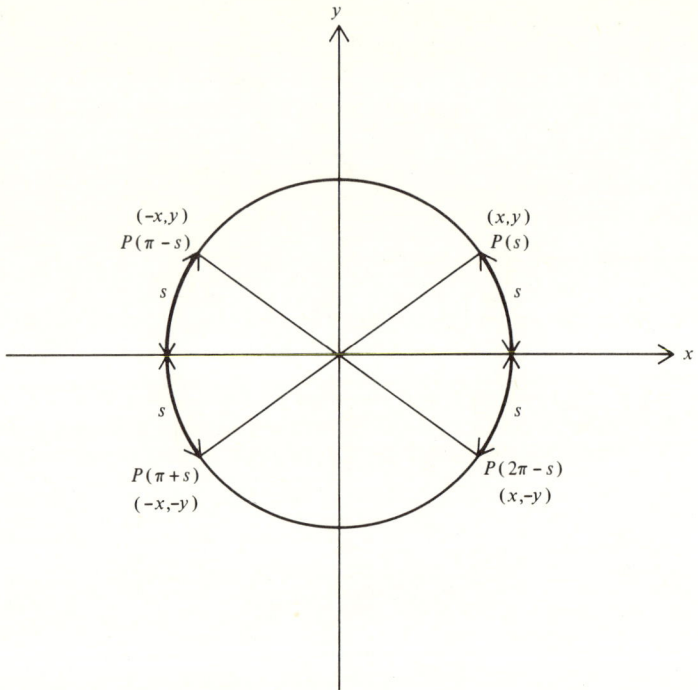

Figure 6.7 Related Quadrantal Values

then

$$(-x, y) = (-\cos s, \sin s) \text{ are the coordinates of } P(\pi - s),$$
$$(-x, -y) = (-\cos s, -\sin s) \text{ are the coordinates of } P(\pi + s),$$

and

$$(x, -y) = (\cos s, -\sin s) \text{ are the coordinates of } P(2\pi - s).$$

Theorem (Complementary Values) If s is any real number, then

$$\sin\left(\frac{\pi}{2} - s\right) = \cos s$$

and

$$\cos\left(\frac{\pi}{2} - s\right) = \sin s.$$

Since in a circle equal arcs subtend equal central angles, triangles $OQP(\pi/2 - s)$ and $ORP(s)$ in Figure 6.8 are congruent (two right triangles with

NATURAL TRIGONOMETRIC FUNCTIONS

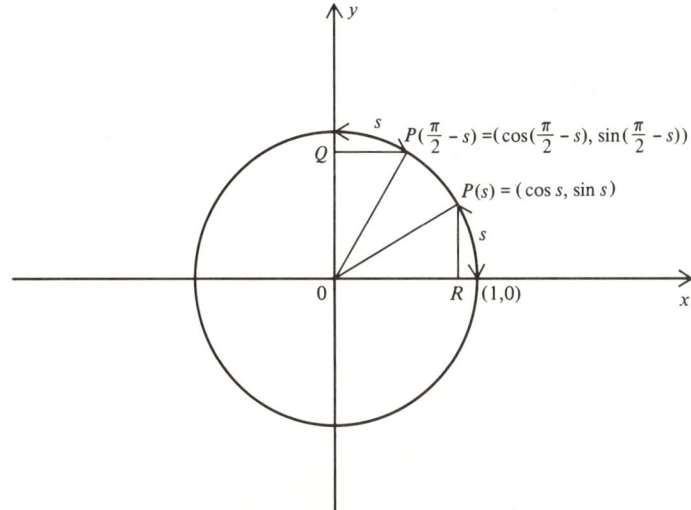

Figure 6.8 Complementary Values

hypotenuse, acute angle equal, respectively, to hypotenuse, acute angle). Therefore their corresponding parts are equal, or

$$\overrightarrow{OQ} = \overrightarrow{OR} \quad \text{and} \quad \overrightarrow{QP\left(\frac{\pi}{2} - s\right)} = \overrightarrow{RP(s)}$$

and

$$\sin\left(\frac{\pi}{2} - s\right) = \cos s \quad \text{and} \quad \cos\left(\frac{\pi}{2} - s\right) = \sin s.$$

6.3.3 Quadrantal Values

The x- and y-axes divide the circle into four parts called quadrants that are numbered counterclockwise as illustrated in Figure 6.9. Listed below are the sets of real numbers corresponding to the four subdivision points on the circle, $A(1, 0)$, $B(0, 1)$, $C(-1, 0)$, and $D(0, -1)$.

$$A(1, 0): \{\ldots, -4\pi, -2\pi, 0, 2\pi, 4\pi, \ldots\} = \{2k\pi \mid k \in I\},$$

$$B(0, 1): \left\{\ldots, \frac{-7\pi}{2}, \frac{-3\pi}{2}, \frac{\pi}{2}, \frac{5\pi}{2}, \frac{9\pi}{2}, \ldots\right\} = \left\{\frac{(4k + 1)\pi}{2} \mid k \in I\right\},$$

$$C(-1, 0): \{\ldots, -3\pi, -\pi, \pi, 3\pi, 5\pi, \ldots\} = \{(2k - 1)\pi \mid k \in I\},$$

$$D(0, -1), \left\{\ldots, \frac{-5\pi}{2}, \frac{-\pi}{2}, \frac{3\pi}{2}, \frac{7\pi}{2}, \frac{11\pi}{2}, \ldots\right\} = \left\{\frac{(4k + 3)\pi}{2} \mid k \in I\right\}.$$

EXPONENTIAL, LOGARITHMIC, AND TRIGONOMETRIC FUNCTIONS

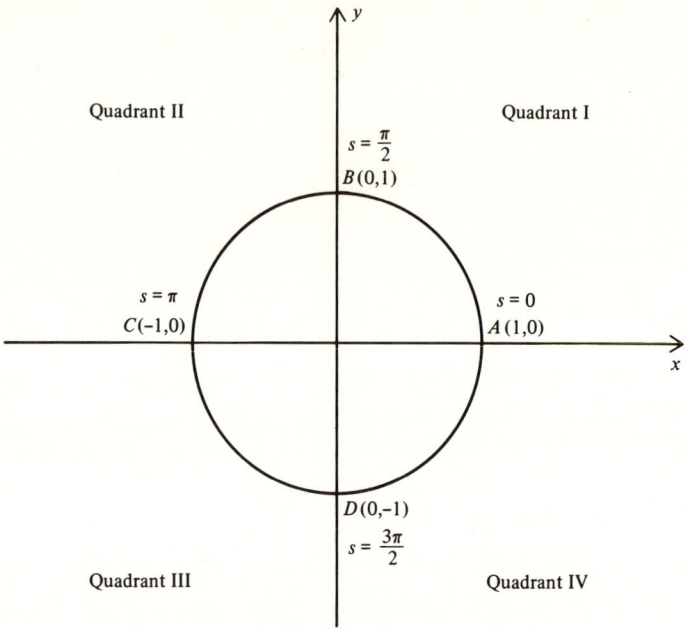

Figure 6.9 Quadrantal Values

The quadrantal values summarized in the table below follow directly from the fact that $\sin s = y$ and $\cos s = x$ for $P(s) = (x, y)$.

s	$P(s)$	$\sin s$	$\cos s$
0	$(1, 0)$	0	1
$\dfrac{\pi}{2}$	$(0, 1)$	1	0
π	$(-1, 0)$	0	-1
$\dfrac{3\pi}{2}$	$(0, -1)$	-1	0
2π	$(1, 0)$	0	1

6.3.4 $\pi/4 + k\pi$ Values

The $y = x$ line divides the arc of the circle in quadrant I into two equal arcs, each having length $\pi/4$. (See Figure 6.10.) The point of subdivision

$$P\left(\frac{\pi}{4}\right) = \left(\cos\frac{\pi}{4}, \sin\frac{\pi}{4}\right), \text{ and } \cos\frac{\pi}{4} = \sin\frac{\pi}{4} \text{ since } y = x.$$

NATURAL TRIGONOMETRIC FUNCTIONS

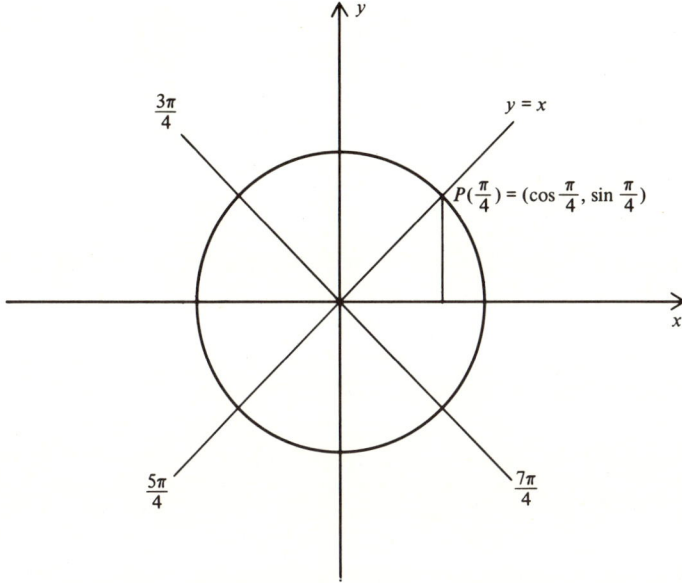

Figure 6.10 $\frac{\pi}{4} + k\pi$ Values

Therefore,

$$\left(\sin\frac{\pi}{4}\right)^2 + \left(\sin\frac{\pi}{4}\right)^2 = 1$$

$$2\left(\sin\frac{\pi}{4}\right)^2 = 1$$

$$\cos\frac{\pi}{4} = \sin\frac{\pi}{4} = \frac{1}{\sqrt{2}} \quad \text{since} \quad \sin\frac{\pi}{4} > 0.$$

The values in the following table are obtained by applying the theorem on related values.

s	$\sin s$	$\cos s$
$\frac{\pi}{4}$	$\frac{1}{\sqrt{2}}$	$\frac{1}{\sqrt{2}}$
$\frac{3\pi}{4}$	$\frac{1}{\sqrt{2}}$	$\frac{-1}{\sqrt{2}}$
$\frac{5\pi}{4}$	$\frac{-1}{\sqrt{2}}$	$\frac{-1}{\sqrt{2}}$
$\frac{7\pi}{4}$	$\frac{-1}{\sqrt{2}}$	$\frac{1}{\sqrt{2}}$

6.3.5 $\pi/6 + k\pi$ and $\pi/3 + k\pi$ Values

The well-known geometric construction for dividing a circle into six equal arcs (by using the radius of the given circle as the radii of circles for the subdivision) enables a regular hexagon and six equilateral triangles to be constructed within the circle. Using the fact that an altitude of an equilateral triangle bisects the corresponding base and placing an equilateral triangle as illustrated in Figure 6.11, values of sine and cosine for $s = \pi/6$ can be determined.

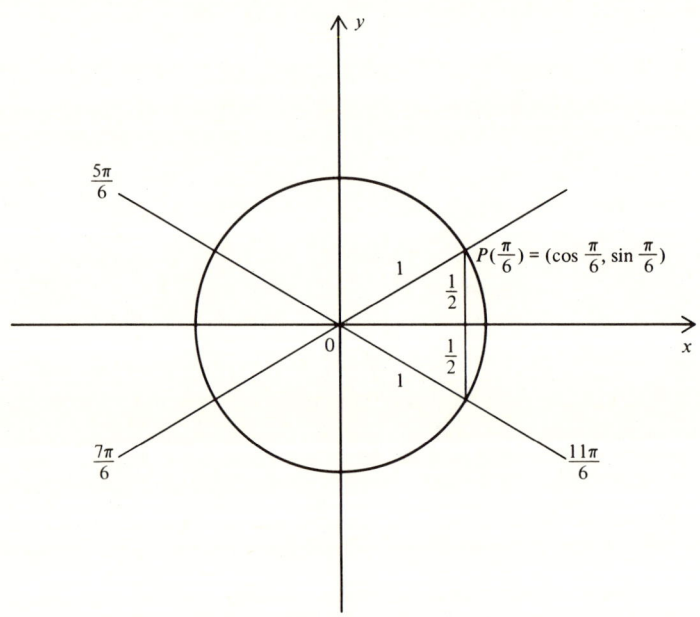

Figure 6.11 $\dfrac{\pi}{6} + k\pi$ Values

Since $(\sin \pi/6)^2 + (\cos \pi/6)^2 = 1$ and since $\sin \pi/6 = \frac{1}{2}$,

$$\left(\frac{1}{2}\right)^2 + \left(\cos \frac{\pi}{6}\right)^2 = 1$$

$$\left(\cos \frac{\pi}{6}\right)^2 = \frac{3}{4}$$

$$\cos \frac{\pi}{6} = \frac{\sqrt{3}}{2} \quad \text{since} \quad \cos \frac{\pi}{6} > 0.$$

NATURAL TRIGONOMETRIC FUNCTIONS

Placing the equilateral triangle as illustrated in Figure 6.12, it is possible to determine the values of sine and cosine for $s = \pi/3$.

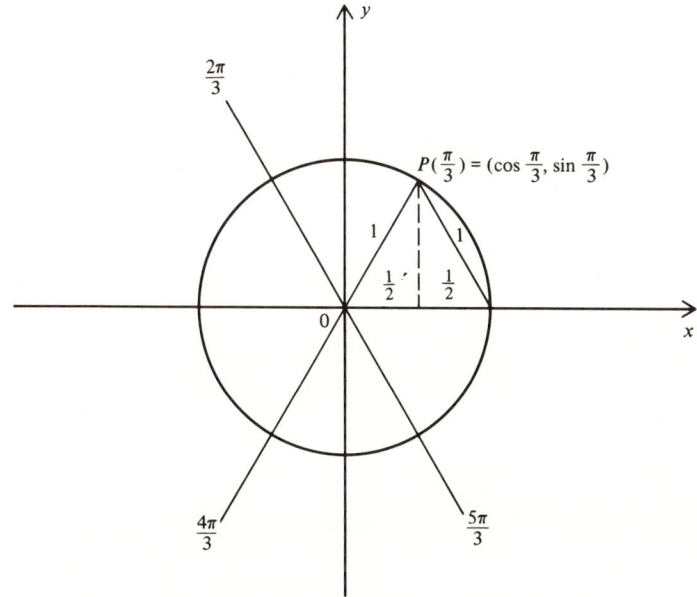

Figure 6.12 $\dfrac{\pi}{3} + k\pi$ Values

Since $(\sin \pi/3)^2 + (\cos \pi/3)^2 = 1$ and since $\cos \pi/3 = \frac{1}{2}$, it follows that $\sin \pi/3 = \sqrt{3}/2$.

The tables following are obtained by using the theorem on related values.

s	sin s	cos s
$\dfrac{\pi}{6}$	$\dfrac{1}{2}$	$\dfrac{\sqrt{3}}{2}$
$\dfrac{5\pi}{6}$	$\dfrac{1}{2}$	$\dfrac{-\sqrt{3}}{2}$
$\dfrac{7\pi}{6}$	$\dfrac{-1}{2}$	$\dfrac{-\sqrt{3}}{2}$
$\dfrac{11\pi}{6}$	$\dfrac{-1}{2}$	$\dfrac{\sqrt{3}}{2}$

s	sin s	cos s
$\dfrac{\pi}{3}$	$\dfrac{\sqrt{3}}{2}$	$\dfrac{1}{2}$
$\dfrac{2\pi}{3}$	$\dfrac{\sqrt{3}}{2}$	$\dfrac{-1}{2}$
$\dfrac{4\pi}{3}$	$\dfrac{-\sqrt{3}}{2}$	$\dfrac{-1}{2}$
$\dfrac{5\pi}{3}$	$\dfrac{-\sqrt{3}}{2}$	$\dfrac{1}{2}$

6.3.6 Tables and Graphs of sin s and cos s

The increasing and decreasing properties of the sine and cosine functions follow directly from their definitions and are summarized in the following table.

QUADRANT	I	II	III	IV
s	$0 < s < \dfrac{\pi}{2}$	$\dfrac{\pi}{2} < s < \pi$	$\pi < s < \dfrac{3\pi}{2}$	$\dfrac{3\pi}{2} < s < 2\pi$
sin s	Increases from 0 to 1	Decreases from 1 to 0	Decreases from 0 to -1	Increases from -1 to 0
cos s	Decreases from 1 to 0	Decreases from 0 to -1	Increases from -1 to 0	Increases from 0 to 1

The values of sin s and cos s for s not in the interval $0 \leq s \leq 2\pi$ are obtained by using the property that sin s and cos s repeat their values every 2π interval. Then, by using the theorem on related values, the values of sin s and cos s can be derived from the values for $0 \leq s \leq \pi/2$. The special values obtained in the previous sections are summarized in the table that follows:

s	sin s	cos s
$0.0000 = 0$	0	1
$0.5236 \approx \dfrac{\pi}{6}$	$\dfrac{1}{2}$	$\dfrac{\sqrt{3}}{2}$
$0.7854 \approx \dfrac{\pi}{4}$	$\dfrac{\sqrt{2}}{2}$	$\dfrac{\sqrt{2}}{2}$
$1.0472 \approx \dfrac{\pi}{3}$	$\dfrac{\sqrt{3}}{2}$	$\dfrac{1}{2}$
$1.5708 \approx \dfrac{\pi}{2}$	1	0

Values of sin s and cos s for values of s between those listed in the table above may be read from a larger table, such as Table III in the Appendix, where s is listed under the column headed radians.

By letting $x = s$, the graphs of $y = \sin x$ and $y = \cos x$ may be obtained from the tabular values and the properties of these two functions. These graphs are shown in Figure 6.13.

NATURAL TRIGONOMETRIC FUNCTIONS

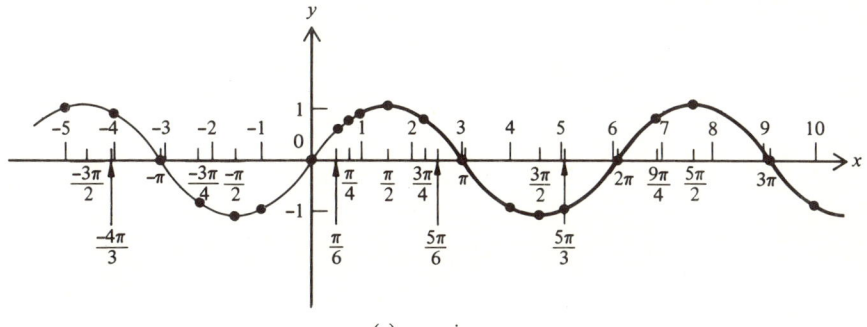

(a) $y = \sin x$

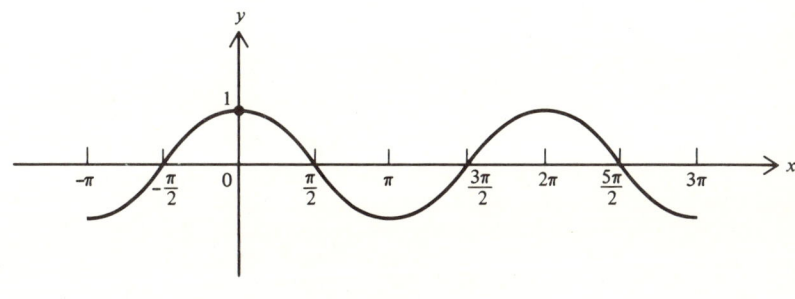

(b) $y = \cos x$

Figure 6.13

EXERCISES 6.3

1–10. For each of the following numbers, s, state the circular coordinates $(\cos s, \sin s)$, and the quadrant in which the corresponding point, $P(s)$, is located

1. $7\pi/6$
2. $\pi/3$
3. $3\pi/4$
4. $5\pi/3$
5. $-5\pi/3$
6. $-3\pi/2$
7. $-8\pi/3$
8. $19\pi/2$
9. $-46\pi/4$
10. 16π

11–15. If s is a real number, determine the quadrant or quadrants where $P(s)$ is located for each of the following:

11. $\cos s > 0$
12. $\sin s < 0$
13. $\sin s > 0$ and $\cos s < 0$
14. $\sin s > 0$ and $\cos s > 0$
15. $\sin s < 0$ and $\cos s < 0$

16–20. If $0 \le s \le 2\pi$, for what values of s is each of the following true?

16. $\sin s$ is an increasing function.
17. $\sin s$ is a decreasing function.
18. $\cos s$ is an increasing function.
19. $\cos s$ is a decreasing function.
20. $(\sin s + \cos s)$ is an increasing function.

21–30. Using the approximation $\pi \approx 3.14$, determine the quadrant in which each of the following points is located and state the algebraic sign of the corresponding values for $\sin s$ and $\cos s$.

21. $P(1)$
22. $P(4)$
23. $P(-13/2)$
24. $P(3.5)$
25. $P(0.4)$
26. $P\left(-\dfrac{6\pi}{5}\right)$
27. $P(10)$
28. $P(-10)$
29. $P\left(-\dfrac{12\pi}{5}\right)$
30. $P(2 + 2\pi)$

31–50. Let $P(s)$ be $(\tfrac{3}{5}, y)$ where $y > 0$. Evaluate each of the following:

31. $\sin s$
32. $\cos s$
33. $\sin(-s)$
34. $\cos(-s)$
35. $\sin(\pi - s)$
36. $\cos(\pi - s)$
37. $\sin(\pi + s)$
38. $\cos(\pi + s)$
39. $\sin(2\pi - s)$
40. $\cos(2\pi - s)$
41. $\sin(2\pi + s)$
42. $\cos(2\pi + s)$
43. $\sin\left(\dfrac{\pi}{2} - s\right)$
44. $\cos\left(\dfrac{\pi}{2} - s\right)$
45. $\sin\left(s - \dfrac{\pi}{2}\right)$
46. $\cos\left(s - \dfrac{\pi}{2}\right)$
47. $\sin(s - \pi)$
48. $\cos(s - \pi)$
49. $\sin(s - 2\pi)$
50. $\cos(s - 2\pi)$

51–72. Let t be a nonnegative real number. For each of the following, find s so that $0 \le s < 2\pi$:

51. $P(s) = (t, 0.5)$
52. $P(s) = (-0.5, t)$
53. $P(s) = (\sqrt{3}/2, t)$
54. $P(s) = (t, \sqrt{2}/2)$
55. $P(s) = (-1, t)$
56. $P(s) = (-t, 0)$
57. $P(s) = (-1/2, -t)$
58. $P(s) = (t, -\sqrt{3}/2)$
59. $P(s) = (-\sqrt{2}/2, t)$
60. $P(s) = (-\sqrt{2}/2, -t)$

NATURAL TRIGONOMETRIC FUNCTIONS

61. $P(s) = (t, -1/2)$
62. $P(s) = (-\sqrt{3}/2, -t)$
63. $P(s) = (-t, \sqrt{3}/2)$
64. $P(s) = (t, -1)$
65. $P(s) = (0, -t)$
66. $P(s) = (t, t)$
67. $P(s) = (t, -t)$
68. $P(s) = (-t, t)$
69. $P(s) = (-t, -t)$
70. $P(s) = (t, 1-t)$
71. $P(s) = (1+t, t)$
72. $P(s) = (1-t, t)$

6.4 NATURAL TRIGONOMETRIC FUNCTIONS (Continued)

6.4.1 Addition, Double-Value, Half-Value Theorems

The theorems in this section are derived by using the distance formula stated below.

Distance Formula If $P_1(x_1, y_1)$ and $P_2(x_2, y_2)$ are two points on a rectangular coordinate plane, then the distance between P_1 and P_2 is denoted by $|P_1P_2|$ and

$$|P_1P_2| = \sqrt{(x_2 - x_1)^2 + (y_2 - y_1)^2}.$$

(See Figure 6.14.)

Since equal arcs subtend equal chords in a circle, in Figure 6.15 chord $AP(s-t)$ equals chord $P(t)P(s)$.

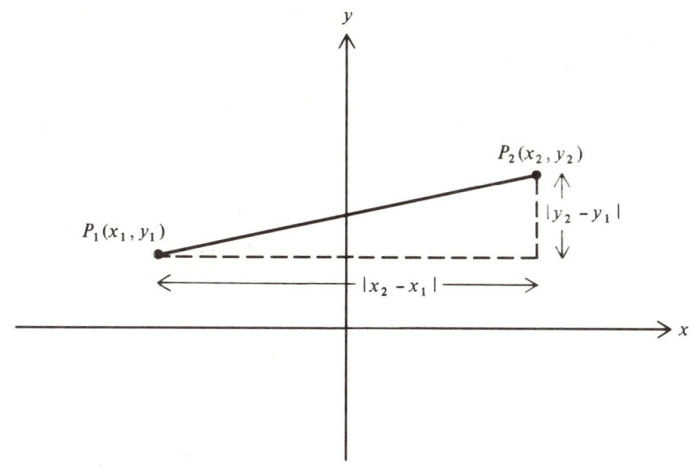

Figure 6.14 Distance between P_1 and $P_2 = \sqrt{(x_2 - x_1)^2 + (y_2 - y_1)^2}$

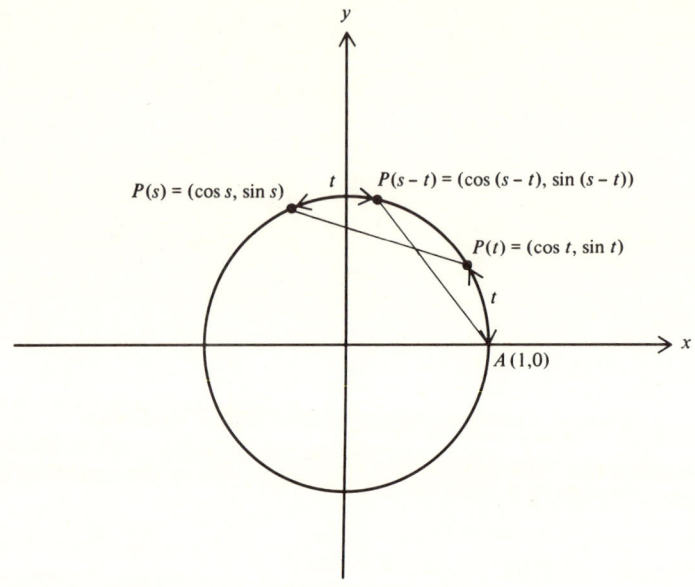

Figure 6.15

Using the distance formula,

$$(\cos(s-t) - 1)^2 + (\sin(s-t))^2 = (\cos s - \cos t)^2 + (\sin s - \sin t)^2$$
$$\cos^2(s-t) + \sin^2(s-t) - 2\cos(s-t) + 1 = \cos^2 s + \sin^2 s + \cos^2 t + \sin^2 t$$
$$- 2\cos s \cos t - 2\sin s \sin t$$
$$2 - 2\cos(s-t) = 2 - 2\cos s \cos t - 2\sin s \sin t$$
$$\cos(s-t) = \cos s \cos t + \sin s \sin t.$$

Now by replacing s and t by the values indicated at the right of each of the following equations, these equations are derived as theorems.

Addition Theorems

$\cos(a-b) = \cos a \cos b + \sin a \sin b$	$s = a, t = b$
$\cos(a+b) = \cos a \cos b - \sin a \sin b$	$s = a, t = -b$, and negative value theorems
$\sin(a+b) = \sin a \cos b + \cos a \sin b$	$s = \dfrac{\pi}{2} - a, t = b$, and complementary theorems
$\sin(a-b) = \sin a \cos b - \cos a \sin b$	$s = \dfrac{\pi}{2} - a, t = -b$, and complementary and negative value theorems

NATURAL TRIGONOMETRIC FUNCTIONS

Double-Value Theorems

$\cos 2a = \cos^2 a - \sin^2 a$ $\qquad s = a, t = -a$

$\sin 2a = 2 \sin a \cos a$ $\qquad s = \dfrac{\pi}{2} - a, t = a$

Half-Value Theorems

$$\sin^2 a = \frac{1 - \cos 2a}{2}$$

$$\cos^2 a = \frac{1 + \cos 2a}{2}$$

The last two theorems are derived from $\cos 2a = \cos^2 a - \sin^2 a$. Replacing $\cos^2 a$ by $1 - \sin^2 a$ since $\sin^2 a + \cos^2 a = 1$ for all a,

$$\cos 2a = 1 - 2 \sin^2 a \text{ and thus } \sin^2 a = \frac{1 - \cos 2a}{2}.$$

Replacing $\sin^2 a$ by $1 - \cos^2 a$ in $\cos 2a = \cos^2 a - \sin^2 a$,

$$\cos 2a = 2 \cos^2 a - 1 \text{ and thus } \cos^2 a = \frac{1 + \cos 2a}{2}.$$

6.4.2 Other Trigonometric Functions

Definition of Tangent

$$\text{tangent} = \left\{ (s, \tan s) \,\Big|\, \tan s = \frac{\sin s}{\cos s} \text{ and } \cos s \neq 0 \right\}$$

Definition of Cotangent

$$\text{cotangent} = \left\{ (s, \cot s) \,\Big|\, \cot s = \frac{\cos s}{\sin s} \text{ and } \sin s \neq 0 \right\}$$

Definition of Secant

$$\text{secant} = \left\{ (s, \sec s) \,\Big|\, \sec s = \frac{1}{\cos s} \text{ and } \cos s \neq 0 \right\}$$

Definition of Cosecant

$$\text{cosecant} = \left\{ (s, \csc s) \,\Big|\, \csc s = \frac{1}{\sin s} \text{ and } \sin s \neq 0 \right\}$$

The following theorems are readily derived from the definitions and the theorems for the sine and cosine functions:

(1) $\tan^2 s + 1 = \sec^2 s$

(2) $\cot^2 s + 1 = \csc^2 s$

(3) $\tan(s + t) = \dfrac{\tan s + \tan t}{1 - \tan s \tan t}$

(4) $\tan 2s = \dfrac{2 \tan s}{1 - \tan^2 s}$

(5) $\tan \dfrac{s}{2} = \dfrac{1 - \cos s}{\sin s} = \dfrac{\sin s}{1 + \cos s}$

(6) $\tan(-s) = -\tan s$

(7) $\tan\left(\dfrac{\pi}{2} - s\right) = \cot s$

(8) $\tan(\pi - s) = -\tan s$

(9) $\tan(\pi + s) = \tan s$

(10) $\tan(2\pi - s) = -\tan s$

(11) $\tan(s + 2k\pi) = \tan s$

6.4.3 Trigonometric Functions of Angles

Definition of Angle An angle is an ordered pair of rays that have a common endpoint.

The first member of an ordered pair of rays forming an angle is called the **initial side** of the angle and the second member is called the **terminal side**. The common endpoint is called the **vertex** of the angle. (See Figure 6.16.)

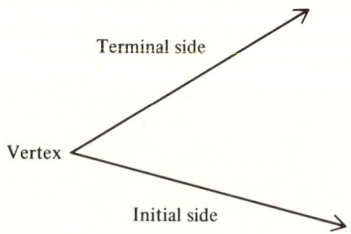

Figure 6.16 Angle

MEASURE OF AN ANGLE

The measure of an angle is determined by placing the angle in standard position on a rectangular coordinate system; that is, the vertex is placed at the origin and the initial side is placed along the positive x-axis. The

NATURAL TRIGONOMETRIC FUNCTIONS

terminal side then determines a unique point $P(s)$ on the unit circle where $0 \leq s \leq 2\pi$. Assigning this arc length s as the measure of the angle, the angle is said to be measured in **radians**. Angles may be assigned a negative measure or a measure of more than 2π by considering the arc length s as measured in the clockwise direction or by considering the arc length as greater than one complete rotation.

The **degree**, °, is another commonly used unit of measurement for angles. The circumference of the circle is considered as divided into 360° so that 2π radians = 360 degrees, and

$$x \text{ radians} = \frac{360}{2\pi} x \text{ degrees} = \frac{180}{\pi} x \text{ degrees}$$

and

$$x \text{ degrees} = \frac{\pi}{180} x \text{ radians}.$$

If θ represents an angle in standard position and if $P(x, y)$ is any point on its terminal side with the distance $|OP| = r$, then it follows that $\cos \theta / 1 = x/r$ and $\sin \theta / 1 = y/r$ by using the fact that corresponding sides of similar triangles are proportional. See Figure 6.17.

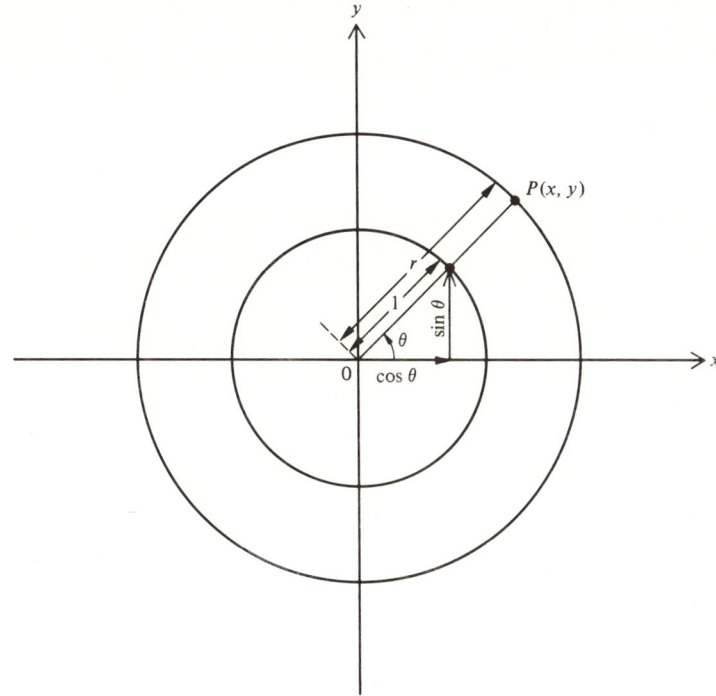

Figure 6.17 Angle in Standard Position

Definition of Trigonometric Functions of Angles Let $P(x, y)$ be a point on the terminal side of angle θ in standard position, with $|OP| = r$. Then

$$\sin \theta = \frac{y}{r} \qquad \csc \theta = \frac{r}{y} \qquad y \neq 0$$

$$\cos \theta = \frac{x}{r} \qquad \sec \theta = \frac{r}{x} \qquad x \neq 0$$

$$\tan \theta = \frac{y}{x} \quad x \neq 0 \qquad \cot \theta = \frac{x}{y} \qquad y \neq 0.$$

Definition α is the *reference angle* of an angle θ in standard position if and only if α is the acute angle that the terminal side of θ makes with the x-axis. (See Figure 6.18.)

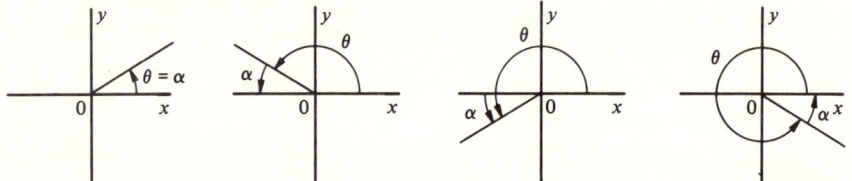

Figure 6.18 Reference Angle, α

Following directly from the theorem on related quadrantal values, $\sin \theta = \pm \sin \alpha$, $\cos \theta = \pm \cos \alpha$, and $\tan \theta = \pm \tan \alpha$. Thus the trigonometric functional values of an angle θ are readily derived from the functional values for its reference angle. The positive and negative qualities of these values are summarized in Figure 6.19.

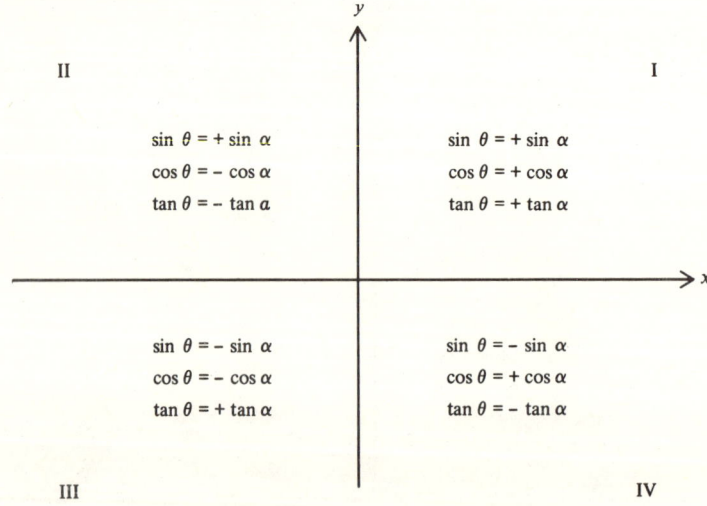

Figure 6.19 Quadrantal Signs

NATURAL TRIGONOMETRIC FUNCTIONS

It may easily be shown that the theorems for the trigonometric functions of real numbers are also valid for the trigonometric functions of angles.

EXERCISES 6.4

1–10. Using only the definitions, theorems, and the facts that $\sin \pi/6 = 1/2$ and $\sin \pi/4 = \sqrt{2}/2$, evaluate each of the following:

1. $\sin \dfrac{2\pi}{3}$
2. $\tan \dfrac{\pi}{6}$
3. $\sec \dfrac{\pi}{6}$
4. $\cot \dfrac{\pi}{6}$
5. $\tan \dfrac{\pi}{4}$
6. $\cos \dfrac{5\pi}{12}$
7. $\sin \dfrac{5\pi}{12}$
8. $\csc \dfrac{3\pi}{4}$
9. $\cot \dfrac{\pi}{12}$
10. $\sin \dfrac{\pi}{12}$

11–30. Without using tables, evaluate each of the following for

$$\tan \theta = 7/24 \text{ and } \sin \theta > 0.$$

11. $\sin \theta$
12. $\cos \theta$
13. $\cot \theta$
14. $\sec \theta$
15. $\csc \theta$
16. $\sin 2\theta$
17. $\cos 2\theta$
18. $\tan 2\theta$
19. $\sin \dfrac{\theta}{2}$
20. $\cos \dfrac{\theta}{2}$
21. $\tan \dfrac{\theta}{2}$
22. $\cot \dfrac{\theta}{2}$
23. $\sec \dfrac{\theta}{2}$
24. $\csc \dfrac{\theta}{2}$
25. $\tan\left(\dfrac{\pi}{4} + \theta\right)$
26. $\csc\left(\dfrac{\pi}{6} - \theta\right)$
27. $\sin\left(\dfrac{\pi}{3} + \theta\right)$
28. $\sec\left(\dfrac{\pi}{3} + \theta\right)$
29. $\cot\left(\dfrac{\pi}{4} - \theta\right)$
30. $\cos 3\theta$

31–60. Without using tables, evaluate each of the following:

31. $\sin^2 5 + \cos^2 5$
32. $\sec^2 2.3 - \tan^2 2.3$
33. $\tan^2 \dfrac{\pi}{7} - \sec^2 \dfrac{\pi}{7}$
34. $\sin^2 \dfrac{\pi}{5} + \sin^2\left(\dfrac{\pi}{2} - \dfrac{\pi}{5}\right)$
35. $\cos^2 \dfrac{\pi}{7} + \cos^2\left(\dfrac{\pi}{2} - \dfrac{\pi}{7}\right)$
36. $\cot^2 3 - \csc^2 3.$
37. $\sec^2 \dfrac{\pi}{9} - \cot^2\left(\dfrac{\pi}{2} - \dfrac{\pi}{9}\right)$
38. $\csc^2 \dfrac{\pi}{9} - \tan^2\left(\dfrac{\pi}{2} - \dfrac{\pi}{9}\right)$
39. $\sin \dfrac{2\pi}{5} - \cos \dfrac{\pi}{10}$
40. $\sin \dfrac{\pi}{7} + \cos \dfrac{9\pi}{14}$
41. $\sin \dfrac{3\pi}{7} \cos \dfrac{4\pi}{7} + \cos \dfrac{3\pi}{7} \sin \dfrac{4\pi}{7}$
42. $\cos \dfrac{7\pi}{24} \cos \dfrac{5\pi}{24} - \sin \dfrac{7\pi}{24} \sin \dfrac{5\pi}{24}$
43. $\cos \dfrac{7\pi}{12} \cos \dfrac{5\pi}{12} + \sin \dfrac{7\pi}{12} \sin \dfrac{5\pi}{12}$
44. $\cos \dfrac{7\pi}{12} \sin \dfrac{5\pi}{12} - \cos \dfrac{5\pi}{12} \sin \dfrac{7\pi}{12}$
45. $\cos^2 \dfrac{\pi}{8}$
46. $\sin^2 \dfrac{\pi}{12}$
47. $\sin^2 \dfrac{\pi}{8}$
48. $\cos^2 \dfrac{\pi}{12}$
49. $\dfrac{\tan(2\pi/9) + \tan(\pi/9)}{1 - \tan(2\pi/9)\tan(\pi/9)}$
50. $\dfrac{\tan(3\pi/8) - \tan(5\pi/24)}{1 + \tan(3\pi/8)\tan(5\pi/24)}$
51. $\dfrac{2\tan(\pi/8)}{1 - \tan^2(\pi/8)}$
52. $\dfrac{2\tan(\pi/12)}{1 - \tan^2(\pi/12)}$
53. $2 \sin \dfrac{\pi}{8} \cos \dfrac{\pi}{8}$
54. $2 \sin \dfrac{3\pi}{8} \cos \dfrac{3\pi}{8}$
55. $\cos^2 \dfrac{\pi}{12} - \sin^2 \dfrac{\pi}{12}$
56. $\sin^2 \dfrac{\pi}{12} - \cos^2 \dfrac{\pi}{12}$
57. $\dfrac{1 - \cos(\pi/3)}{\sin(\pi/3)}$
58. $\dfrac{\sin(\pi/2)}{1 + \cos(\pi/2)}$
59. $\tan \dfrac{5\pi}{4} - \sec \dfrac{5\pi}{6} + \cot \dfrac{7\pi}{4}$
60. $\cos \dfrac{\pi}{3} \sin \dfrac{11\pi}{6} \tan \dfrac{3\pi}{4}$

61–75. Verify each of the following identities:

61. $\tan^2 s - \sin^2 s = \sin^2 s \tan^2 s$
62. $\dfrac{2}{1 + \cos 2s} = \sec^2 s$
63. $\sin 2s = \dfrac{2 \tan s}{1 + \tan^2 s}$

64. $(1 - \cos 2s)/(1 - \cos^2 s) = 2$

65. $\dfrac{2}{\sin 2s} = \tan s + \cot s$

66. $\cos 2s = \dfrac{1 - \tan^2 s}{1 + \tan^2 s}$

67. $\dfrac{\sin 3s}{\sin s} - \dfrac{\cos 3s}{\cos s} = 2$

68. $\tan^2 \dfrac{s}{2} = \dfrac{1 - \cos s}{1 + \cos s}$

69. $\dfrac{2 \tan(s/2)}{1 + \tan^2(s/2)} = \sin s$

70. $\cos s + \cos 2s + \cos 3s = \dfrac{\cos 2s \sin(3s/2)}{\sin(s/2)}$

71. $\dfrac{\sin 2s}{1 - \cos 2s} = \cot s$

72. $\sin 3s = 4 \cos^2 s \sin s - \sin s$

73. $(\sin s + \cos s)^2 = 1 + \sin 2s$

74. $\dfrac{\sin s - \sin t}{\sin s + \sin t} = \tan\left(\dfrac{s-t}{2}\right) \cot\left(\dfrac{s+t}{2}\right)$

75. $\cos 3s = 4 \cos^3 s - 3 \cos s$

76–80. If the point $(-0.8, 0.6)$ is on the terminal side of an angle in standard position with measure θ, find each of the following:

76. $\sin \theta$
77. $\cos \theta$
78. $\tan \theta$
79. $\sec \theta$
80. $\csc \theta$

81–85. If $P(\theta)$ is on the terminal side of an angle in standard position with measure θ, find $\sin \theta$ and $\cos \theta$ for each of the following:

81. $P(\theta): (3, 4)$
82. $P(\theta): (-1, 7)$
83. $P(\theta): (-4, 3)$
84. $P(\theta): (6, 0)$
85. $P(\theta): (0, -2)$

6.5 INVERSE TRIGONOMETRIC FUNCTIONS

6.5.1 Definitions

The increasing and decreasing properties of the sine, cosine, and tangent functions are summarized in the table that follows.

EXPONENTIAL, LOGARITHMIC, AND TRIGONOMETRIC FUNCTIONS

s	$-\frac{\pi}{2}$ to 0	0 to $\frac{\pi}{2}$	$\frac{\pi}{2}$ to π	π to $\frac{3\pi}{2}$	$\frac{3\pi}{2}$ to 2π
$\sin s$	-1 to 0	0 to 1	1 to 0	0 to -1	-1 to 0
$\cos s$	0 to 1	1 to 0	0 to -1	-1 to 0	0 to 1
$\tan s$[1]	$-\infty$ to 0	0 to ∞	$-\infty$ to 0	0 to ∞	$-\infty$ to 0

Since for each real value of $\sin s$, $\cos s$, or $\tan s$, there are infinitely many real values for s, the sine, cosine, and tangent functions are not one-to-one. However, if the domain of each of these functions is restricted so that for each functional value, there will be exactly one value in the domain, then these new functions so defined will be one-to-one and therefore, each will have an inverse function. From the preceding table, it may be seen that $\sin s$ takes on all its values exactly once in increasing order for $-\pi/2 \leq s \leq \pi/2$, $\cos s$ in decreasing order for $0 \leq s \leq \pi$, and $\tan s$ in increasing order for $-\pi/2 < s < \pi/2$. Capital letters are used to distinguish the functions defined for these restricted domains from the natural trigonometric functions.

Definition of Sine $\text{Sine} = \left\{(s, \sin s) \,\middle|\, -\frac{\pi}{2} \leq s \leq \frac{\pi}{2}\right\}$

Definition of Cosine $\text{Cosine} = \{(s, \cos s) \mid 0 \leq s \leq \pi\}$

Definition of Tangent $\text{Tangent} = \left\{(s, \tan s) \,\middle|\, -\frac{\pi}{2} < s < \frac{\pi}{2}\right\}$

Since Sine, Cosine, and Tangent are one-to-one functions, their inverse functions exist and are defined as follows:

Definition of Arcsine[2]

$\text{Arcsine} = \left\{(x, \text{Arcsin } x) \,\middle|\, \sin(\text{Arcsin } x) = x \text{ and } -\frac{\pi}{2} \leq \text{Arcsin } x \leq \frac{\pi}{2}\right\}$

Definition of Arccosine[2]

$\text{Arccosine} = \{(x, \text{Arccos } x) \mid \cos(\text{Arccos } x) = x \text{ and } 0 \leq \text{Arccos } x \leq \pi\}$.

Definition of Arctangent[2]

$\text{Arctangent} = \left\{(x, \text{Arctan } x) \,\middle|\, \tan(\text{Arctan } x) = x \text{ and } -\frac{\pi}{2} < \text{Arctan } x < \frac{\pi}{2}\right\}$

---

[1] The ∞ symbol is used to indicate that the absolute values of $\tan s$ become larger and larger without bound. It should be noted that $\tan s$ is undefined for $s = \pi/2 \pm \pi k$.

[2] Another notation commonly used for Arcsin, Arccos, and Arctan is Sin^{-1}, Cos^{-1}, and Tan^{-1}, respectively.

INVERSE TRIGONOMETRIC FUNCTIONS

The following theorems are immediate consequences of the definitions and the general inverse function theorem:

Theorems on Inverse Trigonometric Functions

(1) $y = \text{Arcsin } x$ if and only if $x = \sin y$ and $-\frac{\pi}{2} \leq y \leq \frac{\pi}{2}$

(2) $y = \text{Arccos } x$ if and only if $x = \cos y$ and $0 \leq y \leq \pi$

(3) $y = \text{Arctan } x$ if and only if $x = \tan y$ and $-\frac{\pi}{2} < y < \frac{\pi}{2}$

(4) $\text{Arcsin}(\sin x) = x$ if and only if $-\frac{\pi}{2} \leq x \leq \frac{\pi}{2}$

(5) $\text{Arccos}(\cos x) = x$ if and only if $0 \leq x \leq \pi$

(6) $\text{Arctan}(\tan x) = x$ if and only if $-\frac{\pi}{2} < x < \frac{\pi}{2}$

It is also possible to define the inverse Arcsecant, Arccosecant functions, and Arccotangent by selecting $-\pi/2 \leq s < 0$ and $0 < s \leq \pi/2$ as the domain for Cosecant, by selecting $0 \leq s < \pi/2$ and $\pi/2 < s \leq \pi$ as the domain for Secant, and by selecting $-\pi/2 \leq s < 0$ and $0 < s \leq \pi/2$ as the domain for Cotangent.

The graphs of the Arcsine, Arccosine, and Arctangent functions are illustrated in Figure 6.20.

Example 1 Evaluate $\text{Arcsin } \frac{1}{2}$, $\text{Arccos}\left(-\frac{\sqrt{2}}{2}\right)$, and $\text{Arctan}(-1)$.

Solution

$x = \text{Arcsin } \frac{1}{2}$ if and only if $\sin x = \frac{1}{2}$ and $-\frac{\pi}{2} \leq x \leq \frac{\pi}{2}$.

Thus, $x = \frac{\pi}{6}$.

$x = \text{Arccos}\left(-\frac{\sqrt{2}}{2}\right)$ if and only if $\cos x = -\frac{\sqrt{2}}{2}$ and $0 \leq x \leq \pi$.

Thus, $x = \frac{3\pi}{4}$.

$x = \text{Arctan}(-1)$ if and only if $\tan x = -1$ and $-\frac{\pi}{2} < x < \frac{\pi}{2}$.

Thus, $x = -\frac{\pi}{4}$.

228   EXPONENTIAL, LOGARITHMIC, AND TRIGONOMETRIC FUNCTIONS

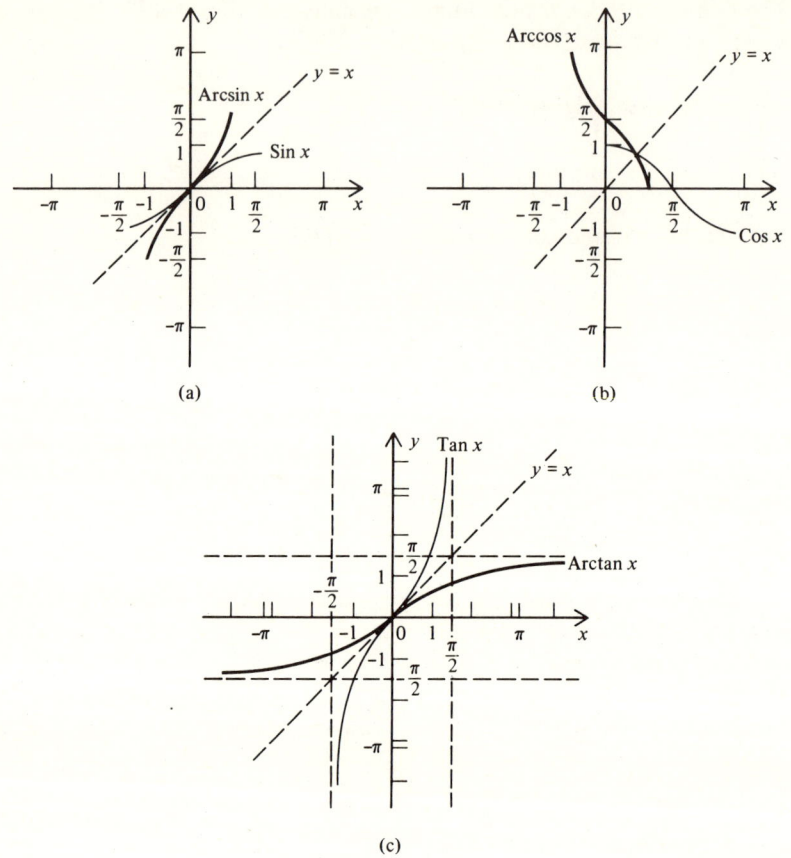

Figure 6.20

Example 2 If $\text{Arcsin}\,\dfrac{3}{5} = k$, find $\sin 2k$ and $\cos \dfrac{k}{2}$.

Solution $\sin k = \dfrac{3}{5}$ and $0 < k < \dfrac{\pi}{2}$

$$\cos k = \sqrt{1 - \left(\dfrac{3}{5}\right)^2} = \dfrac{4}{5}$$

$$\sin 2k = 2 \sin k \cos k = 2\left(\dfrac{3}{5}\right)\left(\dfrac{4}{5}\right) = \dfrac{24}{25} \text{ or } 0.96$$

$$\cos^2 \dfrac{k}{2} = \dfrac{1 + \cos k}{2} = \dfrac{1 + 4/5}{2} = \dfrac{9}{10} \text{ or } 0.9$$

INVERSE TRIGONOMETRIC FUNCTIONS

$$\cos \frac{k}{2} = \sqrt{\frac{9}{10}} = \sqrt{\frac{90}{100}} = \frac{3\sqrt{10}}{10}$$

Example 3 Is sin(Arcsin 3) = 3? Why?

Solution No. The domain of Arcsin x is $-1 \leq x \leq 1$. Since 3 is not in the domain, Arcsin 3 is undefined.

Example 4 Is $\text{Arcsin}\left(\sin \frac{5\pi}{6}\right) = \frac{5\pi}{6}$?

Solution No. $\sin \frac{5\pi}{6} = \frac{1}{2}$ and $\text{Arcsin} \frac{1}{2} = \frac{\pi}{6}$ and $\frac{\pi}{6} \neq \frac{5\pi}{6}$.

Example 5 If $0 \leq x < 2\pi$, solve $2 \sin^2 x + \cos x = 2$.

Solution $2 \sin^2 x + \cos x = 2(1 - \cos^2 x) + \cos x$

$$= 2 - 2\cos^2 x + \cos x$$

$$-2\cos^2 x + \cos x + 2 = 2$$

$$-2\cos^2 x + \cos x = 0$$

$$\cos x(-2\cos x + 1) = 0$$

$\cos x = 0$ or $\cos x = \frac{1}{2}$

$x = \frac{\pi}{2}$ or $\frac{3\pi}{2}$, or $x = \frac{\pi}{3}$ or $\frac{5\pi}{3}$

EXERCISES 6.5

1–10. Evaluate each of the following without using tables:

1. Arccos($\sqrt{3}/2$)
2. Arcsin($-1/2$)
3. Arcsin 0
4. Arctan(-1)
5. Arcsec 2
6. Arccot($-\sqrt{3}$)
7. Arcsin($-\sqrt{3}/2$)
8. Arctan 0
9. Arccos(-1)
10. Arccsc(-2)

11–20. Evaluate each of the following without using tables:

11. $\cos(\text{Arcsin } \sqrt{3}/2)$
12. $\text{Arcsin}(\sin -\pi/6)$
13. $\text{Arctan}(\tan x)$
14. $\text{Arcsin}(\cot 3\pi/4)$
15. $\tan(\text{Arccos } 5/13)$
16. $\text{Arccot}(\tan -\pi/3)$
17. $\text{Arctan}(\cot \pi/3)$
18. $\cos(\text{Arccos } x + \text{Arcsin } y)$
19. $\cos(\text{Arcsin } 1/4 + \text{Arccos } 1/4)$
20. $\tan(\text{Arcsin } \sqrt{2}/2)$

21–30. If $0 \le x \le 2\pi$, solve each of the following:

21. $\sin^2 x + \sin x - 2 = 0$
22. $\sin x + \tan x = 0$
23. $\sin 2x + \cos x = 0$
24. $\tan^2 x - \tan x < 0$
25. $\sin x < \cos x$
26. $\cos 2x + \cos 4x = 0$
27. $\sin x/2 + \cos x = 0$
28. $3 - 3 \sin x - 2 \cos^2 x > 0$
29. $\sin 3x + \sin x = 0$
30. $\sin^2 x < 1/4$

31. If $\text{Arcsin}(4/5) = k$, find $\sin 2k$ and $\cos(k/2)$.
32. If $\text{Arccos}(5/13) = k$, find $\cos 2k$ and $\sin(k/2)$.
33. If $\text{Arctan}(\sqrt{3}) = k$, find $\tan 2k$ and $\cos(k/2)$.
34. Prove: $\text{Arcsin } 1 = \text{Arcsin}(\sqrt{3}/2) + \text{Arcsin}(1/2)$.
35. Prove: $\text{Arctan } 1 = 2 \text{ Arctan}(1/3) + \text{Arctan}(1/7)$.

36–50. Solve for x.

36. $\text{Arccos } x = \text{Arctan } \dfrac{7}{24} - \text{Arcsin } \dfrac{3}{5}$
37. $\text{Arcsin } x = \text{Arctan } 2 + \text{Arccos } 1/\sqrt{10}$
38. $\text{Arcsin}[\cos(\pi - x)] = 2$
39. $\text{Arccos}[\sin(\pi - x)] = \frac{1}{2}$
40. $\text{Arccos } x = \text{Arcsin}(-\frac{2}{3})$
41. $\text{Arctan}\left(\dfrac{\tan x - 1}{\tan x + 1}\right) = \dfrac{\pi}{12}$
42. $\text{Arccos}(\cos x) = \sin(\text{Arcsin } x)$
43. $\text{Arcsin}(\cos x) = \cos(\text{Arcsin } \frac{4}{5})$
44. $\text{Arctan}(\sec x + \tan x) = \dfrac{\pi}{8}$ [*Hint:* First prove $\sec x + \tan x = \tan\left(\dfrac{\pi}{4} + \dfrac{x}{2}\right)$.]
45. $\text{Arcsin } \sqrt{\dfrac{1 - \cos x}{2}} = \dfrac{1}{2}$
46. $\text{Arccos } \sqrt{\dfrac{1 + \cos x}{2}} = -\dfrac{1}{2}$

47. $\text{Arccos}\left(\dfrac{1}{2}\cos x - \dfrac{\sqrt{3}}{2}\sin x\right) = \dfrac{7\pi}{12}$

48. $\text{Arcsin}\dfrac{\sqrt{2}}{2}(\cos x + \sin x) = 1$

49. $\text{Arctan } x + \text{Arccos}\dfrac{x}{\sqrt{1+x^2}} = \dfrac{\pi}{2}$

50. $\text{Arcsin } x - \text{Arctan}\dfrac{x}{\sqrt{1-x^2}} = 0$

51. Is $\sin(\text{Arcsin } 2) = 2$? Why?
52. Is $\text{Arcsin}(\sin 2) = 2$? Why?

CHAPTER SUMMARY

Definition of Exponential Function
$$\text{Exp} = \{(x, b^x) \mid x \in R, b \in R, \text{ and } 0 < b < 1 \text{ or } b > 1\}$$

Definition of Inverse Function, f^{-1} If $f = \{(x, y) \mid f \text{ is one-to-one}\}$, then
$$f^{-1} = \{(y, x) \mid (x, y) \in f\}.$$

Definition of $f^{-1}(x)$ $f^{-1}(x) = y$ if and only if $x = f(y)$

Inverse Function Theorem If f is a one-to-one function, $x \in \text{Dm}$ of f, and $y \in \text{Rn}$ of f, then
$$f^{-1}(f(x)) = x \quad \text{and} \quad f(f^{-1}(y)) = y.$$

Definition of Logarithm Function
$$\log = \{(x, y) \mid x = b^y \text{ where } x > 0 \text{ and } b > 1 \text{ or } 0 < b < 1\}$$

Definition of $\log_b x$ $\log_b x = y$ if and only if $x = b^y$

Inverse Logarithm Theorem $b^{\log_b y} = y$ and $\log_b b^x = x$

Logarithmic Operations Theorem

(1) $\log_b xy = \log_b x + \log_b y$

(2) $\log_b \dfrac{x}{y} = \log_b x - \log_b y$

(3) $\log_b x^y = y(\log_b x)$

Common Logarithm Theorem $\log_{10} x(10^k) = k + \log_{10} x$

Definition of Antilogarithm antilog $y = x$ if and only if $y = \log x$

Change of Logarithm Base Theorem $\log_b x = \dfrac{\log_a x}{\log_a b}$

$\log_{10} e = 0.4343, \qquad \log_e 10 = 2.3026, \qquad e = 2.71828\ldots.$

Definitions (Trigonometric Functions of Real Numbers) If $P(s) = (x, y)$ and $x^2 + y^2 = 1$, then

$$\sin s = y \qquad\qquad \csc s = \frac{1}{y}, \quad y \neq 0$$

$$\cos s = x \qquad\qquad \sec x = \frac{1}{x}, \quad x \neq 0$$

$$\tan s = \frac{y}{x}, \quad x \neq 0 \qquad \cot x = \frac{x}{y}, \quad y \neq 0.$$

Definitions (Trigonometric Functions of Angles) If $P(x, y)$ is any point on the terminal side of an angle θ in standard position and if $r^2 = x^2 + y^2$, then

$$\sin \theta = \frac{y}{r} \qquad\qquad \csc \theta = \frac{r}{y}, \quad y \neq 0$$

$$\cos \theta = \frac{x}{r} \qquad\qquad \sec \theta = \frac{r}{x}, \quad x \neq 0$$

$$\tan \theta = \frac{y}{x}, \quad x \neq 0 \qquad \cot \theta = \frac{x}{y}, \quad y \neq 0.$$

Negative Value Theorems

$$\sin(-s) = -\sin s, \qquad \cos(-s) = \cos s, \qquad \tan(-s) = -\tan s$$

Related Value Theorems

$$\sin\left(\frac{\pi}{2} - s\right) = \cos s \qquad \cos\left(\frac{\pi}{2} - s\right) = \sin s \qquad \tan\left(\frac{\pi}{2} - s\right) = \cot s$$

$$\sin(\pi - s) = \sin s \qquad \cos(\pi - s) = -\cos s \qquad \tan(\pi - s) = -\tan s$$

$$\sin(\pi + s) = -\sin s \qquad \cos(\pi + s) = -\cos s \qquad \tan(\pi + s) = \tan s$$

$$\sin(2\pi - s) = -\sin s \qquad \cos(2\pi - s) = \cos s \qquad \tan(2\pi - s) = -\tan x$$

Pythagorean Identity Theorems

$$\sin^2 s + \cos^2 s = 1 \qquad \text{and} \qquad \tan^2 s + 1 = \sec^2 s$$

CHAPTER SUMMARY

Addition Theorems

$$\sin(s + t) = \sin s \cos t + \cos s \sin t$$
$$\sin(s - t) = \sin s \cos t - \cos s \sin t$$
$$\cos(s + t) = \cos s \cos t - \sin s \sin t$$
$$\cos(s - t) = \cos s \cos t + \sin s \sin t$$
$$\tan(s + t) = \frac{\tan s + \tan t}{1 - \tan s \tan t}$$
$$\tan(s - t) = \frac{\tan s - \tan t}{1 + \tan s \tan t}$$

Double-Value Theorems

$$\sin 2s = 2 \sin s \cos s$$
$$\cos 2s = \cos^2 s - \sin^2 s$$
$$\tan 2s = \frac{2 \tan s}{1 - \tan^2 s}$$

Half-Value Theorems

$$\sin^2 \frac{s}{2} = \frac{1 - \cos s}{2}, \quad \cos^2 \frac{s}{2} = \frac{1 + \cos s}{2}, \quad \tan \frac{s}{2} = \frac{\sin s}{1 + \cos s} = \frac{1 - \cos s}{\sin s}$$

Special Values

s	$\sin s$	$\cos s$	$\tan s$
0	0	1	0
$\pi/6$	$1/2$	$\sqrt{3}/2$	$1/\sqrt{3}$
$\pi/4$	$1/\sqrt{2}$	$1/\sqrt{2}$	1
$\pi/3$	$\sqrt{3}/2$	$1/2$	$\sqrt{3}$
$\pi/2$	1	0	

INVERSE TRIGONOMETRIC FUNCTIONS

(1) $y = \text{Arcsin } x$ if and only if $x = \sin y$ and $-\frac{\pi}{2} \leq y \leq \frac{\pi}{2}$

(2) $y = \text{Arccos } x$ if and only if $x = \cos y$ and $0 \leq y \leq \pi$

(3) $y = \text{Arctan } x$ if and only if $x = \tan y$ and $-\frac{\pi}{2} < y < \frac{\pi}{2}$

REVIEW EXERCISES

1–6. Evaluate each of the following, if possible.

1. $\log_{10} 10$
2. $\log_{10} 1$
3. $\log_{10} 0$
4. $\log_{10}(-10)$
5. $\log_{10} 10^{-3}$
6. $10^{\log_{10} 2}$

7–12. Evaluate each of the following, given that $\log_b 2 = 0.4$.

7. $\log_b 8$
8. $\log_b \frac{1}{2}$
9. $\log_b \sqrt{2}$
10. $\log_b \sqrt[3]{0.25}$
11. $\log_b(\log_b \sqrt{b})$
12. $\dfrac{\log_b 2}{\log_b b^{\sqrt{3}}}$

13–20. Complete each of the following. Do not use tables.

13. If $N = 5^{-4.6}$, then $\log_5 N =$
14. If $\log_5 N = -\frac{1}{2}$, then $N =$
15. If $5^x = 3$, then $x =$
16. If $\log_{10} N = -1 + 0.38$, then $\log_{10} \sqrt[3]{N} =$
17. If $\log_{10} x = 0.4$ and $\log_{10} y = -2 + 0.7$, then $\log_{10} \dfrac{x}{y} =$
18. If $\log x = \log 15 + 3 \log 2 - \frac{1}{4} \log 81$, then $x =$
19. $(10^{\sqrt{5}+3} 10^{\sqrt{5}-3})^{-1/\sqrt{20}} =$
20. If $f(x) = 3^x$, then $f^{-1}(\frac{1}{9}) =$

21. (a) Write the logarithmic equation that would be used to compute
$$\sqrt[3]{\dfrac{(42.21)(23.50)}{5650}}.$$
(b) Using tables, compute the number designated in (a).

22–25. For each of the following numbers, s, state the circular coordinates $(\cos s, \sin s)$ and the quadrant in which the corresponding point, $P(s)$, is located.

22. $-3\pi/4$
23. $5\pi/6$
24. $13\pi/2$
25. $-5\pi/3$

REVIEW EXERCISES

26–30. Use the theorems for trigonometric functions and the fact that $\cos \pi/6 = \sqrt{3}/2$ to evaluate each of the following:

26. $\sin(\pi/3)$
27. $\tan(\pi/6)$
28. $\cos(\pi/12)$
29. $\sin(2\pi/3)$
30. $\tan(\pi/12)$

31–38. Verify each of the following identities:

31. $\tan(\alpha + \theta) = \dfrac{\tan \alpha + \tan \theta}{1 - \tan \alpha \tan \theta}$

32. $\tan 2\theta = \dfrac{2 \tan \theta}{1 - \tan^2 \theta}$

33. $\dfrac{2 \tan(\theta/2)}{1 + \tan^2(\theta/2)} = \sin \theta$

34. $\dfrac{\tan^2 \theta}{1 + \sec \theta} + 1 = \sec \theta$

35. $\dfrac{\sin^3 \theta - \cos^3 \theta}{1 + \sin \theta \cos \theta} = \sin \theta - \cos \theta$

36. $\dfrac{\sin 2\theta}{1 - \cos 2\theta} = \cot \theta$

37. $\tan \theta \tan 2\theta = \sec 2\theta - 1$

38. $\dfrac{2}{1 - \cos 2\theta} = \csc^2 \theta$

39–44. Solve each of the following for all values of θ such that $0 \leq \theta \leq 2\pi$:

39. $\sin \theta - \cos(\theta/2) = 0$
40. $2 \tan \theta - \sec \theta = 1$
41. $\cos 2\theta - \sin \theta = 0$
42. $2 \sin \theta - 2 \cos \theta = \sqrt{2}$
43. $\sin^2 \theta + \sin \theta - 2 = 0$
44. $\cos \theta + \sin^2 \theta = 3$

45–48. Evaluate each of the following without using tables:

45. $\text{Arcsin}(\cos \pi/4)$
46. $\tan(\text{Arcsin} \sqrt{3}/2)$
47. $\cos[1/2(\text{Arctan } 0)]$
48. $\cos[\text{Arcsin}(\sqrt{2}/2) + \text{Arccos}(4/5)]$

49. If the point $(2, -\sqrt{3})$ is on the terminal side of an angle in standard position with measure θ, find $\sin \theta$, $\cos \theta$, $\tan \theta$, $\sec \theta$, $\csc \theta$, and $\cot \theta$.

50. Prove: $\text{Arcsin}(-x) = -\text{Arcsin } x$.

7
THE COMPLEX NUMBERS

It was noted in Chapter 5 that the system of rational numbers had to be extended to the system of real numbers if the equation $x^2 = 2$ was to have a solution. Later it was seen that every positive real number had exactly one positive real square root. On the other hand, there are equations such as $x^2 = -1$ that do not have a solution in the set of real numbers, since the square of a real number cannot be negative.

An awareness of this difficulty was first stated by the Hindu Mahavira, about 850, who wrote "as, in the nature of things, a negative is not a square, it has therefore no square root." Later the Italian Luca Pacioli explained in his *Summa de Arithmetica* of 1494 that the quadratic equation $x^2 + c = bx$ could be solved only if $b^2 \geq 4c$. This meant, in particular, that the solution of an equation such as $x^2 = -a^2$, where a is a real number, was impossible in the set of real numbers.

The origin of the set of complex numbers is credited to the Italian Jerome Cardan. In his *Ars Magna* of 1545, Cardan used the square root of a negative number in a computation. It is interesting to note that Cardan was motivated by investigations concerned with the solution of the cubic equation and *not* the quadratic equation. For certain cases, Cardan's formula exhibited a real solution in terms of square roots of negative squares. Such a situation did not occur for the quadratic eqaution.

THE COMPLEX NUMBERS

Cardan's work is significant because it illustrates how a man, spurred by intellectual curiosity, proceeds by analogy and discovers new results even though he is unable to completely understand them. The famous problem of Cardan is to find two numbers whose sum is 10 and whose product is 40; that is, to solve $x(10-x) = 40$. Cardan says, "This is obviously impossible but nevertheless we will operate." His procedure may be transcribed as follows:

$$x^2 - 10x = -40$$
$$x^2 - 10x + 25 = -40 + 25$$
$$(x-5)^2 = -15$$
$$x = 5 + \sqrt{-15} \quad \text{or} \quad x = 5 - \sqrt{-15}.$$

CHECK $(5 + \sqrt{-15}) + (5 - \sqrt{-15}) = 10$,

$(5 + \sqrt{-15}) \cdot (5 - \sqrt{-15}) = 25 - (-15) = 40$.

At this time, a solution of an equation was interpreted as the length of a line segment. Since this geometric interpretation was not possible for a number such as $\sqrt{-15}$, Cardan wrote "you will *imagine*" this quantity. Finally, Cardan stated that these quantities are "truly sophisticated" and further work with these numbers would be "as subtle as it would be useless."

In 1572, the Italian Rafael Bombelli, also motivated by the solution of the cubic equation, stated the rules for operations with square roots of negatives. With these rules Bombelli could show, for example, that if Cardan's formula yielded

$$x = \sqrt[3]{52 + \sqrt{-2209}} + \sqrt[3]{52 - \sqrt{-2209}},$$

then

$$\sqrt[3]{52 + \sqrt{-2209}} = 4 + \sqrt{-1}$$

and

$$\sqrt[3]{52 - \sqrt{-2209}} = 4 - \sqrt{-1}$$

and thus

$$x = (4 + \sqrt{-1}) + (4 - \sqrt{-1}) = 8.$$

René Descartes, in his *La Géométrie* of 1637, introduced the terms "real" and "imaginary" to describe the kinds of solutions of an equation. The letter "i" to designate $\sqrt{-1}$ was first used in 1748 by the Swiss mathematician Leonhard Euler.

The German mathematician Carl Friedrich Gauss, in 1832, named a number having the form $a + b\sqrt{-1}$ a "complex number" to distinguish it from numbers of the form $a\sqrt{-1}$.

Finally, in 1837, Sir William Rowan Hamilton published his rigorous development of the set of complex numbers as ordered pairs of real numbers.

7.1 THE COMPLEX NUMBER SYSTEM

The system of complex numbers is the set of ordered pairs of real numbers with definitions for equality, addition, and multiplication.

The Set of Complex Numbers, C

$$C = \{(x, y) \mid x \in R \text{ and } y \in R\}$$

Definition of Equality

$$(x_1, y_1) = (x_2, y_2) \text{ if and only if } x_1 = x_2 \text{ and } y_1 = y_2$$

For example, $(\sqrt{18}, \sqrt[3]{8}) = (3\sqrt{2}, 2)$ since $\sqrt{18} = 3\sqrt{2}$ and $\sqrt[3]{8} = 2$. On the other hand, $(0, 1) \neq (1, 0)$ since $0 \neq 1$ and $1 \neq 0$.

Definition of Addition

$$(x_1, y_1) + (x_2, y_2) = (x_1 + x_2, y_1 + y_2)$$

For example,

$$(3, -2) + (4, 3) = (3 + 4, -2 + 3) = (7, 1),$$

and

$$(4, 3) + (-4, -3) = (0, 0).$$

Definition of Multiplication

$$(x_1, y_1) \cdot (x_2, y_2) = (x_1 x_2 - y_1 y_2, x_1 y_2 + x_2 y_1)$$

For example, $(3, -2) \cdot (4, 3) = (12 - (-6), 9 + (-8)) = (18, 1)$, and $(0, 1) \cdot (0, 1) = (-1, 0)$, and $(4, 3) \cdot (4/25, -3/25) = (1, 0)$.

The following theorems show that the field axioms are valid for the system of complex numbers, and thus the complex number system is a field.

Closure Theorem (Addition) There is exactly one complex number
$$(x_1, y_1) + (x_2, y_2).$$

Closure Theorem (Multiplication) There is exactly one complex number
$$(x_1, y_1) \cdot (x_2, y_2).$$

THE COMPLEX NUMBER SYSTEM

Commutative Theorem (Addition)

$$(x_1, y_1) + (x_2, y_2) = (x_2, y_2) + (x_1, y_1)$$

Commutative Theorem (Multiplication)

$$(x_1, y_1) \cdot (x_2, y_2) = (x_2, y_2) \cdot (x_1, y_1)$$

Associative Theorem (Addition)

$$[(x_1, y_1) + (x_2, y_2)] + (x_3, y_3) = (x_1, y_1) + [(x_2, y_2) + (x_3, y_3)]$$

Associative Theorem (Multiplication)

$$[(x_1, y_1) \cdot (x_2, y_2)] \cdot (x_3, y_3) = (x_1, y_1) \cdot [(x_2, y_2) \cdot (x_3, y_3)]$$

Distributive Theorem (Addition)

$$(x_1, y_1) \cdot [(x_2, y_2) + (x_3, y_3)] = (x_1, y_1) \cdot (x_2, y_2) + (x_1, y_1) \cdot (x_3, y_3)$$

Identity Theorem (Addition) There is exactly one complex number $(0, 0)$ so that for all (x, y) in C, $(x, y) + (0, 0) = (x, y)$.

Identity Theorem (Multiplication) There is exactly one complex number $(1, 0)$ so that for all (x, y) in C, $(x, y) \cdot (1, 0) = (x, y)$.

Inverse Theorem (Addition) For each (x, y) in C, there is exactly one element $(-x, -y)$ in C so that $(x, y) + (-x, -y) = (0, 0)$.

Inverse Theorem (Multiplication) For each (x, y) in C such that $(x, y) \neq (0, 0)$, there is exactly one element

$$\left(\frac{x}{x^2 + y^2}, \frac{-y}{x^2 + y^2} \right)$$

in C so that

$$(x, y) \cdot \left(\frac{x}{x^2 + y^2}, \frac{-y}{x^2 + y^2} \right) = (1, 0).$$

The closure, commutative, associative, and distributive theorems follow directly from the definitions for the complex numbers and the properties of the real numbers. Therefore, these proofs are left for the student.

Proof of the Identity Theorem (Addition)

$(x, y) + (0, 0) = (x + 0, y + 0)$	Definition of addition of complex numbers.
$= (x, y).$	Addition identity axiom, real numbers.

If another element (a, b) exists so that $(x, y) + (a, b) = (x, y)$, then $(x + a, y + b) = (x, y)$.	Definition of addition, complex numbers.
$x + a = x$ and $y + b = y$.	Definition of equality, complex numbers.

Thus, $a = 0$ and $b = 0$ since the addition identity axiom for real numbers states that the addition identity is unique.

Proof of the Identity Theorem (*Multiplication*) If there is a complex number (a, b) so that for all (x, y), $(x, y) \cdot (a, b) = (x, y)$, then

$(xa - yb, xb + ya) = (x, y)$	Definition of multiplication, complex numbers.
and	
$xa - yb = x$ and $xb + ya = y$.	Definition of equality, complex numbers.

Then,

$$(a - 1)x - yb = 0 \quad \text{and} \quad xb + (a - 1)y = 0$$
$$[(a - 1)^2 x - b(a - 1)y] + [b^2 x + b(a - 1)y] = 0$$
$$((a - 1)^2 + b^2)x = 0.$$

Since $x \neq 0$ for all x, thus $(a - 1)^2 + b^2 = 0$ by the zero-product theorem for real numbers.

Then since for real numbers x and y, $x^2 + y^2 = 0$ if and only if $x = 0$ and $y = 0$, it follows that $a - 1 = 0$ and $b = 0$ or $a = 1$ and $b = 0$. Therefore, $(1, 0)$ is the only multiplication identity.

Proof of the Inverse Theorem (*Addition*) If there is a complex number (a, b) so that $(x, y) + (a, b) = (0, 0)$, then $(x + a, y + b) = (0, 0)$ and thus $x + a = 0$ and $y + b = 0$.

Therefore, $a = -x$ and $b = -y$ since the addition inverse of a real number is unique.

Consequently, $(-x, -y)$ is the only addition inverse of (x, y).

Proof of the Inverse Theorem (*Multiplication*) If there is a complex number (a, b) so that

$$(x, y) \cdot (a, b) = (1, 0),$$

then

$$(xa - yb, xb + ya) = (1, 0)$$

THE COMPLEX NUMBER SYSTEM

and
$$xa - yb = 1 \quad \text{and} \quad xb + ya = 0$$
$$x(xa - yb) + y(xb + ya) = x(1) + y(0)$$

or
$$x^2 a + y^2 a = x \quad \text{and} \quad a = \frac{x}{x^2 + y^2} \quad \text{since } (x, y) \neq (0, 0).$$

Similarly, $-y(xa - yb) + x(xb + ya) = -y(1) + x(0)$ and
$$y^2 b + x^2 b = -y \quad \text{or} \quad b = \frac{-y}{x^2 + y^2} \quad \text{since } (x, y) \neq (0, 0).$$

Therefore,
$$\left(\frac{x}{x^2 + y^2}, \frac{-y}{x^2 + y^2} \right)$$
is the only multiplication inverse of (x, y) when $(x, y) \neq (0, 0)$.

Since every complex number has an addition inverse and since every complex number different from $(0, 0)$ has a multiplication inverse, subtraction and division may be defined in terms of these inverses.

Definition of Subtraction $\quad (x_1, y_1) - (x_2, y_2) = (x_1, y_1) + (-x_2, -y_2)$

Definition of Division $\quad \dfrac{(x_1, y_1)}{(x_2, y_2)} = (x_1, y_1) \cdot \left(\dfrac{x_2}{x_2^2 + y_2^2}, \dfrac{-y_2}{x_2^2 + y_2^2} \right)$

Since every complex number (x, y) can be expressed as
$$(x, y) = (x, 0) + (0, y)$$
and since
$$(0, y) = (y, 0) \cdot (0, 1),$$
then
$$(x, y) = (x, 0) + (y, 0)(0, 1).$$

It is now possible to identify the subset of C consisting of those numbers having the form $(x, 0)$ as the set of real numbers because all of the axioms for the system of real numbers are valid for this set. The field operations appear as follows:

$$(x, 0) + (y, 0) = (x + y, 0)$$
$$(x, 0) \cdot (y, 0) = (xy, 0)$$
$$(x, 0) - (y, 0) = (x - y, 0)$$
$$(x, 0)/(y, 0) = (x/y, 0) \quad \text{for } y \neq 0.$$

Defining $(x, 0) > (y, 0)$ if and only if $x > y$, it may be verified that the order axioms are valid. The completeness axiom may be verified by identifying the least upper bound of a set $\{(x, 0) \mid x \in S\}$ as the least upper bound of $\{x \mid x \in S\}$.

The special number $(0, 1)$ has the following property:

$$(0, 1)^2 = (0, 1) \cdot (0, 1) = (-1, 0).$$

In other words, the square of $(0, 1)$ is the real number -1. This special number $(0, 1)$ is called the imaginary unit and is designated symbolically by the letter i.

Definition of i $i = (0, 1)$

At this point it is possible to assign a meaning to the symbol $\sqrt{-1}$; that is, $\sqrt{-1} = i$ and $\sqrt{-1}$ is called the principal square root of -1. (*Note:* -1 also has another square root, $-i = (0, -1)$.)

Since the real numbers x and y have been identified as $(x, 0)$ and $(y, 0)$, respectively, the equation

$$(x, y) = (x, 0) + (y, 0)(0, 1)$$

may now be written as

$$(x, y) = x + yi.$$

Definition of Standard Form $x + yi$ is the standard form of the complex number (x, y) where $i = (0, 1)$.

The equation $(0, 1)^2 = (-1, 0)$ can now be expressed as $i^2 = -1$.
A meaning can now be given to $\sqrt{-a^2}$ where a is a real number.

Definition of $\sqrt{-a^2}$ If a is a real number, then $\sqrt{-a^2} = |a| \, i$.

For example, $\sqrt{-16} = 4i$ and $\sqrt{-2} = \sqrt{2}\, i = i\sqrt{2}$.

The real numbers x and y are called, respectively, the **real part** and the **imaginary part** of $x + yi$.

Definition of Real Part of $x + yi$ $\mathscr{R}(x + yi) = x$

Definition of Imaginary Part of $x + yi$ $\mathscr{I}(x + yi) = y$

It is more convenient to work with complex numbers when they are expressed in standard form. Since the field axioms are valid for the set of complex numbers, the usual computations can be performed on the binomials of the form $x + yi$, treating i as a literal constant and replacing i^2 by -1 whenever it occurs.

THE COMPLEX NUMBER SYSTEM 243

Example 1 Simplify $(-3 + 5i) - (1 - 4i)$.

Solution
$$\begin{aligned}(-3 + 5i) - (1 - 4i) &= (-3 + 5i) + (-1 + 4i) \\ &= (-3 - 1) + (5 + 4)i \\ &= -4 + 9i\end{aligned}$$

Example 2 Write in standard form: $(2 + i)^4$.

Solution
$(2 + i)^2 = 4 + 4i + i^2 = 4 + 4i - 1 = 3 + 4i$
$(2 + i)^4 = (3 + 4i)^2 = 9 + 24i + 16i^2 = 9 + 24i - 16 = -7 + 24i$

Example 3 Multiply $(3 + \sqrt{-25})(4 - \sqrt{-9})$.

Solution Since $\sqrt{-25} = 5i$ and $\sqrt{-9} = 3i$,
$$\begin{aligned}(3 + \sqrt{-25})(4 - \sqrt{-9}) &= (3 + 5i)(4 - 3i) \\ &= 12 - 9i + 20i - 15i^2 \\ &= 12 + 11i + 15 \\ &= 27 + 11i.\end{aligned}$$

The division of two complex numbers is most easily accomplished by multiplying the numerator and denominator of the indicated quotient by the *conjugate* of the denominator.

Definition of Conjugates $x - yi$ and $x + yi$ are conjugates of each other.

The notation $\bar{z}$ is used to indicate the conjugate of a complex number z. For example, if $z = 2 + 5i$, then $\bar{z} = 2 - 5i$.

The name "conjugates" (French: conjuguées) was introduced in 1821 by the French mathematician Augustin Cauchy in his *Cours d'Analyse algébrique*.

Example 4 Express in standard form $\dfrac{5 - 3i}{3 + 2i}$.

Solution
$$\begin{aligned}\frac{5 - 3i}{3 + 2i} &= \frac{5 - 3i}{3 + 2i} \cdot \frac{3 - 2i}{3 - 2i} \\ &= \frac{15 - 10i - 9i + 6i^2}{9 - 4i^2} \\ &= \frac{9 - 19i}{13} = \frac{9}{13} + \frac{-19}{13}i\end{aligned}$$

ORDER

Although the system of complex numbers and the system of real numbers are both fields, only the field of real numbers is an ordered field. It is impossible to define an order relation $z > 0$ so that for all z in C, the trichotomy axiom and the positive closure axioms are valid; that is, so that

(1) exactly one is true: $z > 0$ or $z = 0$ or $-z > 0$, and

(2) if $w > 0$ and $z > 0$, then $w + z > 0$ and $wz > 0$.

Note: $i \neq 0$ since $(0, 1) \neq (0, 0)$.

If $i > 0$, then $i^2 > 0$ and $-1 > 0$ by the positive closure axiom. But then, by the positive closure axiom, $i(-1) = -i > 0$, which contradicts the trichotomy axiom. Thus $i \not> 0$.

If $-i > 0$, then $(-i)(-i) = i^2 = -1 > 0$ and $(-i)(-1) = i > 0$ by the positive closure axiom. This contradicts the trichotomy axiom and thus $-i \not> 0$.

Since $i \neq 0$ and $i \not> 0$ and $-i \not> 0$, it is impossible to order the set of complex numbers.

THEOREMS

Those theorems that were derived for the field of real numbers and that did *not* involve the concept of order remain valid for the field of complex numbers. These theorems are conveniently summarized at the end of Chapter 5. The proofs are almost identical to those given in the preceding chapters.

EXERCISES 7.1

1–16. Simplify, leaving the answers in standard form.

1. $(3 + 4i) + (2 + 3i)$
2. $(2 + 3i) - (\sqrt{2} + 5i) + i\sqrt{5}$[1]
3. $(2 + 5i) - (4 - i)$
4. $-(7 - 3i) + (2 + 2i) - (-1)$
5. $(4 + 2i)(3 - 4i)$
6. $(5 + 2i)^3$
7. $(3 - 2i)[(2 - 4i) + (5 + 2i)]$
8. $\dfrac{3 + 4i}{2i}$
9. $\dfrac{2 + 3i}{1 + i\sqrt{2}}$[1]
10. $\dfrac{1}{3i - 5}$
11. $(4 - 3i)^2 (2 + i)^3$
12. $\dfrac{2 + i}{1 + 2i} + \dfrac{2 - i}{1 - 2i}$
13. $\dfrac{2 + 3i}{3 + i} - \dfrac{1 + i}{1 + 2i}$
14. $\dfrac{i\sqrt{3} + 2i}{i\sqrt{2} - 4i}$
15. $\dfrac{\sqrt{-2} + 2\sqrt{-1}}{\sqrt{-3} - 3\sqrt{-1}}$
16. $\dfrac{1 + 2i}{3 + 4i} \div \dfrac{2i}{2 - i}$

[1] The form $i\sqrt{5}$ is preferred over $\sqrt{5}i$ to avoid the error of writing $\sqrt{5i}$.

THE COMPLEX NUMBER SYSTEM

17–20. Determine the real numbers x and y for which each of the following is true:

17. $x + yi = 3 - 4i$
18. $x + yi - (3 + 2i) = 5 - 6i$
19. $(3x + 2y + 1) + (x + 2y)i = 0$
20. $x + 2yi = i + (-2 + 5i)$

21. If a and b are any real numbers, then show that $z^2 = a + bi$ can always be solved in the set of complex numbers.
    Hint: Let $z = x + yi$, then $z^2 = x^2 - y^2 + 2xyi = a + bi$.
    Thus $(x^2 - y^2)^2 = a^2$, $4x^2y^2 = b^2$, and $(x^2 + y^2)^2 = a^2 + b^2$.

$$x^2 = \frac{a + \sqrt{a^2 + b^2}}{2} \geq 0$$

and

$$y^2 = \frac{-a + \sqrt{a^2 + b^2}}{2} \geq 0.$$

22–25. Use the results of Exercise 21 to solve each of the following:

22. $z^2 = i$
23. $z^2 = 3 - 4i$
24. $z^2 = 5 + 12i$
25. $z^2 = 6 - 2i$

26. Show that $\sqrt{1 + \sqrt{-3}} + \sqrt{1 - \sqrt{-3}}$ is a real number.
27. Show that $\sqrt[3]{2 + \sqrt{-121}} + \sqrt[3]{2 - \sqrt{-121}}$ is a real number.
28. Using $\sqrt{-a^2} = |a| i$, find $\sqrt{-16} \sqrt{-25}$.
    Does $\sqrt{-16} \sqrt{-25} = \sqrt{(-16)(-25)}$? Why?
    Does $\sqrt{-a^2} \sqrt{-b^2} = \sqrt{(-a^2)(-b^2)}$? Why? (Assume a and b are real numbers.)

☆29. If possible, define $\sqrt{z}$, where z is any complex number.

☆30. If possible, define $\sqrt{w} \sqrt{z}$, where w and z are any complex numbers.

31–45. Factor each of the following over the set of complex numbers:

31. $x^2 + 1$
32. $x^2 - i$
33. $x^4 + 16$
34. $x^4 + x^2 - 12$
35. $x^3 + 3ix^2 + 3i^2x + i^3$

36. For what complex number z is the addition inverse $(-z)$ equal to the multiplication inverse $(1/z)$?

37. For $z = x + yi$ and $\bar{z} = x - yi$, show that $z + \bar{z} = 2x$; $z - \bar{z} = 2yi$; and $z \cdot \bar{z} = x^2 + y^2$.

38–43. Solve each of the following for z and $\bar{z}$:

38. $z\bar{z} + 2(z - \bar{z}) = 25 - 12i$
39. $z\bar{z} - 3(z - \bar{z}) = 4 + 6i$
40. $z\bar{z} + 2(z + \bar{z}) = 5$
41. $z\bar{z} - 3(z + \bar{z}) = 7$
42. $z\bar{z} + 4(z + \bar{z}) = 4i$
43. $z\bar{z} - 5(z + \bar{z}) = 2i$

44–50. Prove each of the following in the set of complex numbers:

44. The closure theorem for addition.
45. The closure theorem for multiplication.
46. The commutative theorem for addition.
47. The commutative theorem for multiplication.
48. The associative theorem for addition.
49. The associative theorem for multiplication.
50. The distributive theorem.

51–55. Solve each of the following:

51. $(1, 1)(x, y) = (1, 3)$
52. $(2, 5)(x, y) = (1, 17)$
53. $(4, 0)(x, y) = (8, 28)$
54. $(2, 5)(x, y) = (-11, -13)$
55. $(0, 3)(x, y) = (-15, 0)$

☆56. Is $\{x + yi \mid x$ and y are rational numbers$\}$ a field? Prove your answer.

☆57. $J[i]$, the set of Gaussian integers, is defined as $\{m + ni \mid m$ and n are integers$\}$. Determine which field axioms, if any, are valid for this set. Is $J[i]$ an integral domain? Justify your answer.

☆58. If a and b are Gaussian integers where $b \neq 0$, prove that there exist Gaussian integers q and r such that $a = bq + r$, where $|r| \leq |b|$. Use $|x + yi| = \sqrt{x^2 + y^2}$.

7.2 POLAR FORM

7.2.1 Graphic Representation of Complex Numbers

A complex number may be interpreted geometrically as a point in a rectangular coordinate plane with the rectangular coordinates of the point correponding to the real and imaginary parts of the number. It is also possible to think of a complex number as a vector represented by a directed line segment from the origin to the point in the plane. The x-axis is called the **real axis** and the y-axis is called the **imaginary axis**. The plane is called the **complex plane**.

The graphic representation shown in Figure 7.1 is often called the **Argand diagram** or the **Gauss plane** after the mathematicians who used this figure in their works, Argand in 1806 and Gauss in 1831. It should be called the Wessel diagram since the Norwegian surveyor Caspar Wessel (1745–1818) was the first to state it in a paper presented to the Royal Academy of Denmark in 1797.

POLAR FORM

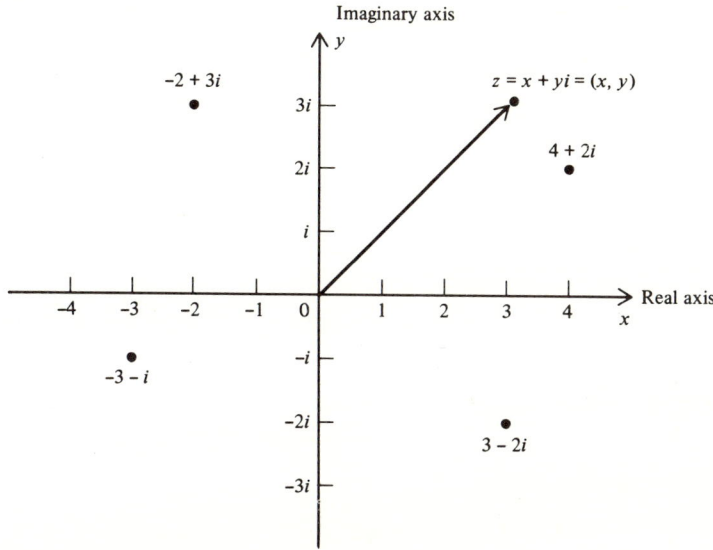

Figure 7.1 The Complex Plane

7.2.2 Polar Form

It is also possible to represent a complex number in terms of the distance r from the origin to the point $P: x + yi$ and the angle θ that the radius vector OP makes with the positive x-axis, as shown in Figure 7.2.

Since $x = r \cos \theta$ and $y = r \sin \theta$,

$$x + yi = (r \cos \theta) + (r \sin \theta)i$$

or

$$x + yi = r(\cos \theta + i \sin \theta).$$

Definition of Polar Form $r(\cos \theta + i \sin \theta)$ is the polar form of $x + yi$ where $r = \sqrt{x^2 + y^2}$, $\cos \theta = x/r$, and $\sin \theta = y/r$. r is called the **modulus** of $x + yi$ (due to Cauchy) or the **absolute value** of $x + yi$, denoted $|x + yi|$ (due to Weierstrass):

$$r = |z| = |x + yi| = \sqrt{x^2 + y^2}.$$

θ is called the **argument** or **amplitude** of $x + yi$.

Example 1 Express $2\left(\cos \dfrac{5\pi}{6} + i \sin \dfrac{5\pi}{6}\right)$ in standard form.

Solution Since $\cos \dfrac{5\pi}{6} = -\dfrac{\sqrt{3}}{2}$ and $\sin \dfrac{5\pi}{6} = \dfrac{1}{2}$,

$$2\left(\cos \dfrac{5\pi}{6} + i \sin \dfrac{5\pi}{6}\right) = 2\left(-\dfrac{\sqrt{3}}{2} + i \cdot \dfrac{1}{2}\right) = -\sqrt{3} + i.$$

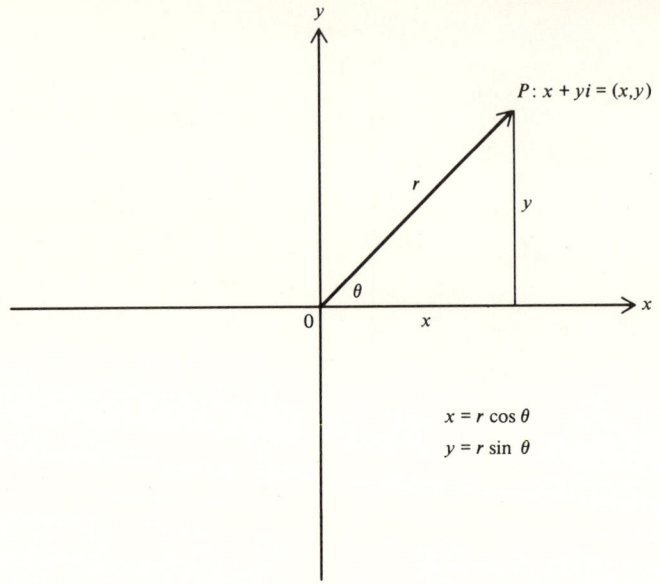

Figure 7.2

Example 2 Express $3 - 3i$ in polar form.

Solution $r^2 = 3^2 + (-3)^2 = 18$ and $r = 3\sqrt{2}$

$$3 - 3i = 3\sqrt{2}\left(\frac{1}{\sqrt{2}} + \frac{-1}{\sqrt{2}}i\right)$$

$$= 3\sqrt{2}\left(\cos\frac{7\pi}{4} + i\sin\frac{7\pi}{4}\right)$$

Example 3 Express -5 in polar form.

Solution $-5 + 0i = 5(-1 + 0i)$
$= 5(\cos\pi + i\sin\pi)$

Example 4 Express $3 + 4i$ in polar form.

Solution $r^2 = 3^2 + 4^2 = 5^2$ and $r = 5$
$3 + 4i = 5(\frac{3}{5} + \frac{4}{5}i)$
$= 5(\cos \text{Arccos } \frac{3}{5} + i\sin \text{Arccos } \frac{3}{5})$

If an approximate value is desired for Arccos $\frac{3}{5}$ = Arccos 0.6, it may be obtained from the tables.

POLAR FORM

Nonuniqueness of Polar Form

Since $\cos(\theta + 2k\pi) = \cos\theta$ and $\sin(\theta + 2k\pi) = \sin\theta$ for any integer k, then

$$r(\cos\theta + i\sin\theta) = r[\cos(\theta + 2k\pi) + i\sin(\theta + 2k\pi)].$$

Therefore, there are infinitely many ways to express a number in polar form.

For example,

$$\frac{5\sqrt{3}}{2} + \frac{5}{2}i = 5\left(\cos\frac{\pi}{6} + i\sin\frac{\pi}{6}\right)$$

$$= 5\left(\cos\frac{13\pi}{6} + i\sin\frac{13\pi}{6}\right)$$

$$= 5\left(\cos\frac{25\pi}{6} + i\sin\frac{25\pi}{6}\right) \quad \text{and so on.}$$

Unless it is otherwise specified, "to express a complex number in polar form" will mean to express it using the smallest nonnegative argument possible.

7.2.3 Multiplication and Division in Polar Form

Since

$$r(\cos a + i\sin a) \cdot R(\cos b + i\sin b)$$
$$= rR[(\cos a\cos b - \sin a\sin b) + i(\sin a\cos b + \sin b\cos a)],$$

it follows that

$$r(\cos a + i\sin a)R(\cos b + i\sin b) = rR[\cos(a+b) + i\sin(a+b)].$$

Therefore, the product of two complex numbers in polar form is obtained by multiplying their moduli to obtain the modulus of the product and by adding their arguments to obtain the argument of the product.

For the quotient, it is first observed that

$$\frac{1}{r(\cos\theta + i\sin\theta)} = \frac{1}{r}\left(\frac{\cos\theta - i\sin\theta}{\cos^2\theta + \sin^2\theta}\right) = \frac{1}{r}(\cos(-\theta) + i\sin(-\theta)).$$

Therefore,

$$\frac{r(\cos a + i\sin a)}{R(\cos b + i\sin b)} = \frac{r}{R}(\cos a + i\sin a)(\cos(-b) + i\sin(-b))$$

$$= \frac{r}{R}(\cos(a-b) + i\sin(a-b)).$$

In other words, the modulus of the quotient is the quotient of the moduli and the argument of the quotient is the difference of the arguments.

Example 5 Express the product in standard form:

$$2(\cos 75° + i \sin 75°) \cdot 5(\cos 60° + i \sin 60°).$$

Solution
$$10(\cos(75° + 60°) + i \sin(75° + 60°))$$
$$= 10(\cos 135° + i \sin 135°)$$
$$= 10\left(-\frac{\sqrt{2}}{2} + i \frac{\sqrt{2}}{2}\right)$$
$$= -5\sqrt{2} + 5i\sqrt{2}.$$

Example 6 Divide and express the quotient in standard form:

$$\frac{6(\cos(\pi/4) + i \sin(\pi/4))}{2(\cos(\pi/12) + i \sin(\pi/12))}.$$

Solution
$$3\left(\cos\left(\frac{\pi}{4} - \frac{\pi}{12}\right) + i \sin\left(\frac{\pi}{4} - \frac{\pi}{12}\right)\right)$$
$$= 3\left(\cos\frac{\pi}{6} + i \sin\frac{\pi}{6}\right)$$
$$= 3\left(\frac{\sqrt{3}}{2} + i\frac{1}{2}\right)$$
$$= \frac{3\sqrt{3}}{2} + \frac{3}{2}i.$$

EXERCISES 7.2

1–15. Change each of the following to standard form:

1. $3(\cos 5\pi/3 + i \sin 5\pi/3)$
2. $2[\cos(-2\pi/3) + i \sin(-2\pi/3)]$
3. $\cos \pi/12 + i \sin \pi/12$
4. $2(\cos 3\pi/8 + i \sin 3\pi/8)$
5. $\frac{1}{2}[\cos(-\pi/3) + i \sin(-\pi/3)]$
6. $2(\cos 7\pi/6 + i \sin 7\pi/6)$
7. $5(\cos 30° + i \sin 30°)$
8. $3(\cos 315° + i \sin 315°)$
9. $\cos(-75°) + i \sin(-75°)$
10. $4(\cos 5\pi/8 + i \sin 5\pi/8)$
11. $4(\cos 225° + i \sin 225°)$
12. $8(\cos 405° + i \sin 405°)$
13. $3[\cos(-\pi/4) + i \sin(-\pi/4)]$

POLAR FORM

14. $3(\cos \text{Arccos } \sqrt{3}/2 + i \sin \text{Arccos } \sqrt{3}/2)$.
15. $2(\cos \text{Arcsin } 3/5 + i \sin \text{Arcsin } 3/5)$.

16–30. Write each of the following complex numbers in polar form where $0 < \theta \le 2\pi$ (do not use tables):

16. $z = 5$
17. $z = 5i$
18. $z = -5$
19. $z = -5i$
20. $z = 2 - 2\sqrt{3}i$
21. $z = -5\sqrt{3} - 5i$
22. $z = -3\sqrt{3} + 3i$
23. $z = -2 - 2i$
24. $z = -1 + i$
25. $z = -4 + 4\sqrt{3}i$
26. $z = -3$
27. $z = -2i$
28. $z = \sqrt{2} - i\sqrt{2}$
29. $z = 6 + 6i$
30. $z = -2\sqrt{2} - 2\sqrt{2}i$

31–38. For each of the following, express in standard form (a) the product $z_1 z_2$; (b) the quotient z_1/z_2 (use tables only when necessary):

31. $z_1 = 2(\cos 3\pi/2 + i \sin 3\pi/2)$, $z_2 = \cos \pi/3 + i \sin \pi/3$
32. $z_1 = 3(\cos 3\pi/4 + i \sin 3\pi/4)$, $z_2 = 4(\cos 5\pi/6 + i \sin 5\pi/6)$
33. $z_1 = 2 (\cos 30° + i \sin 30°)$, $z_2 = 3(\cos 75° + i \sin 75°)$
34. $z_1 = \cos 285° + i \sin 285°$, $z_2 = 2(\cos 135° + i \sin 135°)$
35. $z_1 = 2 - 2i$, $z_2 = -1 + i$
36. $z_1 = 3i$, $z_2 = 2 - 5i$
37. $z_1 = 3(\cos 110° + i \sin 110°)$, $z_2 = 2(\cos 70° + i \sin 70°)$
38. $z_1 = 3(\cos 4\pi/9 + i \sin 4\pi/9)$, $z_2 = 2(\cos 2\pi/9 + i \sin 2\pi/9)$
39. Show that for $\theta = 18°$ (and thus $2\theta = 36°$, $3\theta = 54°$, $4\theta = 72°$, $5\theta = 90°$)
    (a) $\sin 2\theta = \cos 3\theta$
    (b) $4 \sin^2 \theta + 2 \sin \theta - 1 = 0$ using $\sin 2\theta = \cos 3\theta$
    (c) $\sin 18° = \dfrac{-1 + \sqrt{5}}{4} = \cos 72°$ by solving the equation in (b)
    (d) $\cos 18° = \dfrac{\sqrt{10 + 2\sqrt{5}}}{4} = \sin 72°$
    (e) Find the sine and cosine of $36°$ and $54°$
40. Express in standard form, without using tables (use results of Exercise 39):
    (a) $4(\cos 18° + i \sin 18°)$
    (b) $8(\cos 36° + i \sin 36°)$
    (c) $2(\cos 54° + i \sin 54°)$
    (d) $4(\cos 72° + i \sin 72°)$
    (e) $4(\cos 9° + i \sin 9°)$
    (f) $8(\cos 12° + i \sin 12°)$
    [*Hint:* $12° = 30° - 18°$.]

41–48. Change each of the following to polar form:

41. $\sqrt{10 + 2\sqrt{5}} + i(\sqrt{5} - 1)$
42. $\sqrt{5} + 1 + i\sqrt{10 - 2\sqrt{5}}$
43. $(\sqrt{3} + 1) + i(\sqrt{3} - 1)$
44. $(1 - \sqrt{3}) + i(1 + \sqrt{3})$
45. $(\cos a + 1) + i \sin a$
46. $(\cos^2 a - \sin^2 a) + i(1 - 2\sin^2 a)$
47. $(\cos b - \sin b) + i(\cos b + \sin b)$
48. $(\cos a + \cos b) + i(\sin a + \sin b)$

49–58. Locate each of the following points on the complex plane:

49. 1
50. -1
51. i
52. $-i$
53. $1 + i$
54. $1 - i$
55. $-3 - 4i$
56. $2 + 3i$
57. $-3 + 2i$
58. $3 + 4i$

59. What is the geometric location of the points z if $\mathscr{R}(z) = 3$? If $\mathscr{I}(z) = 3$?

60. Illustrate geometrically the relative positions of z and its conjugate $\bar{z}$ for
    (a) $z = 3 + 4i$, (b) $z = 2 - 5i$, (c) $-3 + 2i$.

61. Make a general statement regarding the relative locations of z and $\bar{z}$.

62. Make a general statement regarding the relative locations of
    (a) $c + di$ and $d + ci$, (b) $c + di$ and $-d + ci$.

63. Find z if $|z| = 1$ and $\mathscr{R}(z^2) = 0$.

64. Where are the points z located if
    (a) $|z| = 1$
    (b) $|z| < 1$
    (c) $|z| > 1$
    (d) $1 \le |z| \le 2$.

65. Show that multiplication of $z = a + bi$ by i is geometrically equivalent to a rotation of 90°.

66. What is the geometric significance of z multiplied by $-i$?

67–70. Prove each of the following. Let w and z be any complex numbers.

67. $|wz| = |w| \cdot |z|$
68. $|w + z| \le |w| + |z|$
69. $|z| = \sqrt{z\bar{z}}$
70. $|1 + z| \le 1 + |z|$

71. Show that equal complex numbers have equal moduli.

72. Show that the arguments of equal complex numbers may differ by an integral multiple of 2π.

7.3 POWERS AND ROOTS; DEMOIVRE'S THEOREM

7.3.1 DeMoivre's Theorem

DeMoivre's Theorem If n is a natural number, then
$$[r(\cos\theta + i\sin\theta)]^n = r^n(\cos n\theta + i\sin n\theta).$$

Proof Induction on n. Let S be the set of natural numbers for which DeMoivre's theorem is valid.
PART 1 $n = 1$. $[r(\cos\theta + i\sin\theta)]^1 = r^1(\cos 1 \cdot \theta + i\sin 1 \cdot \theta)$.
Thus, 1 is in S.
PART 2 Let $n = k$. Assume $[r(\cos\theta + i\sin\theta)]^k = r^k(\cos k\theta + i\sin k\theta)$. Then,
$$[r(\cos\theta + i\sin\theta)]^{k+1} = r^k(\cos k\theta + i\sin k\theta)r(\cos\theta + i\sin\theta)$$
$$= r^{k+1}[\cos(k+1)\theta + i\sin(k+1)\theta].$$
Thus, $k + 1$ is in S whenever k is in S. Therefore, $S = N$.

DeMoivre's theorem is also valid when the exponent is a negative integer and $r \neq 0$. Having already observed that
$$z^{-1} = \frac{1}{r(\cos\theta + i\sin\theta)}$$
$$= \frac{1}{r}[\cos(-\theta) + i\sin(-\theta)],$$
then
$$z^{-n} = \left(\frac{1}{z}\right)^n = \left[\frac{1}{r}(\cos(-\theta) + i\sin(-\theta))\right]^n$$
$$= r^{-n}[\cos(-n\theta) + i\sin(-n\theta)].$$
For $n = 0$,
$$[r(\cos\theta + i\sin\theta)]^0 = 1$$
and
$$r^0(\cos 0 + i\sin 0) = 1(1 + 0) = 1.$$
Consequently, DeMoivre's theorem is valid for all integers.

DeMoivre's Theorem Extended If k is any integer, then
$$[r(\cos\theta + i\sin\theta)]^k = r^k(\cos k\theta + i\sin k\theta).$$

Example 1 Express in standard form $(-2 + 2i)^8$.

Solution Let $r(\cos \theta + i \sin \theta) = -2 + 2i$.
Then
$$r = \sqrt{(-2)^2 + 2^2} = 2\sqrt{2}$$

$$(-2 + 2i)^8 = \left[2\sqrt{2}\left(-\frac{1}{\sqrt{2}} + \frac{1}{\sqrt{2}} i\right)\right]^8$$

$$= \left[2\sqrt{2}\left(\cos \frac{3\pi}{4} + i \sin \frac{3\pi}{4}\right)\right]^8$$

$$= (2\sqrt{2})^8 (\cos 6\pi + i \sin 6\pi)$$

$$= 2^8 \, 2^4 (1 + i \cdot 0)$$

$$= 4096 = 4096 + 0i.$$

Example 2 Express $\dfrac{1}{(\sqrt{3} - i)^4}$ in standard form.

Solution Let $r(\cos \theta + i \sin \theta) = \sqrt{3} - i$.
Then
$$r^2 = (\sqrt{3})^2 + (-1)^2 = 4 \quad \text{and} \quad r = 2$$

$$\sqrt{3} - i = 2\left(\frac{\sqrt{3}}{2} - \frac{1}{2} i\right) = 2\left(\cos \frac{-\pi}{6} + i \sin \frac{-\pi}{6}\right)$$

$$r^{-4} = 2^{-4}\left[\cos(-4)\left(\frac{-\pi}{6}\right) + i \sin(-4)\left(\frac{-\pi}{6}\right)\right]$$

$$= 2^{-4}\left(\cos \frac{2\pi}{3} + i \sin \frac{2\pi}{3}\right)$$

$$= \frac{1}{16}\left(-\frac{1}{2} + i \frac{\sqrt{3}}{2}\right) = \frac{-1 + i\sqrt{3}}{32} = -\frac{1}{32} + \frac{\sqrt{3}}{32} i.$$

7.3.2 The Binomial Equation, $z^n = c$

The nth roots of a complex number can be found by applying DeMoivre's theorem to the equation $z^n = c$, where $z = x + yi$ and c is any complex number.
Expressing z and c in polar form,
$$z = r(\cos \theta + i \sin \theta) \quad \text{and} \quad c = s(\cos a + i \sin a),$$

POWERS AND ROOTS; DEMOIVRE'S THEOREM

then
$$[r(\cos \theta + i \sin \theta)]^n = s(\cos a + i \sin a)$$
and
$$r^n(\cos n\theta + i \sin n\theta) = s(\cos a + i \sin a).$$

Since equal complex numbers have equal moduli, $r^n = s$, and since r and s are positive real numbers, then there is exactly one positive real number r such that $r = \sqrt[n]{s}$.

Equal complex numbers may have arguments that differ by an integral multiple of 2π since

$$\sin(\theta + 2k\pi) = \sin \theta \quad \text{and} \quad \cos(\theta + 2k\pi) = \cos \theta.$$

Therefore, $n\theta = a + 2k\pi$ and $\theta = (a + 2k\pi)/n$ for any integer k. Although k may be any integer, there are only n distinct nth roots of a complex number and these may be obtained by selecting $k = 0, 1, 2, \ldots, n - 1$.

Replacing s by r and a by θ, the theorem stating the expressions for the nth roots of a complex number is obtained.

Theorem for nth Roots If $z^n = r(\cos \theta + i \sin \theta)$, then

$$z = \sqrt[n]{r}\left(\cos \frac{\theta + 2k\pi}{n} + i \sin \frac{\theta + 2k\pi}{n}\right), \text{ where } k = 0, 1, 2, \ldots, n-1.$$

Example 3 Find the three cube roots of 125.

Solution $z^3 = 125 = 5^3(\cos 0 + i \sin 0) = 5^3(\cos 2k\pi + i \sin 2k\pi)$

$$z = 5\left(\cos \frac{2k\pi}{3} + i \sin \frac{2k\pi}{3}\right) \quad \text{for } k = 0, 1, 2.$$

For $k = 0$,
$$z_1 = 5(\cos 0 + i \sin 0) = 5.$$

For $k = 1$,
$$z_2 = 5\left(\cos \frac{2\pi}{3} + i \sin \frac{2\pi}{3}\right) = 5\left(-\frac{1}{2} + i\frac{\sqrt{3}}{2}\right) = \frac{-5 + 5i\sqrt{3}}{2}.$$

For $k = 2$,
$$z_3 = 5\left(\cos \frac{4\pi}{3} + i \sin \frac{4\pi}{3}\right) = 5\left(-\frac{1}{2} - i\frac{\sqrt{3}}{2}\right) = \frac{-5 - 5i\sqrt{3}}{2}.$$

Example 4 Solve $z^4 = -4i$.

Solution $z^4 = 4[0 + i(-1)] = 4\left(\cos\dfrac{3\pi}{2} + i\sin\dfrac{3\pi}{2}\right)$

$$z^4 = 4\left[\cos\left(\dfrac{3\pi}{2} + 2k\pi\right) + i\sin\left(\dfrac{3\pi}{2} + 2k\pi\right)\right] \quad k = 0, 1, 2, 3$$

$$z = \sqrt[4]{4}\left(\cos\dfrac{(3+4k)\pi}{8} + i\sin\dfrac{(3+4k)\pi}{8}\right) \quad k = 0, 1, 2, 3$$

For $k = 0$,
$$z_1 = \sqrt[4]{4}\left(\cos\dfrac{3\pi}{8} + i\sin\dfrac{3\pi}{8}\right).$$

For $k = 1$,
$$z_2 = \sqrt[4]{4}\left(\cos\dfrac{7\pi}{8} + i\sin\dfrac{7\pi}{8}\right).$$

For $k = 2$,
$$z_3 = \sqrt[4]{4}\left(\cos\dfrac{11\pi}{8} + i\sin\dfrac{11\pi}{8}\right).$$

For $k = 3$,
$$z_4 = \sqrt[4]{4}\left(\cos\dfrac{15\pi}{8} + i\sin\dfrac{15\pi}{8}\right).$$

Now

$$\sqrt[4]{4} = \sqrt{2} \quad \text{and} \quad \dfrac{3\pi}{8} = \dfrac{1}{2}\left(\dfrac{3\pi}{4}\right) \quad \text{and} \quad \dfrac{\pi}{8} = \dfrac{1}{2}\left(\dfrac{\pi}{4}\right),$$

$$\dfrac{7\pi}{8} = \pi - \dfrac{\pi}{8}, \quad \dfrac{11\pi}{8} = \pi + \dfrac{3\pi}{8}, \quad \text{and} \quad \dfrac{15\pi}{8} = 2\pi - \dfrac{\pi}{8}$$

$$\cos\dfrac{\pi}{8} = \cos\dfrac{1}{2}\left(\dfrac{\pi}{4}\right) = \sqrt{\dfrac{1 + \cos\dfrac{\pi}{4}}{2}} = \sqrt{\dfrac{1 + \dfrac{\sqrt{2}}{2}}{2}} = \dfrac{\sqrt{2 + \sqrt{2}}}{2} = \sin\dfrac{3\pi}{8}.$$

Similarly,

$$\sin\dfrac{\pi}{8} = \dfrac{\sqrt{2 - \sqrt{2}}}{2}, \quad \cos\dfrac{3\pi}{8} = \sin\dfrac{\pi}{8}.$$

Finally,

$$z_1 = \sqrt{2}\left(\frac{\sqrt{2-\sqrt{2}}}{2} + i\frac{\sqrt{2+\sqrt{2}}}{2}\right) = \frac{\sqrt{4-2\sqrt{2}}}{2} + \frac{i\sqrt{4+2\sqrt{2}}}{2}$$

$$z_2 = \sqrt{2}\left(-\frac{\sqrt{2+\sqrt{2}}}{2} + i\frac{\sqrt{2-\sqrt{2}}}{2}\right) = \frac{-\sqrt{4+2\sqrt{2}}}{2} + \frac{i\sqrt{4-2\sqrt{2}}}{2}$$

$$z_3 = \sqrt{2}\left(-\frac{\sqrt{2-\sqrt{2}}}{2} - i\frac{\sqrt{2+\sqrt{2}}}{2}\right) = \frac{-\sqrt{4-2\sqrt{2}}}{2} - \frac{i\sqrt{4+2\sqrt{2}}}{2}$$

$$z_4 = \sqrt{2}\left(\frac{\sqrt{2+\sqrt{2}}}{2} - i\frac{\sqrt{2-\sqrt{2}}}{2}\right) = \frac{\sqrt{4+2\sqrt{2}}}{2} - \frac{i\sqrt{4-2\sqrt{2}}}{2}.$$

At this point it is important to observe that the set of complex numbers is closed with respect to the six algebraic operations: addition, multiplication, subtraction, division, raising to a power, and root extraction.

EXERCISES 7.3

1–10. Compute each of the following in polar form and express in standard form:

1. $[2(\cos 5\pi/12 + i \sin 5\pi/12)]^6$
2. $[3(\cos 3\pi/2 + i \sin 3\pi/2)]^4$
3. $[4(\cos 3\pi/4 + i \sin 3\pi/4)]^4$
4. $[2(\cos \pi/4 + i \sin \pi/4)]^6$
5. $(1 + i)^6$
6. $(1 - \sqrt{3}i)^5$
7. $(-\sqrt{3} + i)^8$
8. $(-2 - 2i)^8$
9. $(-1 + i)^{-4}$
10. $\left(\frac{\sqrt{3}}{2} - \frac{1}{2}i\right)^{-5}$

11–30. Find all roots of each of the following equations:

11. $z^3 = 1$
12. $z^4 = 1$
13. $z^6 = 1$
14. $z^6 = -1$
15. $z^8 = 1$
16. $z^3 = -8$
17. $z^2 = -81$
18. $z^4 = -81$
19. $z^2 = 1 - i\sqrt{3}$
20. $z^2 = 2 + 2i$
21. $z^3 = 1 + i$
22. $z^6 = (\sqrt{3} + 1) + i(1 - \sqrt{3})$
23. $z^3 = -8i$
24. $z^3 = 1 - i$
25. $z^3 = -\frac{1}{2} + \frac{\sqrt{3}}{2}i$
26. $z^4 = -16$
27. $z^4 = -16i$
28. $z^4 = 1 + i$
29. $z^5 = 32$
30. $z^4 = (1 + i)^3$

31. Find the root of $z^{24} = 1$ that has the *smallest positive* argument.
32. Find the root of $z^{10} = 1$ that has the *smallest positive* argument.

33–36. Illustrate the roots of each of the following equations on the unit circle in the complex plane.

33. $z^3 = 1$
34. $z^4 = 1$
35. $z^6 = 1$
36. $z^8 = 1$

☆37. Use the results of Exercises 33–36 to prove or disprove that the problem of dividing a circle into n equal parts is equivalent to solving the equation $z^n = 1$ in terms of the rational operations and square root extractions of positive real numbers.

☆38. Show that $x_1 = 2 \cos 2\pi/9$, $x_2 = 2 \cos 4\pi/9$, and $x_3 = 2 \cos 8\pi/9$ are the solutions of the cubic equation $x^3 - 3x + 1 = 0$.
Hint: $z^9 - 1 = (z^3 - 1)(z^6 + z^3 + 1) = 0$
$$\frac{z^6 + z^3 + 1}{z^3} = z^3 + 1 + 1/z^3. \text{ Let } x = z + 1/z$$

☆39. Show that $x_1 = 2 \cos 2\pi/7$, $x_2 = 2 \cos 4\pi/7$, and $x_3 = 2 \cos 6\pi/7$ are solutions of the cubic equation $x^3 + x^2 - 2x - 1 = 0$.
Hint: $z^7 - 1 = (z - 1)(z^6 + z^5 + z^4 + z^3 + z^2 + z + 1) = 0$
$$\frac{z^6 + z^5 + \ldots + 1}{z^3} = z^3 + z^2 + z + 1 + 1/z + 1/z^2 + 1/z^3$$
Let $x = z + 1/z$.

7.4 GRAPHIC REPRESENTATIONS (Optional)

Many physical concepts, such as force, velocity, and acceleration, are described by two properties: a direction and a magnitude. For example, if the velocity of a car is 35 mph north, then the magnitude of the velocity is 35 mph and the direction is north. A *vector* is a quantity that has both magnitude and direction. Geometrically, a vector may be represented by an arrow, with the length of the arrow indicating the magnitude of the vector and the spearhead indicating the direction. Since a complex number can be interpreted as a vector, this concept furnishes a graphic representation of the sum, difference, product, and quotient of two complex numbers. From the polar form of the complex number, $z = r(\cos \theta + i \sin \theta)$, the magnitude of the vector is identified as $r = \sqrt{x^2 + y^2}$ and the direction is indicated by the angle θ.

7.4.1 Graphic Addition

If $w = a + bi$ and $z = c + di$ are two complex numbers, then their sum is given by
$$w + z = (a + c) + (b + d)i.$$

GRAPHIC REPRESENTATIONS (OPTIONAL)

This sum is illustrated in Figure 7.3.

It may be shown geometrically that the vector $w + z$ is the diagonal of the parallelogram with vectors w and z forming two adjacent sides.

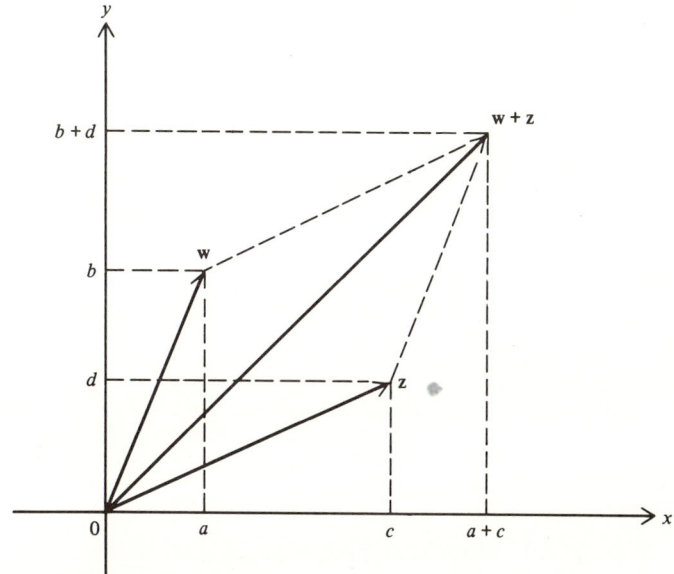

Figure 7.3 Graphic Addition of Complex Numbers

7.4.2 Graphic Subtraction

Since $-z = -c - di$, it follows that the vector $-z$ has the same magnitude as z but the opposite direction.

The magnitude of $-z = \sqrt{(-c)^2 + (-d)^2} = \sqrt{c^2 + d^2}$ and the direction of $-z = \text{Arccos} \dfrac{-c}{r} = \text{Arcsin} \dfrac{-d}{r} = \pi + \theta.$

The difference of two complex numbers using the fact that $w - z = w + (-z)$ is illustrated in Figure 7.4.

7.4.3 Graphic Multiplication

When a vector is considered as an ordered pair of real numbers, then any real number is referred to as a **scalar**. If a is a scalar, then $a(x + yi)$, the product of a scalar and a vector, is a vector having the same direction as the original vector but its magnitude is multiplied by a.

$$\text{Magnitude of } a(x + yi) = \sqrt{(ax)^2 + (ay)^2} = a\sqrt{x^2 + y^2}$$

$$\text{Direction of } a(x + yi) = \text{Arccos} \dfrac{ax}{ar} = \text{Arcsin} \dfrac{ay}{ar} = \theta$$

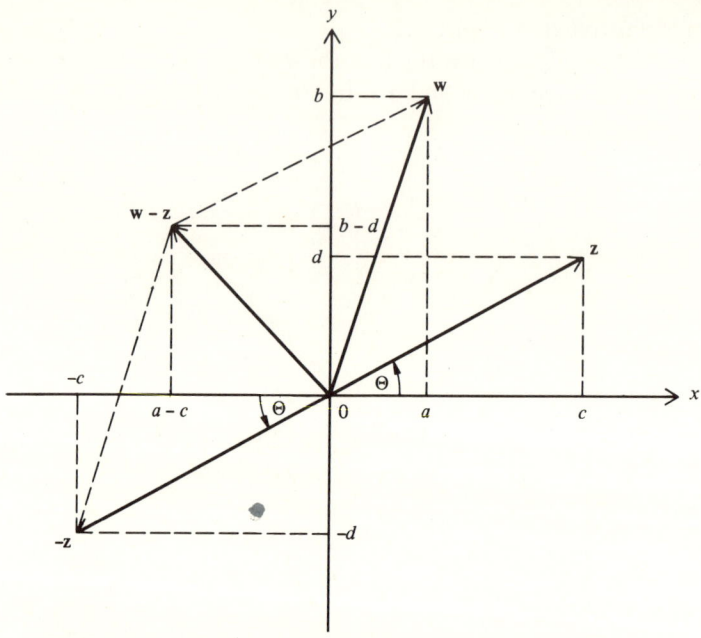

Figure 7.4 Graphic Subtraction of Complex Numbers

The multiplication of a vector or complex number by the imaginary unit i may be interpreted as the rotation of the vector through $\frac{\pi}{2}$ radians or 90°.

$$i(x + yi) = ix + i^2y = -y + xi$$
$$\text{Magnitude of } i(x + yi) = \sqrt{(-y)^2 + x^2} = \sqrt{x^2 + y^2}$$
$$\text{Direction of } i(x + yi) = \frac{\pi}{2} + \theta$$

The multiplication of a vector by ai, $ai(x + yi)$, can now be considered as a rotation through 90° and a multiplication of magnitude by a.

Using these ideas, the product of two complex numbers,

$$(a + bi)(c + di) = a(c + di) + bi(c + di)$$

in illustrated in Figure 7.5.

Another geometric representation of the product of two complex numbers may be obtained from the polar form of the complex numbers. If

$$w = r(\cos a + i \sin a) \quad \text{and} \quad z = s(\cos b + i \sin b),$$

then

$$wz = rs[\cos(a + b) + i \sin(a + b)].$$

The product, then, as illustrated in Figure 7.6, is obtained by adding the angles and multiplying the magnitudes.

GRAPHIC REPRESENTATIONS (OPTIONAL)

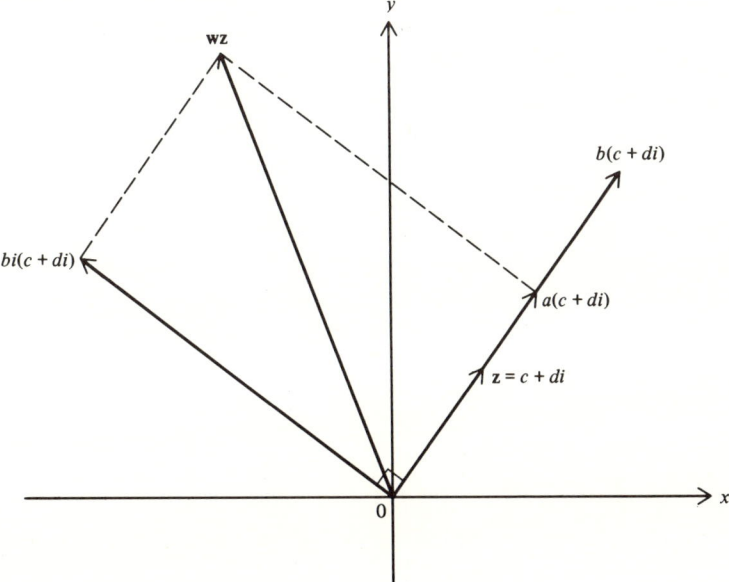

Figure 7.5 Graphic Multiplication of Complex Numbers, Standard Form

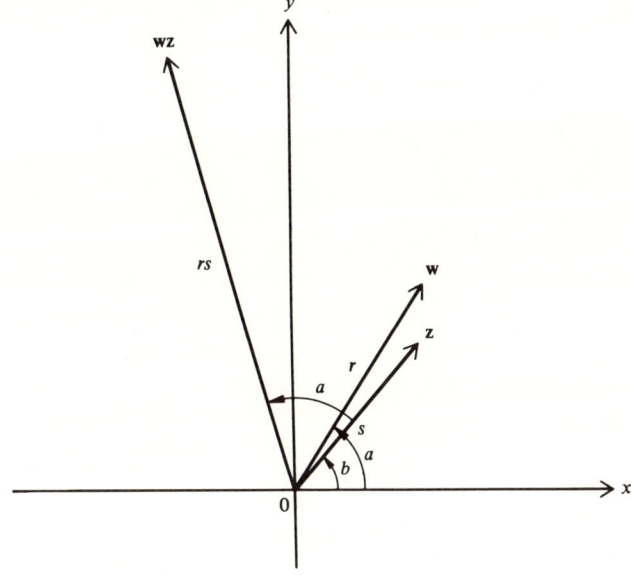

Figure 7.6 Graphic Multiplication of Complex Numbers, Polar Form

7.4.4 Graphic Division

Since for $z = s(\cos b + i \sin b)$, $1/z = (1/s)[\cos(-b) + i \sin(-b)]$, then $w/z = w(1/z) = (r/s)[\cos(a - b) + i \sin(a - b)]$. Thus, the magnitude of the quotient of two complex numbers is obtained by multiplying the magnitude of the numerator by the reciprocal of magnitude of the denominator, and the argument of the quotient is obtained by subtracting the arguments of the two complex numbers. Using this idea, the quotient of two complex numbers is illustrated in Figure 7.7.

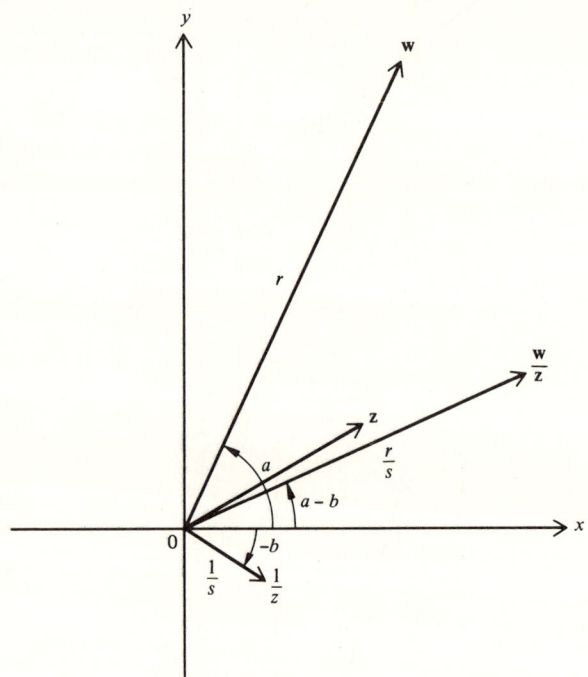

Figure 7.7 Graphic Division of Complex Numbers, Polar Form

If $w = a + bi$, then $w/z = w(1/z) = a(1/z) + bi(1/z)$. Therefore, the quotient of two complex numbers can also be interpreted as the sum of two vectors, one in the direction of $1/z$ with a times its magnitude and the other perpendicular to $1/z$ with b times the magnitude of $1/z$. This interpretation is illustrated in Figure 7.8.

EXERCISES 7.4

1–10. Find the geometric and vector representations of each of the following complex numbers:

GRAPHIC REPRESENTATIONS (OPTIONAL)

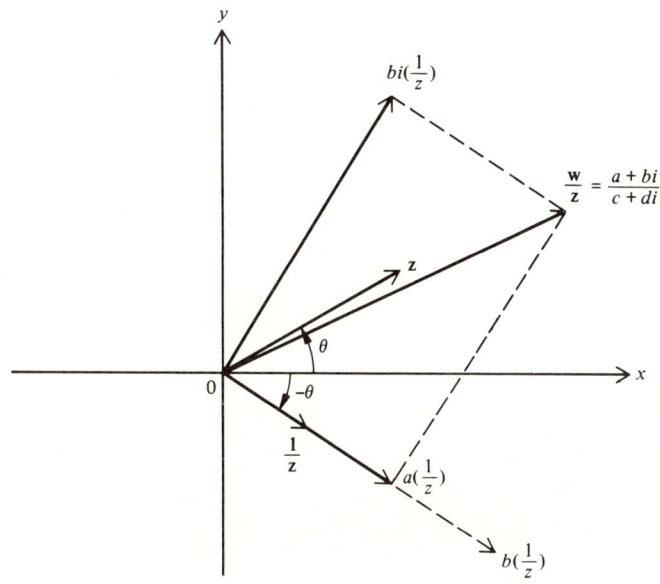

Figure 7.8 Graphic Division of Complex Numbers, Standard Form

1. $3 + 2i$
2. $3 - 2i$
3. $-\sqrt{2} - 3i$
4. $-\sqrt{2} + 3i$
5. $(1 + i)^2$
6. $i(3 + 5i)$
7. $-(2 - 3i)$
8. $3(1 - i)$
9. $i(-1 - 2i)$
10. $(-2)(-1 + 2i)$

11–20. Illustrate the following operations graphically:

11. $(4 + 3i) + (1.5 + 2i)$
12. $(4 + 3i) - (1.5 + 2i)$
13. $(4 + 3i)(1.5 + 2i)$
14. $\dfrac{4 + 3i}{(1.5 + 2i)}$
15. $(-4 + 2i) + (3 + 5i)$
16. $(-4 + 2i) - (3 + 5i)$
17. $(-4 + 2i)(3 + 5i)$
18. $\dfrac{-4 + 2i}{3 + 4i}$
19. $\dfrac{1 + 5i}{1.6 - 1.2i}$
20. $(3 + 4i)^2$

CHAPTER SUMMARY

The Set of Complex Numbers C is the set of ordered pairs of real numbers.

$$C = \{(x, y) \mid x \in R \text{ and } y \in R\}$$

Definition of i $i = (0, 1)$

Definition of $\sqrt{-1}$ $\sqrt{-1} = i$ and $i^2 = -1$

Definition of $\sqrt{-a^2}$, a real $\sqrt{-a^2} = |a|i$

Definition of Standard Form $x + yi$ is the standard form of $(x, y) \in C$

Definition of the Real Part of a Complex Number, z $\mathscr{R}(z) = \mathscr{R}(x + yi) = x$

Definition of the Imaginary Part of a Complex Number, z $\mathscr{I}(z) = \mathscr{I}(x + yi) = y$

Definition of Conjugates $x + yi$ and $x - yi$ are conjugates of each other.

Definition of Equality $a + bi = c + di$ if and only if $a = c$ and $b = d$

Definition of Addition $(a + bi) + (c + di) = (a + c) + (b + d)i$

Definition of Multiplication $(a + bi)(c + di) = (ac - bd) + (ad + bc)i$

Theorem   The system of complex numbers is a field.

For all complex numbers z, w, and v:

CLOSURE, ADDITION
There is exactly one complex number $z + w$.

CLOSURE, MULTIPLICATION
There is exactly one complex number, zw.

COMMUTATIVE, ADDITION
$z + w = w + z$

COMMUTATIVE, MULTIPLICATION
$zw = wz$

CHAPTER SUMMARY

Associative, Addition
$(z + w) + v = z + (w + v)$

Associative, Multiplication
$(zw)v = z(wv)$

Distributive
$z(w + v) = zw + zv$

Identity, Addition
There is exactly one complex number, $0 = (0, 0)$ so that for all z in C, $z + 0 = z$.

Identity, Multiplication
There is exactly one complex number, $1 = (1, 0)$ so that for all z in C, $z \cdot 1 = z$.

Inverse, Addition
For each z in C, there is exactly one complex number, $-z$, so that $z + (-z) = 0$.

Inverse, Multiplication
For each z in C except 0, there is exactly one complex number, $\frac{1}{z}$, so that $z \cdot \frac{1}{z} = 1$.

Definition of Subtraction

$$w - z = w + (-z)$$
$$(a + bi) - (c + di) = (a - c) + (b - d)i$$

Definition of Division

$$\frac{w}{z} = w\left(\frac{1}{z}\right) \quad \text{for} \quad z \neq 0$$

$$\frac{a + bi}{c + di} = (a + bi)\left(\frac{1}{c + di}\right) = \frac{(a + bi)(c - di)}{c^2 + d^2}$$

Definition of Polar Form $r(\cos \theta + i \sin \theta)$ is the polar form of $x + yi$ where

$$r = \sqrt{x^2 + y^2}, \quad \cos \theta = \frac{x}{r}, \quad \text{and} \quad \sin \theta = \frac{y}{r}.$$

Definition of Modulus (Absolute Value) r is the absolute value or modulus of $x + yi$ where $r = |x + yi| = \sqrt{x^2 + y^2}$.

Definition of Argument (Amplitude) θ is the argument or amplitude of the complex number $r(\cos \theta + i \sin \theta)$.

DeMoivre's Theorem If k is any integer, then

$$[r(\cos \theta + i \sin \theta)]^k = r^k(\cos k\theta + i \sin k\theta).$$

Theorem for nth Roots If n is any natural number, and if $z^n = r(\cos \theta + i \sin \theta)$, then

$$z = \sqrt[n]{r}\left(\cos \frac{\theta + 2k\pi}{n} + i \sin \frac{\theta + 2k\pi}{n}\right) \text{ where } k = 0, 1, 2, \ldots, n - 1.$$

Structure of the Set of Complex Numbers

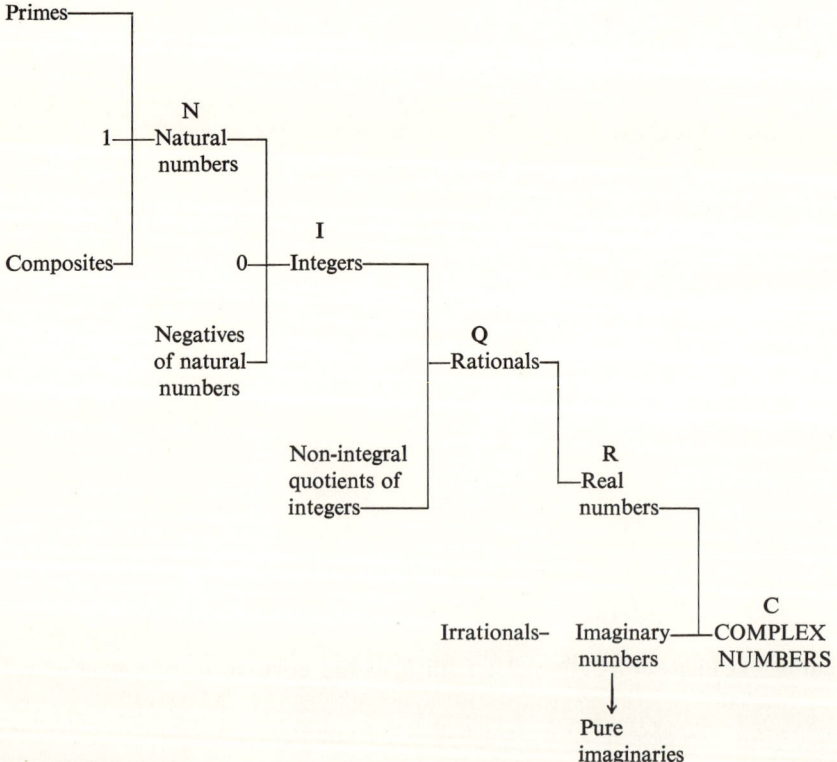

REVIEW EXERCISES

1–10. Simplify, leaving answers in standard form.

1. $(2 + 3i) + (3 - 2i)$
2. $(2 + 3i) - (3 - 2i)$
3. $(2 + 3i)(3 - 2i)$
4. $(2 + 3i)^2$
5. $\dfrac{2 + 3i}{3 - 2i}$
6. $\dfrac{3 + i}{1 + i\sqrt{2}}$
7. $\dfrac{i\sqrt{2} + 2i}{i\sqrt{2} - 3i}$
8. $\dfrac{\sqrt{-3} + 2\sqrt{-1}}{\sqrt{-2} - 3\sqrt{-1}}$
9. $\dfrac{2 + 3i}{1 + 2i} \div \dfrac{3i}{2 - i}$
10. $(3 + 2i)[(2 - 3i) + (3 - 4i)]$

11 and 12. Determine the real numbers x and y for which each of the following is true:

11. $3 + 2i - (x + yi) = 2 - 6i$
12. $2x + 3yi = i - (2 - 5i)$

13–15. Solve each of the following in the set of complex numbers:

13. $z^2 = 2i$
14. $z^2 = 4 - 3i$
15. $z^2 = 3 - i$

16–18. Solve each of the following:

16. $(2, 1)(x, y) = (3, 2)$
17. $(x, y) = (x, 0) + (y, 0)(0, 1)$
18. $(0, 2)(x, y) = (-12, 0)$

19–22. Change each of the following to standard form:

19. $2(\cos 2\pi/3 + i \sin 2\pi/3)$
20. $3[\cos(-7\pi/6) + i \sin(-7\pi/6)]$
21. $4(\cos 75° + i \sin 75°)$
22. $2(\cos \text{Arcsin } \sqrt{3}/2 + i \sin \text{Arcsin } \sqrt{3}/2)$

23–26. Write each of the following complex numbers in polar form:

23. $z = 2$
24. $z = -3i$
25. $z = 3 - 3i$
26. $z = -4 + 4\sqrt{3}i$

27–30. For each of the following problems, write z_1 and z_2 as (a) a product (b) a quotient (z_1/z_2) in standard form:

27. $z_1 = 2(\cos \pi/2 + i \sin \pi/2); z_2 = (\cos 2\pi/3 + i \sin 2\pi/3)$
28. $z_1 = 3(\cos 60° + i \sin 60°); z_2 = 3(\cos 45° + i \sin 45°)$
29. $z_1 = \sqrt{2}/2 - (\sqrt{2}/2)i; z_2 = -1 + i$
30. $z_1 = 2i; z_2 = 2 - 3i$

31–34. Locate each of the following points in the complex plane:

31. 3
32. $-2i$
33. $3 - 2i$
34. $-2 + 3i$

35–38. Compute each of the following in polar form and express in standard form:

35. $[2(\cos 3\pi/2 + i \sin 3\pi/2)]^6$
36. $[3(\cos 2\pi/3 + i \sin 2\pi/3)]^4$
37. $(2 + 3i)^4$
38. $(-2 + i)^{-4}$

39–44. Find all roots of each of the following equations:

39. $z^3 = -1$
40. $z^2 = 2 - i\sqrt{3}$
41. $z^5 = -32i$
42. $z^3 = 1 + i$
43. $z^4 = 16$
44. $z^2 = (1 + i)^3$

45–48. Find the geometric and vector representations of each of the following complex numbers:

45. $2 + 3i$
46. $2 - 3i$
47. $-i(3 + 2i)$
48. $(2 + i)^2$

☆49–52. Illustrate the following operations graphically:

49. $(-2 + 3i) + (3 + 2i)$
50. $(-2 + 3i) - (3 + 2i)$
51. $(-2 + 3i)(3 + 2i)$
52. $\dfrac{-2 + 3i}{3 + 2i}$

8
THEORY OF EQUATIONS

8.1 QUADRATIC EQUATIONS; FUNDAMENTAL THEOREM OF ALGEBRA; ROOTS AND FACTORS

8.1.1 Terminology

A **polynomial** in one variable, z, over the set of complex numbers, C, is an expression of the form

$$a_n z^n + a_{n-1} z^{n-1} + \ldots + a_2 z^2 + a_1 z + a_0$$

where the constants $a_n, a_{n-1}, \ldots, a_2, a_1, a_0$ are elements of C, and n is a natural number.

The constants are called the **coefficients** of the polynomial.

If $a_n \neq 0$, then the polynomial is said to have **degree** n and a_n is called the **leading coefficient**.

Any nonzero constant is a polynomial of degree 0. The number 0 is called the **zero polynomial** but it is not assigned a degree. Polynomials of degrees 1, 2, 3, 4, or 5 are assigned the names **linear, quadratic, cubic, quartic** or **biquadratic,** or **quintic,** respectively.

A polynomial in one variable may be designated by the functional notation, such as $p(z)$, $q(z)$, $r(z)$, and so on.

DEGREE	NAME	GENERAL EXAMPLE	SPECIAL EXAMPLE
1	Linear	$az + b$	$3z - 2$
2	Quadratic	$az^2 + bz + c$	$5z^2 + 2z - 4$
3	Cubic	$az^3 + bz^2 + cz + d$	$8z^3 - 6z + 1$
4	Quartic	$az^4 + bz^3 + cz^2 + dz + e$	$z^4 + z^3 - 3z + 7$
5	Quintic	$az^5 + bz^4 + cz^3 + dz^2 + ez + f$	$7z^5 - z^4 + 3z - 5$

A polynomial equation is an equation of the form $p(z) = 0$ where $p(z)$ is a polynomial.

A **solution** or **root** of a polynomial equation is a constant r such that $p(r) = 0$.

A solution or root of a polynomial equation is also called a *zero* of the polynomial. For example, 2 is a zero of $z^3 - 3z - 2$ since

$$2^3 - 3(2) - 2 = 8 - 6 - 2 = 0.$$

The theory of equations is concerned with finding the roots of polynomial equations with coefficients in the set of complex numbers. One important aspect of this problem is to exhibit the solutions as the results of rational operations or root extractions on the coefficients of the polynomial.

For the linear equation, $az + b = 0$ and $a \neq 0$, the one solution may be expressed as $z = -(b/a)$.

For the quadratic equation, $az^2 + bz + c = 0$ and $a \neq 0$, it may be shown that the two solutions may be expressed as

$$z_1 = \frac{-b + \sqrt{b^2 - 4ac}}{2a} \quad \text{and} \quad z_2 = \frac{-b - \sqrt{b^2 - 4ac}}{2a}.$$

8.1.2 The Quadratic Equation

Theorem (Quadratic Formula) If a, b, and c are any complex numbers and $a \neq 0$, then the roots of $az^2 + bz + c = 0$ are

$$z_1 = \frac{-b + \sqrt{b^2 - 4ac}}{2a} \quad \text{and} \quad z_2 = \frac{-b - \sqrt{b^2 - 4ac}}{2a}.$$

Proof Since $a \neq 0$, $z^2 + \frac{b}{a}z + \frac{c}{a} = 0$.

Adding $(b/2a)^2 - (b/2a)^2$ in order to complete the square,

$$\left(z^2 + \frac{b}{a}z + \frac{b^2}{4a^2}\right) - \left(\frac{b^2}{4a^2} - \frac{c}{a}\right) = 0$$

and

$$\left(z + \frac{b}{2a}\right)^2 - \left(\frac{b^2 - 4ac}{4a^2}\right) = 0.$$

QUADRATIC EQUATIONS; FUNDAMENTAL THEOREM; ROOTS, FACTORS

Factoring the difference of two squares,

$$\left(z + \frac{b}{2a} + \sqrt{\frac{b^2 - 4ac}{4a^2}}\right)\left(z + \frac{b}{2a} - \sqrt{\frac{b^2 - 4ac}{4a^2}}\right) = 0.$$

By the zero product theorem,[1]

$$z + \frac{b}{2a} + \frac{\sqrt{b^2 - 4ac}}{2a} = 0$$

or

$$z + \frac{b}{2a} - \frac{\sqrt{b^2 - 4ac}}{2a} = 0.$$

Consequently,

$$z = \frac{-b \pm \sqrt{b^2 - 4ac}}{2a}.$$

DISCRIMINANT

The radicand $b^2 - 4ac$ is called the **discriminant** of the quadratic equation since it can be used to determine the nature of the roots of the equation.

If $b^2 - 4ac = 0$, then $z_1 = z_2$; that is, the roots are equal.

If $b^2 - 4ac = k^2 \neq 0$ and if a, b, c, and k are rational numbers, then the roots are rational and unequal.

If $b^2 - 4ac > 0$ and a, b, and c are real numbers, then the two roots are real and unequal.

If $b^2 - 4ac < 0$ and a, b, and c are real numbers, then the two roots are imaginary and unequal.

Theorem (Sum and Product of Quadratic Roots) z_1 and z_2 are the two roots of the quadratic equation $az^2 + bz + c = 0$, $a \neq 0$, *if and only if*

$$z_1 + z_2 = -\frac{b}{a} \quad \text{and} \quad z_1 z_2 = \frac{c}{a}.$$

The proof of this theorem is based on the statement that $z^2 + \frac{b}{a}z + \frac{c}{a} = 0$ is equivalent to $z^2 - (z_1 + z_2)z + z_1 z_2 = 0$ if and only if z_1 and z_2 are the roots of $az^2 + bz + c = 0$.

---

[1] The student may verify that if w and z are complex numbers, $wz = 0$ if and only if $w = 0$ or $z = 0$.

Example 1 Find the sum and the product of the roots of $5z^2 + 2z + 3 = 0$.

Solution $z_1 + z_2 = -\dfrac{b}{a} = -\dfrac{2}{5}$

$z_1 z_2 = \dfrac{c}{a} = \dfrac{3}{5}.$

Example 2 Check that $3 + \sqrt{2}$ and $3 - \sqrt{2}$ are the roots of $z^2 - 6z + 7 = 0$.

Solution $z_1 + z_2 = (3 + \sqrt{2}) + (3 - \sqrt{2}) = 6$ and $-\dfrac{b}{a} = -\dfrac{-6}{1} = 6$

$z_1 z_2 = (3 + \sqrt{2})(3 - \sqrt{2}) = 9 - 2 = 7$ and $\dfrac{c}{a} = 7$

Example 3 Solve and check: $4z^2 + 8z + 5 = 0$.

Solution Using $z = \dfrac{-b \pm \sqrt{b^2 - 4ac}}{2a}$,

$z = \dfrac{-8 \pm \sqrt{64 - 80}}{8} = -1 \pm \dfrac{i}{2}$

$z_1 = -1 + \dfrac{i}{2}$ and $z_2 = -1 - \dfrac{i}{2}.$

Checking,

$z_1 + z_2 = -2$ and $-\dfrac{b}{a} = -\dfrac{8}{4} = -2$

$z_1 z_2 = (-1)^2 - \left(\dfrac{i}{2}\right)^2 = 1 + \dfrac{1}{4} = \dfrac{5}{4}$ and $\dfrac{c}{a} = \dfrac{5}{4}.$

Example 4 Solve $z^2 - 2iz - 2 = 0$, where $i^2 = -1$.

Solution $z = \dfrac{-(-2i) \pm \sqrt{4i^2 + 8}}{2} = \dfrac{2i \pm 2}{2} = i \pm 1.$ Thus, $z_1 = -1 + i$ and $z_2 = 1 + i$.

8.1.3 Fundamental Theorem of Algebra

Although the linear and quadratic equations over the set of complex numbers can be solved by means of algebraic operations on their coefficients, it does not follow that all polynomial equations can be solved in this manner. Moreover,

QUADRATIC EQUATIONS; FUNDAMENTAL THEOREM; ROOTS, FACTORS

even though the set of complex numbers is closed with respect to the six algebraic operations, it does not follow that every polynomial equation has even one solution in the set of complex numbers.

The great German mathematician Carl Friedrich Gauss (1777–1855) was the first to prove the theorem that establishes whether or not a polynomial equation has a solution. This theorem is so important that it is called the *Fundamental Theorem of Algebra*. It will be accepted without proof in this text because there is no proof simple enough to present. In fact, all of the known proofs involve concepts from calculus.

The Fundamental Theorem of Algebra   Every polynomial equation with coefficients in the set of complex numbers has at least one solution in the set of complex numbers.

8.1.4 Roots and Factors of Polynomial Equations

Since the set of complex numbers, C, is a field, the field axioms with the exception of the multiplication inverse axiom are valid for the set of polynomials over C and these polynomials may be added, subtracted, and multiplied in the usual way. The set of polynomials over C, designated by $C[z]$, is called the polynomial ring over C. Although the quotient of two polynomials is not in general a polynomial, there is a division algorithm for polynomials over C.

Theorem (Division Algorithm for Polynomials over C)   If $a(z)$ and $b(z)$ are polynomials over C with positive degrees, then there exist unique polynomials $q(z)$ and $r(z)$ over C so that

$$a(z) = b(z)q(z) + r(z)$$

and either

$$r(z) = 0$$

or degree of $r(z)$ < degree of $b(z)$.

This theorem is proved by subtracting suitable cz^k multiples of $b(z)$ from $a(z)$ until either 0 or a remainder is reached (the ordinary division process). The proof is similar to that for integral polynomials and is left for the student.

Corollary   If $a(z)$ and $z - c$ are polynomials over C of positive degrees n and 1, respectively, then there exist a unique polynomial $q(z)$ of degree $n - 1$ and a unique constant r in C so that

$$a(z) = (z - c)q(z) + r.$$

The factor and remainder theorems for the system of polynomials over C follow directly from the division algorithm.

Remainder Theorem for $C[z]$ If $p(z) = (z - a)q(z) + r$, then $p(a) = r$.

Factor Theorem for $C[z]$ $z - a$ is a factor of $p(z)$ if and only if $p(a) = 0$.

The proofs of the factor and remainder theorems are left for the student who is referred to Chapter 3.

Example 5 Find $p(3 + i)$ if $p(z) = z^3 - 6z^2 + 11z + 2$.

Solution Using synthetic division to find the remainder (see Chapter 3),

$$
\begin{array}{r|rrrr}
3+i & 1 & -6 & 11 & 2 \\
 & & 3+i & -10 & 3+i \\
\hline
 & 1 & -3+i & 1 & 5+i
\end{array}
$$

Thus, $p(3 + i) = 5 + i$.

Example 6 Factor $p(z) = z^3 - (4 + 2i)z^2 + (1 + 8i)z - 2i$ over C if $z - 2i$ is a factor.

Solution

$$
\begin{array}{r|rrrr}
2i & 1 & -4-2i & 1+8i & -2i \\
 & & 2i & -8i & 2i \\
\hline
 & 1 & -4 & 1 &
\end{array}
$$

Therefore,

$$p(z) = (z - 2i)(z^2 - 4z + 1)$$
$$z^2 - 4z + 1 = (z^2 - 4z + 4) - 3 = (z - 2)^2 - (\sqrt{3})^2$$
$$= (z - 2 + \sqrt{3})(z - 2 - \sqrt{3}).$$

Finally,

$$p(z) = (z - 2i)(z - 2 + \sqrt{3})(z - 2 - \sqrt{3}).$$

Example 6 may suggest that every polynomial over C of positive degree can be expressed as a product of linear factors with coefficients in C. This is indeed the case as the following theorem indicates.

Theorem (Complete Linear Factorization of Polynomials over C) If $p(z)$ is a polynomial over C of positive degree n, then $p(z)$ can be expressed as the product of n linear factors of the form $z - c_j$ where $c_j \in C$. In symbols, if

$$p(z) = a_n z^n + \ldots + a_1 z + a_0$$

where

$$a_n \neq 0 \text{ and } n > 0, \qquad n \in I$$

then
$$p(z) = a_n(z - c_1)(z - c_2)\ldots(z - c_n).$$

Proof The fundamental theorem of algebra guarantees that $p(z) = 0$ has one solution, call it c_1.

Then $p(z) = (z - c_1)q_1(z)$ by the factor theorem, and moreover, the degree of $q_1(z)$ is $n - 1$ by the division algorithm.

Similarly, $q_1(z)$ has one solution, call it c_2.

Then, $p(z) = (z - c_1)(z - c_2)q_2(z)$ and degree of $q_2(z) = n - 2$.

Continuing in this way, until a polynomial of degree $n - n = 0$ (that is, a constant k) is obtained,
$$p(z) = (z - c_1)(z - c_2)\ldots(z - c_n)k.$$

Since two polynomials are identical if and only if their corresponding coefficients are equal, $k = a_n$.

MULTIPLE ROOTS

While a polynomial over C of positive degree n has n linear factors of the form $z - c$, these factors may not be distinct. If the linear factor $z - c$ occurs exactly once, then c is called a **simple** root or a root of multiplicity 1. If the linear factor occurs exactly twice, then c is called a **double** root or a root of multiplicity 2. If the linear factor occurs exactly three times, then c is called a **triple** root or a root of multiplicity 3.

Definition c is a root of $p(z) = 0$ with *multiplicity* k if and only if the linear factor $z - c$ occurs exactly k times in the factorization of $p(z)$.

For example, if $p(z) = 3(z - 5)(z - 1)^2(z + 2)^3$, then 5, 1, and -2 are simple, double, and triple roots, respectively, of $p(z) = 0$.

Theorem (Uniqueness of the Linear Factorization) If $p(z) = a_n z^n + \ldots + a_1 z + a_0$ is a polynomial over C and if $c_1, c_2, \ldots, c_n$ are complex numbers so that
$$p(z) = a_n(z - c_1)(z - c_2)\ldots(z - c_n),$$
then this linear factorization is unique, disregarding the order of the factors.

Proof $p(z)$ cannot have a factor $z - d$ where d is different from all the c_j's. Otherwise d would be a root of $p(z) = 0$ and
$$p(d) = a_n(d - c_1)(d - c_2)\ldots(d - c_n) = 0.$$
This is impossible since none of the factors is zero.

Now it might be possible that the roots could occur with different multiplicities.

Suppose
$$p(z) = a_n(z-c)^k f(z), \text{ where } f(c) \neq 0$$
and
$$p(z) = a_n(z-c)^j g(z), \text{ where } g(c) \neq 0 \text{ and } k < j.$$
Then, for all z,
$$a_n(z-c)^j g(z) = a_n(z-c)^k f(z)$$
and for all z,
$$(z-c)^{j-k} g(z) = f(z).$$
For $z = c$,
$$(c-c)^{j-k} g(c) = f(c)$$
$$0 = 0 \cdot g(c) = f(c).$$

However, $f(c) \neq 0$ and therefore $k = j$. This means the factorization is unique.

Theorem (Number of Roots of a Polynomial Equation) If $p(z)$ is a polynomial over C of positive degree n, then $p(z) = 0$ has exactly n roots, with a root of multiplicity k counted as k roots.

Proof Since $p(z)$ has a unique linear factorization,
$$p(z) = a_n(z-c_1)(z-c_2)\ldots(z-c_n),$$
there are n roots, $c_1, c_2, \ldots, c_n$, if each is counted with its multiplicity.

EXERCISES 8.1

1–20. Solve each of the following equations and check by using the sum and product of roots theorem:

1. $x^2 - 7x + 6 = 0$
2. $3x^2 + 5x = 0$
3. $3x^2 + 11x - 4 = 0$
4. $x^2 - 5x + 6 = 0$
5. $x^2 - 4x + 2 = 0$
6. $x^2 + 6x + 13 = 0$
7. $2x^2 + 2\sqrt{2}x + 3 = 0$
8. $4x^2 + x - 3 = 0$
9. $(x-3)(x+1) = 1$
10. $\sqrt{2}x^2 - 3x + \sqrt{2} = 0$
11. $2ix^2 + 5x - 2i = 0$
12. $ax^2 + 2bx - c = 0$; a, b, and c are complex numbers, $a \neq 0$
13. $4x^2 - 4x + 3 = 0$
14. $3ix^2 + 4x - 2i = 0$
15. $2ix^2 - 3x + 5i = 0$
16. $x^2 - 9 = -2x^2 - 36$

17. $\dfrac{x^2}{3} = \dfrac{x^2}{6} + 2$

18. $\dfrac{1}{2x-1} - \dfrac{1}{4} = \dfrac{1}{2x+1}$

19. $x(5x - 4) = 2$

20. $\dfrac{3x-2}{3x+2} = \dfrac{4x+1}{8x-7}$

21–25. Determine the value(s) of k that will make the roots of each of the following equations equal:

21. $2x^2 + 8x + k = 0$
22. $9x^2 + kx + 16 = 0$
23. $kx^2 + kx + 2x + 3 = 0$
24. $3x^2 + kx + k = 4$
25. $kx^2 + 6x + k - 8 = 0$

26. Prove that $ax^2 + bx + c$ is a perfect square if and only if $b^2 - 4ac = 0$. ($a \neq 0$.)

27. Prove: For all real numbers a, b, c, and x, $ax^2 + bx + c > 0$ if and only if $a > 0$ and $b^2 - 4ac < 0$.

☆28. Show that $B^2 - 4AC = B'^2 - 4A'C'$ for $Ax^2 + Bxy + Cy^2 + Dx + Ey = 0$ if $x = x' \cos \theta - y' \sin \theta$, and $y = x' \sin \theta + y' \cos \theta$.

29. If $P(x) = 2x^3 - 3x^2 + 2x + 4$, find $P(2)$, $P(i)$, and $P(1 - i)$.

30. If $Q(x) = x^4 - 3x^2 + 6x - 2$, find $Q(2)$, $Q(-3)$, $Q(i)$, and $Q(1 + i)$.

31. If $P(x) = 3x^2 + 4x + k$, find k so that $P(3) = 6$.

32. If $P(x) = 2x^3 - kx^2 + 13x + k$, find k so that $P(4) = 0$.

33. Find k so that $x + 2$ is a factor of $x^4 + 2x^3 + x - k$.

34. Find m so that when $y^3 - m^2y^2 - my - 4$ is divided by $y - 3$, the remainder is 23.

35. Find n such that $x^3 - 4nx^2 + nx + 6$ is exactly divisible by $x - 2$.

36. Find the remainder when $2x^4 - 3x^3 + 5x - 1$ is divided by $x - 4$
    (a) by long division
    (b) by synthetic division
    (c) by the remainder theorem.

37. Use the factor theorem to check if $2 + 4i$ is a root of the equation
$$y^4 - 4y^3 + 18y^2 + 8y - 40 = 0.$$

38. Use the factor theorem to check if $2i$ is a root of the equation
$$x^3 - x^2 + 4x - 4 = 0.$$

39–45. Factor each of the following polynomials over (a) the integers; (b) the real numbers; (c) the complex numbers.

39. $x^4 - 4$
40. $x^4 - x^2 - 2$
41. $x^3 + 1$
42. $x^4 + 4$
43. $x^4 + 3x^3 + 6x^2 + 12x + 8$
44. $x^3 - 4x^2 + 14x - 20$
45. $x^4 + 2x^2 + 1$

46–50. For each of the following, write the equation of lowest degree with integral coefficients, having the given numbers as two of its roots:

46. 2 and $1 + 3i$
47. $1 - \sqrt{3}$ and 2
48. $-3i$ and $\sqrt{3}$
49. -2 and $1 - 2i$
50. $1 + 2i$ and $2 - \sqrt{3}$

51–58. Solve each of the following equations in the set of complex numbers and state the multiplicity of each root.

51. $x^4 - 4x^3 - 5x^2 = 0$
52. $x^4 - 6x^3 + 13x^2 - 24x + 36 = 0$
53. $x^3 - x - 6 = 0$
54. $x^4 + 19x^2 - 150 = 0$
55. $(x^2 + 2x - 3)(x^2 + 5x + 6) = 0$
56. $x^3 - 3x^2 + 3x - 1 = 0$
57. $x^3 + 6x^2 + 12x + 8 = 0$
58. $x^4 - 4x^2 + 4 = 0$

59. Find the remainder when $x^4 - x^3 - 3x^2 + 4x + 4$ is divided by $x^2 - 4$.
60. Find a quadratic equation whose roots are the squares of the roots of $x^2 - bx + c = 0$.
61. Prove that $x^3 + bx^2 + cx + d$ has a factor of the form $x^2 - k^2$ if and only if $d = bc$.
62. Show that for a cubic equation $ax^3 + bx^2 + cx + d = 0$ $(a \neq 0)$, the sum of the roots equals $-b/a$ and the product of the roots equals $-d/a$. What does c/a represent?
63. Show that for a quartic equation $ax^4 + bx^3 + cx^2 + dx + e = 0$ $(a \neq 0)$, the sum of the roots equals $-b/a$ and the product of the roots equals e/a. Generalize for an equation of degree n.

8.2 RATIONAL ROOTS AND SPECIAL PAIRS OF ROOTS

8.2.1 Rational Roots

The rational roots of integral polynomial equations may be found by using the following theorem:

Theorem (Rational Roots of Integral Polynomial Equations) If the coefficients of $p(z) = a_n z^n + \ldots + a_1 z + a_0$ are integers and if r/s is a rational number in simplified form (that is, r and s are integers with no common factor except 1 and -1), and if r/s is a root of

$$a_n z^n + \ldots + a_1 z + a_0 = 0,$$

then the numerator r divides the constant term, a_0, and the denominator s divides the leading coefficient, a_n.

RATIONAL ROOTS AND SPECIAL PAIRS OF ROOTS 279

Proof The number r/s is a root if and only if

$$a_n\left(\frac{r}{s}\right)^n + \ldots + a_1\left(\frac{r}{s}\right) + a_0 = 0.$$

Multiplying by s^n,

$$a_n r^n + a_{n-1} r^{n-1} s + \ldots + a_1 r s^{n-1} + a_0 s^n = 0.$$

Subtracting $a_0 s^n$ from both sides and then factoring the left side,

$$r(a_n r^{n-1} + a_{n-1} r^{n-2} s + \ldots + a_1 s^{n-1}) = -a_0 s^n.$$

Since r is a factor of the left side, r must also be a factor of $a_0 s^n$. However, r is not a factor of s since r/s is in simplified form and therefore r divides a_0.
Similarly, $a_n r^n = -s(a_{n-1} r^{n-1} + \ldots + a_1 r s^{n-2} + a_0 s^{n-1})$ and since s is not a factor of r, therefore s divides a_n.

Example 1 Find the rational roots of $3x^4 + 4x^3 - x^2 + 4x - 4 = 0$ and solve.

Solution Possible numerators: $1, -1, 2, -2, 4, -4$
Possible denominators: $1, 3$
Possible rational roots: $\pm 1, \pm 2, \pm 4, \pm \frac{1}{3}, \pm \frac{2}{3}, \pm \frac{4}{3}$

These possible roots are tested until it is seen that

$$\begin{array}{r|rrrrr} -2 & 3 & 4 & -1 & 4 & -4 \\ & & -6 & 4 & -6 & 4 \\ \hline & 3 & -2 & 3 & -2 & 0 \end{array}$$

$$3x^4 + 4x^3 - x^2 + 4x - 4 = (x+2)(3x^3 - 2x^2 + 3x - 2).$$

Eliminating the possibilities that failed the first time, the possible rational roots of the depressed equation are $-2, \pm \frac{1}{3}, \pm \frac{2}{3}$. ($-2$ must be tried again in case it is a multiple root.)
Finally,

$$\begin{array}{r|rrrr} \frac{2}{3} & 3 & -2 & 3 & -2 \\ & & 2 & 0 & 2 \\ \hline & 3 & 0 & 3 & 0 \end{array}$$

$$\begin{aligned} 3x^4 + 4x^3 - x^2 + 4x - 4 &= (x+2)(x-\tfrac{2}{3})(3x^2+3) \\ &= 3(x+2)(x-\tfrac{2}{3})(x^2+1) \\ &= 3(x+2)(x-\tfrac{2}{3})(x+i)(x-i). \end{aligned}$$

Thus, the rational roots are -2 and $\frac{2}{3}$ and the imaginary roots are i and $-i$.

Example 2 Find the rational roots of $4x^5 - 3x^3 - 3x^2 + 4x - 1 = 0$.

Solution The possible rational roots are $1, -1, \frac{1}{2}, -\frac{1}{2}, \frac{1}{4}, -\frac{1}{4}$. Finally, after testing 1 and -1,

$$\begin{array}{r|rrrrrr} \frac{1}{2} & 4 & 0 & -3 & -3 & 4 & -1 \\ & & 2 & 1 & -1 & -2 & 1 \\ \hline & 4 & 2 & -2 & -4 & 2 & 0 \end{array}$$

Therefore, $\frac{1}{2}$ is a root and the depressed equation is

$$4x^4 + 2x^3 - 2x^2 - 4x + 2 = 2(2x^4 + x^3 - x^2 - 2x + 1) = 0.$$

Since 1 and -1 were ruled out from the original equation, the remaining possibilities are $\frac{1}{2}$ and $-\frac{1}{2}$.

$$\begin{array}{r|rrrrr} \frac{1}{2} & 2 & 1 & -1 & -2 & 1 \\ & & 1 & 1 & 0 & -1 \\ \hline & 2 & 2 & 0 & -2 & 0 \end{array}$$

Thus, $\frac{1}{2}$ is a double root, and

$$4x^5 - 3x^3 - 3x^2 + 4x - 1 = 4(x - \tfrac{1}{2})(x - \tfrac{1}{2})(x^3 + x^2 - 1) = 0.$$

The only possibilities for $x^3 + x^2 - 1 = 0$ are 1 and -1, and since these have been eliminated, the only rational roots are $\frac{1}{2}$ and $\frac{1}{2}$, or $\frac{1}{2}$ is a double root.

8.2.2 Conjugate Imaginary Roots

The following theorem states that the imaginary zeros of a polynomial with *real* coefficients always occur in pairs.

Theorem (Conjugate Imaginary Roots) If $a + bi$ $(b \neq 0)$ is an imaginary root of $p(z) = a_n z^n + \ldots + a_0 = 0$, and if the coefficients of $p(z)$ are *real* numbers with $a_n \neq 0$, then $a - bi$ is also a root.

Proof Let $f(z) = [z - (a + bi)][z - (a - bi)] = z^2 - 2az + a^2 + b^2$.
Then the coefficients of $f(z)$ are real numbers.
Now, by the division algorithm,

$$p(z) = a_n z^n + \ldots + a_1 = (z^2 - 2az + a^2 + b^2)q(z) + cz + d,$$

where the remainder $cz + d$ is linear since its degree is less than the degree of the quadratic divisor. Moreover, the coefficients of $q(z)$ and $cz + d$ are real numbers since only rational operations were performed and the set of real numbers is closed with respect to the rational operations.
Now since $a + bi$ is a root, then

$$p(a + bi) = 0 \cdot q(a + bi) + c(a + bi) + d = 0.$$

RATIONAL ROOTS AND SPECIAL PAIRS OF ROOTS

Thus,
$$(ca + d) + cbi = 0 = 0 + 0i.$$
Therefore,
$$ca + d = 0 \quad \text{and} \quad cb = 0.$$

Since $b \neq 0$, $c = 0$ and as a result, $d = 0$.
Then,
$$p(a - bi) = [(a - bi) - (a + bi)][(a - bi) - (a - bi)]q(a - bi)$$
or
$$p(a - bi) = 0$$
and $a - bi$ is a root of $p(z) = 0$.

Example 3 Solve $z^4 + 12z - 5 = 0$ given that $1 + 2i$ is a root.

Solution Since $1 + 2i$ is a root, $1 - 2i$ is also a root. Subsequently, $[z - (1 + 2i)][z - (1 - 2i)] = z^2 - 2z + 5$ is a factor of $z^4 + 12z - 5$. By division, $z^4 + 12z - 5 = (z^2 - 2z + 5)(z^2 + 2z - 1)$.
Solving $z^2 + 2z - 1 = 0$, $z = -1 \pm \sqrt{2}$.
The solution set is $\{-1 + \sqrt{2}, -1 - \sqrt{2}, 1 + 2i, 1 - 2i\}$.

Example 4 Show that $1 - 3i$ is a root of $z^3 + 3iz^2 - z + 9 + 3i = 0$ but $1 + 3i$ is not a root.

Solution

$$
\begin{array}{r|rrrr}
1 - 3i & 1 & 3i & -1 & 9 + 3i \\
& & 1 - 3i & 1 - 3i & -9 - 3i \\
\hline
1 + 3i & 1 & 1 & -3i & \\
& & 1 + 3i & -7 + 9i & \\
\hline
& 1 & 2 + 3i & -7 + 6i &
\end{array}
$$

Therefore, $1 - 3i$ is a root and $1 + 3i$ is *not* a root.

Note that this example does not contradict the theorem on conjugate imaginary roots since the coefficients of the polynomial were *not* all real numbers.

Example 3 may suggest that if $a + \sqrt{b}$ is an irrational root of the rational polynomial equation $p(z) = 0$, then $a - \sqrt{b}$ is also a root. The next theorem indicates that this is the case.

8.2.3 Conjugate Surd Roots

If a and b are rational numbers and $\sqrt{b}$ is irrational, then $a + \sqrt{b}$ is called a binomial **surd** and $a - \sqrt{b}$ is called its **conjugate surd**. The conjugate surd zeros of polynomials with *rational* coefficients also occur in pairs.

Theorem (Conjugate Surd Roots) If the surd $a + \sqrt{b}$ is a root of $p(z) = 0$ and if the polynomial $p(z)$ has *rational* coefficients, then $a - \sqrt{b}$ is also a root of $p(z) = 0$.

The proof of this theorem is similar to that for conjugate imaginary roots by considering

$$f(z) = [z - (a + \sqrt{b})][z - (a - \sqrt{b})]$$
$$= z^2 - 2az + a^2 - b.$$

The details of the proof are left for the student.

Example 5 Solve $z^4 - 5z^3 + 4z^2 + 3z - 1 = 0$ given that $2 + \sqrt{3}$ is a root.

Solution Since $2 + \sqrt{3}$ is a root and since the coefficients of the polynomial are rational, then $2 - \sqrt{3}$ is a root.
Since

$$(2 + \sqrt{3}) + (2 - \sqrt{3}) = 4$$

and

$$(2 + \sqrt{3})(2 - \sqrt{3}) = 4 - 3 = 1,$$

$z^2 - 4z + 1$ is a factor of $p(z)$ and

$$z^4 - 5z^3 + 4z^2 + 3z - 1 = (z^2 - 4z + 1)(z^2 - z - 1).$$

Solving,

$$z^2 - z - 1 = 0, \qquad z = \frac{1 \pm \sqrt{5}}{2}.$$

The solution set is

$$\left\{ 2 + \sqrt{3},\, 2 - \sqrt{3},\, \frac{1 + \sqrt{5}}{2},\, \frac{1 - \sqrt{5}}{2} \right\}.$$

EXERCISES 8.2

1–20. Find the rational roots of each of the following equations:

1. $x^4 - 2x^2 - 3x - 2 = 0$
2. $2x^4 - x^3 - 3x^2 - 31x - 15 = 0$
3. $2x^3 + x^2 - 7x - 6 = 0$
4. $2x^3 + 3x^2 - 14x - 21 = 0$
5. $4x^4 - 13x^3 - 7x^2 + 41x - 14 = 0$
6. $2x^3 - 7x^2 + 10x - 6 = 0$
7. $2x^4 + x^2 + 2x - 4 = 0$
8. $x^3 - x^2 + 2x - 14 = 0$

RATIONAL ROOTS AND SPECIAL PAIRS OF ROOTS 283

9. $x^4 - x^2 - 2 = 0$
10. $x^3 - 7x^2 + 13x - 3 = 0$
11. $2x^3 - 10x^2 + 13x - 20 = 0$
12. $3x^3 + 5x^2 - 16x - 12 = 0$
13. $x^5 - 12x^3 + x^2 - 4 = 0$
14. $x^3 - 5x^2 + 7x + 13 = 0$
15. $2x^3 - 5x^2 - 28x + 15 = 0$
16. $4x^3 + 15x - 36 = 0$
17. $x^3 + 2x^2 - 21x + 18 = 0$
18. $4x^4 - x^3 - 8x^2 + 18x - 4 = 0$
19. $x^3 + 3x^2 - 5x - 10 = 0$
20. $x^4 - 6x^3 + 10x^2 - 6x + 9 = 0$

21–30. Solve each of the following:

21. $x^4 - 6x^3 + 5x^2 + 12x - 14 = 0$ given $3 + \sqrt{2}$ is a root.
22. $x^4 - 18x^2 + 1 = 0$ given $2 - \sqrt{5}$ is a root.
23. $x^4 + 2x^3 + 2x - 1 = 0$ given $-1 + \sqrt{2}$ is a root.
24. $x^4 + 6x^3 + 8x^2 + 24x + 16 = 0$ given $-3 - \sqrt{5}$ is a root.
25. $x^4 - 7x^2 + 12x + 10 = 0$ given $2 + i$ is a root.
26. $x^4 - 2x^3 + 4x - 4 = 0$ given $1 - i$ is a root.
27. $x^4 - 12x - 5 = 0$ given $-2i - 1$ is a root.
28. $x^4 + 4x^3 + 10x^2 - 12x - 39 = 0$ given $-2 + 3i$ is a root.
29. $x^4 + x^3 + 4x - 16 = 0$ given $2i$ is a root.
30. $x^5 - 7x^3 + x^2 + 10x - 5 = 0$ given $\sqrt{5}$ is a root.

31. If $1 + i\sqrt{5}$ is a root of $x^3 - 6x^2 + cx + d$, where c and d are real numbers, find the other two roots, and also find c and d.
32. If $2 + \sqrt{3}$ is a root of $4x^3 - 15x^2 + cx + d$, where c and d are rational numbers, find the other two roots and also find c and d.
33. Find a cubic equation with real coefficients if two of its roots are 1 and $4 - 3i$.
34. Find a cubic equation with integral coefficients if two of its roots are 1/3 and $1 - \sqrt{2}$.
35. Find a quartic equation with rational coefficients if two of its roots are $1 + \sqrt{5}$ and $-2 + \sqrt{3}i$.
36. Find a quartic equation with integral coefficients if two of its roots are $\sqrt{5}i$ and $3 - \sqrt{7}$.
37. Find a quartic equation with integral coefficients if $1 - \sqrt{3}$ is a double root.
38. Find a quartic equation with integral coefficients if $1 + i$ and $-1 - i$ are two of its roots.
39. Prove that every polynomial with real coefficients can be expressed as a product of linear and quadratic factors with real coefficients.

☆40. Show that if a cubic polynomial with integral coefficients has a double root or a triple root, then it has a rational root.

☆41. Show that if a quartic integral polynomial has a multiple root and the quartic is not a perfect square, then it has a rational root.

42. Use the theorem on rational roots to prove that $\sqrt{3}$ is not rational.

8.3 UPPER AND LOWER BOUNDS; DESCARTES' RULE OF SIGNS

Determining the rational zeros of an integral polynomial may be a tedious task if the constant term or the leading coefficient have many factors. For example, the possible rational zeros of $15z^5 - z^4 - 8z^3 + 6z^2 + 8z - 24$ are

$$\pm \left[1, 2, 3, 4, 6, 8, 12, 24, \frac{1}{3}, \frac{2}{3}, \frac{4}{3}, \frac{8}{3}, \frac{1}{5}, \frac{2}{5}, \frac{3}{5}, \frac{4}{5}, \frac{6}{5}, \frac{8}{5}, \frac{12}{5}, \frac{24}{5}, \frac{1}{15}, \frac{2}{15}, \frac{4}{15}, \frac{8}{15} \right].$$

The theorems in this section provide further information regarding the nature of the roots of a polynomial equation and indicate how to reduce the number of trials for locating the rational roots.

8.3.1 Upper and Lower Bounds

Definition of Upper Bound for Real Roots The real number U is an upper bound for the real roots of $p(z) = 0$ if and only if all real roots of $p(z) = 0$ are less than or equal to U. (If r is a real root of $p(z) = 0$, then $r \leq U$.)

Definition of Lower Bound for Real Roots The real number L is a lower bound for the real roots of $p(z) = 0$ if and only if all real roots of $p(z) = 0$ are greater than or equal to L. (If r is a real root of $p(z) = 0$, then $L \leq r$.)

Theorem (Upper Bounds for Real Roots) If U is a positive real number and if the coefficients of $p(z) = a_n z^n + \ldots + a_1 z + a_0 = 0$ are real numbers with a_n positive and if the third line of the synthetic division for the divisor $z - U$ has no negative numbers, then U is an upper bound for the real roots of $p(z) = 0$.

Proof By the division algorithm,

$$p(z) = (z - U)q(z) + k, \quad \text{where } k \text{ is a constant.}$$

Now, the third line of the synthetic division contains the coefficients of $q(z)$ and the constant k and these are all assumed positive or zero. Suppose $s > U$. Then s is positive since U is positive and

$$p(s) = (s - U)q(s) + k.$$

UPPER AND LOWER BOUNDS; DESCARTES' RULE OF SIGNS

Now $q(s)$ cannot be negative since s is positive and all the coefficients of $q(z)$ are either positive or zero. (By the positive closure axioms for the set of real numbers, sums and products of positive real numbers are positive.) Also $q(s) \neq 0$ since $q(z)$ has at least one positive coefficient, a_n. Since $s - U$ is positive and k is positive or zero, therefore $p(s)$ is positive, $p(s) \neq 0$, and s is not a root of $p(z)$.

A lower bound for the real roots may be obtained by noting that the opposite of a negative real root of $p(z) = 0$ is a positive real root of $p(-z) = 0$.

Theorem (Negative Real Roots) If $-r$ is a negative real root of $p(z) = 0$, then r is a positive real root of $p(-z) = 0$.

Proof $p(z) = (z + r)q(z)$
$p(-z) = (-z + r)q(-z)$
$p(-z) = -(z - r)q(-z)$ or r is a root of $p(-z) = 0$.

By obtaining an upper bound for the positive roots of $p(-z) = 0$, a lower bound is thereby obtained for the negative roots of $p(z) = 0$.

Let L be an upper bound for the roots of $p(-z) = 0$. Then, for all roots r of $p(-z) = 0$, $r \leq L$. This implies that $-r \geq -L$, or $-L \leq -r$.

Example 1 Determine an upper and lower bound for the real roots of

$$z^5 - z^4 - 8z^3 + 6z^2 + 8z - 24 = 0$$

and then solve.

Solution The possible rational roots are $\pm 1, \pm 2, \pm 3, \pm 4, \pm 6, \pm 8, \pm 12, \pm 24$. Trying the possible positive integral roots, in the listed order, finally,

$$\begin{array}{r|rrrrrr} 3 & 1 & -1 & -8 & 6 & 8 & -24 \\ & & 3 & 6 & -6 & 0 & 24 \\ \hline & 1 & 2 & -2 & 0 & 8 & 0 \end{array}$$

Therefore, 3 is a root, but since one of the numbers in the third line is negative, it cannot be concluded that 3 is an upper bound. Trying 4,

$$\begin{array}{r|rrrrrr} 4 & 1 & -1 & -8 & 6 & 8 & -24 \\ & & 4 & 12 & 16 & 88 & 384 \\ \hline & 1 & 3 & 4 & 22 & 96 & 360 \end{array}$$

Thus, 4 is an upper bound, and 4, 6, 8, 12, and 24 cannot be roots. Also, 1 and 2 were eliminated by the first two trials. Thus, 3 is the only positive rational root.

Now the negative roots are investigated.

$p(-z) = (-z)^5 - (-z)^4 - 8(-z)^3 + 6(-z)^2 + 8(-z) - 24$
$p(-z) = -z^5 - z^4 + 8z^3 + 6z^2 - 8z - 24$

Since the theorem on upper bounds requires the leading coefficient to be positive, $p(-z)$ is replaced by $-p(-z)$ by noting that the roots of $-p(-z) = 0$ are the same as the roots of $p(-z) = 0$.

$$-p(-z) = z^5 + z^4 - 8z^3 - 6z^2 + 8z + 24$$

Trying 1, 2, 3, 4, 6, 8, 12, and 24 in order, $-p(-1) = 20$ and -1 is not a root of $p(z) = 0$. Now trying 2,

$$\begin{array}{r|rrrrrr} 2 & 1 & 1 & -8 & -6 & 8 & 24 \\ & & 2 & 6 & -4 & -20 & -24 \\ \hline & 1 & 3 & -2 & -10 & -12 & 0 \end{array}$$

Therefore, -2 is a root of $p(z) = 0$, but it cannot be concluded that -2 is a lower bound for the real roots of $p(z) = 0$ since the third line of the synthetic division contains negative numbers. Trying 3,

$$\begin{array}{r|rrrrrr} 3 & 1 & 1 & -8 & -6 & 8 & 24 \\ & & 3 & 12 & 12 & 18 & 78 \\ \hline & 1 & 4 & 4 & 6 & 26 & 102 \end{array}$$

Consequently, -3 is a lower bound for the real roots of $p(z) = 0$. Combining these results, it is seen for all real roots r of $p(z) = 0$,

$$-3 < r < 4.$$

Now,

$$p(z) = (z - 3)(z^4 + 2z^3 - 2z^2 + 8)$$

and

$$p(z) = (z - 3)(z + 2)(z^3 - 2z + 4).$$

The possible rational roots of $z^3 - 2z + 4$ are $\pm 1, 2, 4$. However, the bounds and the synthetic division trials already made now reduce the possibilities to -2 only. Trying -2,

$$\begin{array}{r|rrrr} -2 & 1 & 0 & -2 & 4 \\ & & -2 & 4 & -4 \\ \hline & 1 & -2 & 2 & 0 \end{array} \quad \text{and } -2 \text{ is a root.}$$

Thus, $p(z) = (z - 3)(z + 2)(z + 2)(z^2 - 2z + 2)$.
Solving $z^2 - 2z + 2 = 0$,

$$z^2 - 2z + 1 = -1$$
$$(z - 1)^2 = -1 \quad \text{and} \quad z = 1 \pm i.$$

Finally,

$$p(z) = (z - 3)(z + 2)(z + 2)(z - 1 - i)(z - 1 + i).$$

The roots of $p(z) = 0$ are $3, -2, -2, 1 + i, 1 - i$ where -2 is a double root.

8.3.2 Descartes' Rule of Signs

If the terms of a polynomial are written in descending order (or ascending order), then the polynomial is said to have a **variation** in sign if the coefficient of two consecutive terms have opposite signs, the missing terms being ignored. For example,

$$x^5 - 3x^3 - 2x + 4$$

has two variations in sign. The number of variations in sign provides information regarding the number of positive and negative real zeros of a polynomial with real coefficients.

Theorem (Descartes' Rule of Signs) If $p(z)$ is a polynomial with real coefficients and if $v^+ =$ the number of variations in sign of $p(z)$, then P, the number of positive real roots of $p(z) = 0$, is either equal to the number of variations or differs from it by an even positive integer; that is

$$P = v^+ - 2k, \quad \text{where } k = 0 \text{ or a positive integer.}$$

If v^- is the number of variations in sign of $p(-z)$, then N, the number of negative real roots of $p(z) = 0$, is either equal to v^- or differs from it by an even positive integer; that is,

$$N = v^- - 2k, \quad \text{where } k = 0 \text{ or a positive integer.}$$

Before proving this theorem, its usefulness will be illustrated by examples.

Example 2 Determine the nature of the roots of $z^3 - 3z^2 - 5 = 0$ by using Descartes' rule of signs.

Solution $p(z) = z^3 - 3z^2 - 5$ and $v^+ = 1$.

Therefore, $p(z)$ has exactly one positive real root. (Note that $1 - 2 = -1$ is impossible since the number of roots must be positive or 0.)

$$p(-z) = -z^3 - 3z^2 - 5 \quad \text{and} \quad v^- = 0.$$

Therefore, $p(z)$ has no negative real roots. Finally, $p(z)$ has one positive real root and two conjugate imaginary roots.

Example 3 Determine the nature of the roots of $z^3 - 3z + 5 = 0$ by using Descartes' rule of signs.

Solution $p(z) = z^3 - 3z + 5$ and $v^+ = 2$.

Therefore, $p(z)$ has either two positive real roots or none at all.

$$p(-z) = -z^3 + 3z + 5 \quad \text{and} \quad v^- = 1.$$

Therefore, $p(z)$ has exactly one negative real root. Finally, $p(z)$ has either one negative real root and two imaginary roots or one negative real root and two positive real roots.

Example 4 Find the real roots of $25z^5 + 21z^3 + 11z - 6 = 0$ making use of Descartes' rule of signs.

Solution $v^+ = 1$ and there is exactly one positive real root. $v^- = 0$ and there are no negative real roots.

$$p(-z) = -25z^5 - 21z^3 - 11z - 6$$

The possible rational roots are now

$$1, 2, 3, 6, \frac{1}{5}, \frac{2}{5}, \frac{3}{5}, \frac{6}{5}, \frac{1}{25}, \frac{2}{25}, \frac{3}{25}, \frac{6}{25}.$$

Trying 1,

$$\begin{array}{r|rrrrrr}
1 & 25 & 0 & 21 & 0 & 11 & -6 \\
& & 25 & 25 & 46 & 46 & 57 \\
\hline
& 25 & 25 & 46 & 46 & 57 & 51
\end{array}$$

Therefore, 1 is an upper bound.
Trying $\frac{1}{5}$, it is seen that $\frac{1}{5}$ is not a root.
Trying $\frac{2}{5}$,

$$\begin{array}{r|rrrrrr}
\frac{2}{5} & 25 & 0 & 21 & 0 & 11 & -6 \\
& & 10 & 4 & 10 & 4 & 6 \\
\hline
& 25 & 10 & 25 & 10 & 15 & 0
\end{array}$$

Thus, $\frac{2}{5}$ is the only real root. No further trials are necessary since there is exactly one positive real root and no negative real roots.

Proof of Descartes' Rule of Signs (Optional) If r is a positive real root of $p(z) = 0$, then $p(z) = (z - r)q(z)$ and $q(z)$ has at least one less variation in sign than $p(z)$.

By considering the synthetic division,

$$\begin{array}{r|cccc}
r & a_n & a_{n-1} \ldots a_1 & a_0 \\
& & ra_n & rq_0 \\
\hline
& a_n & q_{n-1} \ldots q_0 & 0
\end{array}$$

no change in sign can occur in the third line until there is a change of sign in the first line. When there is a change, the numbers in the third line will remain negative or positive until a change of sign again occurs in the first line. Since the last number in the third line is 0, q_0 and a_0 must have opposite signs. Therefore, a sign change had to occur in the first line for which there was no change in the third line. This means that $q(z)$ has at least one less variation in sign than $p(z)$ and perhaps even fewer.

Now suppose $p(z)$ has exactly P positive real roots. Then

$$p(z) = (z - r_1)(z - r_2)\ldots(z - r_P)q_P(z).$$

UPPER AND LOWER BOUNDS; DESCARTES' RULE OF SIGNS

Each successive synthetic division process produces at least one less variation in sign. In all, there are at least P fewer variations in sign in $q_P(z)$ than in $p(z)$. Therefore, since the number of variations in sign of $p(z)$ cannot be negative, the number P of positive real roots is either equal to v^+ or less than v^+.

Since $r_1, r_2, \ldots, r_P$ are the positive real roots of $p(z) = 0$, then $q_P(z)$ must be the product of linear factors producing negative real roots and quadratic factors producing conjugate imaginary roots.

Now it will be shown that $q_P(z)$ has an even number of variations because its first and last coefficients are positive.

The first coefficient is a_n which has been assumed positive.

Now let $-s_1, -s_2, \ldots, -s_N$ be the negative real roots. Then the product $(z + s_1)(z + s_2)\ldots(z + s_N)$ has positive first and last coefficients.

Since imaginary roots occur in conjugate pairs, each quadratic factor $(z - a - bi)(z - a + bi) = z^2 - 2az + a^2 + b^2$ has positive first and last coefficients.

Now, if P, the number of positive real roots, is even, then the first and last coefficients of $p(z)$ are positive and $p(z)$ must have an even number of variations in sign.

If P is odd, then the first coefficient of $p(z)$ is positive and its last coefficient is negative and $p(z)$ must have an odd number of variations in sign.

Since the difference of two even numbers is even and the difference of two odd numbers is even, in either case, the difference $v^+ - P$ is always an even number.

$$v^+ - P = 2k \quad \text{or} \quad P = v^+ - 2k.$$

The second part of the theorem follows directly from the first part and the details of the proof are left for the student.

EXERCISES 8.3

1–10. Find integers which are upper bounds and lower bounds of the real roots of each of the following equations:

1. $x^4 - 2x^3 + 33x^2 - 72x - 108 = 0$
2. $x^5 + x^3 - 2x^2 - 2 = 0$
3. $2x^4 - 7x^3 + x^2 - 5x - 20 = 0$
4. $x^3 + 2x^2 - 5x - 15 = 0$
5. $2x^4 + 2x^3 + 29x^2 + 9x + 90 = 0$
6. $x^4 + x^3 - 2x^2 - 2 = 0$
7. $x^3 + 2x^2 + 8x - 12 = 0$
8. $2x^3 + x^2 + 8x - 12 = 0$
9. $2x^4 - 6x^3 - 8x^2 + 5x - 16 = 0$
10. $4x^4 + x^3 - 7x^2 - 18x - 20 = 0$

11–20. Use Descartes' rule of signs to determine the nature of the roots of each of the following equations:

11. $2x^3 - x - 5 = 0$
12. $x^6 - 3x^5 + 2x^2 - 3x + 1 = 0$
13. $x^5 + 3x^3 - 5x - 16 = 0$
14. $x^6 - 64 = 0$
15. $x^5 + 32 = 0$
16. $x^4 + 3x^2 - 5x - 1 = 0$

17. $x^6 + 2x^4 + 5x^2 + 3 = 0$
18. $3x^5 - 2x^3 + 5x - 8 = 0$
19. $x^6 - 3x^2 + 4x - 1 = 0$
20. $x^8 - 2x^4 + 1 = 0$

21–30. Find the rational roots of each of the following, using the theorems on upper and lower bounds and Descartes' rule of signs to reduce the trials:

21. $x^4 + x^3 + 10x^2 + 12x - 24 = 0$
22. $x^4 - 2x^3 - 15x^2 + 24x + 36 = 0$
23. $48x^5 + 8x^4 - 16x^3 + 162x^2 + 27x - 54 = 0$
24. $x^4 + x^2 - 370x - 1200 = 0$
25. $x^4 - 11x^3 + 32x^2 - 22x + 60 = 0$
26. $x^4 + 6x^3 + 17x^2 + 54x + 72 = 0$
27. $36x^4 - 65x^2 - 36 = 0$
28. $x^4 - 5x^3 + 3x^2 - 11x + 60 = 0$
29. $x^4 + 13x^3 + 52x^2 + 84x + 144 = 0$
30. $50x^4 + 35x^3 + 18x^2 - 16x - 32 = 0$

31–36. Find the real roots of each of the following:

31. $x^6 - 8 = 0$
32. $x^5 + 32 = 0$
33. $x^4 + 81 = 0$
34. $x^6 + 64 = 0$
35. $x^7 - 10^{-7} = 0$
36. $x^8 - 81 = 0$

37. Show that $z^n - 1 = 0$ has exactly one real root if n is odd.
38. Show that $z^n - 1 = 0$ has exactly two real roots if n is even.
39. Show that $z^n + 1 = 0$ has exactly one real root if n is odd.
40. Show that $z^n + 1 = 0$ has no real roots if n is even.
☆41. Prove: If the coefficients of $P(z)$ are real, and if the degree of $P(z) = n$, and if $P(z)$ has n real roots, then the number of positive real roots of $P(z)$ is equal to v^+.
42. Prove that $x^4 + x^3 - 2x^2 - 4 = 0$ has exactly two irrational roots.
43. Prove that $2x^4 - 15x + 3$ has exactly two real and two imaginary roots.

8.4 WEIERSTRASS ZERO THEOREM; APPROXIMATIONS

8.4.1 Weierstrass Zero Theorem

While the rational real roots of a polynomial equation with real coefficients can be readily determined by using the techniques developed in the previous sections, it is a more difficult task to find the irrational real roots. The Weierstrass zero theorem provides a method for locating an irrational real root between two rational numbers.

WEIERSTRASS ZERO THEOREM; APPROXIMATIONS

Theorem (The Weierstrass Zero Theorem) If $p(z)$ is a polynomial with real coefficients and if a and b are real numbers so that $p(a) < 0$ and $p(b) > 0$, then $p(z)$ has a real zero between a and b.

Proof Since the imaginary zeros of a polynomial with real coefficients occur in conjugate pairs such as $c + di$ and $c - di$, and since the product

$$(z - c - di)(z - c + di) = z^2 - 2cz + c^2 + d^2$$

is a polynomial with real coefficients, it follows from the unique factorization of polynomials over C that every polynomial with real coefficients can be expressed as a product of linear factors and prime quadratic factors over R, where each linear factor and each prime quadratic factor have real coefficients.

The prime quadratic factors have the form

$$(z^2 - 2cz + c^2) + d^2 = (z - c)^2 + d^2$$

which is always positive for all real z and $d \neq 0$.

Now if $p(a) < 0$ and $p(b) > 0$ for the real numbers a and b, then the sign change must result from a sign change of one of the linear factors, say $L(z) = z - r$ where r is a real zero of $p(z)$.

There are two possibilities.

CASE 1 $L(a) = a - r < 0$ and $r > a$
            $L(b) = b - r > 0$ and $r < b$

Then $a < r < b$.

CASE 2 $L(a) = a - r > 0$ and $r < a$
            $L(b) = b - r < 0$ and $r > b$

Then, $b < r < a$.
In either case, the real root is between a and b.

Example 1 Show that $z^3 - 5z - 2 = 0$ has a real root between 2 and 3.

Solution $p(2) = 8 - 10 - 2 = -4$
            $p(3) = 27 - 15 - 2 = 10$

Since $p(2)$ is negative and $p(3)$ is positive, there is a real root r between 2 and 3, $2 < r < 3$.

Example 2 Show that $z^3 - 5z^2 + 3 = 0$ has a real root between 0 and 1.

Solution $p(0) = 3$ and $p(1) = 1 - 5 + 3 = -1$.
Therefore, there is a real root between 0 and 1.

Example 3 If $p(x) = 4x^3 - 24x^2 + 45x - 27 = (2x - 3)^2(x - 3)$, show that $p(1) < 0$, $p(2) < 0$, and $p(x)$ has a real root between 1 and 2.

Solution $p(1) = -2$ and $p(2) = -1$. However, $p(\frac{3}{2}) = 0$ and thus $\frac{3}{2}$ is a root and $1 < \frac{3}{2} < 2$.

Example 3 illustrates that the converse of the Weierstrass zero theorem is not valid; that is, there may be a real root between two real numbers a and b even though $p(a)$ and $p(b)$ do not differ in sign.

8.4.2 Approximations

After locating a root by the Weierstrass zero theorem, an approximation to an irrational root may be obtained by the process of linear interpolation, where it is assumed that the corresponding differences in z and $p(z)$ are proportional.

If $a < b$ and $p(a)$ and $p(b)$ are opposite in sign, then for $r = a + d$, and assuming $p(r) = 0$,

$$\frac{d}{b-a} = \frac{-p(a)}{p(b) - p(a)}$$

by observing the differences from the following format:

$$\begin{array}{cc} z & p(z) \\ \hline \end{array}$$

$$b - a \quad d \begin{bmatrix} a & p(a) \\ r & 0 \\ b & p(b) \end{bmatrix} \begin{array}{c} 0 - p(a) \\ \\ p(b) - p(a) \end{array}$$

Example 3 Find, correct to the nearest hundredth, the real root of

$$z^3 - 5z - 2 = 0 \text{ between 2 and 3.}$$

Solution Since for $a = 2$, $p(a) = -4$ and for $b = 3$, $p(b) = 10$,

$$\frac{d}{3-2} = \frac{-(-4)}{10-(-4)} = \frac{2}{7} \quad \text{and} \quad d = 0.3 \text{ to the nearest tenth.}$$

Now try $r = a + d = 2.3$.
By synthetic division,

$$\begin{array}{r|rrrr} 2.3 & 1 & 0 & -5 & -2 \\ & & 2.3 & 5.29 & 0.667 \\ \hline & 1 & 2.3 & 0.29 & -1.333 \end{array},$$

WEIERSTRASS ZERO THEOREM; APPROXIMATIONS

$$\begin{array}{r|rrrr} 2.4 & 1 & 0 & -5 & -2 \\ & & 2.4 & 5.76 & 1.824 \\ \hline & 1 & 2.4 & 0.76 & -0.176 \end{array},$$

$$\begin{array}{r|rrrr} 2.5 & 1 & 0 & -5 & -2 \\ & & 2.5 & 6.25 & 3.125 \\ \hline & 1 & 2.5 & 1.25 & 1.125 \end{array}$$

Therefore, $2.4 < r < 2.5$.
Repeating the process,

$$\frac{d}{2.5 - 2.4} = \frac{0.176}{0.176 + 1.125} \quad \text{or} \quad d = 0.01 \text{ to the nearest hundredth.}$$

Now try $r = 2.4 + 0.01 = 2.41$:

$$\begin{array}{r|rrrr} 2.41 & 1 & 0 & -5 & -2 \\ & & 2.41 & 5.8081 & 1.9475^+ \\ \hline & 1 & 2.41 & 0.8081 & -0.0525 \end{array},$$

$$\begin{array}{r|rrrr} 2.42 & 1 & 0 & -5 & -2 \\ & & 2.42 & 5.8564 & 2.0725^- \\ \hline & 1 & 2.42 & 0.8564 & 0.0725 \end{array}$$

Therefore, $2.41 < r < 2.42$.
Repeating the process,

$$\frac{d}{2.42 - 2.41} = \frac{0.0525}{0.0525 + 0.0725} \quad \text{or} \quad d = 0.004 \text{ to the nearest thousandth.}$$

Now try $r = 2.414$:

$$\begin{array}{r|rrrr} 2.414 & 1 & 0 & -5 & -2 \\ & & 2.414 & 5.8274^- & 1.9973^+ \\ \hline & 1 & 2.414 & 0.8274 & -0.0027 \end{array},$$

$$\begin{array}{r|rrrr} 2.415 & 1 & 0 & -5 & -2 \\ & & 2.415 & 5.8322^+ & 2.0098^- \\ \hline & 1 & 2.415 & 0.8322 & 0.0098 \end{array}$$

Therefore, $2.414 < r < 2.415$.
Correct to the nearest hundredth, $r = 2.41$.

Other approximation techniques exist but the one shown above is comparatively simple to use and easy to remember. Moreover, it is self-checking while providing whatever accuracy is desired.

EXERCISES 8.4

1–5. Show that each of the following equations has a real root, r, such that $a < r < b$, where a and b are integers:

1. $x^3 + x - 15$, $2 < r < 3$
2. $x^3 + 3x^2 - 4x - 5 = 0$, $-4 < r < -3$
3. $x^3 + 3x^2 - 4x - 5 = 0$, $1 < r < 2$
4. $x^3 + 3x^2 - 12 = 0$, $1 < r < 2$
5. $2x^4 - 3x^3 - 7x^2 - 8x + 6 = 0$, $0 < r < 1$

6–10. For each of the following equations, determine consecutive integers a and b such that $a < r < b$, where r is a real root of the given equation:

6. $x^3 - 5x^2 + 4x + 5 = 0$
7. $x^3 + 5x^2 - 3 = 0$
8. $x^3 + 2x - 7 = 0$
9. $x^3 + 3x + 8 = 0$
10. $2x^4 - 3x^3 + x - 8 = 0$

11–15. Find the indicated roots correct to the nearest hundredth for each of the following equations:

11. $x^3 + x^2 - 4x - 15 = 0$,   between 2 and 3
12. $x^3 + x - 5 = 0$,   between 1 and 2
13. $x^3 - 9x - 5 = 0$,   between 3 and 4
14. $x^3 - 3x^2 - 3x + 18 = 0$,   root is negative
15. $x^4 - 3x^3 + x^2 - 7x + 12 = 0$,   between 1 and 2

8.5 ALGEBRAIC SOLUTION OF THE CUBIC EQUATION (Optional)

Attempts to solve the cubic equation can be traced to the Greeks, especially Archimedes (287–212 B.C.) who solved special cases geometrically. Interest in this problem was revived by the Arabs and Persians who algebraically solved certain special cases. The poet Omar Khayyam (c. 1100), who lists thirteen different special cases in his *Algebra*, is noted for his contributions in this area.

The solution of the general case of the cubic equation was accomplished by the sixteenth century Italian algebraists. It was common practice at this time for mathematicians to challenge each other to solve certain problems within a specified time limit. In 1535 Tartaglia had a contest with Florido, a pupil of Scipio del Ferro. Tartaglia stated that he defeated Florido because he discovered only a few days before the contest how to solve equations of the type $x^3 + ax^2 = c$ while Florido could only solve equations of the type $x^3 + bx = c$. A few years later, Tartaglia revealed his method to Cardan, with Cardan pledged to secrecy.

ALGEBRAIC SOLUTION OF THE CUBIC EQUATION (OPTIONAL)

However, in 1545, Cardan published Tartaglia's solution of the cubic in his famous *Ars Magna*. He also showed how to transform equations of the type $x^3 + ax^2 = c$ to the form $x^3 + bx = c$. A heated controversy developed with Cardan's pupil Ferrari claiming that Cardan had obtained his knowledge from del Ferro, just as Tartaglia had.

The method published by Cardan was generalized in 1615 by Viète who derived the solution beginning with the general form $x^3 + ax^2 + bx + c = 0$.

Cardan's Solution of the Cubic The solution presented below is commonly called Cardan's solution although it probably should be called the del Ferro-Tartaglia-Cardan-Viète solution.

Theorem (Cardan's Solution of the Cubic) The three solutions of
$$z^3 + bz^2 + cz + d = 0$$
are
$$z_1 = -\frac{b}{3} + \sqrt[3]{-\frac{q}{2} + \sqrt{\frac{q^2}{4} + \frac{p^3}{27}}} + \sqrt[3]{-\frac{q}{2} - \sqrt{\frac{q^2}{4} + \frac{p^3}{27}}},$$

$$z_2 = -\frac{b}{3} + \omega\sqrt[3]{-\frac{q}{2} + \sqrt{\frac{q^2}{4} + \frac{p^3}{27}}} + \omega^2\sqrt[3]{-\frac{q}{2} - \sqrt{\frac{q^2}{4} + \frac{p^3}{27}}},$$

$$z_3 = -\frac{b}{3} + \omega^2\sqrt[3]{-\frac{q}{2} + \sqrt{\frac{q^2}{4} + \frac{p^3}{27}}} + \omega\sqrt[3]{-\frac{q}{2} - \sqrt{\frac{q^2}{4} + \frac{p^3}{27}}},$$

where
$$\omega = -\frac{1}{2} + \frac{i\sqrt{3}}{2}, \qquad \omega^2 = -\frac{1}{2} - \frac{i\sqrt{3}}{2}, \qquad \omega^3 = 1,$$

and
$$y^3 + py + q = 0 \qquad \text{where} \qquad z = y - \frac{b}{3}.$$

Proof

STEP 1 Let $z = y - \frac{b}{3}$.

Then $z^3 + bz^2 + cz + d = 0$ becomes
$$\left(y - \frac{b}{3}\right)^3 + b\left(y - \frac{b}{3}\right)^2 + c\left(y - \frac{b}{3}\right) + d = 0$$
or
$$y^3 + py + q = 0$$

where
$$p = c - \frac{b^2}{3} \quad \text{and} \quad q = \frac{2b^3}{27} - \frac{bc}{3} + d.$$

This substitution was motivated by an attempt to complete the cube paralleling the completion of the square in the solution of the quadratic equation.

STEP 2 Let $y = x - \frac{p}{3x}$. (This is Viète's substitution.)
Then $y^3 + py + q = 0$ becomes

$$x^6 + qx^3 - \frac{p^3}{27} = 0, \quad \text{a quadratic in } x^3.$$

Thus, $x^3 = -\frac{q}{2} \pm \sqrt{\frac{q^2}{4} + \frac{p^3}{27}}$.

STEP 3 Let A = one of the cube roots of $-\frac{q}{2} + \sqrt{\frac{q^2}{4} + \frac{p^3}{27}}$

and
B = one of the cube roots of $-\frac{q}{2} - \sqrt{\frac{q^2}{4} + \frac{p^3}{27}}$

so that $AB = -p/3$. This is always possible since the product $A^3 B^3$ must equal $-p^3/27$. (The product of the roots of a quadratic equation with leading coefficient 1 is equal to the constant term.)

STEP 4 Let $y_1 = A + B$,
$y_2 = \omega A + \omega^2 B$,
$y_3 = \omega^2 A + \omega B$,

where

$$\omega = -\frac{1}{2} + \frac{i\sqrt{3}}{2}, \quad \omega^2 = -\frac{1}{2} - \frac{i\sqrt{3}}{2};$$

that is, $\omega^3 = 1$ or ω and ω^2 are the two imaginary cube roots of 1.

There are six solutions of the sixth degree equation in x; namely, $A, B, \omega A, \omega B, \omega^2 A,$ and $\omega^2 B$. These can be paired so that $xx' = -\frac{p}{3}$ or $x' = \frac{-p}{3x}$ and thus $y = x + \frac{-p}{3x} = x + x'$.

From $\omega^3 AB = \frac{-p}{3}$, it follows that $\omega^2 B = \frac{-p}{3\omega A}$ and $\omega B = \frac{-p}{3\omega^2 A}$ and thus the pairs are $(A, B), (\omega A, \omega^2 B), (\omega^2 A, \omega B)$.

Therefore, $y_1, y_2,$ and y_3 are the three roots of $y^3 + py + q = 0$.

ALGEBRAIC SOLUTION OF THE CUBIC EQUATION (OPTIONAL)

STEP 5 Let $z_1 = -\dfrac{b}{3} + A + B,$

$$z_2 = -\dfrac{b}{3} + \omega A + \omega^2 B,$$

$$z_3 = -\dfrac{b}{3} + \omega^2 A + \omega B.$$

These are the three roots of $z^3 + bz^2 + cz + d = 0$ since $z = y - b/3$.

Example 1 (One real, two imaginary roots) Solve by Cardan's method:

$$z^3 + 6z^2 + 18z + 22 = 0.$$

Solution

(1) Let $z = y - 2,$ $\qquad \left(y - \dfrac{b}{3} = y - \dfrac{6}{3} = y - 2 \right)$

$$(y-2)^3 + 6(y-2)^2 + 18(y-2) + 22 = 0, \text{ or } y^3 + 6y + 2 = 0.$$

(2) Let $y = x - \dfrac{2}{x},$ $\qquad \left(x - \dfrac{p}{3x} = x - \dfrac{6}{3x} = x - \dfrac{2}{x} \right)$

$$\left(x - \dfrac{2}{x} \right)^3 + 6\left(x - \dfrac{2}{x} \right) + 2 = 0$$

$$x^3 - \dfrac{8}{x^3} + 2 = 0$$

$$x^6 + 2x^3 - 8 = 0$$

$$(x^3 - 2)(x^3 + 4) = 0.$$

Thus, $x^3 = 2$ or $x^3 = -4$.

(3) Let $A = \sqrt[3]{2}$ and let $B = -\sqrt[3]{4}$ $\qquad \left(AB = -2 = -\dfrac{p}{3} \right).$

(4) Let $y_1 = A + B = \sqrt[3]{2} - \sqrt[3]{4}$

$$y_2 = \omega A + \omega^2 B = \left(-\dfrac{1}{2} + \dfrac{i\sqrt{3}}{2} \right)\sqrt[3]{2} + \left(-\dfrac{1}{2} - \dfrac{i\sqrt{3}}{2} \right)(-\sqrt[3]{4})$$

$$y_3 = \omega^2 A + \omega B = \left(-\dfrac{1}{2} - \dfrac{i\sqrt{3}}{2} \right)\sqrt[3]{2} + \left(-\dfrac{1}{2} + \dfrac{i\sqrt{3}}{2} \right)(-\sqrt[3]{4}).$$

(5) Letting $z = y - 2 = -2 + y$, and simplifying,

$$z_1 = -2 - \sqrt[3]{4} + \sqrt[3]{2},$$

$$z_2 = -2 + \frac{\sqrt[3]{4} - \sqrt[3]{2}}{2} + \frac{\sqrt{3}(\sqrt[3]{4} + \sqrt[3]{2})}{2} i,$$

$$z_3 = -2 + \frac{\sqrt[3]{4} - \sqrt[3]{2}}{2} - \frac{\sqrt{3}(\sqrt[3]{4} + \sqrt[3]{2})}{2} i.$$

Example 2 (Three real roots, at least one rational) Solve by Cardan's method: $z^3 - 15z + 4 = 0$.

Solution

(1) $z = y$ or $y^3 - 15y + 4 = 0$ $\left(y - \frac{b}{3} = y - \frac{0}{3} = y\right)$.

(2) $y = x + \frac{5}{x}$ $\left(x - \frac{p}{3x} = x - \frac{-15}{3x} = x + \frac{5}{x}\right)$

$$\left(x + \frac{5}{x}\right)^3 - 15\left(x + \frac{5}{x}\right) + 4 = 0,$$

$$x^6 + 4x^3 + 125 = 0,$$

$$x^3 = -2 + 11i \quad \text{or} \quad x^3 = -2 - 11i.$$

(3) Let $A = \sqrt[3]{-2 + 11i}$ and $B = \sqrt[3]{-2 - 11i}$.[2]

(4) Then $y_1 = \sqrt[3]{-2 + 11i} + \sqrt[3]{-2 - 11i}$,

$$y_2 = \left(-\frac{1}{2} + \frac{i\sqrt{3}}{2}\right)\sqrt[3]{-2 + 11i} + \left(-\frac{1}{2} - \frac{i\sqrt{3}}{2}\right)\sqrt[3]{-2 - 11i},$$

$$y_3 = \left(-\frac{1}{2} - \frac{i\sqrt{3}}{2}\right)\sqrt[3]{-2 + 11i} + \left(-\frac{1}{2} + \frac{i\sqrt{3}}{2}\right)\sqrt[3]{-2 - 11i}.$$

However, $z^3 - 15z + 4 = 0$ has a rational root -4 and

$$z^3 - 15z + 4 = (z + 4)(z^2 - 4z + 1) = 0$$

yielding

$$z_1 = -4, \; z_2 = 2 + \sqrt{3}, \; \text{and} \; z_3 = 2 - \sqrt{3}.$$

Since a polynomial of degree three has exactly three roots, these three real roots must be the same as those expressed in terms of the imaginary unit i.

---

[2] If $b \neq 0$, $\sqrt[3]{a + bi}$ shall mean the cube root of $a + bi$ with the smallest positive argument.

ALGEBRAIC SOLUTION OF THE CUBIC EQUATION (OPTIONAL)

It may be shown that $\sqrt[3]{-2 + 11i} = -2 + i$ and $\sqrt[3]{-2 - 11i} = -2 - i$ by assuming that $\sqrt[3]{-2 + 11i} = a + bi$, that $\sqrt[3]{-2 - 11i} = a - bi$, and that $a = z_1/2 = -4/2 = -2$ and then solving $(-2 + bi)^3 = -2 + 11i$ for b.

Then

$$y_1 = \sqrt[3]{-2 + 11i} + \sqrt[3]{-2 - 11i} = (-2 + i) + (-2 - i) = -4,$$
$$y_2 = \omega(-2 + i) + \omega^2(-2 - i) = 2 - \sqrt{3},$$
$$y_3 = \omega^2(-2 + i) + \omega(-2 - i) = 2 + \sqrt{3}.$$

This example illustrates that the general algebraic solution yields the three real roots as expressions involving the imaginary unit i and not as three real numbers. However, if one of the real roots is rational, then it is possible to exhibit the three real roots without involving the imaginary unit i.

Example 3 (Three real roots, none rational—the irreducible case) Solve $z^3 - 3z + 1 = 0$ by Cardan's method.

Solution Let $z = x + \dfrac{1}{x}$ $\left(x - \dfrac{p}{3x} = x - \dfrac{-3}{3x} = x + \dfrac{1}{x} \right).$

$$\left(x + \frac{1}{x} \right)^3 - 3\left(x + \frac{1}{x} \right) + 1 = 0$$

$$x^6 + x^3 + 1 = 0$$

$$x^3 = \frac{-1 \pm \sqrt{-3}}{2};$$

that is, $x^3 = \omega$ or $x^3 = \omega^2$.
Then,

$$z_1 = \sqrt[3]{\omega} + \sqrt[3]{\omega^2}$$
$$z_2 = \omega \sqrt[3]{\omega} + \omega^2 \sqrt[3]{\omega^2}$$
$$z_3 = \omega^2 \sqrt[3]{\omega} + \omega \sqrt[3]{\omega^2}.$$

This equation has three real roots, but Cardan's solution expresses them in terms of the imaginary unit. There are no rational roots since 1 and -1 are the only possibilities and these are not roots. In this case, the irreducible case, it is not possible to express any of the roots in exact form as the result of algebraic operations involving real numbers only.

Attempting to solve $\sqrt[3]{\omega} = x + yi$,

$$(x + yi)^3 = \omega = -\frac{1}{2} + \frac{i\sqrt{3}}{2}.$$

$$x^3 - 3xy^2 = -\frac{1}{2} \quad \text{and} \quad 3x^2y - y^3 = \frac{\sqrt{3}}{2},$$

$$y^2 = \frac{2x^3 + 1}{6x} \quad \text{and} \quad y\left(3x^2 - \frac{2x^3 + 1}{6x}\right) = \frac{\sqrt{3}}{2}$$

or

$$y = \frac{3\sqrt{3}\,x}{16x^3 - 1}.$$

Thus,

$$y^2 = \frac{2x^3 + 1}{6x} = \frac{27x^2}{(16x^3 - 1)^2}$$

and

$$(2x)^9 + 3(2x)^6 - 24(2x)^3 + 1 = 0.$$

Now, letting $(2x)^3 = 3z - 1$,

$$(3z - 1)^3 + 3(3z - 1)^2 - 24(3z - 1) + 1 = 0$$

or

$$27z^3 - 27(3z) + 27 = 0$$

or

$$z^3 - 3z + 1 = 0.$$

This, however, is the original equation whose solutions were to be found. Although the three real roots of the irreducible case of the cubic cannot be expressed exactly as the results of algebraic operations involving only real numbers, approximations to these roots can be found by using the trigonometric functions. This will be discussed in the next section.

Definition The *discriminant*, D, of the cubic equation

$$y^3 + py + q = 0 \quad \text{is} \quad D = \frac{q^2}{4} + \frac{p^3}{27}.$$

Noting that

$$y_1 = \sqrt[3]{-\frac{q}{2} + \sqrt{D}} + \sqrt[3]{-\frac{q}{2} - \sqrt{D}},$$

ALGEBRAIC SOLUTION OF THE CUBIC EQUATION (OPTIONAL)

$$y_2 = \omega \sqrt[3]{-\frac{q}{2} + \sqrt{D}} + \omega^2 \sqrt[3]{-\frac{q}{2} - \sqrt{D}},$$

$$y_3 = \omega^2 \sqrt[3]{-\frac{q}{2} + \sqrt{D}} + \omega \sqrt[3]{-\frac{q}{2} - \sqrt{D}},$$

it may be shown that for real numbers p and q,

if $D > 0$, there is one real root and two imaginary roots,
if $D < 0$, there are three real unequal roots,
if $D = 0$ and $p^2 + q^2 \neq 0$, there are three real roots with a double root,
if $D = 0$ and $p = q = 0$, then there is one real triple root.

Since there are three real roots if and only if $D < 0$ or $D = 0$ and since $D = q^2/4 + p^3/27$, it follows that p must be negative in this case when $q \neq 0$.

The irreducible case, three unequal irrational roots, can occur, then, only when $p < 0$ and $q^2/4 + p^3/27 < 0$, and $q \neq 0$.

EXERCISES 8.5

1–15. (a) Solve each of the following cubic equations by Cardan's method if possible. (b) Solve by factoring, if there is a rational root. (c) Reconcile the results obtained in (a) and (b).

1. $y^3 + 6y + 2 = 0$
2. $y^3 + 6y - 8 = 0$
3. $y^3 + 3y^2 - 9y + 5 = 0$
4. $y^3 + 6y^2 + 18y + 16 = 0$
5. $y^3 + 3y - 2i = 0$
6. $y^3 - 6y^2 + 9y - 2 = 0$
7. $y^3 - 3y^2 + y + 1 = 0$
8. $y^3 + 3\omega y - 4 = 0$, where $\omega^3 = 1$
9. $y^3 - 3\omega y - 2 = 0$, where $\omega^3 = 1$
10. $y^3 - 2y^2 - y + 2 = 0$
11. $y^3 + 63y - 316 = 0$
12. $y^3 + 3y^2 - 6 = 0$
13. $y^3 - 6y^2 + 11y - 6 = 0$
14. $y^3 - 63y - 162 = 0$
15. $y^3 + 6y^2 + 3y + 18 = 0$

16. Show that for any complex number k, $y^3 - 3k^2 y + 2k^3 = 0$ always has a double root.

17. Cardan, in his *Ars Magna*, solved $z^3 + 6z - 20 = 0$. Find the solutions by Cardan's method.

18. The following problem is taken from the works of Fibonacci: Solve $z^3 + 2z^2 + 10z - 20 = 0$.

19–20. These two problems were presented to Tartaglia, who at first could not solve them, but later discovered the general solution. Solve these equations.

19. $z^3 + 3z^2 = 5$
20. $z^3 + 6z^2 + 8z = 1000$

8.6 TRIGONOMETRIC SOLUTION OF THE CUBIC EQUATION (Optional)

If the cubic equation has three real roots ($p < 0$ and $q^2/4 + p^3/27 < 0$), then these roots can be expressed in terms of the trigonometric functions and approximated by using the tables. The solution is based on the identity

$$4 \cos^3 \theta - 3 \cos \theta = \cos 3\theta.$$

Theorem (Trigonometric Solution of the Cubic Equation) If

$$x^3 - px + q = 0 \quad \text{and} \quad p > 0 \quad \text{and} \quad q^2/4 - p^3/27 < 0,$$

then

$$x = 2\sqrt{\frac{p}{3}} \cos\left(\theta + \frac{2k\pi}{3}\right), \quad \text{where } k = 0, 1, \text{ or } 2$$

and

$$\cos 3\theta = \frac{-3q}{2p} \sqrt{\frac{3}{p}}.$$

Proof Let $x = t \cos \theta$. Then $t^3 \cos^3 \theta - pt \cos \theta = -q$. Multiplying both sides by $4/t^3$,

$$4 \cos^3 \theta - \frac{4p}{t^2} \cos \theta = \frac{-4q}{t^3}.$$

Since $4 \cos^3 \theta - 3 \cos \theta = \cos 3\theta$, select $4p/t^2 = 3$.
Then

$$t^2 = \frac{4p}{3} \quad \text{and} \quad t = 2\sqrt{\frac{p}{3}},$$

choosing the positive root.
Then

$$\cos 3\theta = \cos(3\theta + 2k\pi) = \frac{-3q}{2p} \sqrt{\frac{3}{p}} \quad \text{and} \quad k = 0, 1, \text{ or } 2.$$

Example 1 Solve $x^3 - 3x + 1 = 0$ trigonometrically. (Note that this is the same equation as that in Example 3, Section 8.5.)

Solution

(1) Let $x = t \cos \theta$.

$$t^3 \cos^3 \theta - 3t \cos \theta + 1 = 0$$

TRIGONOMETRIC SOLUTION OF THE CUBIC EQUATION (OPTIONAL) 303

(2) Subtract 1 from both sides and then multiply both sides by $4/t^3$.

$$4\cos^3\theta - \frac{12}{t^2}\cos\theta = \frac{-4}{t^3}$$

(3) Let $12/t^2 = 3$. Then $t^2 = 4$ and choose $t = 2$. Then $4\cos^3\theta - 3\cos\theta = -\frac{1}{2}$.

(4) Solve $\cos 3\theta = \cos(3\theta + 360°k) = -\frac{1}{2}$ for $k = 0, 1, 2$.

If $k = 0$, $3\theta = 120°$ and $\theta = 40°$.
If $k = 1$, $3\theta = 480°$ and $\theta = 160°$.
If $k = 2$, $3\theta = 840°$ and $\theta = 280°$.

(5) Since $x = t\cos\theta = 2\cos\theta$, and $\cos 160° = -\cos 20°$ and $\cos 280° = \cos 80°$,

$$x_1 = 2\cos 40° \quad (=) \quad 1.5320$$
$$x_2 = -2\cos 20° \quad (=) \quad -1.8794$$
$$x_3 = 2\cos 80° \quad (=) \quad 0.3472.$$

Example 2   Solve $2x^3 - 3x + 1 = 0$ trigonometrically.

Solution

(1) Let $x = t\cos\theta$.

$$2t^3\cos^3\theta - 3t\cos\theta = -1$$

(2) Multiply both sides by $2/t^3$.

$$4\cos^3\theta - \frac{6}{t^2}\cos\theta = \frac{-2}{t^3}$$

(3) Let $6/t^2 = 3$, so $t^2 = 2$ and choose $t = \sqrt{2}$. Then

$$\frac{-2}{t^3} = \frac{-1}{\sqrt{2}}.$$

(4) Solving $\cos(3\theta + 360°k) = \frac{-1}{\sqrt{2}}$ for $k = 0, 1, 2$,

$$3\theta = 135° \text{ or } 225° \text{ or } 495°,$$
$$\theta = 45° \text{ or } 75° \text{ or } 165°.$$

(5)

$$\cos 75° = \cos\frac{150°}{2} = \sqrt{\frac{1+\cos 150°}{2}} = \sqrt{\frac{1-\sqrt{3}/2}{2}} = \frac{\sqrt{2-\sqrt{3}}}{2}$$

$$\cos 165° = \cos(90° + 75°) = -\sin 75° = \frac{-\sqrt{2+\sqrt{3}}}{2}$$

(6) Since $x = \sqrt{2} \cos \theta$,

$$x_1 = \sqrt{2} \cos 45° = \sqrt{2}\left(\frac{1}{\sqrt{2}}\right) = 1,$$

$$x_2 = \sqrt{2} \cos 75° = \frac{\sqrt{4 - 2\sqrt{3}}}{2},$$

$$x_3 = \sqrt{2} \cos 165° = \frac{-\sqrt{4 + 2\sqrt{3}}}{2}.$$

Observing that 1 is a root, then

$$2x^3 - 3x + 1 = (x - 1)(2x^2 + 2x - 1).$$

Solving $2x^2 + 2x - 1 = 0$,

$$x = \frac{-1 \pm \sqrt{3}}{2}.$$

Therefore,

$$x_1 = 1$$

$$x_2 = \frac{-1 + \sqrt{3}}{2} = \frac{\sqrt{4 - 2\sqrt{3}}}{2},$$

$$x_3 = \frac{-1 - \sqrt{3}}{2} = \frac{-\sqrt{4 + 2\sqrt{3}}}{2},$$

since there are exactly three roots.

Simplifications such as this will always be possible when one of the roots is rational.

EXERCISES 8.6

1–15. Solve each of the following cubic equations trigonometrically:

1. $4x^3 - 3x + 1 = 0$
2. $4x^3 - 3x - 1 = 0$
3. $8x^3 - 6x + 1 = 0$
4. $8x^3 - 6x - 1 = 0$
5. $8x^3 - 6x + \sqrt{2} = 0$
6. $8x^3 - 6x - \sqrt{2} = 0$
7. $8x^3 - 6x + \sqrt{3} = 0$
8. $8x^3 - 6x - \sqrt{3} = 0$
9. $x^3 - 3x + 2 = 0$
10. $x^3 - 3x - 2 = 0$
11. $x^3 - 3x + 1 = 0$
12. $x^3 - 27x - 27 = 0$
13. $x^3 - 12x + 8 = 0$
14. $x^3 - 12x + 16 = 0$
15. $4x^3 - 75x - 125 = 0$

TRIGONOMETRIC SOLUTION OF THE CUBIC EQUATION (OPTIONAL)

16. Why can't $x^3 + 6x + 2 = 0$ be solved trigonometrically?
17. Why can't $8x^3 - 6x + 1 = 0$ be solved by Cardan's method?
18. Solve $z^3 - 6z + 4 = 0$:
    (a) trigonometrically, simplifying the results
    (b) by Cardan's method.
    (c) reconcile the results obtained by the two methods.

☆19. (a) Solve $z^3 = 1 + i$ algebraically.
    (*Hint:* Let $z = x + yi$ to obtain $(x^3 - 3xy^2) + (3x^2y - y^3)i = 1 + i$ and $x^3 - 3xy^2 = 1$ and $3x^2y - y^3 = 1$.)
    Subtracting, $x^3 - 3xy^2 + y^3 - 3x^2y = 0$.
    Factor and solve, noting that $x + y$ is a factor.
    (b) Solve $z^3 = 1 + i$ trigonometrically.
    (c) Reconcile that (a) yields $\sin 15° = \dfrac{1}{\sqrt[6]{2}\sqrt[3]{20 + 12\sqrt{3}}}$ and (b) yields
    $\sin 15° = \dfrac{\sqrt{2 - \sqrt{3}}}{2}$.

☆20. Solve $z^4 = i$ algebraically and trigonometrically. Reconcile the fact that the algebraic solution yields $\cos \pi/8 = \dfrac{\sqrt[4]{6 + 4\sqrt{2}}}{2}$, and the trigonometric solution yields $\cos \pi/8 = \dfrac{\sqrt{2 + \sqrt{2}}}{2}$.

☆21. (a) Solve $z^5 = i$ algebraically.
    (*Hint:* $x^5 - 10x^3y^2 + 5xy^4 = 0$ and $5x^4y - 10x^2y^3 + y^5 = 1$; $x(x^4 - 10x^2y^2 + 5y^4) = 0$.)
    Solve $x^4 - 10x^2y^2 + 5y^4 = 0$ for x^2 as a function of y^2 and substitute the result in $5x^4y - 10x^2y^3 + y^5 = 1$. Solve for y.
    (b) Solve $y^5 = i$ trigonometrically, leaving solution in terms of sines and cosines.
    (c) Solve $z^5 = i$ by substituting $y^2 = 1 - x^2$ in $x^4 - 10x^2y^2 + 5y^4 = 0$.
    (d) Reconcile the fact that
    (a) yields $\sin 18° = \dfrac{1}{\sqrt[5]{176 + 80\sqrt{5}}}$ while (b) and (c) yield
    $\sin 18° = \dfrac{\sqrt{5} - 1}{4}$.

☆22. Show that the quadratic equation $z^2 - bz + c = 0$, where $b^2 \geq 4c$, can be solved trigonometrically. Let $x = b \cos^2 \theta$, $y = b \sin^2 \theta$; $x + y = b$ and
$xy = b^2(\cos \theta \sin \theta)^2 = \dfrac{b^2 \sin^2 2\theta}{4} = c$; or $\sin 2\theta = \pm \sqrt{\dfrac{4c}{b^2}}$.

8.7 THE QUARTIC EQUATION (Optional)

In 1540 the Italian mathematician da Coi presented Cardan a problem requiring the solution of the quartic (also called biquadratic) equation

$$z^4 + 6z^3 - 6z^2 - 30z - 11 = 0.$$

Although Cardan could not solve this problem, his young pupil Ferrari did and Ferrari's solution was published in Cardan's *Ars Magna*.

Theorem (Ferrari's Solution of the Quartic) The four roots of

$$z^4 + bz^3 + cz^2 + dz + e = 0$$

are obtained from the four roots of

$$\left(y^2 + \frac{x}{2}\right)^2 - \left(y\sqrt{x-p} - \frac{q}{2\sqrt{x-p}}\right)^2 = 0,$$

where $z = y - b/4$ yielding $y^4 + py^2 + qy + r = 0$, and x is a real solution of $x^3 - px^2 - 4rx + (4pr - q^2) = 0$, called the **resolvent cubic**.

Proof Given $z^4 + bz^3 + cz^2 + dz + e = 0$.

STEP 1 Let $z = y - b/4$. (An attempt to complete the 4th power.)

$$\left(y - \frac{b}{4}\right)^4 + b\left(y - \frac{b}{4}\right)^3 + c\left(y - \frac{b}{4}\right)^2 + d\left(y - \frac{b}{4}\right) + e = 0.$$

$$y^4 + \left(c - \frac{3b^2}{8}\right)y^2 + \left(d - \frac{bc}{2} + \frac{2b^3}{16}\right)y + \left(e - \frac{bd}{4} + \frac{b^2c}{16} - \frac{3b^4}{256}\right) = 0.$$

Renaming the coefficients, $y^4 + py^2 + qy + r = 0$.

STEP 2 Add and subtract $y^2 x + x^2/4$.

$$y^4 + y^2 x + \frac{x^2}{4} - y^2 x - \frac{x^2}{4} + py^2 + qy + r = 0.$$

STEP 3 Rearrange terms to obtain the form $A^2 - B^2$.

$$\left(y^2 + \frac{x}{2}\right)^2 - \left[(x - p)y^2 - qy + \left(\frac{x^2}{4} - r\right)\right] = 0.$$

STEP 4 Set $q^2 - 4(x - p)\left(\frac{x^2}{4} - r\right) = 0$.

The number in brackets in Step 3 is a perfect square if and only if its discriminant $B^2 - 4AC = 0$, where $B = -q$, $A = x - p$, and $C = x^2/4 - r$. Then, $x^3 - px^2 - 4rx + (4pr - q^2) = 0$, the resolvent cubic.

THE QUARTIC EQUATION (OPTIONAL)

STEP 5 Let X be any real root of the resolvent cubic equation. Then the quartic becomes

$$\left(y^2 + \frac{X}{2}\right)^2 - \left(y\sqrt{X-p} - \frac{q}{2\sqrt{X-p}}\right)^2 = 0.$$

STEP 6 Factor, using the form $A^2 - B^2 = (A-B)(A+B)$ and then equate the factors to 0. The four roots of the quartic are obtained from the solutions of these two quadratic equations.

Example Solve $z^4 + 4z^3 + 6z^2 + 16z + 8 = 0$ using Ferrari's method.

Solution

(1) Let $z = y - 1$. $\quad \left(y - \frac{b}{4} = y - \frac{4}{4} = y - 1\right)$

$$(y-1)^4 + 4(y-1)^3 + 6(y-1)^2 + 16(y-1) + 8 = 0$$
$$y^4 + 12y - 5 = 0.$$

(2) Add and subtract $y^2 x + \frac{x^2}{4}$.

(3) $\left(y^2 + \frac{x}{2}\right)^2 - \left(xy^2 - 12y + \frac{x^2}{4} + 5\right) = 0$

(4) Set $B^2 - 4AC = 0$, $B = -12$, $A = x$, $C = \frac{x^2}{4} + 5$,

$$(-12)^2 - 4x\left(\frac{x^2}{4} + 5\right) = 0$$
$$x^3 + 20x - 144 = 0.$$

(5) Let $X = 4$ since $(4)^3 + 20(4) - 144 = 0$.
(Always try to find a rational root first)

$$\left(y^2 + \frac{4}{2}\right)^2 - \left(4y^2 - 12y + \frac{16}{4} + 5\right) = 0$$
$$(y^2 + 2)^2 - (4y^2 - 12y + 9) = 0$$

(6) Factor and solve.

$$(y^2 + 2)^2 - (2y - 3)^2 = 0$$
$y^2 + 2 + 2y - 3 = 0 \quad$ or $\quad y^2 + 2 - 2y + 3 = 0$
$y^2 + 2y - 1 = 0 \quad\quad$ or $\quad y^2 - 2y + 5 = 0$
$y = -1 \pm \sqrt{2} \quad\quad$ or $\quad y = 1 \pm 2i$.

(7) Since $z = y - 1$,

$$z_1 = -2 + \sqrt{2}, \quad z_2 = -2 - \sqrt{2}, \quad z_3 = 2i, \quad z_4 = -2i.$$

EQUATIONS OF FIFTH DEGREE AND HIGHER

Among other mathematicians, Euler in 1750 and Lagrange in 1780 attempted to solve the quintic equation but without success. Finally the Norwegian mathematician Niels Abel (1802–1829), when he was 22 years old, showed that the general fifth degree equation could *not* be solved by rational operations and root extractions on the coefficients. Since there are special cases of the quintic that can be solved, the next development was concerned with finding the conditions that determine whether or not a polynomial equation can be solved. At the age of 21, on the evening before the duel in which he was killed, the French mathematician Evariste Galois (1811–1832) wrote a letter to a friend summarizing his discoveries. Galois, who together with Abel, laid the foundations for the theory of groups and the modern theory of equations, showed that each polynomial equation determines a group of substitutions and whether or not the polynomial is solvable depends on the special properties of its (Galois) group.

EXERCISES 8.7

1–10. Solve each of the following quartic equations by Ferrari's method:

1. $2x^4 + 4x + 1 = 0$
2. $x^4 + 12x - 5 = 0$
3. $x^4 - 4x^3 - 2x^2 - 4x + 21 = 0$
4. $x^4 - 5x^2 + 6 = 0$
5. $x^4 + 4x^3 + 7x^2 + 10x + 3 = 0$
6. $x^4 - 14x^2 - 60x - 200 = 0$
7. $x^4 + 2x^3 - x^2 + x - \frac{3}{4} = 0$
8. $x^4 + 4x^2 + 32x - 48 = 0$
9. $x^4 - 4x^3 + 4x^2 - 4 = 0$
10. $2x^4 - x^3 - 14x^2 + 19x - 6 = 0$

11. The following problem was solved by Petri in 1567: Solve
$$z^4 + 6z^3 - 6z^2 - 30z - 11 = 0.$$

12. Solve $z^4 + 2z^3 + 2z + 1 = 13z^2$ by adding $3z^2$ to both sides of the equation. Cardan used this technique for solving this equation.

CHAPTER SUMMARY

Quadratic Formula If a, b, and c are complex numbers and $a \neq 0$, then the roots of $az^2 + bz + c = 0$ are given by

$$z = \frac{-b \pm \sqrt{b^2 - 4ac}}{2a}.$$

Discriminant of Quadratic Equation $b^2 - 4ac$ is the discriminant of the quadratic equation $az^2 + bz + c = 0$, $a \neq 0$.

Sum and Product of Quadratic Roots If r and s are the two roots of

$$az^2 + bz + c = 0, \quad a \neq 0,$$

then

$$r + s = -(b/a) \quad \text{and} \quad rs = c/a.$$

Fundamental Theorem of Algebra Every polynomial equation with coefficients in the set of complex numbers has at least one solution in the set of complex numbers.

Division Algorithm for Polynomials over C If $a(z)$ and $b(z)$ are polynomials over C with positive degree, then there exist unique polynomials $q(z)$ and $r(z)$ over C so that

$$a(z) = b(z)q(z) + r(z)$$

and

either $r(z) = 0$ or degree of $r(z)$ < degree of $b(z)$.

Remainder Theorem If $p(z) = (z - a)q(z) + r$, then $p(a) = r$.

Factor Theorem $z - a$ is a factor of $p(z)$ if and only if $p(a) = 0$.

Unique Linear Factorization of Polynomials over C If $a_n \neq 0$ and $n > 0$ and if $p(z) = a_n z^n + \ldots + a_1 z + a_0$, then the factorization $p(z) = a_n(z - c_1)(z - c_2) \ldots (z - c_n)$ is unique for n complex numbers $c_1, c_2, \ldots, c_n$.

Multiplicity A root r of $p(z) = 0$ has multiplicity k if and only if the linear factor $z - r$ occurs exactly k times in the factorization of $p(z)$.

Number of Roots of a Polynomial Equation If the degree of $P(z) = n$, then $P(z) = 0$ has exactly n roots if each root is counted with its multiplicity.

Rational Roots The simplified rational number r/s is a zero of the integral polynomial $a_n z^n + \ldots + a_1 + a_0$ if and only if r is a factor of a_0 and s is a factor of a_n.

Conjugate Imaginary Roots If $a + bi$ is an imaginary root of a polynomial equation with real coefficients, then $a - bi$ is a root.

Conjugate Surd Roots If $a + \sqrt{b}$ is an irrational real root of a polynomial equation with rational coefficients, then $a - \sqrt{b}$ is a root.

Theorem on Negative Real Roots If $-r$ is a negative root of $p(z) = 0$, then r is a positive root of $p(-z) = 0$.

Upper and Lower Bounds for Real Roots U and L are upper and lower bounds, respectively, for the real roots of $p(z) = 0$ if and only if for all real roots r, $L \le r \le U$.

Theorem on Upper Bounds The positive real number U is an upper bound for the real roots of $p(z) = a_n z^n + \ldots + a_1 z + a_0 = 0$ if $a_n > 0$ and the third line of the synthetic division of $p(z)$ by $z - U$ contains no negative numbers.

Descartes' Rule of Signs The number of positive real roots of $p(z) = 0$ is either equal to the number of variations in sign of $p(z)$ or it differs from it by an even integer.

The number of negative real roots of $p(z) = 0$ is either equal to the number of variations in sign of $p(-z)$ or it differs from it by an even integer.

The Weierstrass Zero Theorem If $p(z)$ is a polynomial with real coefficients and if a and b are real numbers so that $p(a) < 0$ and $p(b) > 0$, then $p(z)$ has a real zero between a and b.

Transformation of the Cubic Equation A cubic equation of the form $z^3 + bz^2 + cz + d = 0$ can be transformed to the form $y^3 + py + q = 0$ by the substitution $z = y - (b/3)$.

Cardan's Solution of the Cubic The three roots of $y^3 + py + q = 0$ are

$$y_1 = A + B, \qquad y_2 = \omega A + \omega^2 B, \qquad \text{and} \qquad y_3 = \omega^2 A + \omega B$$

where

$$A = \sqrt[3]{-\frac{q}{2} + \sqrt{\frac{q^2}{4} + \frac{p^3}{27}}}, \qquad B = \sqrt[3]{-\frac{q}{2} - \sqrt{\frac{q^2}{4} + \frac{p^3}{27}}},$$

$$\omega = \frac{-1 + i\sqrt{3}}{2}, \qquad \text{and} \qquad \omega^2 = \frac{-1 - i\sqrt{3}}{2} \qquad (\omega^3 = 1).$$

Discriminant of the Cubic Equation The discriminant of $y^3 + py + q = 0$ is

$$\frac{q^2}{4} + \frac{p^3}{27}.$$

Trigonometric Solution of the Cubic If $y^3 + py + q = 0$ has three real roots, then it can be solved trigonometrically by letting $y = t \cos \theta$ and by forming the identity $4 \cos^3 \theta - 3 \cos \theta = \cos 3\theta$.

Transformation of the Quartic Equation A quartic equation of the form

$$z^4 + bz^3 + cz^2 + dz + e = 0$$

can be transformed to the form

$$y^4 + py^2 + qy + r = 0$$

by the substitution $z = y - (b/4)$.

Ferrari's Solution of the Quartic The four roots of

$$y^4 + py^2 + qy + r = 0$$

are the four roots of

$$\left(y^2 + \frac{x}{2}\right)^2 - \left(y\sqrt{x-p} - \frac{q}{2\sqrt{x-p}}\right)^2 = 0$$

where x is a real solution of

$$x^3 - px^2 - 4rx + 4pr - q^2 = 0,$$

the resolvent cubic.

The *quintic equation* cannot be solved by algebraic operations on the coefficients.

REVIEW EXERCISES

1–5. Solve each of the following quadratic equations and check by using the sum and product of roots theorem:

1. $2x^2 + 3x - 4 = 0$
2. $x^2 + \sqrt{3}x + 2 = 0$
3. $ix^2 + 3x - 2i = 0$
4. $3x^2 + 2x = 3$
5. $(x + 2)(x - 3) = 1$

6–8. Determine the value(s) of k that will make the roots or each of the following equations equal:

6. $3x^2 + 2x + k = 0$ 7. $2x^2 + kx + 5 = 0$ 8. $x^2 + kx + k = 3$

9. Find k so that $x + 3$ is a factor of $x^3 + 2x^2 + x - k$.

10. Find k so that $x^4 + 2x^3 + x - k$ is exactly divisible by $x - 3$.

11. Use the factor theorem to check if $2 + 3i$ is a root of the equation

$$x^4 - 3x^3 + 12x^2 + 8x - 20 = 0.$$

12–14. Factor each of the following polynomials over (a) the integers (b) the real numbers and (c) the complex numbers:

12. $2x^3 + 3x^2 + 4x + 12$ 13. $x^3 - 2x^2 - 3x + 6$ 14. $x^4 + 2x^2 - 8$

15–20. Find the rational roots of each of the following equations:

15. $x^3 + 3x^2 - 4x - 12 = 0$ 18. $2x^3 - x^2 - 22x - 24 = 0$
16. $x^3 - 7x^2 + 8x + 4 = 0$ 19. $6x^4 - 5x^3 + 7x^2 - 5x + 1 = 0$
17. $x^4 - 2x^3 - 3x^2 + 4x - 12 = 0$ 20. $24x^3 - 10x^2 - 13x + 6 = 0$

21. Find a quartic equation with real coefficients if two of its roots are $3 - 2i$ and $\sqrt{2} - 1$.

22–24. Find integers which are upper and lower bounds of the real roots of each of the following equations:

22. $x^4 - 5x^3 + 4x^2 + 2x - 30 = 0$
23. $6x^3 - 8x^2 - 3x - 12 = 0$
24. $2x^6 + x^5 + 8x^4 - 32x^2 - 16x - 128 = 0$

25–28. Use Descartes' rule of signs to determine the nature of the roots of each of the following equations:

25. $x^3 - 2x^2 + 3x - 7 = 0$ 27. $3x^6 + 2x^5 - x^2 + 2x + 3 = 0$
26. $x^4 - 3x^2 - 7 = 0$ 28. $x^5 + 3x^3 + 2x = 0$

29–32. Determine consecutive integers a and b such that a real root r of each of the following equations is located between a and b.

29. $x^3 + 5x^2 - 3 = 0$ 31. $2x^4 + x^3 - 3x + 1 = 0$
30. $x^3 - 6x^2 + 11x - 6 = 0$ 32. $x^5 - 9x^3 + 3x^2 + 16x - 4 = 0$

33. Determine to the nearest hundredth the positive real root of

$$2x^3 + x^2 + 3x - 4 = 0.$$

34. Determine to the nearest hundredth the real root of $x^4 - 3x^3 + 7x - 8 = 0$ between -2 and -1.

REVIEW EXERCISES

35–40. Solve each of the following cubic equations by the most appropriate of these methods: Cardan's, trigonometric, or factoring.

35. $x^3 - 3x - 1 = 0$
36. $x^3 - 9x - 12 = 0$
37. $3x^3 - 5x^2 + 4x + 2 = 0$
38. $x^3 - 7x + 6 = 0$
39. $x^3 - 3x - 2i = 0$
40. $x^3 - 12x - 8\sqrt{2} = 0$

9
PLANE ANALYTIC GEOMETRY

The concept of a coordinate system probably originated with the ancient Egyptian surveyors. The hieroglyphic symbol used to designate the districts into which Egypt was divided was a grid symbol. Records indicate that the Greeks used the ideas of longitude and latitude to locate points in the sky and on the earth. The Romans, who were noted for their surveying techniques, arranged the streets of their cities on a rectangular coordinate system.

The Arab and Persian mathematicians were the first to use geometric figures for algebraic problems. Examples are found in the works of the Arab al-Khowarizmi (*c.* 825) and the Persian Omar Khayyam (*c.* 1100). This usage is again found in the writings of Fibonacci (1220), Pacioli (1494), and Cardan (1545).

René Descartes (1596–1650) is credited with the invention of analytic geometry, since he used a rectangular coordinate system to establish a relationship between equations and curves. Descartes' best-known work is a philosophical treatise, published in 1637, called *A Discourse on the Method of Rightly Conducting the Reason and Seeking Truth in the Sciences*. This publication had three appendices, the third of which is the famous *La Géométrie*. In this appendix Descartes deals with such topics as the solution of equations of degree greater than 2 and

RECTANGULAR COORDINATES; DISTANCE; MIDPOINT

the use of exponents to indicate powers of numbers. But what Descartes is best remembered for is his introduction of modern analytic geometry in *La Géométrie*, and the Cartesian coordinate system is named in his honor. Although Descartes is credited with the development of analytic geometry, Pierre de Fermat, another great French mathematician, also formulated coordinate geometry at the same time and made a considerable contribution in this field. The modern terms *coordinates*, *abscissa*, and *ordinate* were contributed by the German mathematician, Gottfried Wilhelm Leibniz, in 1692.

9.1 RECTANGULAR COORDINATES; DISTANCE; MIDPOINT

9.1.1 Coordinate Lines

A **coordinate line** is a line whose points have been placed in one-to-one correspondence with the set of real numbers, R.

Two distinct points, O and U, on the line are assigned the numbers 0 and 1, and a positive direction is assigned as the direction from point O to point U.

The point O is called the **origin**, the point U is called the **unit point**, and the line segment OU is called the **unit segment**.

The real numbers are assigned to the points on the coordinate line so that point a is to the right of point b if and only if $a > b$. In this way, there is exactly one real number associated with each point on the coordinate line and exactly one point associated with each real number. The real number p assigned to point P is called the **coordinate** of P and the point will be designated symbolically by $P:p$.

9.1.2 Distance on a Coordinate Line

If b is greater than a, then the difference $b - a$ is positive, and if $B:b$ is in the positive direction from $A:a$, then $b - a$ is the length of the line segment AB. However, if $B:b$ is in the negative direction from $A:a$, then b is less than a and the difference $b - a$ is negative. In either case, $|b - a|$ is never negative and can be assigned as the distance between A and B.

Definition (Distance between Two Points on a Coordinate Line) If $A:a$ and $B:b$ are two points on a coordinate line, then the distance, $|AB|$, between A and B is defined as follows:

$$|AB| = |b - a|.$$

9.1.3 Directed Distance

It is convenient to define a directed distance from one point to another point with a positive number indicating that the second point is in the positive direction from the first point and a negative number indicating that the second point is in the negative direction from the first point.

Definition (The Directed Distance) $\overrightarrow{AB}$, from $A:a$ to $B:b$ is defined as follows:

$$\overrightarrow{AB} = b - a.$$

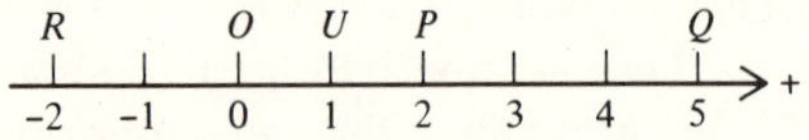

```
   R         O  U  P           Q
   |---|---|---|---|---|---|---|--->+
  -2  -1   0   1   2   3   4   5
```

Figure 9.1 A Coordinate Line

In Figure 9.1,

$$\text{distance, } |OP| = |OR|$$
$$|2 - 0| = |-2 - 0| = 2,$$
$$\text{distance, } |PQ| = |QP|$$
$$|5 - 2| = |2 - 5| = 3,$$
$$\text{directed distance, } \overrightarrow{PQ} = 5 - 2 = 3,$$
$$\text{directed distance, } \overrightarrow{QP} = 2 - 5 = -3.$$

9.1.4 Rectangular Coordinate Systems

A **coordinate plane** is a plane whose points have been placed in one-to-one correspondence with the set of ordered pairs of real numbers, $R \times R$.

A **rectangular coordinate system** is obtained by taking two perpendicular coordinate lines, called the **axes**, intersecting at their origins.

The point of intersection of the axes is called the **origin** of the coordinate system.

It is customary to select one axis horizontal, called the **x-axis**, with its positive direction to the right, and the other axis vertical, called the **y-axis**, with its positive direction upward.

Unless it is specified otherwise, the unit segment on the x-axis is selected congruent to the unit segment on the y-axis; that is, the same scale is used for both axes.

The axes separate the plane into four regions, called **quadrants**, that are numbered consecutively starting with the upper right quadrant and proceeding counterclockwise (see Figure 9.2).

To each ordered pair (a, b) in $R \times R$ is associated a unique point P, the point of intersection of a vertical line through point a on the x-axis and a horizontal line through point b on the y-axis.

Conversely, to each point P in the plane is associated a unique ordered pair (a, b) where a is the coordinate of the point of intersection of the vertical line through P and the x-axis and b is the coordinate of the point of intersection of the horizontal line through P and the y-axis.

RECTANGULAR COORDINATES; DISTANCE; MIDPOINT

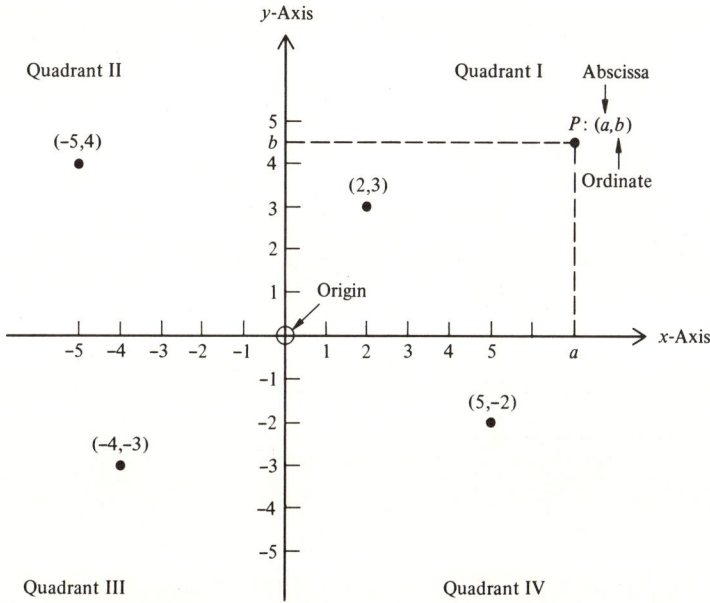

Figure 9.2 A Rectangular Coordinate System

A point a on the x-axis is assigned the ordered pair $(a, 0)$ and a point b on the y-axis is assigned the ordered pair $(0, b)$. The origin is assigned the ordered pair $(0, 0)$.

The numbers of the ordered pair (a, b) are called the **coordinates** of P, with the first component a called the **abscissa** and the second component b called the **ordinate** (see Figure 9.2). A point will be designated symbolically by $P: (a, b)$.

The association in which two perpendicular coordinate lines are used to establish a one-to-one correspondence between the set of points on a plane and the set of ordered pairs of real numbers is called a **rectangular** (or **Cartesian**) **coordinate system**.

9.1.5 Distance Formula

Theorem (The Distance Formula) The distance $|P_1P_2|$ between the points $P_1: (x_1, y_1)$ and $P_2: (x_2, y_2)$ is given by

$$|P_1P_2| = \sqrt{(x_2 - x_1)^2 + (y_2 - y_1)^2}.$$

Referring to Figure 9.3, it may be seen that the distance

$$|P_1Q| = |x_2 - x_1|$$

and the distance

$$|P_2Q| = |y_2 - y_1|$$

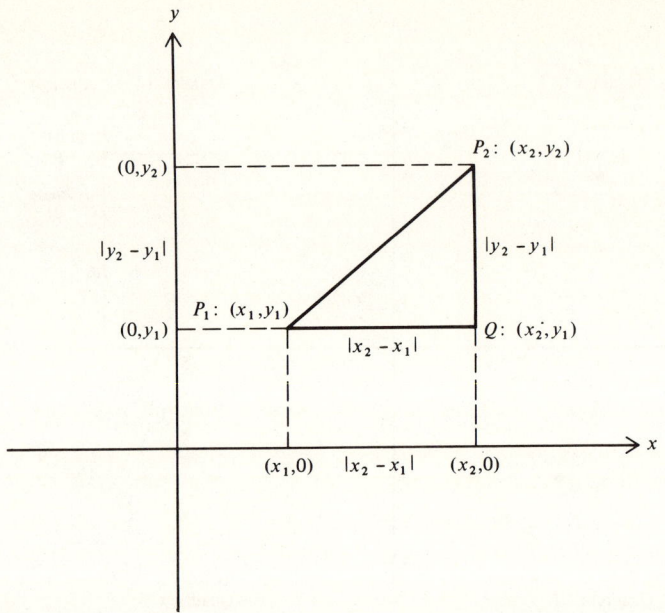

Figure 9.3 Distance Formula: $|P_1 P_2| = \sqrt{(x_2 - x_1)^2 + (y_2 - y_1)^2}$

since the line segments involved are congruent to the corresponding line segments on the x-axis and on the y-axis.

By the theorem of Pythagoras, the square of the length of the hypotenuse of a right triangle is equal to the sum of the squares of the lengths of the legs, and consequently,

$$|P_1 P_2|^2 = (x_2 - x_1)^2 + (y_2 - y_1)^2$$

and

$$|P_1 P_2| = \sqrt{(x_2 - x_1)^2 + (y_2 - y_1)^2}.$$

Example 1 Find the distance between $A: (2, -3)$ and $B: (5, 1)$.

Solution
$$\begin{aligned}|AB| &= \sqrt{(5-2)^2 + (1-(-3))^2} \\ &= \sqrt{3^2 + 4^2} \\ &= 5\end{aligned}$$

9.1.6 Midpoint Formula

Theorem (Midpoint Formula) The midpoint, $M: (x, y)$, of the line segment joining $P_1: (x_1, y_1)$ to $P_2: (x_2, y_2)$ has coordinates

$$x = \frac{x_1 + x_2}{2} \quad \text{and} \quad y = \frac{y_1 + y_2}{2}.$$

RECTANGULAR COORDINATES; DISTANCE; MIDPOINT

Referring to Figure 9.4, and using the geometric theorem that parallel lines intercepting equal segments on one transversal also intercept equal segments on another transversal, it follows that Q is the midpoint of line segment $P_1 R$ and S is the midpoint of line segment $P_2 R$. Thus,

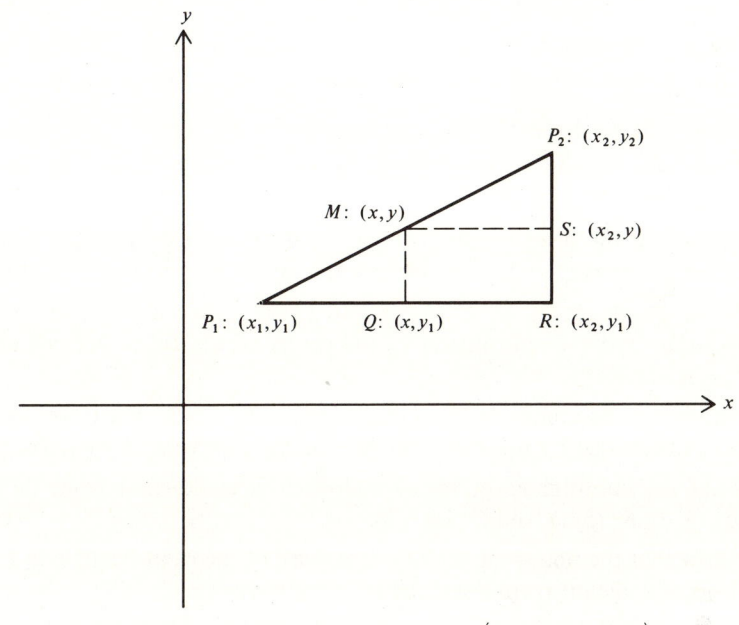

Figure 9.4 Midpoint Formula $M: \left(\dfrac{x_1+x_2}{2}, \dfrac{y_1+y_2}{2}\right)$

$$x - x_1 = x_2 - x \quad \text{and} \quad y - y_1 = y_2 - y$$
$$2x = x_1 + x_2 \qquad\qquad 2y = y_1 + y_2$$
$$x = \frac{x_1 + x_2}{2} \qquad\qquad y = \frac{y_1 + y_2}{2}.$$

Example 2 Find the midpoint of the line segment joining the points (3, 4) and (5, −6).

Solution $x = \dfrac{3 + 5}{2} = 4$ and $y = \dfrac{4 + (-6)}{2} = -1$. The midpoint is (4, −1).

EXERCISES 9.1

1–10. Find the distance $|AB|$ between each of the following pairs of points:

1. $A: (2, -10), B: (-3, 2)$
2. $A: (8, 3), B: (2, -5)$
3. $A: (-1, -4), B: (-2, -5)$
4. $A: (2, 0), B: (6, -2)$
5. $A: (0, -2), B: (-3/2, -4)$
6. $A: (-1/2, -2/3), B: (0, -4/3)$

7. $A: (6, -1)$, $B: (-4, -3)$
8. $A: (a, 0)$, $B: (0, a)$
9. $A: (a, b)$, $B: (1, 2)$
10. $A: (a, b)$, $B: (c, d)$

11–15. State the coordinates of the midpoint of the line segment joining each of the following pairs of points:

11. $(2, 3)$, $(-2, 1)$
12. $(-2, 2)$, $(3, -1)$
13. $(-3, -5)$, $(-1, -4)$
14. $(2, a)$, $(-3, b)$
15. (a, b), (c, d)

16. Show that the points $A: (3, 2)$, $B: (2, 1)$, and $C: (2, 3)$ are the vertices of an isosceles triangle.
17. Show that the three points $A: (-4, 4)$, $B: (6, -8)$, and $C: (7, +3)$ lie on a circle with center $(1, -2)$. What is the radius of this circle?
18. If the points $A: (1, 2)$, $B: (3, 4)$, and $C: (6, 1)$ are the vertices of a right triangle, find the coordinates of the center of the circumscribed circle of this triangle.
19. Show that the points $A: (2, -1)$, $B: (4, 2)$, and $C: (7, 0)$ are the vertices of a right triangle by using the converse of the theorem of Pythagoras.
20. Find the coordinates of the point which is equidistant from the points $A: (3, 2)$, $B: (5, 1)$, and $C: (-1, 3)$.
21. Show that the midpoint of the hypotenuse of the right triangle in Exercise 19 is equidistant from the vertices of the triangle.
22. Show that the line segment joining the midpoints of sides AB and AC of the triangle in Exercise 16 is 1/2 side BC.
23. Find the perimeter of the triangle formed by joining the midpoints of the triangle with vertices $A: (0, 0)$, $B: (0, 6)$, and $C: (3, 4)$.
24. Find the value of k if the distance between the points $(k, 3)$ and $(-3, k)$ is $2\sqrt{17}$ units.
25. Find the value of k if the distance of $(-2, k)$ from $(4, 6)$ is $2\sqrt{10}$.
26. Find $P: (x, y)$ if the point $(-2, 3)$ bisects the line segment having endpoints $P: (x, y)$ and $(5, 4)$.
27. The line segment joining $A: (0, -3)$ and $B: (-5, 0)$ is extended to a point C so that $|AC| = 2|AB|$. Find the coordinates of C.

9.2 SLOPE; PARALLEL AND PERPENDICULAR LINES

9.2.1 Slope

Since an angle is a basic concept of geometry, it is convenient to associate with the angle a measure based on the coordinates of points forming the angle. The concept of *slope* accomplishes this objective.

SLOPE; PARALLEL AND PERPENDICULAR LINES

Definition of Inclination The inclination, θ, of a line not parallel to the x-axis is the smallest positive angle measured counterclockwise from the positive x-axis to the line.

The inclination of a line parallel to the x-axis is 0.

Definition of Slope The **slope**, m, of a line with inclination θ, is defined as follows:

$$m = \tan \theta.$$

From the definition of the inclination θ, it follows that $0° \leq \theta < 180°$. It also follows that a line is horizontal, parallel to the x-axis, if and only if $\theta = 0$ and therefore its slope is 0. A line is vertical, parallel to the y-axis, if and only if $\theta = 90°$. Since $\tan 90°$ is undefined, a vertical line has no slope.

From Figure 9.5 it may be observed that lines with positive slope, $0 < \theta < 90°$, rise toward the right, and lines with negative slope, $90° < \theta < 180°$, fall toward the right.

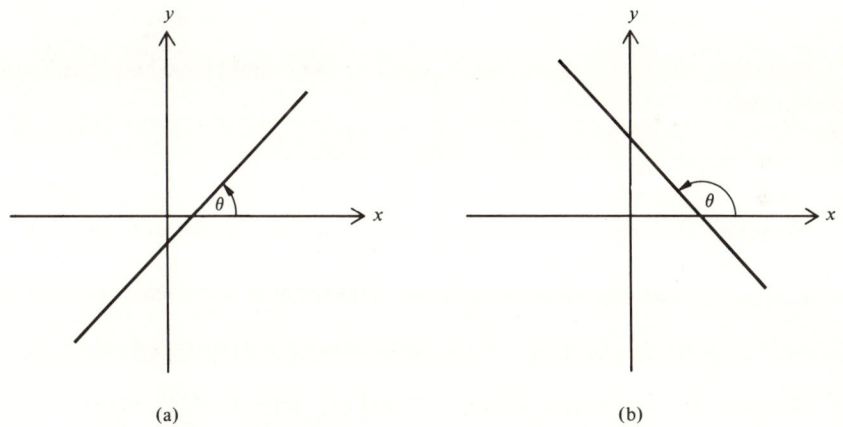

Figure 9.5 (a) A Line with Positive Slope, $0 < \theta < 90°$
(b) A Line with Negative Slope, $90° < \theta < 180°$

Theorem (Slope Formula) If m is the slope of the line on the points $P_1: (x_1, y_1)$ and $P_2: (x_2, y_2)$ and if $x_1 \neq x_2$, then

$$m = \frac{y_2 - y_1}{x_2 - x_1}.$$

By definition, $m = \tan \theta$, and, from Figure 9.6, it may be seen that

$$m = \tan \theta = \frac{\overrightarrow{QP_2}}{\overrightarrow{P_1Q}} = \frac{y_2 - y_1}{x_2 - x_1}.$$

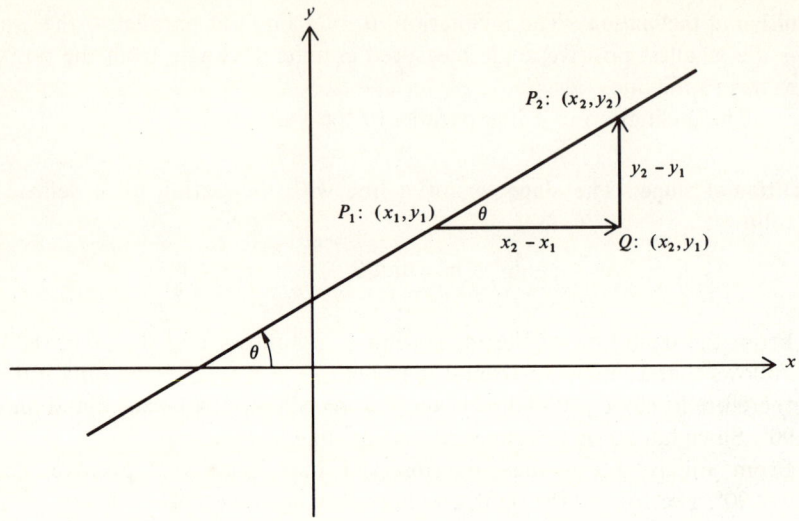

Figure 9.6 Slope Formula: $m = \dfrac{y_2 - y_1}{x_2 - x_1}$

Note that if P_1 and P_2 were interchanged, then Q would have had coordinates (x_1, y_2) and

$$m = \frac{\overrightarrow{QP_1}}{\overrightarrow{P_2Q}} = \frac{y_1 - y_2}{x_1 - x_2} = \frac{y_2 - y_1}{x_2 - x_1}.$$

Therefore, the slope is the same if P_2 is chosen either higher or lower than P_1. If $y_1 = y_2$, then $y_2 - y_1 = 0$, $\theta = 0$, and the line is horizontal. If $x_1 = x_2$, then the slope is undefined and the line is vertical.

Example 1 Find the slope of the line on points $(3, -2)$ and $(-4, 1)$.

Solution Choose $(x_2, y_2) = (3, -2)$ and $(x_1, y_1) = (-4, 1)$. Then

$$m = \frac{y_2 - y_1}{x_2 - x_1} = \frac{-2 - 1}{3 - (-4)} = \frac{-3}{7}.$$

9.2.2 Parallel and Perpendicular Lines

Since parallel lines have equal inclinations and since $\tan \theta_1 = \tan \theta_2$ whenever $\theta_1 = \theta_2$, it follows that nonvertical lines are parallel if and only if their slopes are equal.

Theorem on Parallel Lines If two nonvertical lines, L_1 and L_2, have slopes m_1 and m_2, respectively, then

$$L_1 \parallel L_2 \quad \text{if and only if} \quad m_1 = m_2.$$

(See Figure 9.7.)

SLOPE; PARALLEL AND PERPENDICULAR LINES

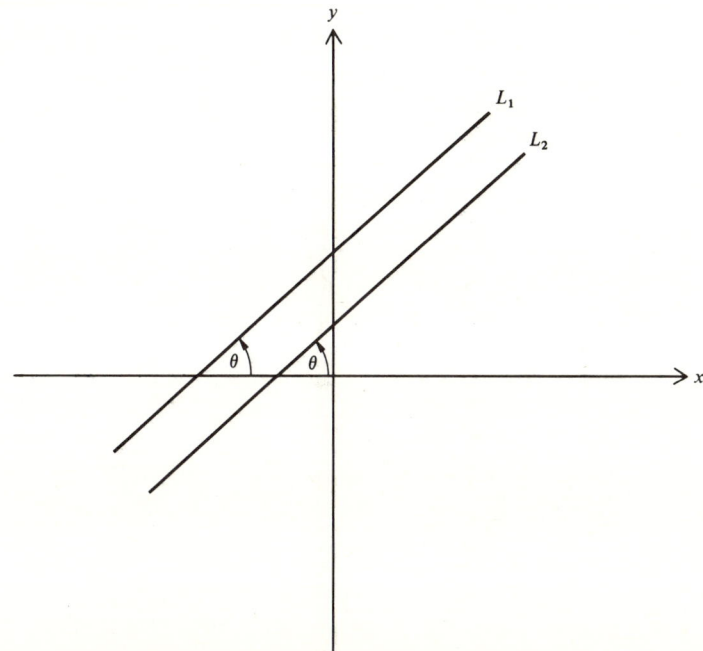

Figure 9.7 Parallel Lines: $m_1 = m_2 = \tan \theta$

If two nonvertical lines are perpendicular, then their inclinations must differ by 90° or $\theta_2 = \theta_1 + 90°$. Then

$$m_2 = \tan(\theta_1 + 90°) = -\cot \theta_1 = \frac{-1}{\tan \theta_1} = \frac{-1}{m_1}.$$

Conversely, if the slope of one line is the negative reciprocal of the slope of the other line, then $\tan \theta_2 = (-1/\tan \theta_1) = -\cot \theta_1$ and the inclinations of the lines must differ by 90° or an odd multiple of 90°. However, since the inclination of a line is between 0° and 180°, there is only one possibility for $\theta_2 > \theta_1$ and $\theta_2 - \theta_1 = 90°$ or the lines are perpendicular.

Theorem on Perpendicular Lines If two nonvertical lines, L_1 and L_2, have slopes m_1 and m_2, respectively, then

$$L_1 \perp L_2 \quad \text{if and only if} \quad m_2 = \frac{-1}{m_1}.$$

(See Figure 9.8.)

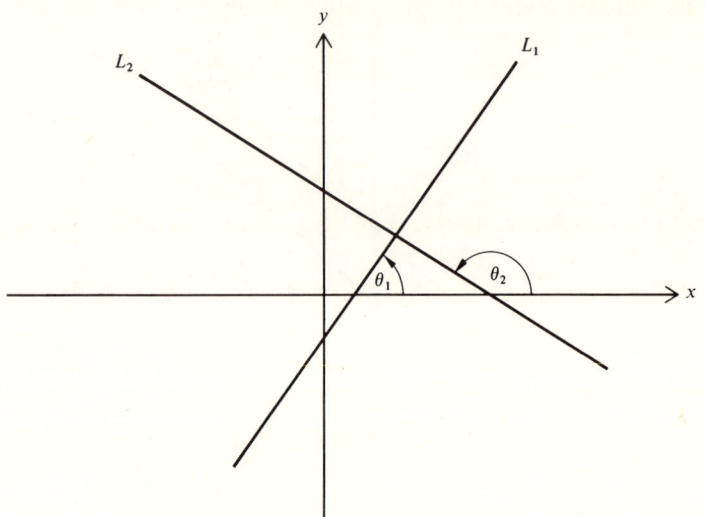

Figure 9.8 Perpendicular Lines: $m_2 = \dfrac{-1}{m_1}$

Example 2 Show that the line on points $(-3, 4)$ and $(2, 6)$ is perpendicular to the line on points $(1, -2)$ and $(-1, 3)$.

Solution $\quad m_1 = \dfrac{6-4}{2-(-3)} = \dfrac{2}{5}$

$$m_2 = \dfrac{3-(-2)}{-1-1} = \dfrac{-5}{2}$$

Since $m_2 = (-1/m_1)$, the lines are perpendicular.

EXERCISES 9.2

1–5. Find the slope of each of the following lines on the given pairs of points:

1. $(2, 1), (3, -2)$
2. $(3, 2), (8, -1)$
3. $(-2, -3), (-4, -5)$
4. $(4, 3), (4, -1)$
5. $(1, -2), (3, -2)$

6–10. Find the slope of each of the following lines with the given inclination:

6. $30°$
7. $3\pi/4$ radians
8. $5\pi/6$ radians
9. $180° - \text{Tan}^{-1} 3$
10. $\pi - \text{Tan}^{-1} 2$

ANALYTIC PROOFS

11–15. Find the inclinations of each of the following lines with given slopes:

11. $m = 1$
12. $m = \sqrt{3}/3$
13. $m = -1$
14. $m = \sqrt{3}$
15. $m = -2$

16. Find the slope of the line representing all points equally distant from the points (2, 3) and (−1, 4).
17. Show that the points A: (4, 3), B: (7, 4), C: (5, 2), and D: (2, 1) are the vertices of a parallelogram.
18. Use the concept of slope to show that the points A: (−1, 2), B: (1, −3), and C: (−9, −7) are the vertices of a right triangle.
19. Prove that the points A: (−1, −5), B: (−4, 1), C: (2, 4), and D: (5, −2) are the vertices of a square.
20. Using only the slope formula, determine whether the points (−1, 0), (−9, −2), and (7, 2) are collinear.
21. Find k so that the line through (7, k) and (2, 3) is
    (a) parallel to a line having slope 5;
    (b) perpendicular to a line having slope 5.
22. Find k so that the line through (k, 3) and (1, 5) is
    (a) parallel to a line having slope $-1/2$;
    (b) perpendicular to a line having slope $-1/2$.
23. Find the slope of a line parallel to the line on the points (4, 5) and (3, 5).
24. Find the slope of a line perpendicular to the line on the points (−2, 1) and (−2, −1).
25. Find the slope of a line
    (a) parallel to the line on the points (−2, 6) and (2, 9);
    (b) perpendicular to the line on the points (−2, 6) and (2, 9).
26. Find the slope of a line
    (a) parallel to the line on the points (−1, 3) and (1, −2);
    (b) perpendicular to the line on the points (−1, 3) and (1, −2).
27. The inclination of the line joining (5, 6) and (k, −3) is 45°. Find k.
28. The inclination of the line joining (−2, 4) and (7, k) is 120°. Find k.

9.3 ANALYTIC PROOFS

Many geometric theorems have simple algebraic proofs obtained by establishing a coordinate system on the plane of the geometric figure and by using the numerical concepts developed in the preceding sections. Such a proof is called an **analytic proof**. A summary comparing geometric and algebraic concepts is given in Table 9.1.

Table 9.1 Comparison of Geometric and Algebraic Concepts

GEOMETRIC	ALGEBRAIC
Point, P	(x, y)
Line segment	$\|P_1P_2\| = \sqrt{(x_2 - x_1)^2 + (y_2 - y_1)^2}$
Midpoint	$\left(\dfrac{x_1 + x_2}{2}, \dfrac{y_1 + y_2}{2}\right)$
Angle, θ	$m = \tan \theta = \dfrac{y_2 - y_1}{x_2 - x_1}$
Nonvertical Parallel Lines	$m_1 = m_2$
Nonvertical Perpendicular Lines	$m_2 = \dfrac{-1}{m_1}$

Usually an analytic proof is simplest when the geometric figure is placed in a preferential position with respect to the coordinate axes. This can often be accomplished by selecting the origin as one of the essential points or by placing a side of a figure along one of the axes.

Example 1 Prove analytically that the line segment joining the midpoints of two sides of a triangle is parallel to the third side and equal to one-half its length.

Solution
(1) Place triangle OAB with one vertex O at the origin and another vertex on the x-axis at $A: (a, 0)$. Then the third vertex can be selected anywhere not on the x-axis, say at $B: (b, c)$. (See Figure 9.9.)
(2) Let M and N be the midpoints of sides OB and AB, respectively. Then, using the midpoint formula,

$$M = \left(\frac{b}{2}, \frac{c}{2}\right) \quad \text{and} \quad N = \left(\frac{a+b}{2}, \frac{c}{2}\right).$$

(3) Using the distance formula,

$$|MN| = \sqrt{\left(\frac{a+b}{2} - \frac{b}{2}\right)^2 + \left(\frac{c}{2} - \frac{c}{2}\right)^2} = \frac{a}{2}$$

$$|OA| = a.$$

Therefore, $|MN| = \frac{1}{2}|OA|$.

ANALYTIC PROOFS

(4) Using the slope formula,

$$\text{slope of } MN = \frac{\frac{c}{2} - \frac{c}{2}}{\frac{a+b}{2} - \frac{b}{2}} = 0$$

$$\text{slope of } OA = \frac{0-0}{a-0} = 0.$$

Therefore, $MN \parallel OA$.

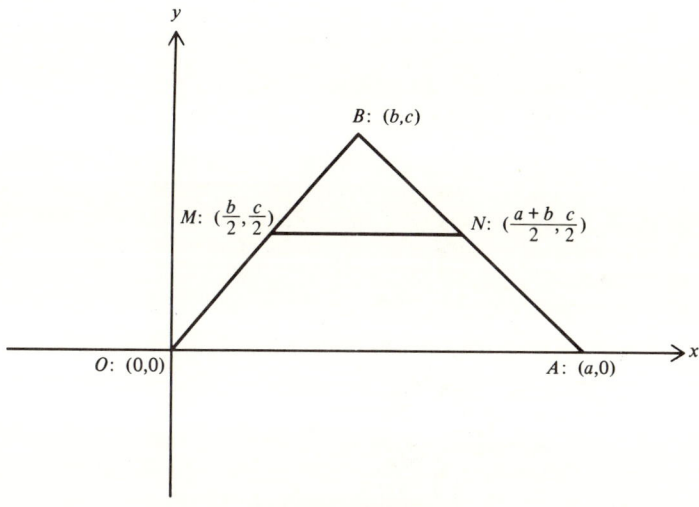

Figure 9.9

Example 2 Prove analytically that the diagonals of a rhombus are perpendicular.

Solution
(1) Place the rhombus with one vertex at the origin $O: (0, 0)$ and one vertex at $C: (c, 0)$. Let $A: (a, b)$ and $B: (x, y)$ be the other two vertices.
(2) Determine the coordinates of B and C so that $OABC$ is a rhombus. If $OC = OA$, then $c = \sqrt{a^2 + b^2}$ by the distance formula. If $AB \parallel OC$, then $y = b$ by using the slope formula. If $AB = OC$, then $x - a = \sqrt{a^2 + b^2}$ or $x = a + \sqrt{a^2 + b^2}$.

(3) Now find the slopes of the diagonals.

$$\text{slope of } OB = m_1 = \frac{b}{a + \sqrt{a^2 + b^2}}$$

$$\text{slope of } AC = m_2 = \frac{b - 0}{a - \sqrt{a^2 + b^2}} = \frac{b(a + \sqrt{a^2 + b^2})}{a^2 - (a^2 + b^2)} = \frac{a + \sqrt{a^2 + b^2}}{-b}$$

Since $m_2 = (-1/m_1)$, the diagonals are perpendicular.

(See Figure 9.10.)

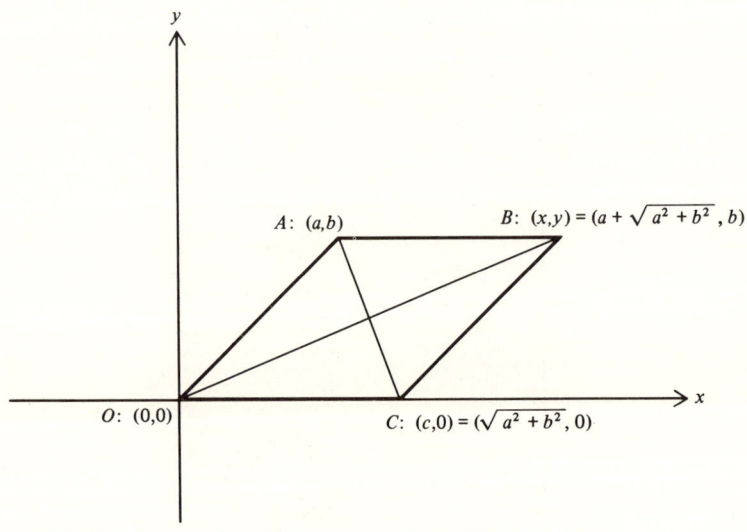

Figure 9.10

EXERCISES 9.3

1–10. Prove each of the following analytically:

1. If two lines in the same plane are perpendicular to a given line, then the two lines are parallel.
2. In an isosceles triangle, the medians drawn to the congruent sides are equal in measure.
3. The median of an isosceles trapezoid is equal to one-half the sum of the bases.
4. The diagonals of an isosceles trapezoid are equal in measure.
5. The diagonals of a parallelogram bisect each other.
6. An angle inscribed in a semicircle is a right angle.

7. The lines joining successively the midpoints of any quadrilateral form a parallelogram.
8. If the adjacent sides of a parallelogram are congruent, then the diagonals of the parallelogram are perpendicular.
9. If the diagonals of a parallelogram are equal in measure, then the parallelogram is a rectangle.
10. An equilateral triangle is equiangular.

9.4 THE STRAIGHT LINE

Definition (Graph of a Relation) The graph of a relation is the set of points whose coordinates are the ordered pairs of the relation.

Definition (Graph of an Equation in Two Variables) The graph of an equation in two variables is the graph of the relation defined by the equation.

The coordinates of a point on the graph of an equation are said to *satisfy the equation* or to be *a solution of the equation*.

For example, consider the relation defined by the equation $2x + 3y = 8$. The graph of this equation is the set of points corresponding to the set, $\{(x, y) | 2x + 3y = 8\}$.

The point $(7, -2)$ is on this graph since its coordinates satisfy the relation; that is, $2(7) + 3(-2) = 14 - 6 = 8$.

Theorem (Point-Slope Form of a Line) If $P: (x_1, y_1)$ is a point on a line whose slope is m, then the point-slope form of the equation of this line is

$$y - y_1 = m(x - x_1).$$

Proof Let (x, y) be the coordinates of any point Q on the line different from P. Then Q is on the line if and only if

$$m = \frac{y - y_1}{x - x_1} \quad \text{or} \quad y - y_1 = m(x - x_1).$$

P is also on this line since

$$y_1 - y_1 = m(x_1 - x_1) = 0.$$

Theorem (Horizontal Lines) A horizontal line on the point (x_1, y_1) has the equation $y = y_1$.

This follows since a point $Q: (x, y)$ is on this line if and only if $y = y_1$.

Theorem (Vertical Lines) A vertical line on the point (x_1, y_1) has the equation $x = x_1$.

This follows since a point $Q: (x, y)$ is on this line if and only if $x = x_1$.

Example 1 Find the equation of the line on the points $(2, -5)$ and $(4, 3)$.

Solution The slope $m = \dfrac{3 - (-5)}{4 - 2} = \dfrac{8}{2} = 4.$

Since
$$y - y_1 = m(x - x_1),$$
$$y - (-5) = 4(x - 2)$$
or
$$y + 5 = 4(x - 2).$$

9.4.1 Slope-Intercept Form of a Line

If a line intersects the x-axis at $(a, 0)$, then a is called the *x-intercept* of the line.

If a line intersects the y-axis at $(0, b)$, then b is called the *y-intercept* of the line.

By selecting the point (x_1, y_1) on the line as the point $(0, b)$, then from the point-slope form of the line,
$$y - b = m(x - 0)$$
or
$$y = mx + b.$$

Since the slope m and the y-intercept b are read directly from this equation, it is called the slope-intercept form of the line.

Theorem (Slope-Intercept Form of a Line) If a line has slope m and a y-intercept b, then the slope-intercept form of its equation is
$$y = mx + b.$$

Example 2 Find the equation of a line whose slope is -2 and whose y-intercept is 5.

Solution $m = -2$ and $b = 5$. Using
$$y = mx + b,$$
$$y = -2x + 5.$$

THE STRAIGHT LINE

9.4.2 General Form of a Line

A first degree equation of the form $Ax + By + C = 0$ where A and B are not both zero is called a linear equation since it can be shown that its graph is a straight line.

$$\text{If } B \neq 0, \text{ then } y = \frac{-A}{B} x + \frac{-C}{B}.$$

This is the slope-intercept form of a line with slope $m = -A/B$ and y-intercept $b = -C/B$.

$$\text{If } B = 0 \text{ but } A \neq 0, \text{ then } Ax + C = 0 \text{ and } x = \frac{-C}{A}.$$

This is the equation of a vertical line.
Therefore, the graph of any equation having the form

$$Ax + By + C = 0 \qquad (A^2 + B^2 \neq 0)$$

is a straight line. Conversely, it is readily seen that every straight line has an equation of this form, called the **general form of a straight line.**

Example 3 Find the general form of the equation of a line on the point (1, 2) and perpendicular to $2x - 3y = 6$.

Solution $\quad -3y = -2x + 6$

$$y = \tfrac{2}{3}x - 2$$

Thus, the slope of the given line is $\tfrac{2}{3}$. The slope of a perpendicular line is $-3/2$.

Using

$$y - y_1 = m(x - x_1),$$

$$y - 2 = \frac{-3}{2}(x - 1),$$

$$2y - 4 = -3x + 3.$$

Finally, $3x + 2y - 7 = 0$, the general form.

EXERCISES 9.4

1–10. Find the general form of the equation for the line determined by each of the following pairs of points:

1. (3, 2), (4, 1)
2. (2, −1), (1, 5)
3. (−3, −2), (6, 4)
4. (−3, −2), (−6, −4)
5. (5, 0), (2, 3)
6. (1, 3), (3, 1)

7. $(a, 0), (0, b), a \neq 0, b \neq 0$
8. $(0, 0), (0, 2)$
9. $(3, 1), (3, 4)$
10. $(2, -1), (-2, 3)$

11–15. For each of the following find the general form of the equation for the line with the given slope and passing through the given point:

11. $m = \frac{1}{2}, (2, 3)$
12. $m = -2, (1, -3)$
13. $m = -\frac{3}{4}, (-2, 4)$
14. $m = \frac{2}{3}, (-6, -2)$
15. $m = -\frac{7}{8}, (0, -2)$

16. State an equation of the line which passes through the point $(2, -1)$ and is parallel to the line $3x + 2y + 5 = 0$.

17. State an equation of the line which passes through the point $(4, 3)$ and is perpendicular to the line $4x - 3y + 2 = 0$.

18. State an equation of the line passing through the point $(2, 3)$ and having an inclination of $60°$.

19. State an equation of the line which passes through the origin and is perpendicular to the line through the points $(1, 2)$ and $(-3, 5)$.

20. Find an equation of the perpendicular bisector of the line segment joining the points $(2, -1)$ and $(5, 4)$.

21. Find the coordinates of the point on the line $2x + 3y + 13 = 0$ which is equidistant from the points $(-4, 7)$ and $(2, 3)$.

22. Show that if $(a, 0)$ and $(0, b)$, $a \neq 0$ and $b \neq 0$ are points on a line, then its equation can be expressed as $\frac{x}{a} + \frac{y}{b} = 1$. This is called the intercept form of the line.

23. Show that if θ is the inclination of a line, then its equation may be expressed as
$$x \sin \theta - y \cos \theta = d,$$
where $|d|$ is the perpendicular distance from the origin to the line.

24. Show that if two intersecting lines L_1 and L_2 with slopes m_1 and m_2, respectively, form an angle θ, then
$$\tan \theta = \frac{m_2 - m_1}{1 + m_2 m_1}$$
provided that $\theta \neq 90°$ and L_2 has the greater inclination.

25–29. Use Exercise 24 to find to the nearest degree the angle between the lines with the given slopes.

25. $\frac{3}{4}$ and $\frac{1}{3}$
26. $-\frac{1}{2}$ and $-\frac{2}{3}$
27. 2 and $-\frac{1}{4}$
28. $-\frac{1}{2}$ and $\frac{1}{3}$
29. $\frac{1}{4}$ and $\frac{1}{6}$
30. 3 and $-\frac{1}{5}$

INTERCEPTS; EXTENT; SYMMETRY 333

31. Calculate to the nearest degree the angles of a triangle whose vertices are $(-1, 0)$, $(2, -1)$, $(4, 1)$.
32. The inclination of a line is 150°. Find the slope and inclination of a line that makes a 45° angle with this line.
33. Find the angle that the line on the points $(12, 3)$ and $(7, -2)$ makes with a line having an inclination of 150°.
34. Prove analytically that the lines joining a vertex of an equilateral triangle to the points of trisection of the opposite side do *not* trisect this vertex angle.
35. Find the area of the triangle with vertices at $(3, 1)$, $(4, 9)$ and $(7, -2)$ by using the formula area $= \frac{1}{2}ab \sin C$, where C is the included angle of the two sides having measures a and b.
36. What does each of the following statements imply about the coefficients A, B, and C of the equation $Ax + By + C = 0$?
   (a) the slope of the line is $\frac{3}{4}$; (b) the x-intercept is 4; (c) The inclination of the line is 90°; (d) the line is perpendicular to $3y = 4x - 5$; (e) the point $(0, 3)$ is on the line; (f) the line is parallel to the x-axis.
37. Let $ABCD$ be a parallelogram with M the midpoint of side CD. Prove that the diagonal BD intersects the line segment AM at a point of trisection of the diagonal BD.
38. State the equation of the line that bisects the angle between the lines $x - 3y = 0$ and $x + 2y = 0$.

9.5 INTERCEPTS; EXTENT; SYMMETRY

While the graph of a linear equation is easily obtained by selecting any two points on the graph and joining them with a straight line, the graphs of other equations are more difficult to determine. Useful concepts for this purpose are the *intercepts*, *extent*, and *symmetry* of the graph.

9.5.1 Intercepts

Definition of x-intercept An x-intercept of a curve is the x-coordinate of a point of intersection of the curve and the x-axis.

Definition of y-intercept A y-intercept of a curve is the y-coordinate of a point of intersection of the curve and the y-axis.

Example 1 Find the x- and y-intercepts of the graph of $x^2 + 4y^2 = 4$.

Solution

(1) x-intercepts: Let $y = 0$ and solve for x. $x^2 + 0 = 4$, $x^2 = 4$, and thus $x = 2$ or $x = -2$; 2 and -2 are the x-intercepts.
(2) y-intercepts: Let $x = 0$ and solve for y. $0^2 + 4y^2 = 4$, $y^2 = 1$, and $y = 1$ or $y = -1$; 1 and -1 are the y-intercepts.

9.5.2 Extent

The *extent* of a curve is the region of the plane where the curve lies. It is determined by the *domain* and *range* of the relation defined by the equation of the curve, with the domain and range restricted to the set of real numbers.

Example 2 Determine the extent of the graph of $4x^2 + 9y^2 = 36$.

Solution

(1) Find the domain:
Solving for y, $9y^2 = 36 - 4x^2 = 4(9 - x^2)$.
Thus, $y = \pm\frac{2}{3}\sqrt{9 - x^2}$.
If y is to be real, then $9 - x^2 \geq 0$ or $x^2 \leq 9$.
The domain is $|x| \leq 3$ or $-3 \leq x \leq 3$.

(2) Find the range:
Solving for x, $4x^2 = 36 - 9y^2 = 9(4 - y^2)$.
Thus, $x = \pm\frac{3}{2}\sqrt{4 - y^2}$.
If x is to be real, then $4 - y^2 \geq 0$ or $y^2 \leq 4$.
The range is $|y| \leq 2$ or $-2 \leq y \leq 2$.

Finally, the curve is located within or on a rectangle having the boundary lines: $x = -3$, $x = 3$, $y = -2$, and $y = 2$. (See Figure 9.11.)

Example 3 Find the domain and range of the graph of $4x^2 - 9y^2 = 36$.

Solution

DOMAIN $y = \pm\frac{2}{3}\sqrt{x^2 - 9}$.
If $x^2 - 9 \geq 0$, then $x^2 \geq 9$ or $|x| \geq 3$. The domain is the union of $x \geq 3$ and $x \leq -3$.

RANGE $x = \pm\frac{3}{2}\sqrt{y^2 + 4}$.
If $y^2 + 4 \geq 0$, then y can be any real number. The range is all real y.

INTERCEPTS; EXTENT; SYMMETRY   335

Therefore, the curve is located outside the vertical strip bounded by the lines $x = -3$ and $x = 3$, or on these lines. (See Figure 9.12.)

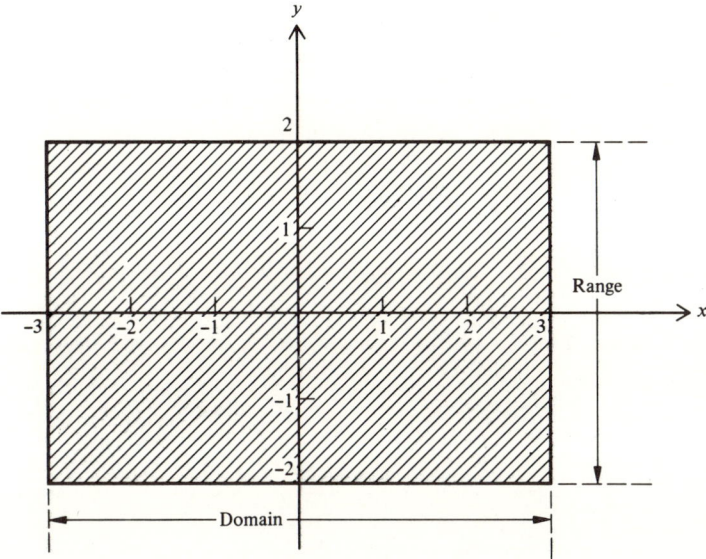

Figure 9.11

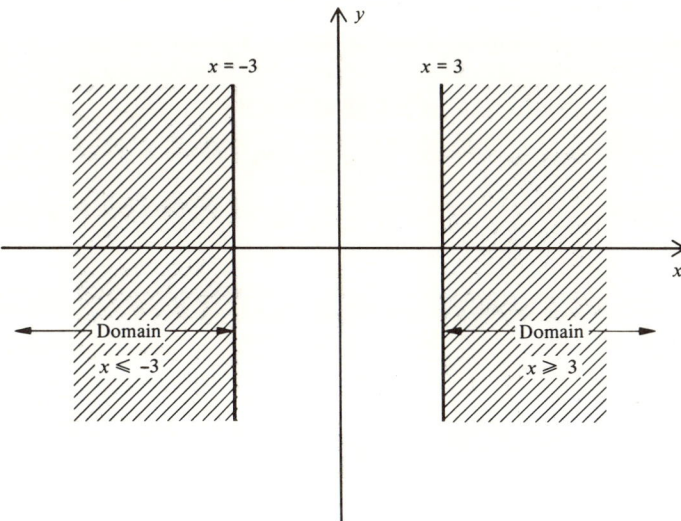

Figure 9.12

9.5.3 Symmetry

Definition of Point Symmetry A curve is symmetric to a point P if and only if for every point A on the curve, there is another point B on the curve so that P is the midpoint of the line segment AB.

Definition of Line Symmetry A curve is symmetric to a line L if and only if for every point A on the curve, there is another point B on the curve so that the line L is the perpendicular bisector of the line segment AB. (See Figure 9.13.)

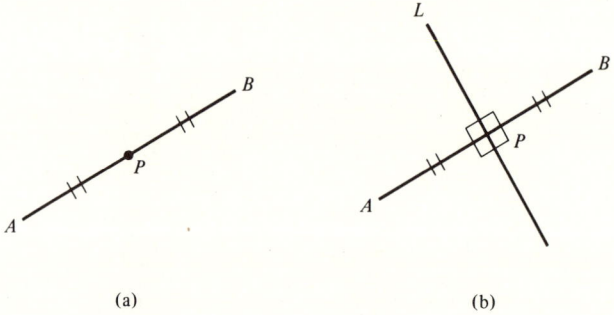

(a) $|BP| = |PA|$; A and B are symmetric to point P

(b) $|BP| = |PA|$ and $L \perp AB$; A and B are symmetric to line L

Figure 9.13

By referring to Figure 9.14, it may be seen that the following statements follow directly from the definitions of point and line symmetry.

(1) A curve is *symmetric to the origin* if and only if the point $(-x, -y)$ is on the curve whenever the point (x, y) is.

(2) A curve is *symmetric to the x-axis* if and only if the point $(x, -y)$ is on the curve whenever the point (x, y) is.

(3) A curve is *symmetric to the y-axis* if and only if the point $(-x, y)$ is on the curve whenever the point (x, y) is.

Symmetry is useful in graphing because it reduces the number of points that need to be tabulated in order to sketch the curve. Points in other quadrants may be obtained as the mirror image of the quadrant I points depending on the symmetry.

(1) To determine symmetry with respect to the origin, replace x by $-x$ and y by $-y$ in the equation of the curve. If the resulting equation is equivalent to the original one, the curve is symmetric to the origin.

INTERCEPTS; EXTENT; SYMMETRY

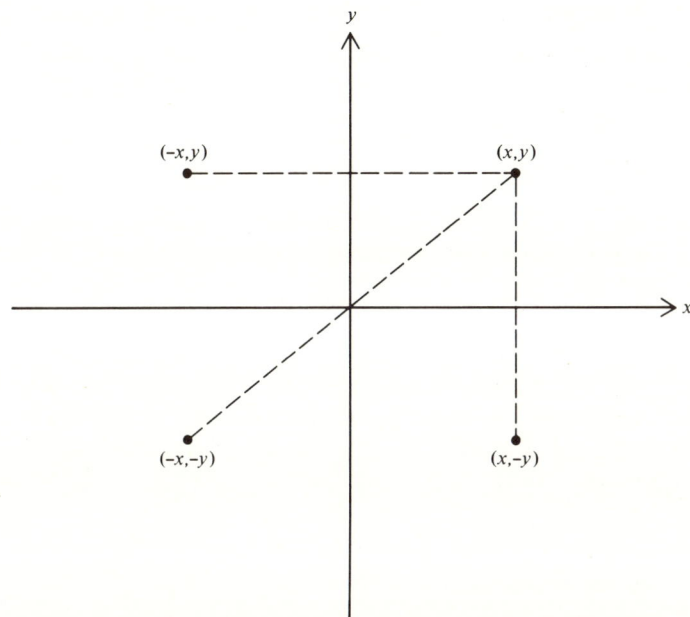

Figure 9.14 (x, y) and $(-x, -y)$ are symmetric to the origin
(x, y) and $(x, -y)$ are symmetric to the x-axis
(x, y) and $(-x, y)$ are symmetric to the y-axis

(2) To determine symmetry with respect to the x-axis, replace y by $-y$. If the resulting equation is equivalent to the original one, then the curve is symmetric to the x-axis.

(3) To determine symmetry with respect to the y-axis, replace x by $-x$. If the resulting equation is equivalent to the original one, then the curve is symmetric to the y-axis.

Example 4 Discuss the symmetry of the graph of $x^2 + 4y^2 = 4$.

Solution

(1) Origin: Replacing x by $-x$ and y by $-y$, $(-x)^2 + 4(-y)^2 = 4$ or $x^2 + 4y^2 = 4$. Thus, the curve is symmetric to the origin.

(2) x-axis: Replacing y by $-y$, $x^2 + 4(-y)^2 = 4$ or $x^2 + 4y^2 = 4$. Thus, the curve is symmetric to the x-axis.

(3) y-axis: Replacing x by $-x$, $(-x)^2 + 4y^2 = 4$ or $x^2 + 4y^2 = 4$. Thus, the curve is symmetric to the y-axis.

Example 5 Discuss the symmetry of the graph of $xy = 6$.

Solution

(1) Origin: $(-x)(-y) = 6$ is the same as $xy = 6$. Thus, the curve is symmetric to the origin.

(2) x-axis: $x(-y) = 6$ or $-xy = 6$. This is not the same as $xy = 6$. Thus, the curve is *not* symmetric to the x-axis.

(3) y-axis: $(-x)y = 6$ or $-xy = 6$. Thus, the curve is *not* symmetric to the y-axis.

Example 6 Determine the symmetry of the graph of $y^2 + 4x = 0$.

Solution

(1) Origin: $(-y)^2 + 4(-x) = 0$ or $y^2 - 4x = 0$. Thus, the curve is *not* symmetric to the origin.

(2) x-axis: $(-y)^2 + 4x = 0$ or $y^2 + 4x = 0$. Thus, the curve is symmetric to the x-axis.

(3) y-axis: $y^2 + 4(-x) = 0$ or $y^2 - 4x = 0$. Thus, the curve is *not* symmetric to the y-axis.

(See Figure 9.15.)

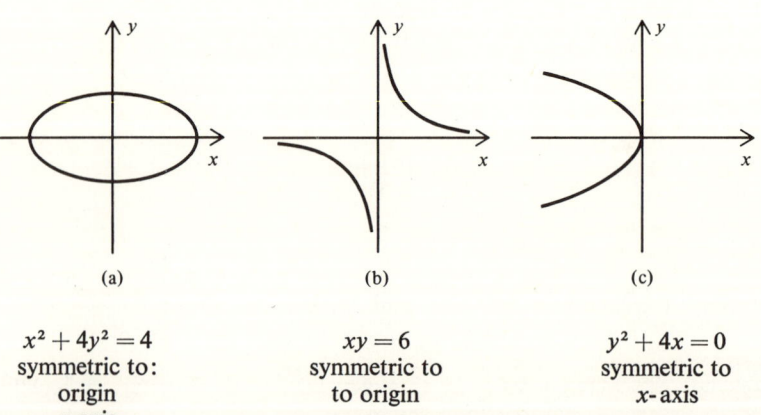

(a)
$x^2 + 4y^2 = 4$
symmetric to:
  origin
  x-axis
  y-axis

(b)
$xy = 6$
symmetric to origin

(c)
$y^2 + 4x = 0$
symmetric to x-axis

Figure 9.15

CIRCLES AND STANDARD PARABOLAS

EXERCISES 9.5

1–20. For each of the following, determine the intercepts, the extent, and the symmetry with respect to the origin, the x-axis, and the y-axis:

1. $4x^2 + y^2 = 9$
2. $xy = -6$
3. $y^2 - x^2 = 9$
4. $y^2 - 4y + 4x = 12$
5. $x^2 - 4y^2 = 0$
6. $x^2 + y^2 = 100$
7. $4x^2 + 3y^2 - 12y = 36$
8. $y^2 = 8x + 1$
9. $x^2 - y^2 = 16$
10. $xy = 12$
11. $yx^2 = 25$
12. $y^2 = x(x^2 - 4)$
13. $y^2(x^2 + 1) = x^2 - 1$
14. $y = x^3 - 4x$
15. $y^2 = (x + 2)(x - 1)(x - 3)$
16. $y(x^2 - 1) = x^2 + 1$
17. $y(x + 1)(x - 2) = x^2$
18. $xy = y^2 - 9$
19. $xy - 3x - 2y - 6 = 0$
20. $y^2 = x^2(y^2 - 25)$

9.6 CIRCLES AND STANDARD PARABOLAS

Definition of Circle A circle is the set of points in a plane that are equidistant from a fixed point, called the **center** of the circle.

The distance between any point on the circle and the center of the circle is called the **radius** of the circle.

9.6.1 Equation of Circle in Standard Form

$$(x - a)^2 + (y - b)^2 = r^2$$

If $P: (x, y)$ designates any point on the circle and if $C: (a, b)$ is the center of the circle whose radius is r, then the equation of the circle is obtained by using the distance formula. Thus,

$$\sqrt{(x - a)^2 + (y - b)^2} = r.$$

Squaring both sides of the above equation yields the standard form of the equation of the circle.

The standard form for the equation of the circle is useful because the radius and the coordinates of the center can be read directly from the equation. The circle is symmetric to every line through its center. (See Figure 9.16.)

Example 1 Write the equation of the circle with center $(3, -4)$ and radius 5.

Solution $(x - 3)^2 + (y + 4)^2 = 25$

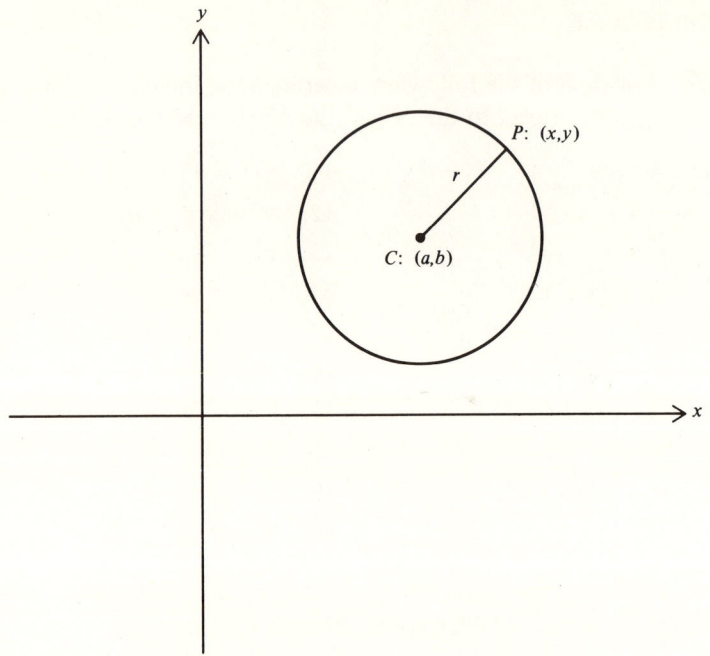

Figure 9.16 Circle: $(x-a)^2 + (y-b)^2 = r^2$

Example 2 Show that $x^2 + y^2 + 8x - 10y + 5 = 0$ is the equation of a circle, and then determine its center and radius.

Solution Completing the squares for x and for y,

$$(x^2 + 8x) + (y^2 - 10y) = -5$$
$$(x^2 + 8x + 16) + (y^2 - 10y + 25) = -5 + 16 + 25$$
$$(x+4)^2 + (y-5)^2 = 36.$$

Thus, the equation is that for a circle with center $(-4, 5)$ and with radius $= 6$.

Example 3 Draw the graph of $(x+2)^2 + (y-3)^2 = 16$.

Solution Draw the circle by using a compass with center at $(-2, 3)$ and with radius $= 4$. (See Figure 9.17.)

Definition of Parabola A parabola is the set of points in a plane that are equidistant from a fixed point called the **focus** and a fixed line called the **directrix**.

CIRCLES AND STANDARD PARABOLAS

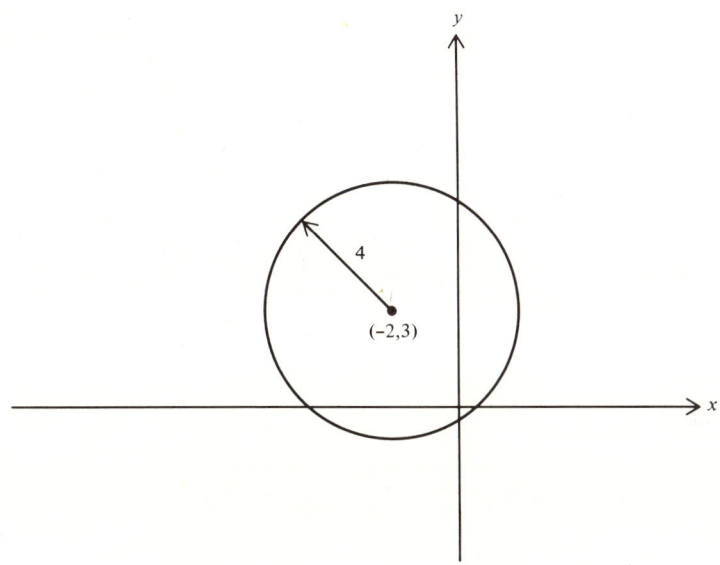

Figure 9.17 $(x+2)^2 + (y-3)^2 = 16$

Definition of Standard Parabola A standard parabola is a parabola with either
(1) its focus at $(0, p)$ and directrix $y = -p$ where $p \neq 0$; or
(2) its focus at $(p, 0)$ and directrix $x = -p$ where $p \neq 0$.

Theorem (Equations of Standard Parabolas) A standard parabola is the graph of either $x^2 = 4py$ or $y^2 = 4px$ where $p \neq 0$.

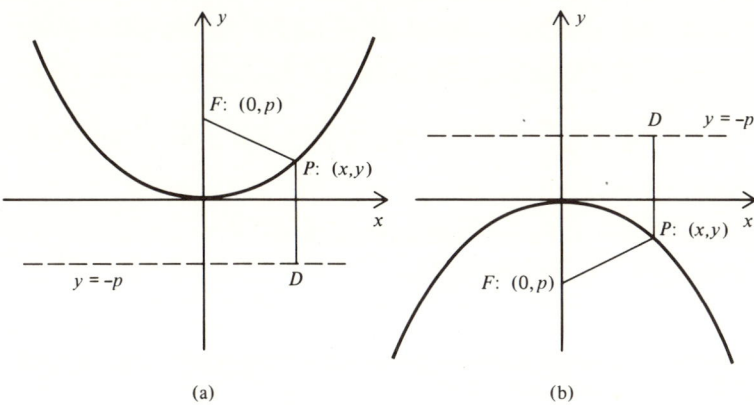

(a)   (b)

Parabola: $x^2 = 4py$, $p > 0$ Parabola: $x^2 = 4py$, $p < 0$

Figure 9.18

Proof

(1) Focus $(0, p)$ and directrix $y = -p$.

By referring to Figure 9.18, it may be seen that a point $P: (x, y)$ is on the parabola if and only if $|PF| = |PD|$; that is,

$$\sqrt{(x-0)^2 + (y-p)^2} = |y - (-p)|$$
$$x^2 + y^2 - 2py + p^2 = y^2 + 2py + p^2$$
$$x^2 = 4py.$$

(2) Focus $(p, 0)$ and directrix $x = -p$.

By referring to Figure 9.19, it may be seen that a point $P: (x, y)$ is on the parabola if and only if $|PF| = |PD|$; that is,

$$\sqrt{(x-p)^2 + (y-0)^2} = |x - (-p)|$$
$$x^2 - 2px + p^2 + y^2 = x^2 + 2px + p^2$$
$$y^2 = 4px.$$

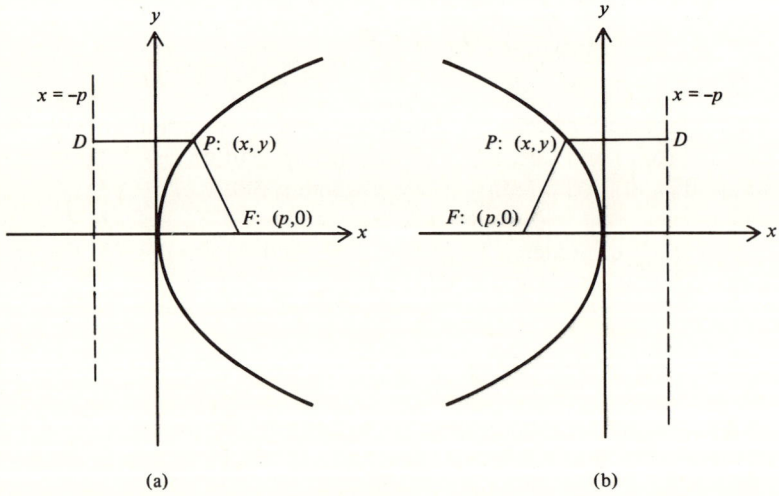

(a) Parabola: $y^2 = 4px$, $p > 0$

(b) Parabola: $y^2 = 4px$, $p < 0$

Figure 9.19

It is readily seen that parabolas with equations of the form $x^2 = 4py$ are symmetric to the y-axis and those with equations of the form $y^2 = 4px$ are symmetric to the x-axis. The line of symmetry of a parabola is called the **axis** of the parabola. The **vertex**, or turning point, of a parabola is the special point on the parabola that is also on its axis. Each of the standard parabolas has its vertex at the origin.

CIRCLES AND STANDARD PARABOLAS

Example 4 Find the coordinates of the focus, the equation of the directrix, and sketch the parabola $x^2 = -12y$.

Solution Since $x^2 = 4py$, $4p = -12$ and $p = -3$. The focus is $(0, p)$ or $(0, -3)$. The directrix is $y = -p$ or $y = 3$.

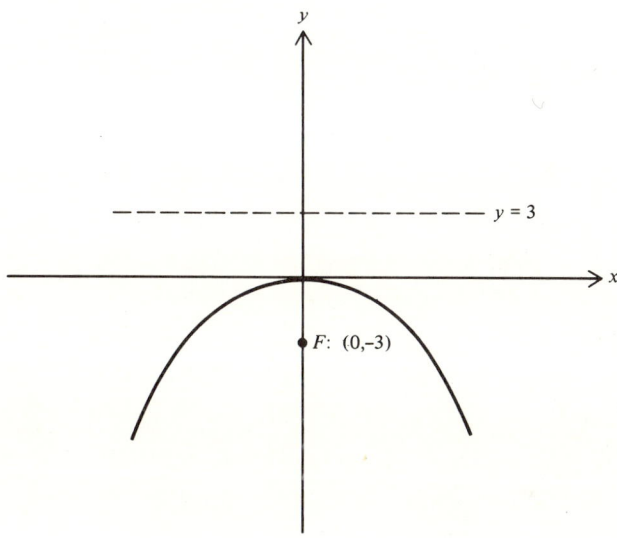

Figure 9.20 $x^2 = -12y$

Example 5 Find the equation of the standard parabola if the equation of its directrix is $x - 5 = 0$.

Solution Since $x = 5$ and since $x = -p$ are the equations of the directrix, $p = -5$. The equation of the parabola is $y^2 = 4px$ or $y^2 = -20x$.

EXERCISES 9.6

1–5. Determine whether each of the following is a circle. Find the center and radius of each circle and draw its graph:

1. $x^2 + y^2 - 4x + 6y - 3 = 0$
2. $2x^2 + 2y^2 + 20x - 12y + 18 = 0$
3. $4x^2 + 4y^2 - 28x - 8y + 37 = 0$
4. $x^2 + y^2 - 14y = 0$
5. $x^2 + y^2 - 6x - 10y + 34 = 0$

6–16. Write an equation of each of the following circles satisfying the given conditions and draw its graph:

6. Center at $(2, -3)$ and radius 4.

7. Center at $(-1, 2)$ and radius 3.
8. Center at $(3, 4)$ and passing through the origin.
9. Center at $(2, -1)$ and passing through the point $(4, 2)$.
10. Center at $(3, 2)$ and tangent to the line $y = 5$.
11. Center at $(5, 4)$ and tangent to the line $3x - 4y + 6 = 0$.
12. Passing through the points $(2, 3)$, $(10, -1)$, and $(4, 5)$.
13. Passing through the points $(1, 2)$ and $(4, 5)$ and having radius 3 (two solutions).
14. Endpoints of a diameter are $(-4, 5)$ and $(10, 3)$.
15. Endpoints of a diameter are $(0, 12)$ and $(8, 0)$.
16. Endpoints of a diameter are $(5, 5)$ and $(-3, -3)$.

17–24. For each of the following parabolas, find the coordinates of the focus, the equation of the directrix, and sketch the curve:

17. $y^2 = -2x$
18. $x^2 = 3y$
19. $y^2 - 3x = 0$
20. $3x^2 + 2y = 0$
21. $x^2 + 10y = 0$
22. $y^2 = -5x$
23. $4x^2 - 3y = 0$
24. $y^2 - 8x = 0$

25–38. Write the equation of each of the following standard parabolas satisfying the given conditions:

25. Directrix is $x + 2 = 0$
26. Directrix is $y + 2 = 0$
27. Focus $(3, 0)$
28. Focus $(0, -2)$
29. Directrix is $x = -3/7$
30. Directrix is $y = \frac{1}{4}$
31. Focus $(0, \frac{1}{4})$
32. Directrix is $2x - 1 = 0$
33. Directrix is $4y + 3 = 0$
34. Focus is $(-5/2, 0)$
35. Passing through the point $(2, 4)$ (two solutions).
36. Passing through the point $(6, 3)$ (two solutions).
37. Passing through the point $(4, -8)$ (two solutions).
38. Passing through the point $(-1, -2)$ (two solutions).

39–44. Graph:

39. $(x - 3)^2 + (y - 4)^2 \leq 16$
40. $(x - 3)^2 + (y - 4)^2 > 16$
41. $y^2 \geq 8x$
42. $x^2 + 10y < x$
43. $\{(x, y) \mid x^2 + y^2 \leq 9\} \cap \{(x, y) \mid y \geq 4 - x^2\}$
44. $\{(x, y) \mid x^2 + y^2 > 9\} \cap \{(x, y) \mid x^2 + y^2 < 25\}$

ELLIPSES AND HYPERBOLAS

9.7 ELLIPSES AND HYPERBOLAS

Definition of Ellipse An ellipse is the set of points in a plane such that the sum of the distances from any point on the ellipse to two fixed points, called the **foci**, is a constant, $2a$.

Definition of Standard Ellipse A standard ellipse is an ellipse having foci either at F_1: $(c, 0)$ and F_2: $(-c, 0)$ or at F_1: $(0, c)$ and F_2: $(0, -c)$, where $a > c > 0$.

Theorem (Equations of Standard Ellipses) A standard ellipse is the graph of either

$$\frac{x^2}{a^2} + \frac{y^2}{b^2} = 1$$

or

$$\frac{x^2}{b^2} + \frac{y^2}{a^2} = 1,$$

where $b^2 = a^2 - c^2$.

Proof By definition, P: (x, y) is on the standard ellipse if and only if $|PF_1| + |PF_2| = 2a$. Therefore,

$$\sqrt{(x-c)^2 + y^2} + \sqrt{(x+c)^2 + y^2} = 2a$$

or

$$\sqrt{x^2 + (y-c)^2} + \sqrt{x^2 + (y+c)^2} = 2a.$$

Rationalizing and simplifying these equations yields

$$\frac{x^2}{a^2} + \frac{y^2}{a^2 - c^2} = 1 \quad \text{or} \quad \frac{x^2}{a^2 - c^2} + \frac{y^2}{a^2} = 1.$$

The equations of the standard ellipses are obtained by letting $b^2 = a^2 - c^2$.

9.7.1 Graphing the Ellipse $\frac{x^2}{a^2} + \frac{y^2}{b^2} = 1$

(1) **Intercepts:** Let $x = 0$. Then $y^2 = b^2$ and $y = b$ or $y = -b$. The y-intercepts are b and $-b$.

Let $y = 0$. Then $x^2 = a^2$ and $x = a$ or $x = -a$. The x-intercepts are a and $-a$.

(2) Extent: To find the domain, solve the equation for y.

$$y = \pm \frac{b}{a}\sqrt{a^2 - x^2}$$

y is real if $x^2 \leq a^2$; that is, $|x| \leq a$ or $-a \leq x \leq a$. The domain is $\{x \mid -a \leq x \leq a\}$. To find the range, solve the equation for x.

$$x = \pm \frac{a}{b}\sqrt{b^2 - y^2}$$

x is real if $y^2 \leq b^2$; that is, $|y| \leq b$ or $-b \leq y \leq b$. The range is $\{y \mid -b \leq y \leq b\}$.

(3) Symmetry: Since the point $(-x, -y)$ is on the graph whenever (x, y) is, the standard ellipse is symmetric to the origin. Moreover, the point $(-x, y)$ is on the curve whenever (x, y) is, and the point $(x, -y)$ is on the curve whenever (x, y) is. Therefore the standard ellipse is also symmetric to both axes.

Figure 9.21 is a useful graphing aid. It also illustrates the graph of $\frac{x^2}{b^2} + \frac{y^2}{a^2} = 1$ which is obtained in a similar way to that described earlier.

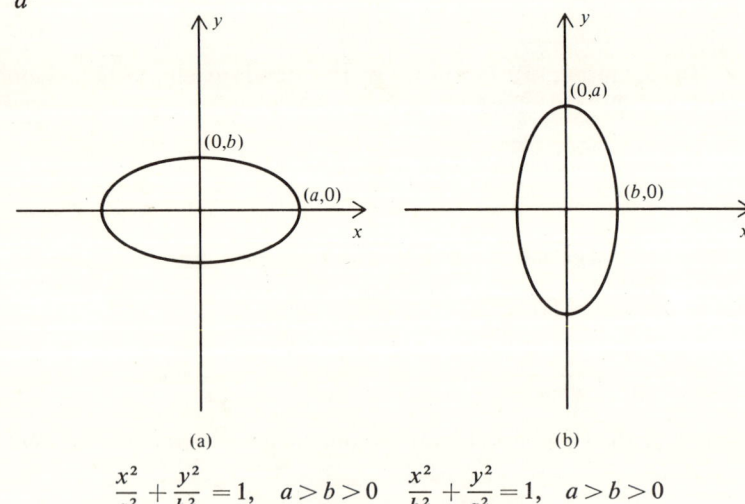

(a) $\quad \frac{x^2}{a^2} + \frac{y^2}{b^2} = 1, \quad a > b > 0 \qquad$ (b) $\quad \frac{x^2}{b^2} + \frac{y^2}{a^2} = 1, \quad a > b > 0$

Figure 9.21 Standard Ellipses

Example 1 Graph $4x^2 + 9y^2 = 36$.

Solution

(1) Identify the curve by obtaining the equation of the standard ellipse. Dividing both sides by 36,

ELLIPSES AND HYPERBOLAS

$$\frac{x^2}{9} + \frac{y^2}{4} = 1.$$

(2) x-intercepts: If $y = 0$, then $x = 3$ or $x = -3$.
(3) y-intercepts: If $x = 0$, then $y = 2$ or $y = -2$.
(4) Domain: $y = \pm \frac{2}{3}\sqrt{9 - x^2}$. Thus, $-3 \leq x \leq 3$.
(5) Range: $x = \pm \frac{3}{2}\sqrt{4 - y^2}$. Thus, $-2 \leq y \leq 2$.
(6) Symmetry: origin, x-axis, y-axis.
(7) Graph: See Figure 9.22.

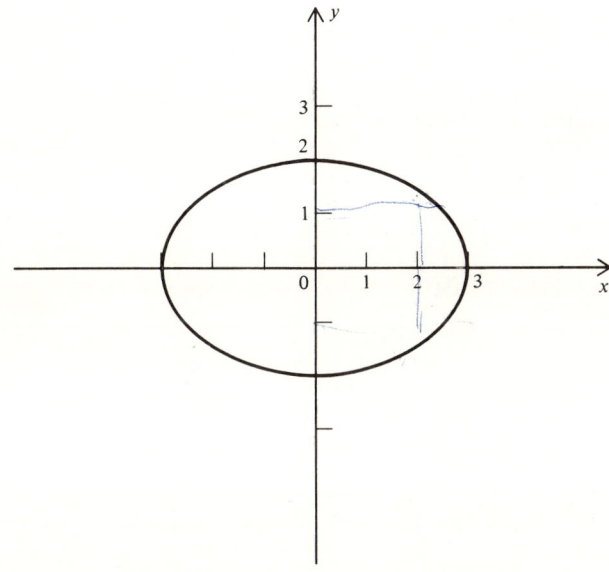

Figure 9.22 Ellipse: $\frac{x^2}{9} + \frac{y^2}{4} = 1$

Definition of Hyperbola   A hyperbola is the set of points in a plane such that the difference of the distances from any point on the hyperbola to two fixed points, called the **foci**, is a constant, $2a$.

Definition of Standard Hyperbola   A standard hyperbola is a hyperbola having foci either at F_1: $(c, 0)$ and F_2: $(-c, 0)$ or at F_1: $(0, c)$ and F_2: $(0, -c)$, where $c > a > 0$.

Theorem (Equations of Standard Hyperbolas)   A standard hyperbola is the graph of either

(1) $$\frac{x^2}{a^2} - \frac{y^2}{b^2} = 1$$

or

(2) $$\frac{y^2}{a^2} - \frac{x^2}{b^2} = 1, \quad \text{where } b^2 = c^2 - a^2.$$

Proof By definition, $P: (x, y)$ is on the standard hyperbola if and only if $|PF_2| - |PF_1| = 2a$. Therefore,

$$\sqrt{(x+c)^2 + y^2} - \sqrt{(x-c)^2 + y^2} = 2a$$

or

$$\sqrt{x^2 + (y+c)^2} - \sqrt{x^2 + (y-c)^2} = 2a.$$

Rationalizing and simplifying these equations yield

$$\frac{x^2}{a^2} - \frac{y^2}{c^2 - a^2} = 1 \quad \text{or} \quad \frac{y^2}{a^2} - \frac{x^2}{c^2 - a^2} = 1.$$

The equations of the standard hyperbolas are obtained by letting $b^2 = c^2 - a^2$.

9.7.2 Graphing the Hyperbola

(1) *Intercepts:* For Equation (1), if $y = 0$, then $x = a$ or $x = -a$, and if $x = 0$, then $y^2 = -b^2$, and there is no real solution for y.

Thus, for Equation (1), the x-intercepts are a and $-a$ and there are no y-intercepts.

For Equation (2), it may be seen similarly that a and $-a$ are the y-intercepts and there are no x-intercepts.

(2) *Extent:* To find the domain for Equation (1),

$$y = \pm \frac{b}{a}\sqrt{x^2 - a^2}.$$

Thus, $x^2 \geq a^2$ and $x \geq a$ or $x \leq -a$. For Equation (2),

$$y = \pm \frac{a}{b}\sqrt{x^2 + b^2}$$

and x is any real number.

To find the range for Equation (1),

ELLIPSES AND HYPERBOLAS

$$x = \pm \frac{a}{b}\sqrt{y^2 + b^2}$$

and y is any real number. For Equation (2),

$$x = \pm \frac{b}{a}\sqrt{y^2 - a^2}.$$

Thus, $y^2 \geq a^2$ and $y \geq a$ or $y \leq -a$.

(3) *Symmetry:* From the equations it is readily seen that any standard hyperbola is symmetric to the origin and to both axes.

(4) *Asymptotes:* The lines obtained by replacing the constant term 1 by 0 are called the asymptotes of the hyperbola.

For $\frac{x^2}{a^2} - \frac{y^2}{b^2} = 1$, the asymptotes are obtained from $\frac{x^2}{a^2} - \frac{y^2}{b^2} = 0$.

Thus, the asymptotes of this hyperbola are the lines

$$y = \frac{b}{a}x \quad \text{and} \quad y = -\frac{b}{a}x.$$

A hyperbola has two branches and these branches lie entirely within two of the regions formed by the asymptotes as shown in Figures 9.23 and

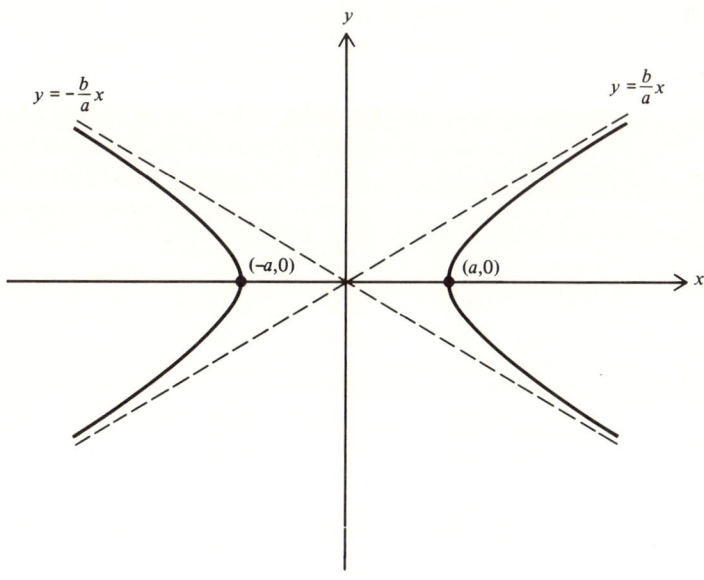

Figure 9.23 Hyperbola: $\frac{x^2}{a^2} - \frac{y^2}{b^2} = 1$

9.24. For larger and larger values of x, the distance between a branch of the hyperbola and its asymptote becomes smaller and smaller. Thus, in graphing the hyperbola, the asymptotes can be drawn to serve as useful guide lines.

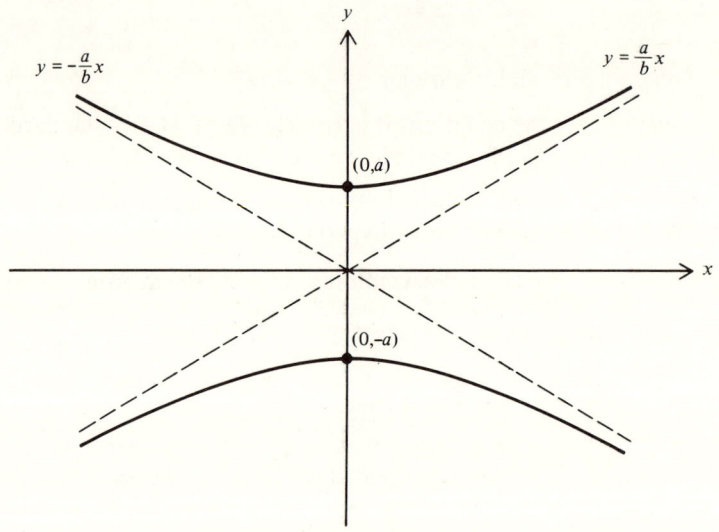

Figure 9.24 Hyperbola: $\dfrac{y^2}{a^2} - \dfrac{x^2}{b^2} = 1$

Example 2 Discuss and sketch the graph of $25x^2 - 16y^2 = 400$. (See Figure 9.25.)

Solution

(1) Identify the curve. Dividing both sides by 400, $\dfrac{x^2}{16} - \dfrac{y^2}{25} = 1$, a standard hyperbola.

(2) x-intercepts: 4 and -4.

(3) y-intercepts: none.

(4) Symmetry: origin, x-axis, y-axis.

(5) Domain: $y = \pm \tfrac{5}{4}\sqrt{x^2 - 16}$. Thus, $|x| \geq 4$.

(6) Range: $x = \pm \tfrac{4}{5}\sqrt{y^2 + 25}$. Thus, all real y.

(7) Asymptotes: $y = \tfrac{5}{4}x$ and $y = -\tfrac{5}{4}x$.

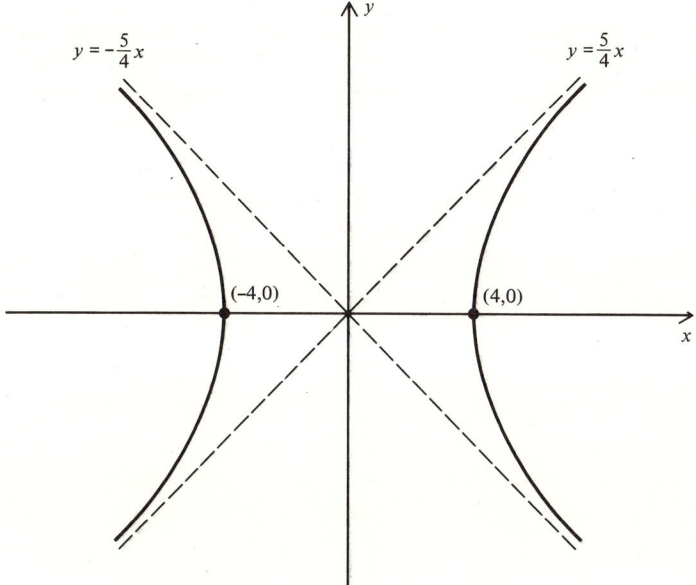

Figure 9.25 Hyperbola: $\dfrac{x^2}{16} - \dfrac{y^2}{25} = 1$

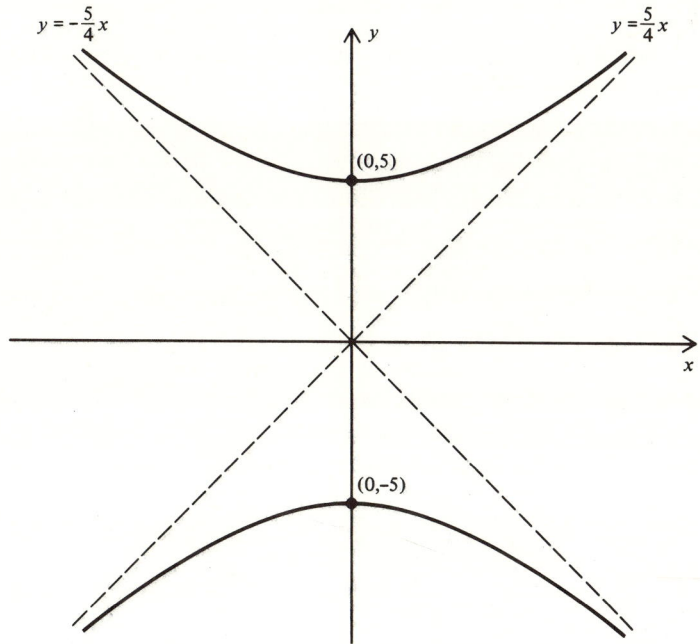

Figure 9.26 Hyperbola: $\dfrac{y^2}{25} - \dfrac{x^2}{16} = 1$

Example 3 Discuss and sketch the graph of $25x^2 - 16y^2 + 400 = 0$.

See Figure 9.26.

Solution

(1) $\dfrac{y^2}{25} - \dfrac{x^2}{16} = 1$. The curve is a standard hyperbola.
(2) x-intercepts: none.
(3) y-intercepts: 5 and -5.
(4) Symmetry: origin, x-axis, y-axis.
(5) Domain: all real x $(y = \pm\frac{5}{4}\sqrt{x^2 + 16})$.
(6) Range: $|y| \geq 5$ $(x = \pm\frac{4}{5}\sqrt{y^2 - 25})$.
(7) Asymptotes: $y = \frac{5}{4}x$ and $y = -\frac{5}{4}x$.

EXERCISES 9.7

1–8. Discuss and graph each of the following equations:

1. $x^2 + 25y^2 = 100$
2. $16x^2 = 1600 - 25y^2$
3. $5x^2 + 2y^2 - 50 = 0$
4. $9x^2 + y^2 = 9$
5. $2x^2 - 2y^2 + 18 = 0$
6. $9x^2 = 16y^2 - 576$
7. $9x^2 - 16y^2 = 144$
8. $9y^2 = 16x^2 - 36$

9–14. For each of the following, find the equation of the standard ellipse that satisfies the stated conditions:

9. One focus at $(2, 0)$, $a = 3$
10. One focus at $(0, 3)$, $a = 5$
11. $a = 4$, $b = 1$ (two solutions)
12. $a = 5$, $b = 2$ (two solutions)
13. $a = 5$, $c = 3$ (two solutions)
14. $b = 5$, $c = 12$ (two solutions)

15–20. For each of the following, find the equation of the standard hyperbola that satisfies the stated conditions:

15. One focus at $(0, 2)$, $a = 1$
16. One focus at $(5, 0)$, $a = 3$
17. $a = 6$, $b = 5$ (two solutions)
18. $a = \sqrt{2}$, $b = 1$ (two solutions)
19. $a = \sqrt{5}$, $c = 3$ (two solutions)
20. $b = 3/2$, $c = 5/2$ (two solutions)

ELLIPSES AND HYPERBOLAS

21. An ellipse may also be defined as the set of points, P, on a plane, whose distance from a fixed point, F, is the product of a constant e and the distance from a fixed line, where $0 < e < 1$. The fixed point F is a focus of the ellipse, the fixed line is the corresponding directrix, and the constant e is called the **eccentricity**. In symbols,

$$|FP| = e|DP|.$$

Use this definition to find the equation of the ellipse with focus $F: (0, 2)$, with corresponding directrix $y = 8$, and eccentricity $e = \frac{1}{2}$.

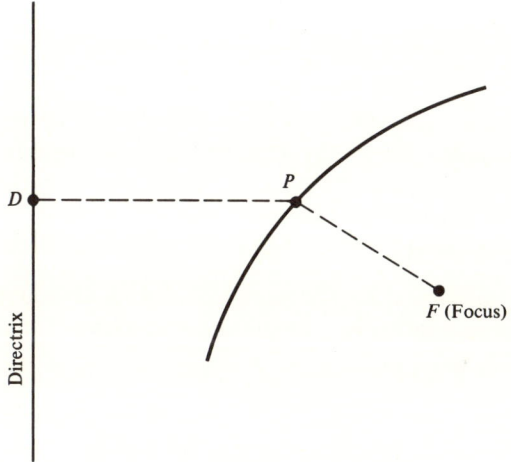

Figure 9.27 $|FP| = e|DP|$

22–23. Use the definition of Exercise 21 to write the equation of each of the following ellipses satisfying the given conditions:

22. A focus at $(-1, 0)$, directrix $x = -9$, and $e = \frac{1}{3}$.
23. A focus at $(0, 4)$, directrix $y = 7$, and $e = \frac{1}{4}$.

24–26. The standard ellipse $\dfrac{x^2}{a^2} + \dfrac{y^2}{b^2} = 1$ has x-intercepts equal to a and $-a$ and y-intercepts equal to b and $-b$. If $a > b$, then the points $(-a, 0)$ and $(a, 0)$ are called **vertices** of the ellipse, and the line segment joining the vertices is called the **major axis**. The **minor axis** is the line segment joining the points $(0, b)$ and $(0, -b)$.

For $a > b$ and the equation $\dfrac{x^2}{b^2} + \dfrac{y^2}{a^2} = 1$, the major axis is the line segment joining the vertices of this ellipse; that is, the points $(0, a)$ and $(0, -a)$. The minor axis is the line segment joining the points $(b, 0)$ and $(-b, 0)$.

For example, for the ellipse $\dfrac{x^2}{9} + \dfrac{y^2}{4} = 1$, the vertices are $(-3, 0)$ and $(3, 0)$; the length of the major axis is six units and the length of the minor axis is four units.

24. Find the equation of the standard ellipse with one vertex at $(-5, 0)$ and with minor axis six units.

25. Find the equation of the standard ellipse with one vertex at $(0, 3)$ and with minor axis four units.

26. Find the equation of the standard ellipse with major axis eight units and with minor axis four units.

27. A hyperbola may be defined in the same manner as the ellipse in Exercise 21, except $e > 1$. Thus,

$$|FP| = e|DP|, \quad \text{where } e > 1,$$

F is a fixed point (the focus), P is any point on the curve, and DP is the distance from the point P to a fixed line (the directrix), defines a hyperbola.

Find the equation of the hyperbola with focus $(0, 4)$, directrix $y = 1$, and eccentricity $e = 2$.

28–29. Use the definition of Exercise 27 to write the equation of each of the following hyperbolas satisfying the given conditions:

28. A focus at $(9, 0)$, corresponding directrix $x = 1$, and $e = 3$.

29. A focus at $(0, -2)$, corresponding directrix $y = -1/2$, and $e = 4/3$.

30. The **vertices** of the standard hyperbola $\dfrac{x^2}{a^2} - \dfrac{y^2}{b^2} = 1$ are the points $(-a, 0)$ and $(a, 0)$ and its **transverse axis** is the line segment joining the two vertices. The line segment joining the points $(0, b)$ and $(0, -b)$ is called the **conjugate axis**.

For the standard hyperbola $\dfrac{y^2}{a^2} - \dfrac{x^2}{b^2} = 1$, the vertices are $(0, a)$ and $(0, -a)$; the transverse axis is the line segment joining the vertices, and the conjugate axis is the line segment joining the points $(-b, 0)$ and $(b, 0)$.

Find the coordinates of the vertices, and the transverse and conjugate axes for the hyperbola $\dfrac{x^2}{16} - \dfrac{y^2}{25} = 1$.

31–33. The definition used in Exercises 21 and 27, $|FP| = e|DP|$ is often referred to as the definition of a *conic* or a *conic section*. If $0 < e < 1$, then the conic is an ellipse. If $e = 1$, the conic is a parabola, and if $e > 1$, the conic is a hyperbola. The eccentricity, e, can be thought of as a ratio, $\dfrac{|FP|}{|DP|} = e$. Identify each of the following conics by finding this ratio e:

31. Focus at (2, 5), directrix $y = 3$, passing through the point (4, 6).
32. Focus at (6, −2), directrix $x = 2$, passing through the point (6, 2).
33. Focus at (−3, 1), directrix $y + 2 = 0$, passing through the point (5, 4).
34. A roadway 400 feet long is supported by a parabolic cable. The cable is 100 feet above the roadway at the ends and 4 feet above at the center. Find the lengths of the vertical supporting cables at 50 foot intervals along the roadway. (See Figure 9.28.)

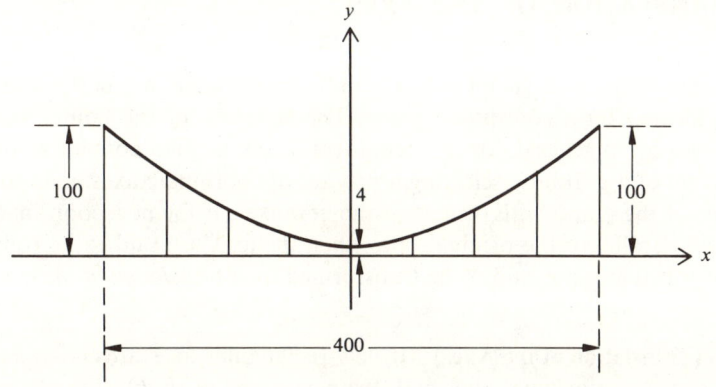

Figure 9.28

35. An arch of a bridge across a river is in the shape of half an ellipse. The span of the arch at water level is 100 feet, and the greatest height of the arch above the water level is 30 feet. Find the equation of the arch. (See Figure 9.29.)

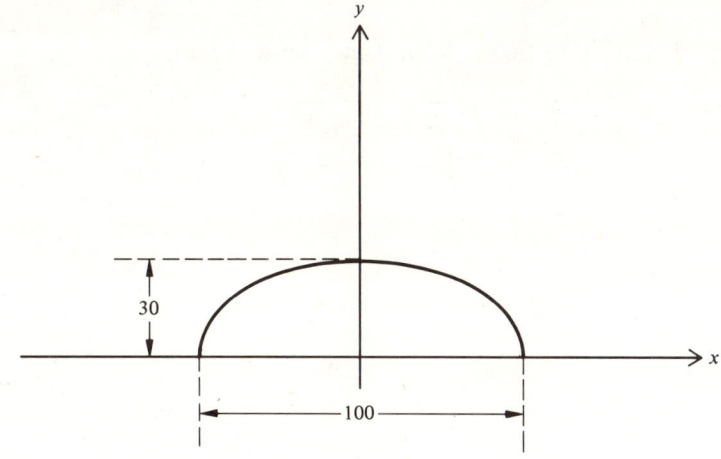

Figure 9.29

36. Referring to the definitions used in Exercises 21 and 27, show that for any ellipse or hyperbola, $c = ae$ and $d = \dfrac{a}{e}$ where c is the distance from the center of the conic (its point of symmetry) to any focus and d is the distance from the center of the conic to any directrix.

9.8 TRANSLATION OF THE AXES

The size and shape of a geometric figure remain the same no matter where the figure is located on a coordinate plane. The equation of the figure, however, will be changed, in general, for different locations. It is often possible to simplify the equation of a graph by selecting a new set of coordinate axes and renaming the points of the graph with respect to the new axes. If the new coordinate axes are chosen parallel to the original ones, then the axes are said to be translated and the coordinates are said to be transformed by a *translation of the axes*.

Theorem (Translation of the Axes) If new rectangular axes are chosen parallel, respectively, to the original ones with the new origin at (h, k), then the new coordinates (x', y') of a point on the plane are related to the old coordinates (x, y) of the point by the equations

$$x' = x - h \qquad x = x' + h$$
$$\text{or}$$
$$y' = y - k \qquad y = y' + k.$$

Proof Referring to Figure 9.30, it may be seen that

$$\overrightarrow{AP} = \overrightarrow{AB} + \overrightarrow{BP}$$

or

$$x = h + x'$$

and

$$x' = x - h.$$

Similarly,

$$y' = y - k \quad \text{and} \quad y = y' + k.$$

TRANSLATION OF THE AXES

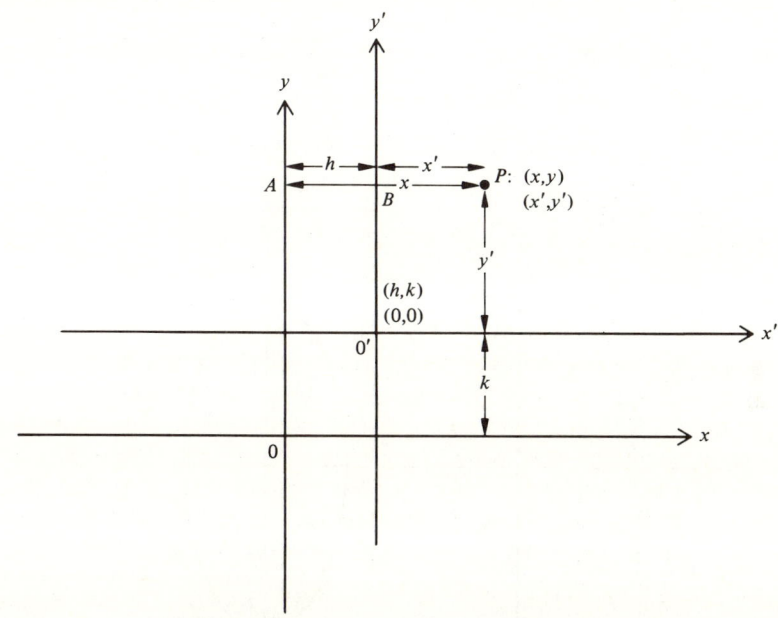

Figure 9.30 Translation of the Axes

Example 1 Simplify by a translation of the axes and sketch the graph with respect to both sets of axes:

$$4x^2 + y^2 - 24x - 4y + 36 = 0.$$

Solution

(METHOD A) Let $x = x' + h$ and $y = y' + k$. Then

$$4(x' + h)^2 + (y' + k)^2 - 24(x' + h) - 4(y' + k) + 36 = 0$$
$$4x'^2 + y'^2 + (8h - 24)x' + (2k - 4)y' + 4h^2 + k^2 - 24h - 4k + 36 = 0.$$

The linear x' and y' terms can be eliminated by selecting their coefficients to be 0; that is,

$$8h - 24 = 0 \quad \text{and} \quad 2k - 4 = 0.$$

Thus,

$$h = 3 \quad \text{and} \quad k = 2.$$

Then,

$$4h^2 + k^2 - 24h - 4k + 36 = 36 + 4 - 72 - 8 + 36 = -4$$

The equation now becomes

$$4x'^2 + y'^2 - 4 = 0$$

or

$$\frac{x'^2}{1} + \frac{y'^2}{4} = 1, \quad \text{a standard ellipse.}$$

(See Figure 9.31.)

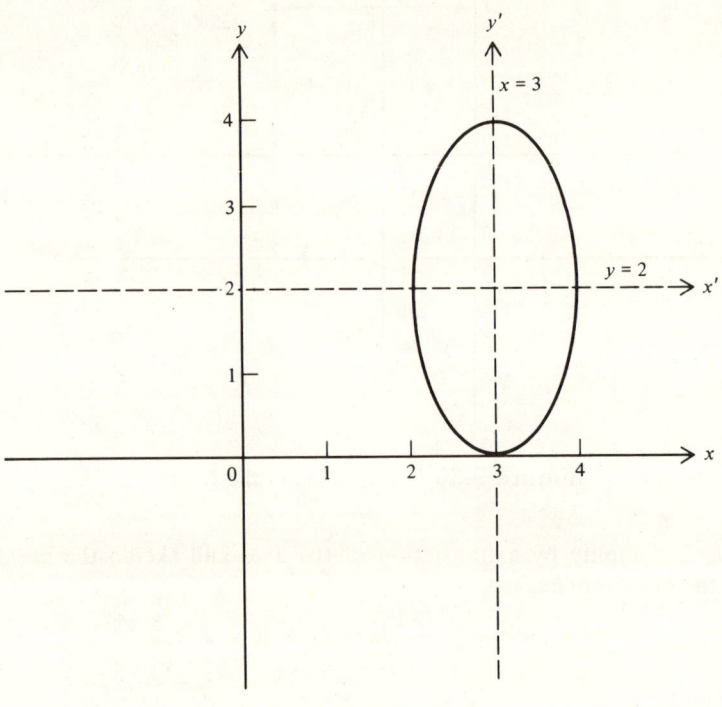

Figure 9.31 $\quad \dfrac{(x-3)^2}{1} + \dfrac{(y-2)^2}{4} = 1$

The coordinates of the center and the vertices in both coordinate systems can now be readily obtained using the equations of the translation, $x = x' + 3$ and $y = y' + 2$.

	CENTER	VERTICES			
New (x', y')	(0, 0)	(1, 0)	(−1, 0)	(0, 2)	(0, −2)
Old (x, y)	(3, 2)	(4, 2)	(2, 2)	(3, 4)	(3, 0)

Example 2 Simplify by a translation of the axes and sketch the graph with respect to both sets of axes:

$$x^2 - 9y^2 + 8x + 18y - 2 = 0.$$

TRANSLATION OF THE AXES

Solution

(METHOD B) Complete the squares with respect to the x and y terms.

$$(x^2 + 8x +) - 9(y^2 - 2y +) = 2$$
$$(x^2 + 8x + 16) - 9(y^2 - 2y + 1) = 2 + 16 - 9$$
$$(x + 4)^2 - 9(y - 1)^2 = 9$$

Now, let $x' = x + 4$ and $y' = y - 1$

$$x'^2 - 9y'^2 = 9$$

or

$$\frac{x'^2}{9} - \frac{y'^2}{1} = 1, \quad \text{a standard hyperbola.}$$

(See Figure 9.32.)

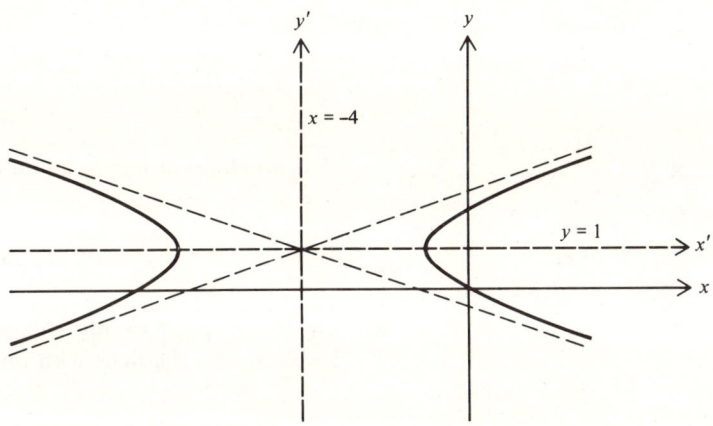

Figure 9.32 $\quad \dfrac{(x+4)^2}{9} - \dfrac{(y-1)^2}{1} = 1$

EXERCISES 9.8

1–15. Simplify each of the following by a translation of the axes and sketch the graph with respect to both axes:

1. $x^2 + y^2 - 4x + 6y - 3 = 0$
2. $4x^2 + 9y^2 - 8x - 36y + 4 = 0$
3. $x^2 + 6x - 16y - 7 = 0$
4. $x^2 + y^2 + 10x - 6y + 18 = 0$
5. $4x^2 + 16x + y^2 - 5y + 2 = 0$
6. $x^2 + 2y + 8x + 10 = 0$
7. $x^2 + y^2 = x + y$
8. $x^2 + 4x = y^2 + 2y - 5$
9. $4x^2 - 5y^2 - 16x + 10y - 29 = 0$
10. $3x^2 - 4y^2 - 6x - 8y - 4 = 0$

11. $x^2 - 4y^2 - 6x + 16y = 23$
12. $x^2 - y^2 = x - y$
13. $4x^2 - y^2 + 8x + 4y = 0$
14. $x^2 + 2y^2 - 8x - 8y + 24 = 0$
15. $3x^2 + y^2 - 6x + 10y + 28 = 0$
16. Show that the x^2 term of $y = ax^3 + bx^2 + cx + d$ can be removed by a translation of the axes.
17. Simplify $y = x^3 + 3x^2 - x + 4$ by a translation of the axes so that the x^2 term and the constant term are removed.
18. Show that the slope of a line does not change as a result of translation of axes.
19. Show that the distance between two points remains unchanged after a translation of axes.
20. Find the new equation obtained from $4x^2 + y^2 - 24x + 4y + 36 = 0$ after a translation with a new origin at $(3, -2)$. Sketch the graph with respect to both sets of axes.

21–23. Prove each of the following theorems:

21. The graph of $\dfrac{(x-h)^2}{a^2} - \dfrac{(y-k)^2}{b^2} = \pm 1$ is a hyperbola.

22. The graph of $\dfrac{(x-h)^2}{a^2} + \dfrac{(y-k)^2}{b^2} = 1$ is an ellipse if $a \neq b$, and a circle if $a = b$.

23. The graphs of either $y - k = a(x - h)^2$ or $x - h = a(y - k)^2$ are parabolas with vertex at (h, k).

24. Translate the axes so that the new origin is the point of intersection of the lines $2x - y - 5 = 0$ and $x - 2y + 2 = 0$. Sketch the lines with respect to both coordinate systems.

25–39. Find the equations of the lines of symmetry of the graphs of the equations in Exercises 1–15.

9.9 ROTATION OF THE AXES

If new perpendicular coordinate axes are chosen so that they have the same origin as the original ones, then the axes are said to be rotated and the coordinates are said to be transformed by a *rotation of the axes*. If θ is the angle from the positive half of the original x-axis to the positive half of the new x'-axis, then the axes are said to be *rotated through an angle θ*.

Theorem (Rotation of the Axes) If new rectangular axes are chosen with the same origin as the original ones, and if θ is the angle from the positive half of the original x-axis to the positive half of the new x'-axis, then the new coordinates (x', y') of a point on the plane are related to the old coordinates (x, y) of the point by the equations

$$x = x' \cos \theta - y' \sin \theta$$
$$y = x' \sin \theta + y' \cos \theta.$$

Proof Referring to Figure 9.33, and by selecting $r = |OP|$, it is seen that

$$x' = r \cos \emptyset \quad \text{and} \quad y' = r \sin \emptyset$$

and

$$x = r \cos(\theta + \emptyset) \quad \text{and} \quad y = r \sin(\theta + \emptyset).$$

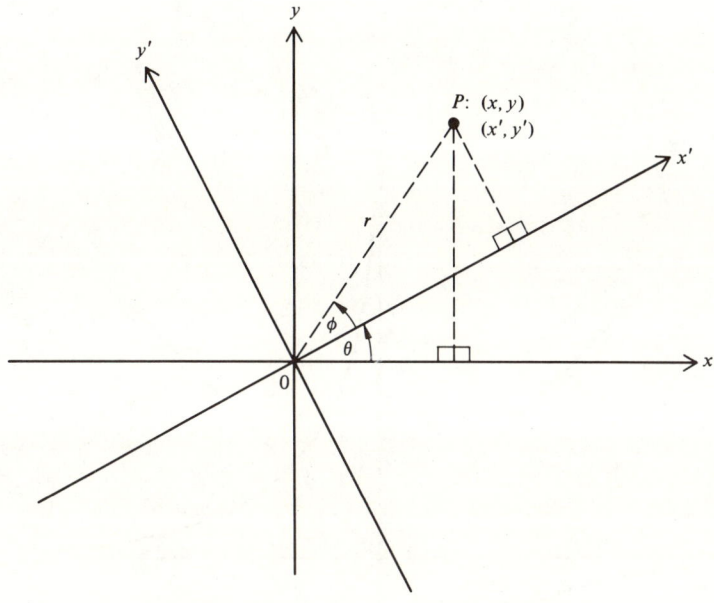

Figure 9.33 Rotation of the Axes

Using the trigonometric addition identity,

$$x = r(\cos \theta \cos \emptyset - \sin \theta \sin \emptyset)$$
$$x = (r \cos \emptyset)\cos \theta - (r \sin \emptyset)\sin \theta.$$

Thus,

$$x = x' \cos \theta - y' \sin \theta.$$

Similarly,

$$y = x' \sin \theta + y' \cos \theta.$$

Example 1 By using a rotation of the axes through 45°, find the new equation of $xy = 2$. Sketch the graph with respect to both axes.

Solution Since $x = x' \cos \theta - y' \sin \theta$, $y = x' \sin \theta - y' \cos \theta$, $\theta = 45°$, $\cos \theta = \sin \theta = \sqrt{2}/2$,

$$x = \frac{\sqrt{2}}{2} x' - \frac{\sqrt{2}}{2} y' = \frac{\sqrt{2}}{2} (x' - y')$$

$$y = \frac{\sqrt{2}}{2} x' + \frac{\sqrt{2}}{2} y' = \frac{\sqrt{2}}{2} (x' + y').$$

Then $xy = 2$ becomes

$$\left(\frac{\sqrt{2}}{2}\right)\left(\frac{\sqrt{2}}{2}\right)(x' - y')(x' + y') = 2$$

or

$$x'^2 - y'^2 = 4.$$

(See Figure 9.34.)

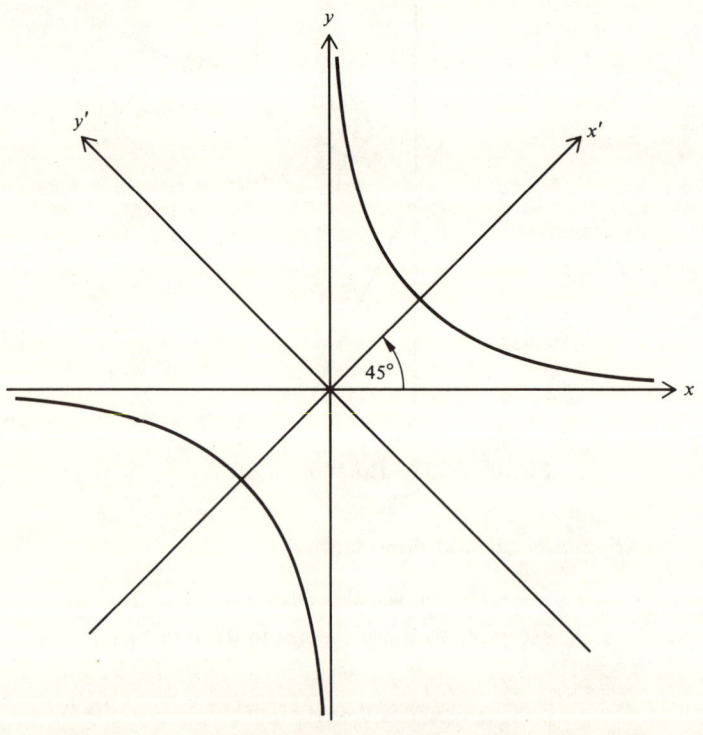

Figure 9.34 $xy = 2$
$x'^2 - y'^2 = 4$

ROTATION OF THE AXES

Example 2 By using a rotation of the axes to remove the xy-term, simplify and sketch the graph with respect to both axes:

$$2x^2 + 4\sqrt{3}\,xy + 6y^2 + \sqrt{3}\,x - y = 0.$$

Solution Let $x = x'\cos\theta - y'\sin\theta$ and $y = x'\sin\theta + y'\cos\theta$.

$2(x'\cos\theta - y'\sin\theta)^2 + 4\sqrt{3}(x'\cos\theta - y'\sin\theta)(x'\sin\theta + y'\cos\theta)$
$\quad + 6(x'\sin\theta + y'\cos\theta)^2 + \sqrt{3}(x'\cos\theta - y'\sin\theta) - (x'\sin\theta + y'\cos\theta) = 0.$

$(2\cos^2\theta + 4\sqrt{3}\cos\theta\sin\theta + 6\sin^2\theta)x'^2$
$\quad + (2\sin^2\theta - 4\sqrt{3}\sin\theta\cos\theta + 6\cos^2\theta)y'^2$
$\quad + (-4\sin\theta\cos\theta + 4\sqrt{3}\cos^2\theta - 4\sqrt{3}\sin^2\theta + 12\sin\theta\cos\theta)x'y'$
$\quad + (\sqrt{3}\cos\theta - \sin\theta)x' + (-\sqrt{3}\sin\theta - \cos\theta)y' = 0.$

Setting the coefficient of $x'y'$ equal to 0,

$$8\sin\theta\cos\theta + 4\sqrt{3}(\cos^2\theta - \sin^2\theta) = 0.$$

Using the double angle identities,

$$4\sin 2\theta + 4\sqrt{3}\cos 2\theta = 0$$

or

$$\tan 2\theta = -\sqrt{3}.$$

Selecting the smallest positive solution, $2\theta = 120°$ and $\theta = 60°$.

Since $\sin 60° = \dfrac{\sqrt{3}}{2}$ and $\cos 60° = \dfrac{1}{2}$, the equation now becomes

$$\left(\frac{1}{2} + 3 + \frac{9}{2}\right)x'^2 + \left(\frac{3}{2} - 3 + \frac{3}{2}\right)y'^2 + \left(\frac{\sqrt{3}}{2} - \frac{\sqrt{3}}{2}\right)x' + \left(-\frac{3}{2} - \frac{1}{2}\right)y' = 0$$

or

$$y' = 4x'^2, \quad \text{a standard parabola.}$$

(See Figure 9.35.)

Theorem The xy term can always be removed from the general quadratic equation

$$Ax^2 + Bxy + Cy^2 + Dx + Ey + F = 0, \quad B \neq 0$$

by a rotation of the axes through an angle θ where

$$\cot 2\theta = \frac{A - C}{B}.$$

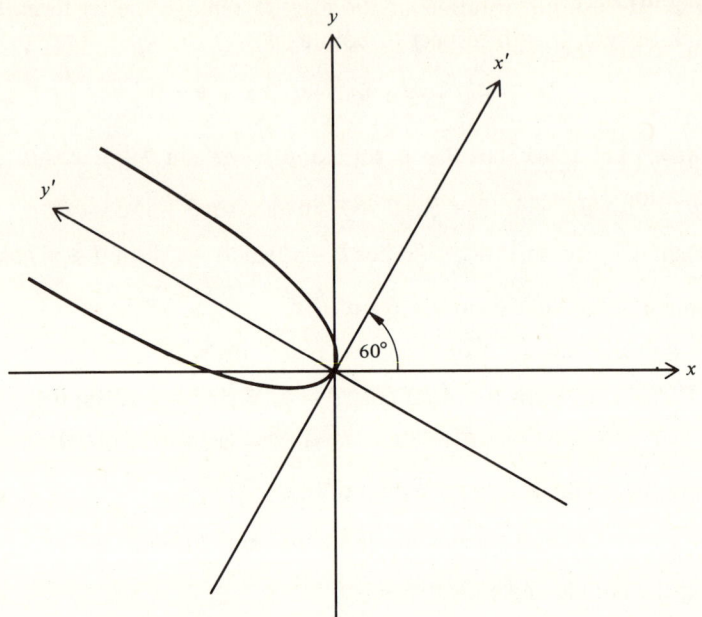

Figure 9.35 $2x^2 + 4\sqrt{3}\,xy + 6y^2 + \sqrt{3}\,x - y = 0$
$y' = 4x'^2$

Proof Letting $x = x'\cos\theta - y'\sin\theta$ and $y = x'\sin\theta + y'\cos\theta$,

$Ax^2 = A(x'^2 \cos^2\theta - 2x'y' \sin\theta \cos\theta + y'^2 \sin^2\theta)$
$Bxy = B(x'^2 \sin\theta \cos\theta + (\cos^2\theta - \sin^2\theta)x'y' - y'^2 \sin\theta \cos\theta)$
$Cy^2 = C(x'^2 \sin^2\theta + 2x'y' \sin\theta \cos\theta + y'^2 \cos^2\theta)$
$Dx = D(x' \cos\theta - y' \sin\theta)$
$Ey = E(x' \sin\theta + y' \cos\theta)$
$F = F.$

Combining like terms, it is seen that the coefficient of the $x'y'$ term is

$$-A(2\sin\theta\cos\theta) + B(\cos^2\theta - \sin^2\theta) + C(2\sin\theta\cos\theta).$$

Setting this equal to zero and using the double angle identities,

$$-A\sin 2\theta + B\cos 2\theta + C\sin 2\theta = 0$$

or

$$B\cos 2\theta = (A - C)\sin 2\theta \quad \text{and} \quad \cot 2\theta = \frac{A - C}{B}.$$

Since $B \neq 0$, this equation can always be solved for θ.

ROTATION OF THE AXES

In simplifying an equation by a rotation of the axes, it is usually more convenient to find the angle first by using the relation $\cot 2\theta = \dfrac{A - C}{B}$.

Example 3 Given $x^2 - 2xy + y^2 - 8x - 8y + 16 = 0$. First simplify by a rotation of the axes to remove the xy term and then simplify by a translation of the axes to remove as many linear terms as possible.

Solution

(1) Since $\cot 2\theta = \dfrac{A - C}{B}$,

$$\cot 2\theta = \dfrac{1 - 1}{-2} = 0$$

and

$$2\theta = 90° \quad \text{and} \quad \theta = 45°.$$

Now let

$$x = x' \cos\theta - y' \sin\theta = \dfrac{1}{\sqrt{2}}(x' - y')$$

$$y = x' \sin\theta + y' \cos\theta = \dfrac{1}{\sqrt{2}}(x' + y')$$

and

$$\tfrac{1}{2}(x' - y')^2 - 2(\tfrac{1}{2})(x'^2 - y'^2) + \tfrac{1}{2}(x' + y')^2 - \dfrac{8}{\sqrt{2}}(2x') + 16 = 0$$

or

$$2y'^2 - 8\sqrt{2}\, x' + 16 = 0.$$

(2) $y'^2 = 4\sqrt{2}(x' - \sqrt{2})$. Let

$$x'' = x' - \sqrt{2} \quad \text{and} \quad y'' = y'.$$

Then the equation becomes

$$y''^2 = 4\sqrt{2}\, x'', \quad \text{a standard parabola.}$$

EXERCISES 9.9

1–4. Find the new equation of each of the following if the axes are rotated counterclockwise through the indicated angle:

1. $x^2 - y^2 = 9$; $45°$
2. $7x^2 - 6\sqrt{3}\, xy + 13y^2 = 16$; $30°$
3. $x^2 - 5xy + y^2 = 3$; $\pi/4$ radians
4. $xy = -6$; $\pi/4$ radians

5–11. Eliminate the xy term in each of the following equations by an appropriate rotation of the axes. Identify the curve.

5. $8x^2 + 5xy - 4y^2 + 2 = 0$
6. $9x^2 + 4xy + 6y^2 = 20$
7. $2x^2 + 9xy + 14y^2 - 5 = 0$
8. $x^2 - 2xy + y^2 - 5x - 5y = 0$
9. $3x^2 - 10xy + 3y^2 + 32 = 0$
10. $2x^2 - 24xy - 5y^2 + 9 = 0$
11. $3x^2 - 2\sqrt{3}\,xy + y^2 - 2x - 2\sqrt{3}\,y = 0$

12–17. For any conic $Ax^2 + Bxy + Cy^2 + Dx + Ey + F = 0$, $B^2 - 4AC$ is called the **discriminant**.

If $B^2 - 4AC < 0$, the conic is an ellipse or a circle.

If $B^2 - 4AC = 0$, the conic is a parabola.

If $B^2 - 4AC > 0$, the conic is a hyperbola.

Use the discriminant to determine the nature of each of the following conics:

12. $3x^2 + 2y^2 - 12x + 8y + 9 = 0$
13. $25x^2 - 9y^2 + 50x + 90y + 25 = 0$
14. $xy + x - 2y + 5 = 0$
15. $16x^2 + 24xy + 9y^2 - 30x + 10y = 0$
16. $25x^2 - 20xy + 4y^2 + 2x - 5y = 0$
17. $x^2 - 2xy + 2y^2 - 2x + 4y = 0$
18. Show that for any conic $Ax^2 + Bxy + Cy^2 + Dx + Ey + F = 0$, $B^2 - 4AC = B'^2 - 4A'C'$ after a rotation and after a translation.

19–23. For each of the following conics perform an appropriate rotation and translation to reduce the conic to standard form. Sketch the curve, showing all sets of axes.

19. $6x^2 - 5xy - 6y^2 + 78x + 52y + 26 = 0$
20. $34x^2 - 24xy + 41y^2 + 20x + 140y + 100 = 0$
21. $xy + y - 2x - 2 = 0$
22. $3x^2 - 8xy - 3y^2 - 2\sqrt{5}\,x - 4\sqrt{5}\,y = 0$
23. $x^2 - 2xy + y^2 - 12x - 4y + 16 = 0$
24. If the slope of a line is 1, find its new slope after a rotation through (a) 60°, (b) 30°, (c) 15°.
25. If the slope of a line is m, find its new slope after a rotation through θ°.
26. Given point $P: (3, 4)$. Find the new coordinates of P after
    (a) first a rotation through 30° and then a translation to a new origin at $(1, 2)$,

POLAR COORDINATES

(b) first a translation to a new origin at (1, 2) and then a rotation through 30°.

27. Show that the distance between two points does not change after a rotation.

9.10 POLAR COORDINATES

9.10.1 Polar Coordinate Systems

The equations of the graphs of certain geometric figures can be expressed more simply by using coordinates different from the rectangular coordinates. One convenient system is the **polar coordinate system**. The positive half of a coordinate line together with the zero point are selected. The zero point is called the **pole** and the nonnegative part of the coordinate line is called the **polar axis**.

For a given angle θ, whose initial side is the polar axis, and for a given real number r, the directed distance from the pole measured along the line determined by the terminal side of θ, a unique point P is determined. The members of the ordered pair (r, θ) are called the polar **coordinates** of P and P is designated as $P: (r, \theta)$.

If r is positive, then r is measured along the terminal side of θ.

If r is negative, then $|r|$ is measured along the extension through the pole of the terminal side of θ. See Figure 9.36.

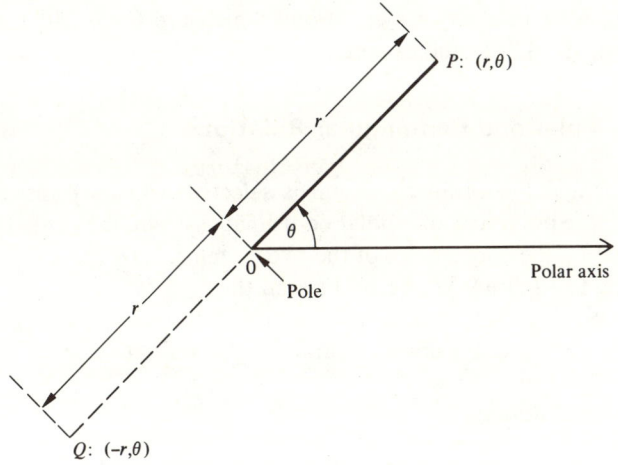

Figure 9.36 A Polar Coordinate Plane

Although for a given ordered pair of real numbers, (r, θ), exactly one point is determined on the plane, a given point does *not* have a unique set of polar coordinates.

For example, (5, 60°) determines a unique point P (see Figure 9.37), but P

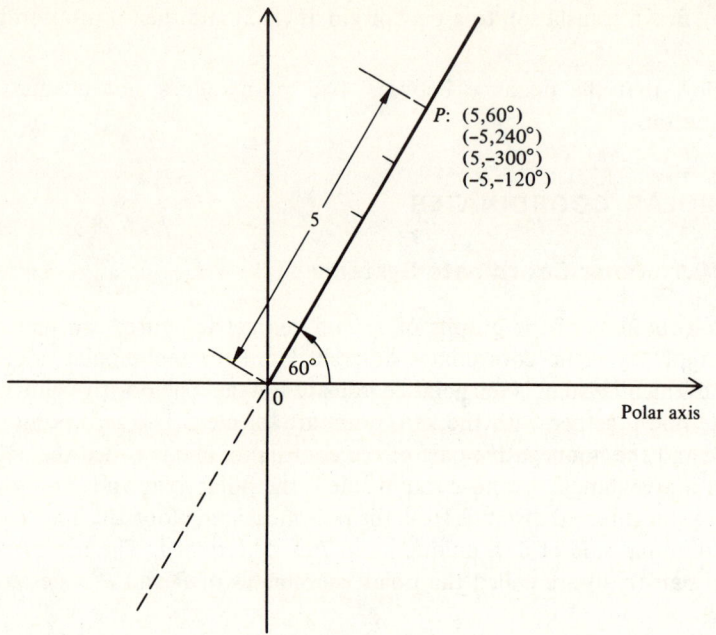

Figure 9.37

can be designated in many ways, among which are $(-5, 240°)$, $(5, -300°)$, $(-5, -120°)$, $(5, 420°)$, and so on.

9.10.2 Polar and Rectangular Relations

If a rectangular coordinate system is selected with its positive x-axis coincident with the polar axis of a polar coordinate system, then equations may be obtained relating the coordinates of the two systems.

Referring to Figure 9.38, it may be seen that

$$x = r \cos \theta \quad \text{and} \quad y = r \sin \theta.$$

Squaring and adding,

$$x^2 + y^2 = r^2 \quad \text{or} \quad r^2 = x^2 + y^2.$$

Dividing, for $x \neq 0$,

$$\frac{y}{x} = \tan \theta \quad \text{or} \quad \tan \theta = \frac{y}{x}.$$

POLAR COORDINATES

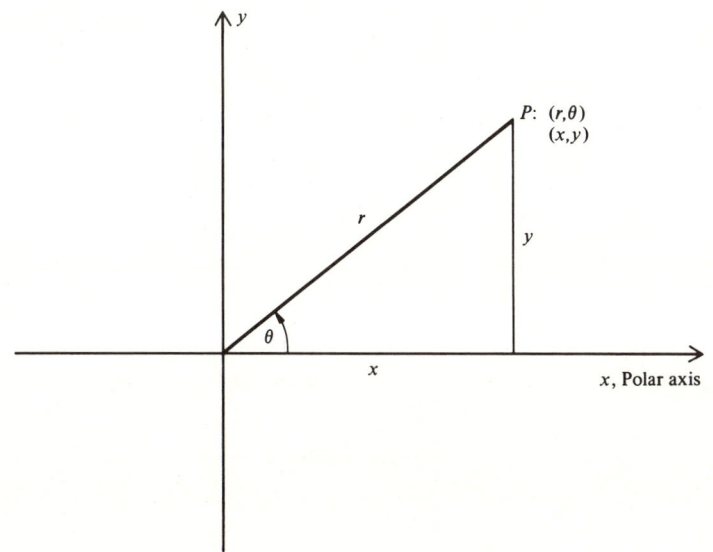

Figure 9.38

9.10.3 Polar Graphs

Definition The graph of an equation in the variables r and θ is the set of points in the plane such that for each point at least one pair of its polar coordinates satisfies the equation.

By using the equations relating the coordinates of the polar and rectangular systems, it may be seen that the graph of $r = c$, where c is a constant, is a circle with center at the pole since $r^2 = c^2 = x^2 + y^2$. (See Figure 9.39(a).)

Also, $\theta = c$, c a constant, is a straight line through the pole since $\tan c = y/x$ or $y = (\tan c)x$. (See Figure 9.39(b).)

The graphs of many polar equations may be readily obtained by first finding the equivalent rectangular equation.

Example 1 Identify and graph $r = 4 \cos \theta$.

Solution Multiplying both sides by r,

$$r^2 = 4r \cos \theta$$

and

$$x^2 + y^2 = 4x$$

or

$$x^2 - 4x + 4 + y^2 = 4$$

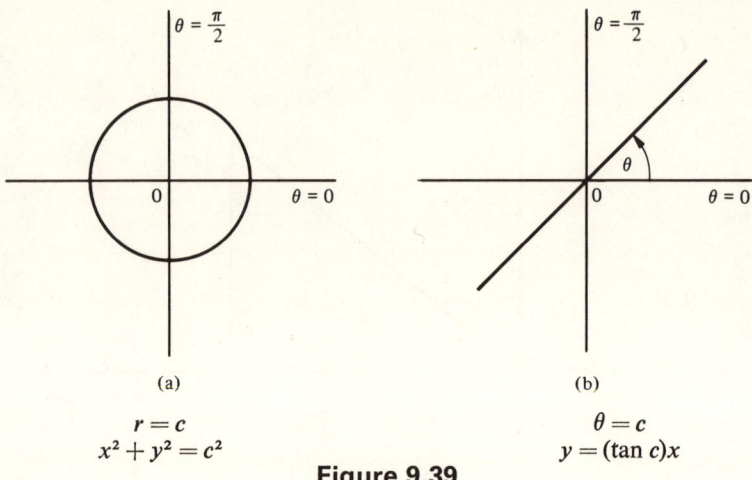

(a)

$r = c$
$x^2 + y^2 = c^2$

(b)

$\theta = c$
$y = (\tan c)x$

Figure 9.39

and thus

$$(x - 2)^2 + y^2 = 4.$$

Therefore, $r = 4 \cos \theta$ is a circle with center $(2, 0°)$ and with radius 2. (See Figure 9.40.)

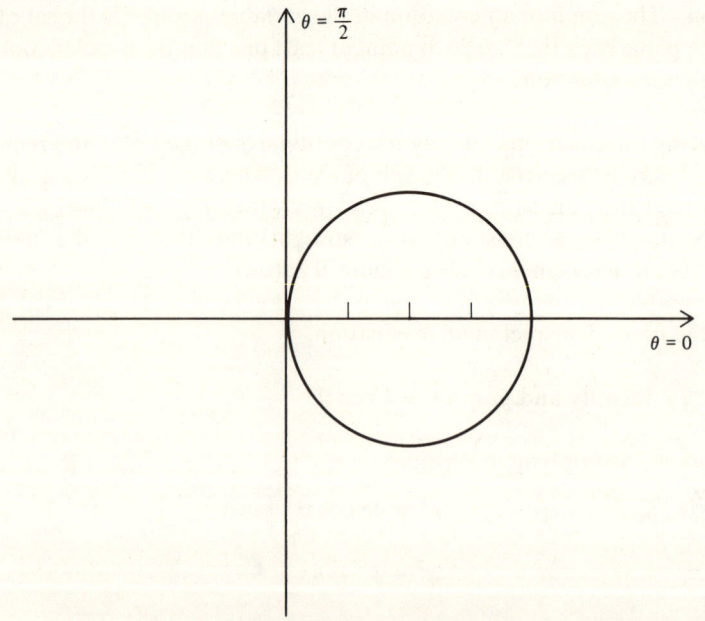

Figure 9.40 $r = 4 \cos \theta$, $(x - 2)^2 + y^2 = 4$

POLAR COORDINATES

Example 2 Graph $r = \dfrac{2}{1 + \sin \theta}$ and identify.

Solution $r + r \sin \theta = 2$

$\sqrt{x^2 + y^2} + y = 2$

$x^2 + y^2 = (2 - y)^2$

$x^2 = -4(y - 1)$, a parabola.

(See Figure 9.41.)

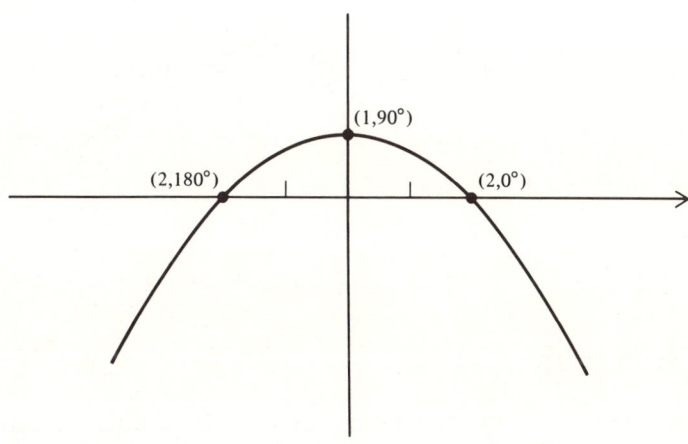

Figure 9.41 $r = \dfrac{2}{1 + \sin \theta}$, $x^2 = -4(y - 1)$

9.10.4 General Polar Graphing Techniques

It is also desirable to graph directly from the polar equation without using the rectangular equation, especially when the polar equation is much simpler. As a general rule, the curve must be graphed for values of θ ranging from 0° to 360° excluding those values that do not produce a real number for r.

Symmetry may be used to limit the values that need to be tabulated. By referring to Figure 9.42, the following observations may be made:

The graph of a polar equation is symmetric

(1) to the origin if $(-r, \theta)$ or $(r, \pi + \theta)$ is on the graph whenever (r, θ) is,

(2) to the polar axis if $(r, -\theta)$ is on the graph whenever (r, θ) is,

(3) to the $\dfrac{\pi}{2}$ line if $(r, \pi - \theta)$ is on the graph whenever (r, θ) is, or if $(-r, -\theta)$ is on the graph whenever (r, θ) is.

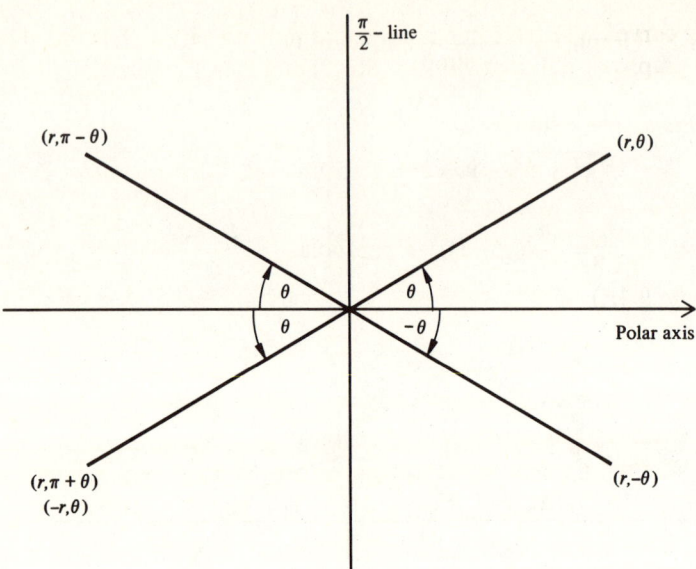

Figure 9.42

Example 3 Graph $r^2 = 4\cos 2\theta$.

Solution This time the rectangular equation is of little assistance since from $r^4 = 4(r^2 \cos^2\theta - r^2 \sin^2\theta)$ is obtained $(x^2 + y^2)^2 = 4(x^2 - y^2)$ which is not as simple as the polar equation.

The graph may be obtained from a table of values noting the symmetry and by also noting that $\cos 2\theta$ must be nonnegative and thus there are no values when $90° < 2\theta < 270°$ or $45° < \theta < 135°$.

Since $\cos 2(-\theta) = \cos 2\theta$, the equation is unchanged if θ is replaced by $-\theta$ or if r is replaced by $-r$. Thus, the curve is symmetric with respect to the pole, the polar axis, and the $\pi/2$ or 90° line.

Table of Values

$\theta°$	2θ	$\cos 2\theta$	$4\cos 2\theta$	r	$\theta^{(r)}$
0	0	1	4	± 2	0
30	60	$\tfrac{1}{2}$	2	$\pm\sqrt{2}$	$\pi/6$
45	90	0	0	0	$\pi/4$
(45° to 90°) No graph ($\pi/4$ to $\pi/2$)					

Other quadrants obtained by symmetry

POLAR COORDINATES

The curve is illustrated on polar coordinate paper which is useful for sketching the graphs of polar equations. (See Figure 9.43.)

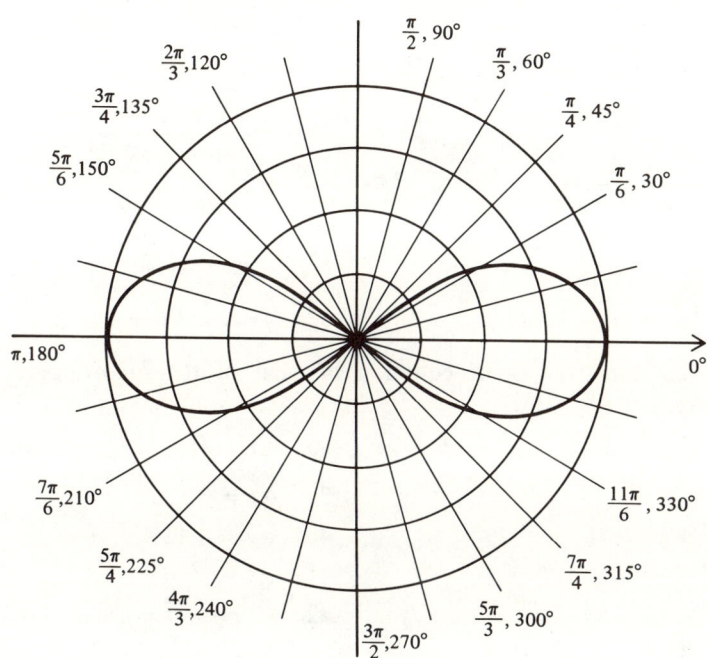

Figure 9.43 $r^2 = 4 \cos 2\theta$ (Lemniscate)

EXERCISES 9.10

1–10. Plot each of the following points with the given polar coordinates:

1. $(1, \pi/3)$
2. $(2, \pi/4)$
3. $(-3, 7\pi/6)$
4. $(-5, 2\pi/3)$
5. $(-2, -\pi/4)$
6. $(-2, \pi/4)$
7. $(3, 5\pi/4)$
8. $(1, 7\pi/6)$
9. $(-1, -7\pi/6)$
10. $(3, 11\pi/6)$

11–20. Plot each of the following points with given rectangular coordinates and give two sets of polar coordinates for each point ($0 \leq \theta < 2\pi$):

11. $(2, 2)$
12. $(-2, 2)$
13. $(2\sqrt{3}, 2)$
14. $(-2\sqrt{3}, -2)$
15. $(\sqrt{3}, 1)$
16. $(1, \sqrt{3})$
17. $(-1, -1)$
18. $(-\sqrt{3}, 1)$
19. $(-2, 0)$
20. $(0, 3)$

21–32. For each of the following equations find an equivalent rectangular equation and graph:

21. $r = 2 \cos \theta$
22. $r = 3 \sin \theta$
23. $r = \dfrac{4}{1 + \sin \theta}$
24. $4 = \dfrac{6}{3 - 2 \cos \theta}$
25. $r = 2 \cos \theta - 2 \sin \theta$
26. $r = \dfrac{6}{2 \cos \theta - 3 \sin \theta}$
27. $r = 2 \sec \theta$
28. $r = -3 \csc \theta$
29. $r = \dfrac{5}{\sin \theta + \cos \theta}$
30. $r^2 = 16 \csc 2\theta$
31. $r^2 = 4 \sec 2\theta$
32. $r = 6 \cos \theta + 4 \sin \theta$

33–42. Find the polar equation of each of the following rectangular equations:

33. $x = 4$
34. $y = 2$
35. $x^2 + y^2 = 36$
36. $x^2 + y^2 - 6x = 0$
37. $x^2 + y^2 - 8y = 0$
38. $x^2 + y^2 - 4x + 6y = 0$
39. $xy = 6$
40. $xy = -12$
41. $x^2 - y^2 = 9$
42. $(x^2 + y^2)^2 = 18(x^2 - y^2)$

43–52. Graph each of the following equations:

43. $r = 4(3 + 2 \cos \theta)$
44. $r = \dfrac{2}{1 - \cos \theta}$
45. $r = 1 - 2 \cos \theta$ (limaçon)
46. $r = 5 \sin 3\theta$ (three-leaved rose)
47. $r = 3 \cos 3\theta$ (three-leaved rose)
48. $r = 4(1 - \cos \theta)$ (cardioid)
49. $r^2 = 4 \sin 2\theta$ (lemniscate)
50. $r = 6 \sin 2\theta$ (four-leaved rose)
51. $r = 4 + 2 \sin \theta$ (limaçon)
52. $r = 3\theta$ (spiral of Archimedes)

53. Find the polar equation of the circle with center $(2, \pi)$ and radius 2.
54. Find the polar equation of the line with slope 3, which intersects the polar axis at $(2, 0)$.
55. Find the polar equation of the circle with center at the pole and radius 5.
56. Find the polar equation of the circle with center at $(2, \pi/6)$ and passing through the pole.

PARAMETRIC EQUATIONS

57. Find the polar equation of the circle whose rectangular equation is $x^2 + y^2 - 6x - 8y = 0$. What are the rectangular coordinates and the polar coordinates of the center of this circle?

58. Find the polar equation of the line that passes through the origin and whose slope is 1.

59. Find the polar equation of the line that passes through the origin and whose slope is -1.

60. Show that the distance between two points, P_1 and P_2, whose polar coordinates are $(r_1, \cos \theta_1)$ and $(r_2, \cos \theta_2)$, is

$$|P_1 P_2| = \sqrt{r_1^2 + r_2^2 - 2r_1 r_2 \cos(\theta_2 - \theta_1)}.$$

61–64. Find the distance between each of the following pairs of points:

61. $(2, \pi/12)$ and $(4, 5\pi/12)$
62. $(3, \pi/2)$ and $(6, \pi/6)$
63. $(1, 4\pi/3)$ and $(\sqrt{3}, 7\pi/6)$
64. $(\sqrt{2}, \pi/8)$ and $(3, 3\pi/8)$

☆65. Show that $r = \dfrac{k}{1 \pm e \cos \theta}$ and $r = \dfrac{k}{1 \pm e \sin \theta}$ are equations of

(a) a parabola with a focus at the pole if $e = 1$,
(b) an ellipse with a focus at the pole if $0 < e < 1$, or
(c) a hyperbola with a focus at the pole if $e > 1$.

9.11 PARAMETRIC EQUATIONS

Instead of defining a relation by a single equation relating the components of the ordered pairs, it is also possible to use two equations whereby each component is expressed as a function of a third variable, say t, called the **parameter**.

Definition of Parametric Equations If f and g are functions with a common domain, Dm, then the set of equations

$$x = f(t)$$
$$y = g(t)$$

is called **parametric equations** of the relation

$$\{(x, y) \mid x = f(t) \text{ and } y = g(t) \text{ and } t \in \text{Dm}\}.$$

The variable t is called a **parameter**.

Example 1 Find the rectangular equation of the relation defined by the parametric equations: $x = 2 - 2 \sin t$, $y = 1 + 2 \cos t$. (This process is called **eliminating the parameter**.)

Solution $x - 2 = -2 \sin t$ and $y - 1 = 2 \cos t$

$$(x - 2)^2 + (y - 1)^2 = 4 \sin^2 t + 4 \cos^2 t = 4$$

Therefore, $(x - 2)^2 + (y - 1)^2 = 4$ is the rectangular equation.

Parametric equations are valuable because it is often more convenient to express the variables x and y in terms of one variable than it is to use the single equation relating x and y. In some cases a parametric representation may be found when it is very difficult, if not impossible, to determine a rectangular or polar equation. For example, to obtain the rectangular equation for

$$x = t^3 - 3t + 2$$
$$y = t^5 + t^2 - 1,$$

one must solve a cubic equation.

In many physical applications it is desirable to express the motion of a particle in terms of the time, where the parameter t is used to indicate units of time. For example, consider a projectile fired with an initial velocity v_0 at an angle θ from the horizontal. If forces such as air resistance and friction are neglected and only the force of gravity is considered, then the laws of mechanics lead to the following equations expressing the x- and y-coordinates of the position of the projectile as functions of the time, t. The letter g designates the magnitude of the gravitational force which is approximately 32 when distance is measured in feet and time in seconds.

$$x = (v_0 \cos \theta)t$$
$$y = (v_0 \sin \theta)t - \tfrac{1}{2}gt^2.$$

(See Figure 9.44.)

Example 2 A projectile is shot upward with an initial velocity of 800 ft/sec at an angle of 30° from the horizontal.

(1) Find the equation of its path.
(2) Find its horizontal range (the horizontal distance it travels before striking the earth) and the time of its flight.

Solution

(1) $x = (v_0 \cos \theta)t = (800 \cos 30°)t$

$y = (v_0 \sin \theta)t - \tfrac{1}{2}gt^2 = (800 \sin 30°)t - \tfrac{1}{2}(32)t^2$

Therefore, $x = 400\sqrt{3}\, t$ and $y = 400t - 16t^2$.

PARAMETRIC EQUATIONS

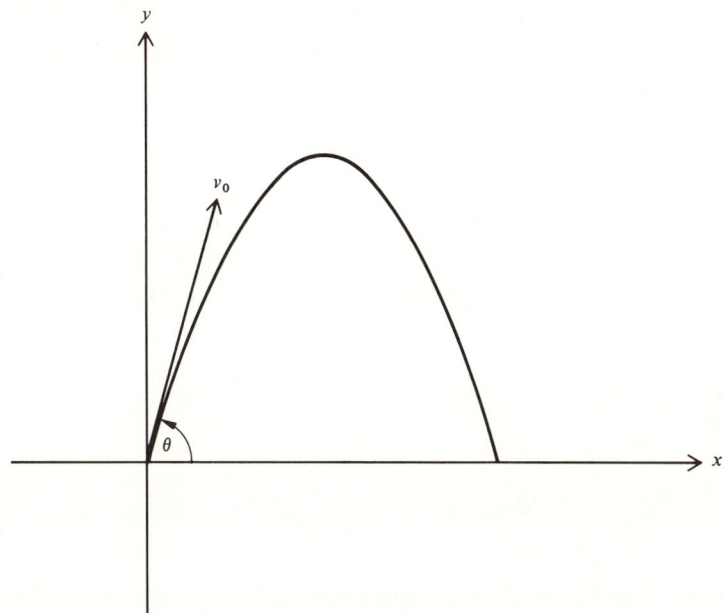

Figure 9.44 Path of Projectile: $x = (v_0 \cos \theta)t, \quad y = (v_0 \sin \theta)t - \tfrac{1}{2}gt^2$

Eliminating the parameter t,

$$y = 400\left(\frac{x}{400\sqrt{3}}\right) - 16\left(\frac{x}{400\sqrt{3}}\right)^2$$

$$y = \frac{\sqrt{3}}{3}x - \frac{1}{30{,}000}x^2.$$

(2) To find the time of the flight, set $y = 0$; $400t - 16t^2 = 0$ and thus $t = 0$ or $t = 25$. $t = 25$ sec, the time of the flight.
To find the horizontal range, find x for $t = 25$:
$x = 400\sqrt{3}(25) = 10{,}000\sqrt{3}$ ft.

A parametric representation for a given rectangular or polar equation is not unique. Many interpretations of the parameter are possible. In addition, care must be exerted in the choice of the parameter so that every ordered pair defined by the single equation is also given by the parametric equations, and the parametric representation should not yield points not defined by the single equation.

For example, suppose that $x = \sin t$, $y = \sin^2 t$ is proposed as a parametric representation of $y = x^2$. Since $-1 \le \sin t \le 1$, x is restricted so that $-1 \le x \le 1$.

However, $y = x^2$ is defined for all real values of x and the set of parametric equations is not equivalent to the rectangular equation but represents only part of the parabola.

Example 3 Find a parametric representation for the semicircle $x^2 + y^2 = a^2$, $y \geq 0$, by letting

(1) $t = x$
(2) $t = \theta$, the polar angle.

Solution Note that the domain is $-a \leq x \leq a$ and the range is $0 \leq y \leq a$.

(1) $x = t$
$y = \sqrt{a^2 - t^2}$ and $-a \leq t \leq a$

(2) Since $x = r \cos \theta$ and $y = r \sin \theta$, for $r = a$ and $t = \theta$,
$x = a \cos t$
$y = a \sin t$ and $0 \leq t \leq \pi$

Example 4 Find a parametric representation for $x^3 + y^3 = 3xy$ by letting $t = y/x$ (that is, $y = xt$).

Solution $x^3 + x^3 t^3 = 3x^2 t$
$x^3(1 + t^3) = 3x^2 t$,
$$x = \frac{3t}{1 + t^3} \quad \text{and} \quad y = \frac{3t^2}{1 + t^3}$$

Exercises 9.11

1–10. Find a rectangular equation for each of the following parametric equations and sketch:

1. $x = t - 2$
   $y = 3t + 1$

2. $x = 2 \cos t$
   $y = 3 \sin t$

3. $x = 3 \sec t$
   $y = 4 \tan t$

4. $x = 1 - \sin t$
   $y = 1 - \cos t$

5. $x = e^t$
   $y = e^{-t}$

6. $x = \sin 2t$
   $y = \sin^2 t$

7. $x = t^3 + t$
   $y = t^3 - t$

8. $x = t$.
   $y = \dfrac{1}{t^2 + 1}$

9. $x = \sin t$
   $y = \cos 2t$

10. $x = a \cos^3 t$
    $y = a \sin^3 t$

PARAMETRIC EQUATIONS

11–15. Find a polar equation for each of the following parametric equations and sketch each one.

11. $r = \cos t$
    $\theta = \tfrac{1}{2}t$
12. $r = 1 + 4t$
    $\theta = \text{Arcsin } t$
13. $r = \sin 2t$
    $\theta = 2t$
14. $r = 2t$
    $\theta = (1/3)t$
15. $r = 1 - 2t$
    $\theta = \text{Arccos } t$

16–20. Determine whether each of the following represents all or part of the parabola $y = x^2$. Sketch each curve.

16. $x = t$
    $y = t^2$
17. $x = e^t$
    $y = e^{2t}$
18. $x = t^2$
    $y = t^4$
19. $x = 2 \sin \theta$
    $y = 2 - 2 \cos 2\theta$
20. $x = \tan t$
    $y = \sec^2 t - 1$

21–30. Eliminate the parameter for each of the following, identify the curve as a line, circle, ellipse, parabola, or hyperbola, determine whether the equation represents part or all of this conic, and sketch.

21. $x = \dfrac{3t}{\sqrt{1 + t^2}}$
    $y = \dfrac{3}{\sqrt{1 + t^2}}$
22. $x = 3 + 2 \cos t$
    $y = 2 + 3 \cos t$
23. $x = 3 + 2 \sec t$
    $y = 2 + 3 \tan t$
24. $x = 2 + \sin t$
    $y = \cos t$
25. $x = t - \dfrac{1}{t}$
    $y = t + \dfrac{1}{t}$
26. $x = \dfrac{3t}{t^2 + 1}$
    $y = \dfrac{6}{t^2 + 1}$
27. $x = 2 + 2 \cos t$
    $y = 4 \sin t$
28. $x = 4t$
    $y = 3t - 16t^2$
29. $x = \dfrac{4t^2 + 12}{(t^2 + 2)^2}$
    $y = \dfrac{2t^2 + 8}{t^2 + 2}$
30. $x = 4 + 2 \tan t$
    $y = 4 - \tan t$

31–40. Find a pair of parametric equations for the following rectangular equations, using the suggested substitutions where t is the parameter.

31. $(x + 3)^2 + 3(y - 2)^2 = 9$; let $y = 2 + \sqrt{3} \cos t$
32. $x^2 - y^2 = 16$; let $y = 4 \tan t$
33. $4x^2 + y^2 = 1$; let $x = \frac{1}{2} \sin t$
34. $x^2 + y^2 - 4y = 0$; let $t = y/x$
35. $x^3 - y^2 = 0$; let $x = t^2$
36. $\sqrt{x} + \sqrt{y} = 2$; let $x = 4 \sin^4 t$
37. $x^{2/3} + y^{2/3} = 4$; let $x = 8 \cos^3 t$
38. $y^2(6 - x) = x^3$; let $x = 6 \sin^2 t$
39. $x^5 + y^5 = 5x^2 y^2$; let $y = xt$
40. $x^3 - 3x^2 y - y^2 = 0$; let $y = xt$

41. Show that $x = 4 + t$ and $y = 6 + 4t$ are parametric equations of the line through the points $A: (4, 6)$ and $B: (3, 2)$.
42. Find parametric equations for the line through the points $A: (2, 5)$ and $B: (-3, 2)$.
43. Find parametric equations for the straight line through the points $P_1: (x_1, y_1)$ and $P_2: (x_2, y_2)$ if t is the ratio of the directed distances,
$$\frac{\overrightarrow{P_1 P}}{\overrightarrow{P_1 P_2}},$$
and P is any point on the line.

44–45. Using the parameter t as the ratio described in Exercise 43, find parametric equations for the straight line through the given points.

44. $P_1: (3, 4)$ and $P_2: (5, -2)$
45. $P_1: (-2, 1)$ and $P_2: (1, -5)$

9.12 TWO-DIMENSIONAL VECTORS

Many quantities such as force, velocity, and acceleration are characterized by having two important properties, magnitude and direction. The concept of vector furnishes a useful description of such quantities.

Definition of Two-Dimensional Vector A two-dimensional vector, $\mathbf{v}$, over the set of real numbers is an ordered pair of real numbers. In symbols,

$$\mathbf{v} = (v_x, v_y),$$

where the real numbers v_x and v_y are called the **components** of the vector.

Definition of Scalar A scalar is any real number.

Definition of Equal Vectors Two vectors are equal if and only if their corresponding components are equal. In symbols, if

$$\mathbf{a} = (a_x, a_y) \quad \text{and} \quad \mathbf{b} = (b_x, b_y),$$

then
$$\mathbf{a} = \mathbf{b} \text{ if and only if } a_x = b_x \text{ and } a_y = b_y.$$

Definition of Vector Addition If $\mathbf{a} = (a_x, a_y)$ and $\mathbf{b} = (b_x, b_y)$, then
$$\mathbf{a} + \mathbf{b} = (a_x + b_x, a_y + b_y).$$

With these definitions it may be proved that vector addition satisfies the closure, commutative, associative, identity, and inverse properties. This is stated below as a theorem with the proof left for the student.

Theorem (Algebra of Vector Addition) For all vectors $\mathbf{a}$, $\mathbf{b}$, and $\mathbf{c}$ over R,

(1) Closure $\mathbf{a} + \mathbf{b} \in R \times R$.
(2) Commutative $\mathbf{a} + \mathbf{b} = \mathbf{b} + \mathbf{a}$.
(3) Associative $(\mathbf{a} + \mathbf{b}) + \mathbf{c} = \mathbf{a} + (\mathbf{b} + \mathbf{c})$.
(4) Identity   There exists a unique vector which is called the **zero vector**, $\mathbf{0} = (0, 0)$, so that for all $\mathbf{a}$, $\mathbf{a} + \mathbf{0} = \mathbf{a}$.
(5) Inverse   For each $\mathbf{a}$ there exists a unique vector $-\mathbf{a} = (-a_x, -a_y)$ so that $\mathbf{a} + (-\mathbf{a}) = \mathbf{0}$.

Definition of Vector Subtraction $\mathbf{a} - \mathbf{b} = \mathbf{a} + (-\mathbf{b})$.

Definition of Magnitude The magnitude, $|\mathbf{v}|$, of a vector $\mathbf{v} = (v_x, v_y)$ is defined as follows:
$$|\mathbf{v}| = \sqrt{v_x^2 + v_y^2}.$$

Definition of Direction The direction, θ, of a nonzero vector $\mathbf{v} = (v_x, v_y)$ is defined as follows:
$$\cos\theta = \frac{v_x}{\sqrt{v_x^2 + v_y^2}} \quad \text{and} \quad \sin\theta = \frac{v_y}{\sqrt{v_x^2 + v_y^2}}.$$

The zero vector is not assigned a direction.

Geometrically, a two-dimensional vector may be interpreted as a directed line segment, $\overrightarrow{PQ}$, in a coordinate plane, where any point $P: (x, y)$ may be chosen as the initial point and the terminal point is then $Q: (x + v_x, y + v_y)$. The length of the directed line segment is identified as the magnitude of the vector and the angle of inclination of the line $\overrightarrow{PQ}$ is identified as the direction of the vector, See Figure 9.45.

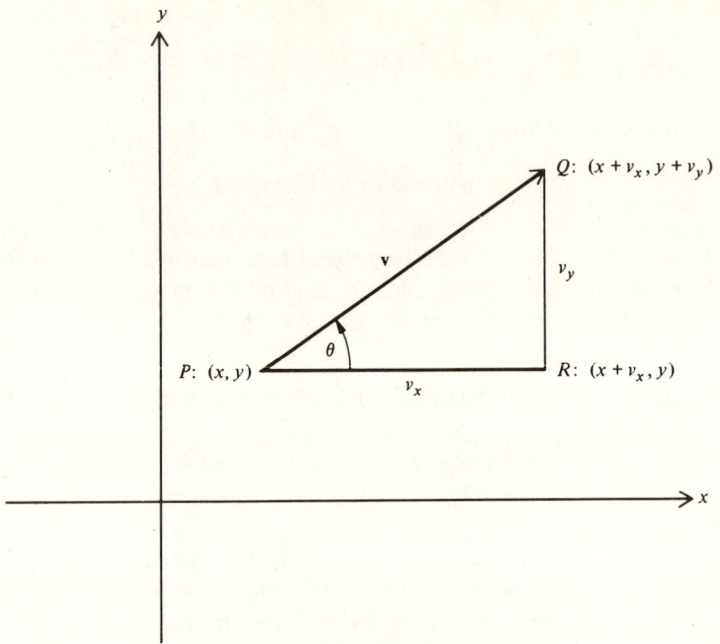

Figure 9.45
Vector Interpreted as Directed Line Segment
$$\mathbf{v} = \overrightarrow{PQ}$$

Example 1 Find the magnitude and direction of $\mathbf{v} = (2, -1)$.

Solution $|\mathbf{v}| = \sqrt{2^2 + (-1)^2} = \sqrt{5}$

$$\cos \theta = \frac{2}{\sqrt{5}} \quad \text{and} \quad \sin \theta = \frac{-1}{\sqrt{5}}$$

The sum of two vectors may be interpreted as the directed line segment forming the third side of a triangle whose other two sides are determined by the two vectors with the initial point of one coincident with the terminal point of the other. See Figure 9.46.

If the two vectors are represented as two directed line segments whose initial points coincide, then the sum of the two vectors can be interpreted as the directed diagonal of a parallelogram. The vector sum is referred to as the **resultant**. See Figure 9.47.

If a vector is represented as a directed line segment with its initial point at the origin, then it is called the **position vector**.

TWO-DIMENSIONAL VECTORS 383

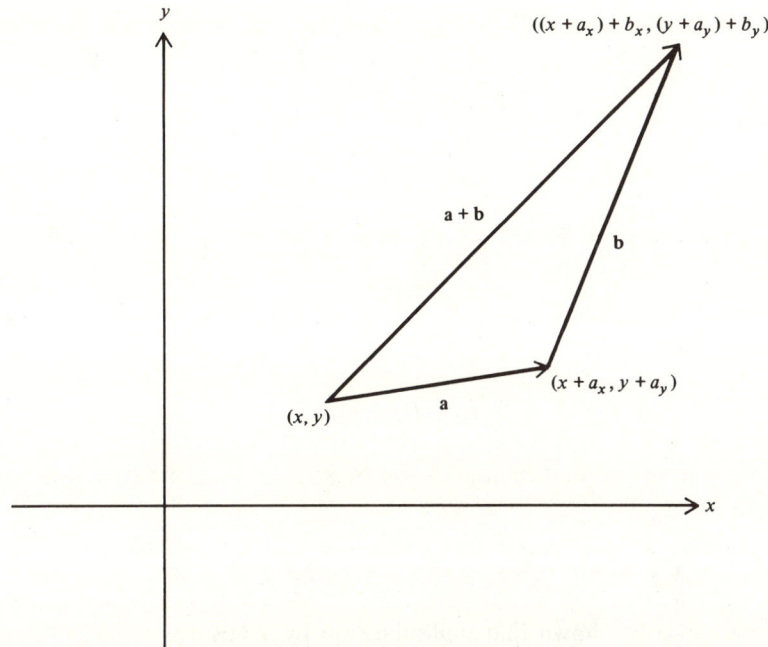

Figure 9.46 Vector Addition: **a** + **b**

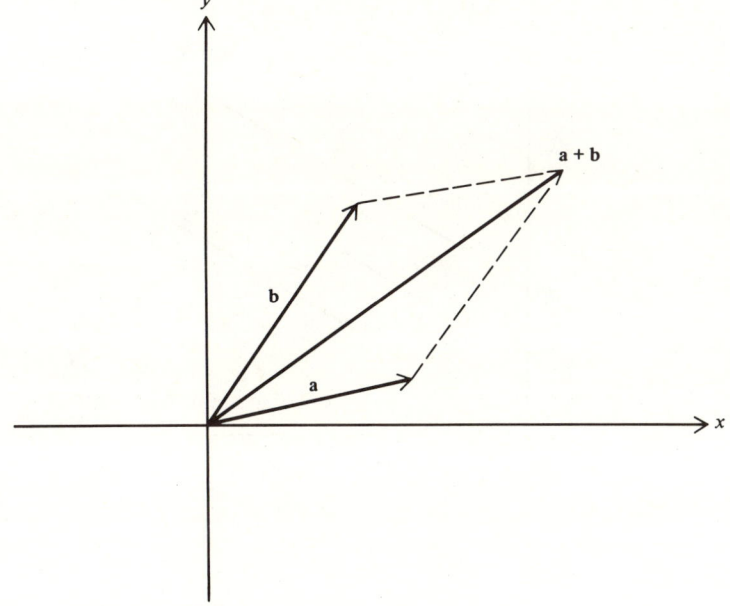

Figure 9.47 Vector Addition: **a** + **b**

Example 2 Find the rectangular equation of a curve determined by the position vector

$$\overrightarrow{OP} = (2 \cos t, 3 \sin t).$$

Solution Let $x = 2 \cos t$ and let $y = 3 \sin t$. Then

$$(3x)^2 + (2y)^2 = 36 \cos^2 t + 36 \sin^2 t = 36 \quad \text{or} \quad 9x^2 + 4y^2 = 36.$$

Definition of Multiplication by a Scalar If k is a scalar and $\mathbf{v} = (v_x, v_y)$ is a vector, then

$$k\mathbf{v} = (kv_x, kv_y).$$

It may be shown that multiplication by a scalar is distributive over vector addition:

$$k(\mathbf{a} + \mathbf{b}) = (ka_x + kb_x, ka_y + kb_y) = k\mathbf{a} + k\mathbf{b}.$$

It may also be shown that multiplication by a positive scalar produces a new vector whose magnitude is k times that of the original one but whose direction is the same. Multiplication by a negative scalar reverses the direction of the original vector. See Figure 9.48.

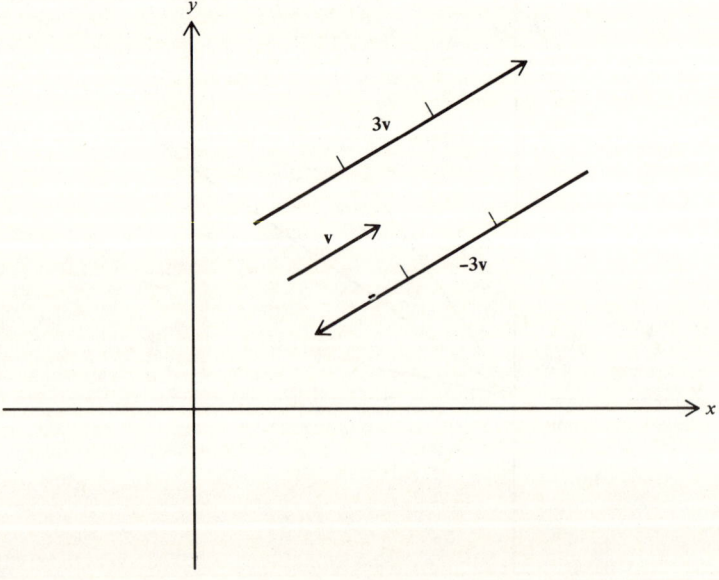

Figure 9.48 Multiplication by a Scalar

TWO-DIMENSIONAL VECTORS

9.12.1 Dot Product (or Inner Product)

A useful product because of its many applications is the **dot product**, also called the **inner product**.

Definition of Dot Product If $\mathbf{a} = (a_x, a_y)$ and $\mathbf{b} = (b_x, b_y)$, then

$$\mathbf{a} \cdot \mathbf{b} = a_x b_x + a_y b_y.$$

By dividing both sides of the equation defining the dot product by the magnitudes of the two vectors,

$$\frac{\mathbf{a} \cdot \mathbf{b}}{|\mathbf{a}| |\mathbf{b}|} = \frac{a_x b_x}{|\mathbf{a}| |\mathbf{b}|} + \frac{a_y b_y}{|\mathbf{a}| |\mathbf{b}|}$$

$$= \cos \alpha \cos \beta + \sin \alpha \sin \beta$$

$$= \cos(\alpha - \beta),$$

where α is the direction of $\mathbf{a}$ and β is the direction of $\mathbf{b}$. Thus,

$$\mathbf{a} \cdot \mathbf{b} = |\mathbf{a}| |\mathbf{b}| \cos(\alpha - \beta).$$

Theorem The dot product of two vectors is equal to the product of the magnitudes of the two vectors and the cosine of the angle between them. In symbols,

$$\mathbf{a} \cdot \mathbf{b} = |\mathbf{a}| |\mathbf{b}| \cos(\alpha - \beta)$$

or

$$a_x b_x + a_y b_y = |\mathbf{a}| |\mathbf{b}| \cos(\alpha - \beta).$$

Geometrically, the dot product may be interpreted as the projection of one vector on a second multiplied by the magnitude of the second. See Figure 9.49.

Example 3 Find the angle between the vectors $\mathbf{a} = (0, 3)$ and $\mathbf{b} = (2, 2\sqrt{3})$.

Solution $\mathbf{a} \cdot \mathbf{b} = 0(2) + 3(2\sqrt{3}) = 6\sqrt{3}$

$|\mathbf{a}| = 3$ and $|\mathbf{b}| = \sqrt{4 + 12} = 4$

$$\cos \theta = \frac{\mathbf{a} \cdot \mathbf{b}}{|\mathbf{a}| \cdot |\mathbf{b}|} = \frac{6\sqrt{3}}{12} = \frac{\sqrt{3}}{2}$$

$$\theta = \frac{\pi}{6} \text{ radians}$$

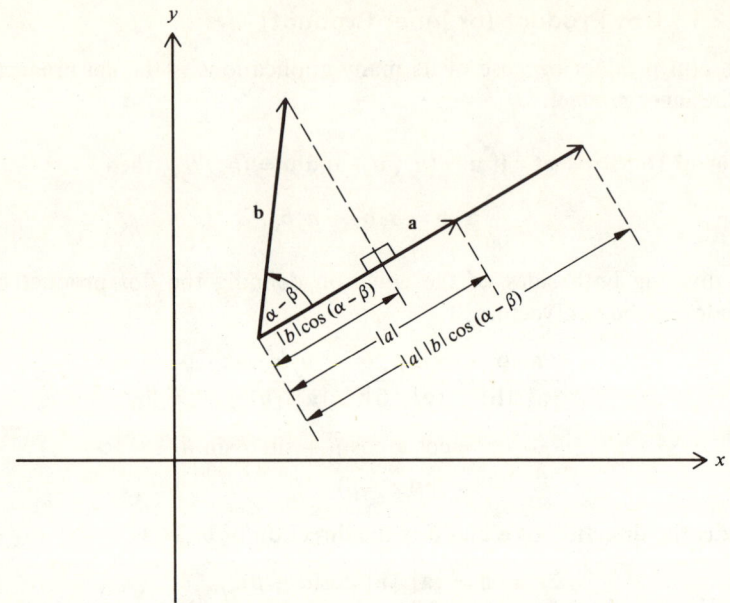

Figure 9.49 Dot Product a · b

Example 4 Find the projection of **a** on **b** if **a** = (2, −1) and **b** = (3, 4).

Solution Projection of **a** on **b** = $|\mathbf{a}| \cos(\alpha - \beta) = \dfrac{\mathbf{a} \cdot \mathbf{b}}{|\mathbf{b}|}$

$$\frac{\mathbf{a} \cdot \mathbf{b}}{|\mathbf{b}|} = \frac{2(3) + (-1)(4)}{\sqrt{3^2 + 4^2}} = \frac{2}{5}$$

EXERCISES 9.12

1–5. Sketch each of the following pairs of vectors whose initial points are at the origin and whose terminal points are indicated, find their sum, and sketch the resultant.

1. (2, 3), (1, 1)
2. (−1, 0), (3, −4)
3. (π, $\sqrt{3}$), (−1, −2)
4. (1, 3), (3, 1)
5. ($\sqrt{2}/2$, 1), (−$\sqrt{2}$, −1)

6–10. For each of the pairs of vectors in Exercises 1–5, subtract the second vector from the first and sketch this difference.

11–15. In each of the following exercises, the endpoints of the directed line segment $\overrightarrow{PQ}$ are given. Find the vector represented by $\overrightarrow{PQ}$, and sketch $\overrightarrow{PQ}$.

TWO-DIMENSIONAL VECTORS

11. $P:(3, 2)$, $Q:(5, 4)$
12. $P:(0, \sqrt{2})$, $Q:(2, 3\sqrt{2})$
13. $P:(-4, 3)$, $Q:(0, 0)$
14. $P:(3, -4)$, $Q:(2, -5)$
15. $P:(\sqrt{3}/2, 1)$, $Q:(\sqrt{2}/2, 1)$

16–25. Find the magnitude and direction of each of the following vectors:

16. $(2, 2)$
17. $(4\sqrt{3}, 4)$
18. $(5, -5\sqrt{3})$
19. $(-1, 1)$
20. $(-3, 4)$
21. $(2, -1.5)$
22. $(0, -3)$
23. $(-7, 0)$
24. $(\sqrt{2} - \sqrt{2}, \sqrt{2} + \sqrt{2})$
25. $(\sqrt{6} + \sqrt{2}, \sqrt{6} - \sqrt{2})$

26–35. (a) Find the dot product for each of the following vectors; (b) Find the cosine of the angle between each pair of vectors; (c) Find the projection of the first vector on the second vector of each pair.

26. $(4, 3)$ and $(2, 2)$
27. $(1, 2)$ and $(2, 1)$
28. $(3, -2)$ and $(4, -6)$
29. $(-2, -4)$ and $(9, -3)$
30. $(-1, \sqrt{3})$ and $(2, -2\sqrt{3})$
31. $(-3, \sqrt{3})$ and $(-\sqrt{3}, -1)$
32. $(2, -3)$ and $(-6, -4)$
33. $(-4, 5)$ and $(10, 8)$
34. $(5, 12)$ and $(2, -1.5)$
35. $(6, 4.5)$ and $(2, -2)$

36–40. If **i** and **j** are unit vectors (magnitude = 1) from the origin along the x-axis and the y-axis, respectively, find the magnitude of each of the following vectors:

36. $\mathbf{v} = 2\mathbf{i} + 3\mathbf{j}$
37. $\mathbf{v} = 3\mathbf{i} - 4\mathbf{j}$
38. $\mathbf{v} = -5\mathbf{i} + 2\mathbf{j}$
39. $\mathbf{v} = -2\mathbf{i} - 4\mathbf{j}$
40. $\mathbf{v} = a\mathbf{i} + b\mathbf{j}$

41–50. Find the rectangular equation of the curve determined by each of the position vectors stated below:

41. $(2 - 5t, 3 + 2t)$
42. $(1 + 2t, 2 + 4t)$
43. $(5 \cos t, 5 \sin t)$
44. $(4 \cos t, 5 \sin t)$
45. $(5 \cos t, -3 \sin t)$
46. $(3 \sin t, -3 \cos t)$
47. $(2 \sec t, 3 \tan t)$
48. $(5 \tan t, -\sec t)$
49. $(2t - 3, 12t - 4t^2)$
50. $(12t - 9t^2, 3t - 2)$

51–55. Prove each of the following theorems for all vectors **a**, **b**, and **c** over R, the set of real numbers:

51. $\mathbf{a}+\mathbf{b} \in R \times R$ (closure)
52. $\mathbf{a}+\mathbf{b}=\mathbf{b}+\mathbf{a}$ (commutativity)
53. $(\mathbf{a}+\mathbf{b})+\mathbf{c}=\mathbf{a}+(\mathbf{b}+\mathbf{c})$ (associativity)
54. There exists a unique vector, called the zero vector, $\mathbf{0}=(0,0)$ so that for all $\mathbf{a}$, $\mathbf{a}+\mathbf{0}=\mathbf{a}$. (Identity)
55. For each $\mathbf{a}$, there exists a unique vector $-\mathbf{a}=(-a_x, -a_y)$ so that $\mathbf{a}+(-\mathbf{a})=\mathbf{0}$.

CHAPTER SUMMARY

Distance Formula $|P_1P_2| = \sqrt{(x_2-x_1)^2 + (y_2-y_1)^2}$

Midpoint Formula $M = \left(\dfrac{x_1+x_2}{2}, \dfrac{y_1+y_2}{2}\right)$

Slope Formula $m = \dfrac{y_2-y_1}{x_2-x_1}$

Parallel Lines Two nonvertical lines are parallel if and only if $m_1 = m_2$.

Perpendicular Lines Two nonvertical lines are perpendicular if and only if $m_2 = \dfrac{-1}{m_1}$.

Point-Slope Form of Straight Line $y - y_1 = m(x - x_1)$

Slope-Intercept Form of Straight Line $y = mx + b$

General Form of Straight Line $Ax + By + C = 0$, where $A^2 + B^2 \neq 0$

Equation of Circle in Standard Form $(x-a)^2 + (y-b)^2 = r^2$; Center: (a, b); radius: r.

Equations of Standard Parabolas

$$x^2 = 4py \text{ with focus } (0, p) \text{ and directrix } y = -p$$
$$y^2 = 4px \text{ with focus } (p, 0) \text{ and directrix } x = -p$$

Equations of Standard Ellipses

$$\dfrac{x^2}{a^2} + \dfrac{y^2}{b^2} = 1 \text{ with foci } (c, 0) \text{ and } (-c, 0) \text{ and } b^2 = a^2 - c^2$$

$$\dfrac{x^2}{b^2} + \dfrac{y^2}{a^2} = 1 \text{ with foci } (0, c) \text{ and } (0, -c) \text{ and } b^2 = a^2 - c^2$$

CHAPTER SUMMARY

Equations of Standard Hyperbolas

$$\frac{x^2}{a^2} - \frac{y^2}{b^2} = 1 \text{ with foci } (c, 0) \text{ and } (-c, 0) \text{ and } b^2 = c^2 - a^2$$

$$\frac{y^2}{a^2} - \frac{x^2}{b^2} = 1 \text{ with foci } (0, c) \text{ and } (0, -c) \text{ and } b^2 = c^2 - a^2$$

Translation of the Axes $x' = x - h$ and $y' = y - k$

Rotation of the Axes $x = x' \cos\theta - y' \sin\theta$
$y = x' \sin\theta + y' \cos\theta$

Removal of xy-term The xy-term can be removed from $Ax^2 + Bxy + Cy^2 + Dx + Ey + F = 0$, $B \neq 0$ by a rotation of the axes if θ is selected so that $\cot 2\theta = \dfrac{A - C}{B}$.

Polar-Rectangular Relations $x = r \cos\theta$, $y = r \sin\theta$
$r^2 = x^2 + y^2$, $\tan\theta = y/x$

Parametric Equations The set of equations $x = f(t)$, $y = g(t)$ are parametric equations of the relation

$\{(x, y) \mid x = f(t) \text{ and } y = g(t) \text{ and } t \text{ is in the intersection of the domains of } f \text{ and } g\}$.

Equal Vectors $\mathbf{a} = \mathbf{b}$ if and only if $a_x = b_x$ and $a_y = b_y$

Vector Addition $\mathbf{a} + \mathbf{b} = (a_x + b_x, a_y + b_y)$

Magnitude of Vector $|\mathbf{v}| = \sqrt{v_x^2 + v_y^2}$

Direction of Vector $\cos\theta = \dfrac{v_x}{|\mathbf{v}|}$ and $\sin\theta = \dfrac{v_y}{|\mathbf{v}|}$

Multiplication by Scalar $k\mathbf{v} = (kv_x, kv_y)$

Dot Product $\mathbf{a} \cdot \mathbf{b} = a_x b_x + a_y b_y = |\mathbf{a}| \, |\mathbf{b}| \cos(\alpha - \beta)$

REVIEW EXERCISES

1–5. Find (a) the distance between each of the following two points; (b) the slope of the line joining the points; (c) the equation of this line; (d) the coordinates of the midpoint of line segment AB.

1. $A: (3, 2), B: (4, -1)$
2. $A: (-2, 3), B: (-1, -1)$
3. $A: (\frac{1}{2}, -\frac{1}{3}), B: (-\frac{1}{4}, -2)$
4. $A: (3, 0), B: (0, -4)$
5. $A: (-1, -2), B: (-3, -4)$

6. Prove analytically that the line segments joining the midpoints of the opposite sides of any quadrilateral bisect each other.

7. Prove analytically that the diagonals of an isosceles trapezoid are congruent.

8. Write an equation for the line through the point $(-2, 3)$ perpendicular to the line $4x + 2y - 3 = 0$.

9–12. Determine the intercepts, the extent, and the symmetry with respect to the origin, the x-axis, and the y-axis for each of the following:

9. $x^2y^2 + 4x^2 - 4y^2 = 0$
10. $x^2y + 3y - 6 = 0$
11. $xy^2 + x + 5y^2 = 5$
12. $y^3 - x - 3y = 0$

13–14. Determine whether the graphs of the following equations have symmetry with respect to a horizontal or a vertical line.

13. $(x - 2)^2 + (y - 3)^2 - 2 = 0$
14. $3x - 2(y - 4)^2 = 12$

15–18. Identify each of the following conics:

15. $x^2 + y^2 - x - y = 0$
16. $x^2 - 2y^2 - 3x + 2 = 0$
17. $5x^2 + 6xy + 3y^2 - 6 = 0$
18. $x^2 - 2xy + y^2 + x + 2y + 1 = 0$

19–23. Identify each of the following conics, use a translation or rotation whenever necessary to simplify the equation, and sketch each curve, labeling all important points.

19. $5x^2 + 4xy + 8y^2 - 36 = 0$
20. $x^2 + y^2 - 4x - 12y + 40 = 0$
21. $16x^2 - 24xy + 9y^2 + 90x - 120y = 0$
22. $x^2 + xy + y^2 + 3\sqrt{2}y = 0$
23. $3x^2 + 10xy + 3y^2 - 4\sqrt{2}x - 12\sqrt{2}y = 8$
24. Find the equation of the parabola with vertex at $(3, -1)$ and focus at $(3, -2)$.

REVIEW EXERCISES

25. State the coordinates of the vertices, the lengths of major and minor axes, coordinates of the foci, and the eccentricity $\left(=\dfrac{c}{a}\right)$ of the ellipse $25x^2 + 169y^2 = 81$.

26. State the coordinates of the vertices and foci, the lengths of the transverse and conjugate axes, and the eccentricity $\left(=\dfrac{c}{a}\right)$ of the hyperbola $9y^2 - 16x^2 = 144$.

27–30. Graph each of the following equations:

27. $r = 2(3 + \cos \theta)$
28. $r = \dfrac{5}{2 - 3 \cos \theta}$
29. $r = 3 \cos \theta - 3 \sin \theta$
30. $r = 1 - 2 \cos \theta$

31. Find the polar equation of the line with slope -2, which intersects the polar axis at $(3, 0)$.

32–37. Eliminate the parameter in each of the following and sketch the curve:

32. $x = 3 \cos t - 2$
    $y = 5 \sin t + 2$
33. $x = 3 - 4t$
    $y = 2 + t$
34. $x = \sin t$
    $y = 1 + \cos t$
35. $x = 2t^2$
    $y = 2t$
36. $x = 2 \tan t$
    $y = 3 \sec t$
37. $2x = e^t + e^{-t}$
    $2y = e^t - e^{-t}$

38–42. For each of the following pairs of vectors, **a** and **b**, find the sum **a** + **b**, the difference **a** − **b**, the dot product **a** · **b**, and the magnitude of **a** and **b**.

38. **a** = 3**i** + 2**j**,   **b** = 2**i** + 3**j**
39. **a** = −3**i** − 2**j**, **b** = **i** + 4**j**
40. **a** = **i** − 3**j**,   **b** = 5**i** − 7**j**
41. **a** = −2**i** + **j**,   **b** = **i** − 2**j**
42. **a** = 6**i** − 2**j**,   **b** = −3**i** − 5**j**

43–47. For each of the following pairs of vectors, (a) find the angle between the two vectors, and sketch; (b) find the projection of the first vector on the second, and sketch.

43. $(-4, 4\sqrt{3})$ and $(2, 2)$
44. $(2, -2)$ and $(-3, -3)$
45. $(1, \sqrt{3})$ and $(5, 0)$
46. $(1, \sqrt{3})$ and $(0, 5)$
47. $(2, -1)$ and $(-2, 4)$

10
SOLID ANALYTIC GEOMETRY

10.1 RECTANGULAR COORDINATES; DISTANCE; MIDPOINT

10.1.1 Rectangular Coordinates

A three-dimensional rectangular coordinate system is an association whereby the set of points in three-dimensional space are placed in one-to-one correspondence with the set of ordered triples of real numbers, the triple Cartesian product, $R \times R \times R$.

Three mutually perpendicular coordinate lines (not all on one plane) having the same scale and intersecting at their origins are selected. These three coordinate lines are called the **x-axis**, the **y-axis**, and the **z-axis**. Their common point of intersection is called the **origin**. If the rotation from the positive x-axis to the positive y-axis to the positive z-axis is counterclockwise, then the system is called a right-handed coordinate system. If the aforesaid rotation is clockwise, then the system is called a left-handed coordinate system. See Figure 10.1.

The three axes determine three planes:

> the x and y axes determine a plane called the **xy-plane**,
> the x and z axes determine a plane called the **xz-plane**, and
> the y and z axes determine a plane called the **yz-plane**.

RECTANGULAR COORDINATES; DISTANCE; MIDPOINT

These three planes divide the space into eight regions called **octants**. Only the first octant, the octant determined by the positive halves of the x, y, and z axes, is referred to by number. See Figure 10.2.

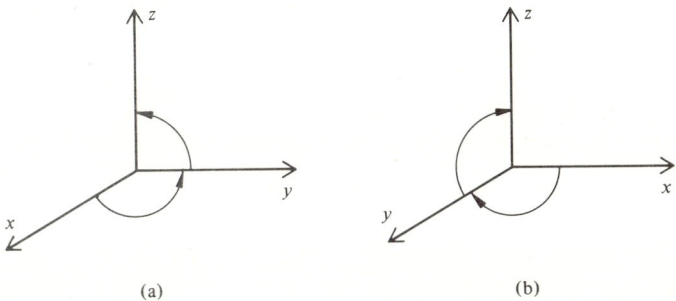

(a)   (b)

Figure 10.1 Right-handed Coordinate System   Left-handed Coordinate System

With each ordered triple of real numbers (a, b, c) is associated a unique point P in space obtained as the point of intersection of

> the plane through point a on the x-axis parallel to the yz-plane,
> the plane through point b on the y-axis parallel to the xz-plane, and
> the plane through point c on the z-axis parallel to the xy-plane.

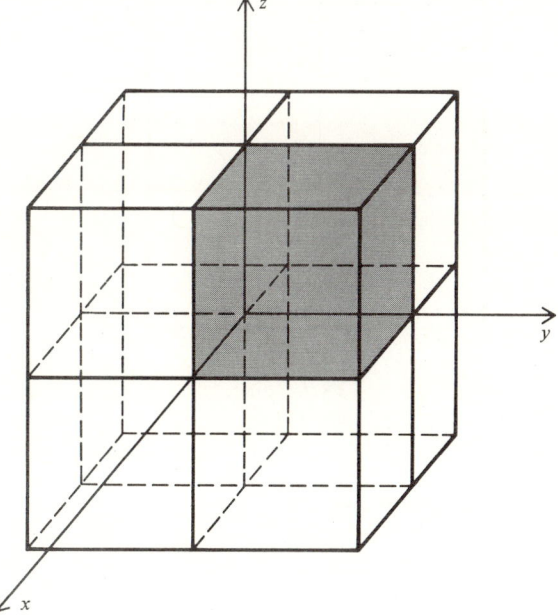

Figure 10.2 Finite Portion of First Octant, Shaded

Conversely, with each point in space is associated a unique ordered triple (a, b, c) where

> a is the point on the x-axis intersected
> by the plane through P parallel to the yz-plane,
> b is the point on the y-axis intersected
> by the plane through P parallel to the xz-plane, and
> c is the point on the z-axis intersected
> by the plane through P parallel to the xy-plane.

Points on the x-axis are assigned triples of the form $(a, 0, 0)$. Points on the y-axis are assigned triples of the form $(0, b, 0)$. Points on the z-axis are assigned triples of the form $(0, 0, c)$. The origin is assigned the triple $(0, 0, 0)$.

A general point in space will be designated as $P: (x, y, z)$ where the real numbers x, y, and z are called, respectively, the **x-coordinate**, the **y-coordinate**, and the **z-coordinate** of point P. See Figure 10.3.

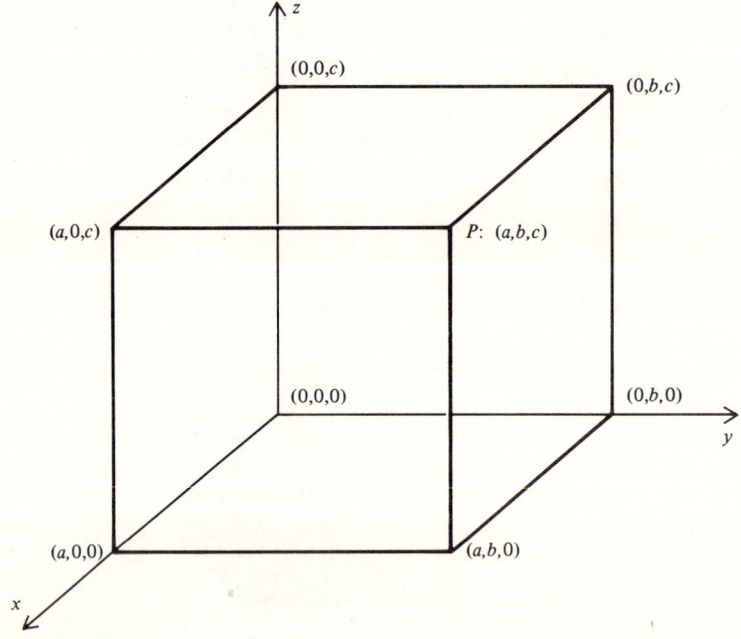

Figure 10.3 Three-dimensional Rectangular Coordinate System

10.1.2 Distance Formula

Theorem (Distance Formula) The distance, $|P_1P_2|$, between the points $P_1: (x_1, y_1, z_1)$ and $P_2: (x_2, y_2, z_2)$ is

$$|P_1P_2| = \sqrt{(x_2 - x_1)^2 + (y_2 - y_1)^2 + (z_2 - z_1)^2}.$$

RECTANGULAR COORDINATES; DISTANCE; MIDPOINT

Proof Planes are constructed through P_1 and P_2 parallel to the coordinate planes and forming a rectangular parallelepiped (box) as shown in Figure 10.4. Then,

$$|QR| = |x_2 - x_1|, \quad |P_2 Q| = |z_2 - z_1|, \quad \text{and} \quad |RP_1| = |y_2 - y_1|$$

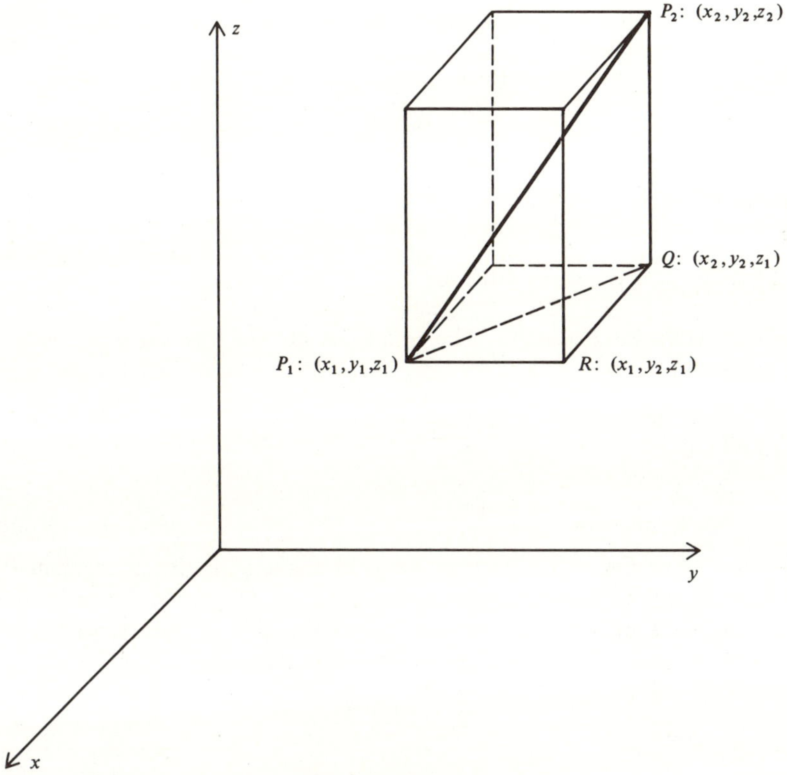

Figure 10.4

since each of these line segments is on a line parallel to one of the coordinate axes.

Now triangle $P_1 R Q$ is a right triangle with right angle at R and triangle $P_1 P_2 Q$ is a right triangle with right angle at Q since a line $(P_2 Q)$ perpendicular to a plane $(P_1 QR)$ is perpendicular to every line $(P_1 Q)$ in the plane passing through the foot of the perpendicular.

Therefore, by the theorem of Pythagoras,

$$\begin{aligned}|P_1 P_2|^2 &= |P_2 Q|^2 + |P_1 Q|^2 \\ &= |P_2 Q|^2 + |P_1 R|^2 + |RQ|^2 \\ &= (z_2 - z_1)^2 + (y_2 - y_1)^2 + (x_2 - x_1)^2.\end{aligned}$$

Finally,
$$|P_1P_2| = \sqrt{(x_2 - x_1)^2 + (y_2 - y_1)^2 + (z_2 - z_1)^2}.$$

Example 1 Find the distance between $A: (-4, 5, 3)$ and $B: (8, 1, 6)$.

Solution Using the distance formula,
$$|AB| = \sqrt{(-4 - 8)^2 + (5 - 1)^2 + (3 - 6)^2}$$
$$= \sqrt{144 + 16 + 9}$$
$$= \sqrt{169}.$$
Thus, $|AB| = 13$.

10.1.3 Midpoint Formula

Theorem (Midpoint Formula) The midpoint of the line segment with endpoints $P_1: (x_1, y_1, z_1)$ and $P_2: (x_2, y_2, z_2)$ is
$$\left(\frac{x_1 + x_2}{2}, \frac{y_1 + y_2}{2}, \frac{z_1 + z_2}{2}\right).$$

Proof

Referring to Figure 10.5, let $M: (x, y, z)$ be the midpoint of segment P_1P_2. Consider the plane passing through the midpoint M parallel to the yz-plane. Then this plane also passes through the midpoint S of line segment R_1R_2. Moreover, M and S have the same x-coordinate; namely,
$$x = \frac{x_1 + x_2}{2}.$$
In a similar way it can be shown that $y = (y_1 + y_2)/2$ and $z = (z_1 + z_2)/2$.

Example 2 Find the midpoint of line segment AB for $A: (5, -1, 2)$ and $B: (-3, 7, 6)$.

Solution Let M be (x, y, z), the midpoint. Then
$$x = \frac{5 + (-3)}{2} = 1$$
$$y = \frac{-1 + 7}{2} = 3$$
$$z = \frac{2 + 6}{2} = 4.$$
Therefore, the midpoint is $M: (1, 3, 4)$.

RECTANGULAR COORDINATES; DISTANCE; MIDPOINT

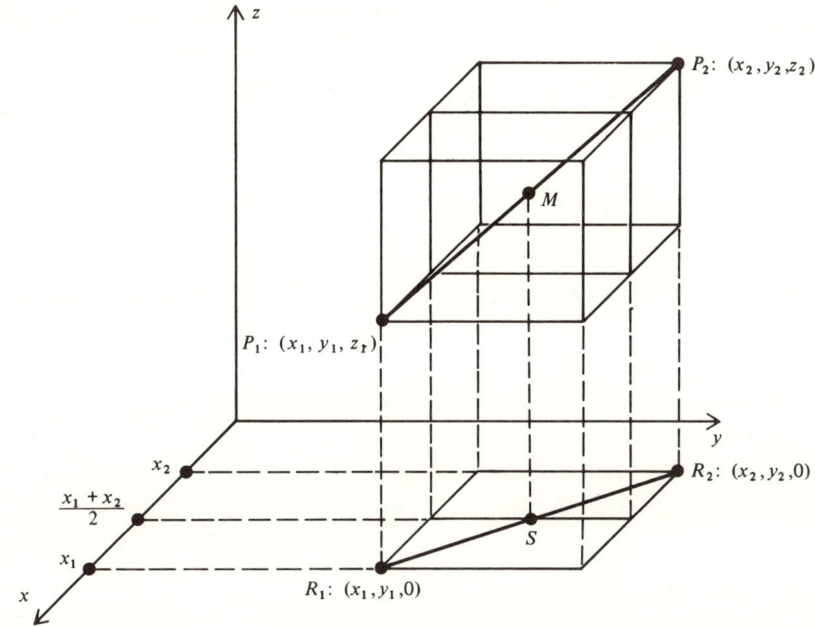

Figure 10.5

EXERCISES 10.1

1–10. Find the distance between each of the following pairs of points.

1. $A: (-1, 2, 0)$, $B: (2, 4, 2)$
2. $A: (1, -2, -1)$, $B: (3, 0, -2)$
3. $A: (3, 2, 4)$, $B: (-1, -2, -3)$
4. $A: (4, 1, 3)$, $B: (-2, 2, -2)$
5. $A: (1, 1, 1)$, $B: (-2, 2, -2)$
6. $A: (-1, 5, 7)$, $B: (-5, -7, 4)$
7. $A: (a, b, c)$, $B: (-a, -b, -c)$
8. $A: (a, 0, c)$, $B: (0, b, 0)$
9. $A: (-3, 2, 6)$, $B: (6, -3, -6)$
10. $A: (1, -1, -1)$, $B: (2, -2, -2)$

11–20. Find the coordinates of the midpoints of the line segments determined by A and B in each of the Exercises 1–10.

21. Find a point on the x-axis that is equidistant from $A: (2, 1, 3)$ and $B: (3, -2, -3)$.

22. Find a point on the y-axis that is equidistant from A: $(4, -2, 5)$ and B: $(1, -4, 2)$.

23–26. Determine if the triangle having the following sets of points as vertices is isosceles, equilateral, or right-angled.

23. $(4, 7, 4)$, $(7, 1, 5)$, $(1, 5, 1)$
24. $(3, 1, -2)$, $(-3, 0, 0)$, $(1, 4, -4)$
25. $(5, -6, -4)$, $(1, -1, 1)$, $(4, 1, -4)$
26. $(3, 2, 0)$, $(1, 5, -6)$, $(-2, -1, -8)$
27. Find a point on line segment AB whose distance from A is $\frac{1}{3}$ of the distance from A: $(2, 5, 3)$ to B: $(4, 7, 0)$.
28. Find a point on line segment AB whose distance from A is $\frac{1}{4}$ of the distance from A: $(3, -2, 6)$ to B: $(1, -6, -8)$.
29. Show that the point on line segment $P_1 P_2$ whose distance from P_1 is $1/n$ of the distance from P_1: (x_1, y_1, z_1) to P_2: (x_2, y_2, z_2) has coordinates

$$\left(\frac{x_1 + x_2}{n}, \frac{y_1 + y_2}{n}, \frac{z_1 + z_2}{n} \right).$$

30. Find the coordinates of two points on the line determined by P_1: (x_1, y_1, z_1) and P_2: (x_2, y_2, z_2) such that the distance of either of the two points from P_1 is $1/n$ of the distance from P_1 to P_2.

10.2 LINES IN SPACE

In two-dimensional space the slope of a line provided a measure of the direction of the line. However, the concept of slope cannot be extended to three-space. On the other hand, since the slope is a function of an angle the line makes with the positive x-axis, this suggests considering in three-space the three angles that a line makes with the three positive coordinate axes.

Definition of Direction Angles and Direction Cosines Let QP be a half-line with endpoint at Q. Then, if Q is the origin, let α, β, and γ be the smallest non-negative angles between the half-line OP and the positive x-axis, the positive y-axis, and the positive z-axis, respectively. Each angle, therefore, is between $0°$ and $180°$. The angles α, β, and γ are called the **direction angles** of the half-line, and $\cos \alpha$, $\cos \beta$, and $\cos \gamma$ are called the **direction cosines** of the half-line. See Figure 10.6.

If a half-line QP with endpoint Q does not go through the origin, then half-lines emanating from Q are constructed having the same directions as the positive coordinate axes. Then the **direction angles** of the half-line QP are the angles

LINES IN SPACE

α, β, and γ between the half-line and the positive x'-, y'-, and z'-axes, respectively, as shown in Figure 10.7. The **direction cosines** are the cosines of α, β, and γ.

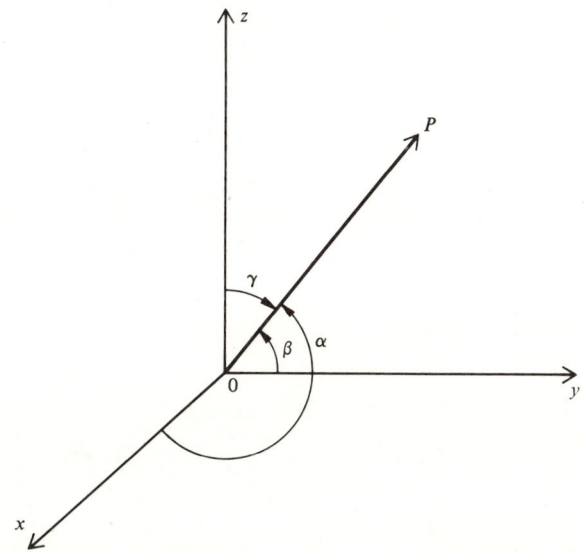

Figure 10.6 Direction Angles of Half-line through Origin

Now consider the half-line QP', the extension of QP through Q. Then, by referring to Figure 10.8, it may be seen that the direction angles of this half-line are the supplements of those for QP. As a result, the direction cosines for half-line QP' are the negatives of those for half-line QP.

$$\alpha + \alpha' = 180°, \qquad \beta + \beta' = 180°, \qquad \gamma + \gamma' = 180°$$

and

$$\cos \alpha' = -\cos \alpha, \qquad \cos \beta' = -\cos \beta, \qquad \cos \gamma' = -\cos \gamma.$$

As a direct consequence of the definition of the direction angles of a half-line, two half-lines are parallel if and only if their direction cosines are equal, respectively, or if the direction cosines of one are equal, respectively, to the negatives of the direction cosines of the other. Two parallel half-lines, then, either have the same direction or opposite directions.

Since, for each line there are two sets of direction angles and two corresponding sets of direction cosines, the following theorem can now be stated.

Theorem on Parallel Lines If (a_1, b_1, c_1) are direction cosines for one line and (a_2, b_2, c_2) are direction cosines for another line, then the two lines are parallel if and only if either

$$a_1 = a_2, \qquad b_1 = b_2, \quad \text{and} \quad c_1 = c_2$$

or

$$a_1 = -a_2, \qquad b_1 = -b_2, \quad \text{and} \quad c_1 = -c_2.$$

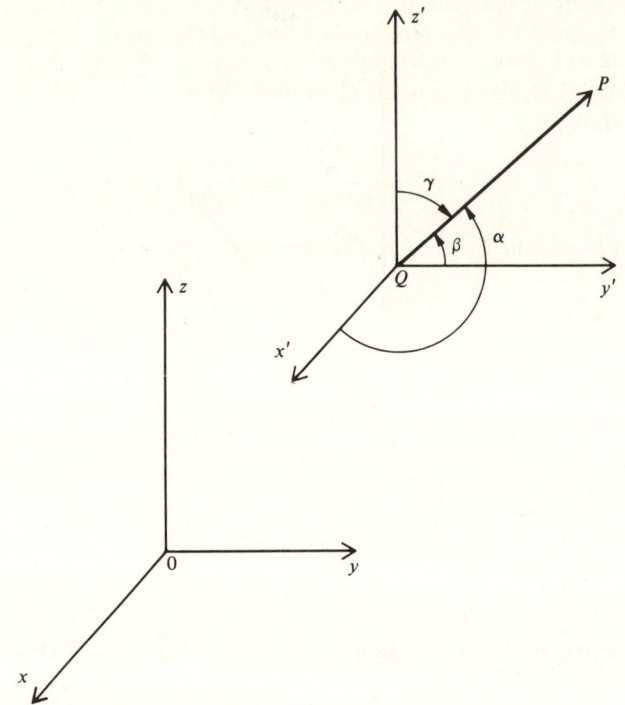

Figure 10.7 Direction Angles of Arbitrary Half-line

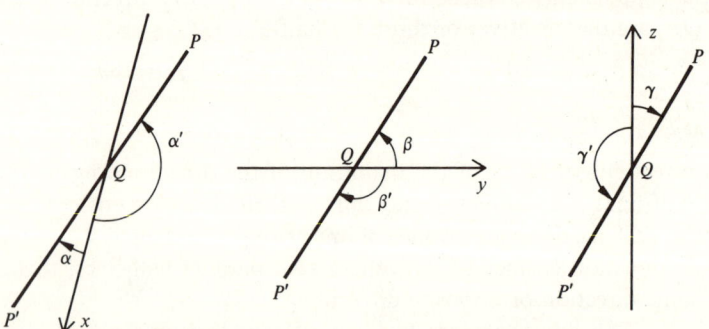

Figure 10.8 The Two Sets of Direction Angles for a Line

Theorem (Equations for the Direction Cosines) The direction cosines of the half-line from $P_1: (x_1, y_1, z_1)$ to $P_2: (x_2, y_2, z_2)$ are

$$\cos \alpha = \frac{x_2 - x_1}{d}, \quad \cos \beta = \frac{y_2 - y_1}{d}, \quad \cos \gamma = \frac{z_2 - z_1}{d},$$

where d is the distance between P_1 and P_2.

Proof

Referring to Figure 10.9, triangle $P_1 P_2 Q$ has a right angle at Q since $P_1 Q$ is perpendicular to the plane through Q parallel to the yz-plane and $P_2 Q$ is in this plane. Therefore,

$$\cos \alpha = \frac{\overrightarrow{P_1 Q}}{|P_1 P_2|} = \frac{x_2 - x_1}{d}.$$

The rest of the theorem is proved similarly.

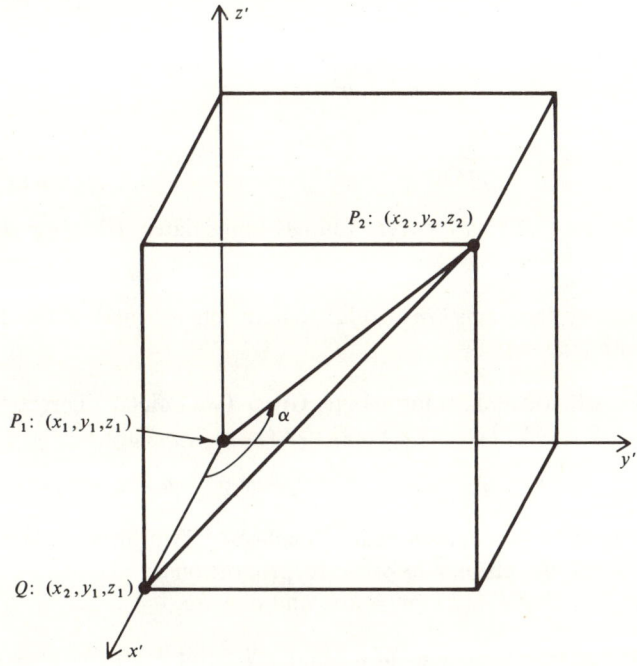

Figure 10.9

Theorem (Direction Cosine Dependence) If $\cos \alpha$, $\cos \beta$, and $\cos \gamma$ are the direction cosines of the half-line from $P_1: (x_1, y_1, z_1)$ through $P_2: (x_2, y_2, z_2)$, then

$$\cos^2 \alpha + \cos^2 \beta + \cos^2 \gamma = 1.$$

Proof $\cos^2 \alpha + \cos^2 \beta + \cos^2 \gamma = \left(\frac{x_2 - x_1}{d}\right)^2 + \left(\frac{y_2 - y_1}{d}\right)^2 + \left(\frac{z_2 - z_1}{d}\right)^2$

$$= \frac{d^2}{d^2} = 1$$

Definition of Direction Numbers l, m, and n are direction numbers of a line if and only if $\cos \alpha$, $\cos \beta$, and $\cos \gamma$ are direction cosines of any half-line of the given line and there exists a real number t so that

$$l = t \cos \alpha, \quad m = t \cos \beta, \quad \text{and} \quad n = t \cos \gamma.$$

Theorem (Direction Cosines from Direction Numbers) If l, m, and n are direction numbers of a line, the direction cosines of the line are

either
$$\cos \alpha = \frac{l}{\sqrt{l^2 + m^2 + n^2}}$$
$$\cos \beta = \frac{m}{\sqrt{l^2 + m^2 + n^2}}$$
$$\cos \gamma = \frac{n}{\sqrt{l^2 + m^2 + n^2}}$$

or

$$\cos \alpha = \frac{-l}{\sqrt{l^2 + m^2 + n^2}}$$
$$\cos \beta = \frac{-m}{\sqrt{l^2 + m^2 + n^2}}$$
$$\cos \gamma = \frac{-n}{\sqrt{l^2 + m^2 + n^2}}.$$

The proof of this theorem follows immediately from the definition of direction numbers.

The next two theorems are also immediate consequences of the definition of direction numbers.

Theorem (Direction Numbers for a Line, Given Two Points) Direction numbers for the line on $P_1: (x_1, y_1, z_1)$ and $P_2: (x_2, y_2, z_2)$ are

$$l = x_2 - x_1, \quad m = y_2 - y_1, \quad n = z_2 - z_1.$$

Theorem (Parallel Lines and Direction Numbers) Two lines are parallel if and only if their direction numbers are proportional; that is, there is a real number k so that $l' = kl$, $m' = km$, and $n' = kn$.

Theorem (Parametric Equations of a Line) A point $P: (x, y, z)$ is on the line through $P_1: (x_1, y_1, z_1)$ and $P_2: (x_2, y_2, z_2)$ if and only if

$$x = x_1 + (x_2 - x_1)t$$
$$y = y_1 + (y_2 - y_1)t$$
$$z = z_1 + (z_2 - z_1)t,$$

where t is any real number. These equations are called **parametric equations** of the line.

Proof A point $P: (x, y, z)$ is on the line through P_1 and P_2 if and only if there is a real number t so that

$$x - x_1 = (x_2 - x_1)t, \quad y - y_1 = (y_2 - y_1)t, \quad \text{and} \quad z - z_1 = (z_2 - z_1)t;$$

that is, their direction numbers are proportional.

LINES IN SPACE

Solving these three equations for x, y, and z, respectively, yields the parametric equations. Clearly, (x_1, y_1, z_1) is on the line for $t = 0$ and (x_2, y_2, z_2) is on the line for $t = 1$. Points on the line between P_1 and P_2 are obtained by selecting values for t between 0 and 1. Points in the direction of P_2 from P_1 correspond to values of $t > 0$ and points in the opposite direction to values of $t < 0$.

Corollary Equations for the line on P_1: (x_1, y_1, z_1) with direction numbers l, m, and n are
$$x = x_1 + lt, \qquad y = y_1 + mt, \qquad z = z_1 + nt.$$

Definition (Angle between Two Lines in Space) The angle θ between two lines in space is the angle of least nonnegative measure, $0° \leq \theta < 180°$, between two half-lines emanating from the origin and parallel to the given lines.

Theorem (Formula for Angle between Two Lines) If θ is the angle between two given lines and if (a_1, b_1, c_1) and (a_2, b_2, c_2) are the direction cosines of the two half-lines emanating from the origin parallel to the given lines, then
$$\cos \theta = a_1 a_2 + b_1 b_2 + c_1 c_2.$$

Proof (See Figure 10.10.)

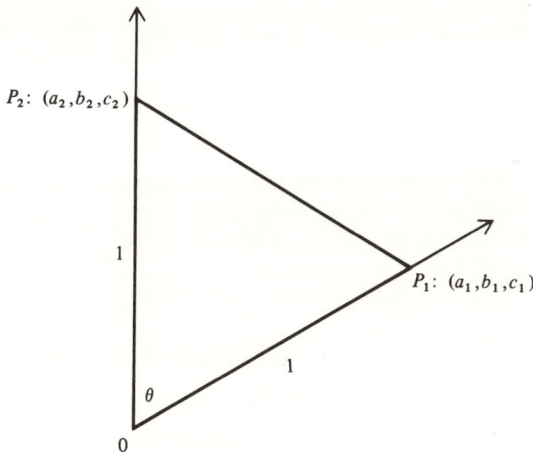

Figure 10.10

Let P_1: (a_1, b_1, c_1) and P_2: (a_2, b_2, c_2) be the two points, one on each of the half-lines, at a distance of one unit from the origin. Then, by the law of cosines,[1]

---

[1] The law of cosines: In triangle ABC, where side a is opposite angle A, and b and c are the other two sides,
$$a^2 = b^2 + c^2 - 2bc \cos A.$$

$$|P_1P_2|^2 = |OP_1|^2 + |OP_2|^2 - 2|OP_1||OP_2|\cos\theta$$
$$|P_1P_2|^2 = 1 + 1 - 2\cos\theta = 2(1 - \cos\theta)$$

and

$$\cos\theta = 1 - \tfrac{1}{2}|P_1P_2|^2$$
$$= 1 - \tfrac{1}{2}[(a_2 - a_1)^2 + (b_2 - b_1)^2 + (c_2 - c_1)^2]$$
$$= a_1a_2 + b_1b_2 + c_1c_2.$$

Theorem (Perpendicular Lines) If one line has direction numbers (l_1, m_1, n_1) and another line has direction numbers (l_2, m_2, n_2), then the two lines are perpendicular if and only if $l_1l_2 + m_1m_2 + n_1n_2 = 0$.

Proof Direction cosines for half-lines on the given lines may be taken as follows, where $d_1 = \sqrt{l_1^2 + m_1^2 + n_1^2}$ and $d_2 = \sqrt{l_2^2 + m_2^2 + n_2^2}$,

$$a_1 = \frac{l_1}{d_1}, \quad b_1 = \frac{m_1}{d_1}, \quad c_1 = \frac{n_1}{d_1},$$

$$a_2 = \frac{l_2}{d_2}, \quad b_2 = \frac{m_2}{d_2}, \quad c_2 = \frac{n_2}{d_2}.$$

Then, the lines are perpendicular if and only if $\cos\theta = 0$, where θ is the angle between the lines. Using the formula for the angle between two lines,

$$\cos\theta = \frac{l_1l_2}{d_1d_2} + \frac{m_1m_2}{d_1d_2} + \frac{n_1n_2}{d_1d_2} = 0$$

or

$$l_1l_2 + m_1m_2 + n_1n_2 = 0.$$

Example 1 Find direction numbers, direction cosines, and direction angles of the half-line from $P_1: (3\sqrt{2}, 2, 5)$ to $P_2: (\sqrt{2}, 4, 3)$.

Solution Direction numbers are $(3\sqrt{2} - \sqrt{2}, 2 - 4, 5 - 3)$ or $(2\sqrt{2}, -2, 2)$

$$d^2 = (2\sqrt{2})^2 + (-2)^2 + (2)^2 = 8 + 4 + 4 = 16$$
$$d = 4.$$

Direction cosines are $\left(\dfrac{\sqrt{2}}{2}, -\dfrac{1}{2}, \dfrac{1}{2}\right).$

LINES IN SPACE

$$\text{Direction angles are } \cos \alpha = \frac{\sqrt{2}}{2}, \quad \alpha = 45°$$

$$\cos \beta = -\frac{1}{2}, \quad \beta = 120°$$

$$\cos \gamma = \frac{1}{2}, \quad \gamma = 60°.$$

Example 2 Find parametric equations of the line through $P: (3, 2, -1)$ and $Q: (1, -4, 1)$.

Solution $x = 3 + (1 - 3)t \qquad x = 3 - 2t.$
$y = 2 + (-4 - 2)t \qquad y = 2 - 6t.$
$z = -1 + (1 + 1)t \qquad z = -1 + 2t$

Example 3 Find the angle between two lines having direction numbers $(2, -2, 1)$ and $(2, -6, -3)$.

Solution

(1) Direction cosines for $(2, -2, 1)$ are $(2/d, -2/d, 1/d)$ where $d^2 = 2^2 + (-2)^2 + 1^2 = 9$. Thus, direction cosines are $(2/3, -2/3, 1/3)$.

(2) Direction cosines for $(2, -6, -3)$ are $(2/d, -6/d, -3/d)$ where $d^2 = 2^2 + (-6)^2 + (-3)^2 = 49$. Thus, direction cosines are $(2/7, -6/7, -3/7)$.

(3) $\cos \theta = a_1 a_2 + b_1 b_2 + c_1 c_2$
$\cos \theta = (2/3)(2/7) + (-2/3)(-6/7) + (1/3)(-3/7)$
$\cos \theta = 13/21$ and $\theta = \text{Arccos } 13/21$.

Example 4 If the direction numbers of a line are $(5, 3, -2)$, then determine which of the following sets are direction numbers of a line parallel to the given line and which are direction numbers of a line perpendicular to the given line.

(a) $(-15, -9, 6)$
(b) $(3, -1, 6)$
(c) $(10, -6, 4)$

Solution

(a) Parallel since for $k = -3$, $-15 = -3(5)$, $-9 = -3(3)$ and $6 = -3(-2)$.
(b) Perpendicular since $5(3) + 3(-1) + (-2)(6) = 0$.
(c) Neither parallel nor perpendicular since $5(10) + 3(-6) + (-2)(4) = 24 \neq 0$ and if $5k = 10$, then $k = 2$, but $3(2) \neq -6$.

EXERCISES 10.2

1–6. Find the direction cosines and direction angles of each of the following half-lines drawn from the origin to the given point P.

1. $P: (-2, 1, 2)$
2. $P: (3, 2, 1)$
3. $P: (-\sqrt{2}, -1, 1)$
4. $P: (2, -2, 0)$
5. $P: (0, 0, -5)$
6. $P: (\sqrt{3}, 0, \sqrt{2})$

7–12. Find direction numbers of the lines through each of the following pairs of points.

7. $(-1, 2, 3), (1, 4, 2)$
8. $(-1, -3, -1), (2, -3, -5)$
9. $(2, -4, -7), (-4, -2, 2)$
10. $(2, 7, 8), (1, 3, 6)$
11. $(3, -1, 1), (1, 3, -1)$
12. $(3, 2, 6), (0, -4, 8)$

13–18. Find the direction cosines of each of the lines in Exercises 7–12.

19–24. Find parametric equations for each of the lines of Exercises 7–12.

25. The direction numbers of two lines are $(-3, 2, c)$ and $(6, c, 8)$, respectively. Find c so that the two lines are perpendicular.

26. In Exercise 25, find c so that the two lines are parallel.

27. If $\pi/3$ and $3\pi/4$ are the radian measures of two direction angles of a ray emanating from the origin, find the third direction angle.

28. If α, β, and $5\pi/6$ are the radian measures of the direction angles of a line and if $\alpha = \beta$, find the direction cosines of the line.

29. Show that the three points $(-4, -13, 4)$, $(5, -1, -2)$ and $(2, -5, 0)$ are collinear.

30. If the direction numbers for two lines are $(-3, 2, -1)$ and $(6, -4, 2)$, are the lines parallel or perpendicular? Justify.

31–38. Find the acute angle between each of the following lines.

31. Direction numbers for the two lines are $(2, -3, -4)$ and $(-\frac{1}{2}, -\frac{3}{4}, 1)$.

32. Direction numbers for the two lines are $(2, 1, -2)$ and $(3, -6, 2)$.

33. One line passes through the points $(3, 2, 1)$ and $(-2, 1, 3)$ and the other line passes through the points $(-3, 2, 3)$ and $(2, 1, 5)$.

34. Line AB for $A: (-2, -2, 8)$ and $B: (4, 1, 6)$ and line PQ for $P: (2, 3, 5)$ and $Q: (6, -2, 2)$.

35. Sides PQ and QR of the triangle with vertices $P: (2, 0, -4)$, $Q: (10, 2, -2)$, and $R: (4, 2, 4)$.

36. Sides PQ and PR of the triangle described in Exercise 35.

37. Sides PQ and PR of the triangle with vertices $P: (6, 5, 3)$, $Q: (4, 8, -3)$, and $R: (1, 2, -5)$.

THREE-DIMENSIONAL VECTORS

38. Sides PQ and QR of the triangle described in Exercise 37.

39–42. Find parametric equations for a line parallel to each of the following lines and passing through the point $(2, 1, -3)$.

39. $x = 5 - 2t,\ y = -2 + 3t,\ z = 6 + t$
40. $x = 1 - t,\ y = 5 - 2t,\ z = -4 + t$
41. $x = t,\ y = 2,\ z = 1 - t$
42. $x = 3,\ y = -t,\ z = 2 + t$

43–46. Find parametric equations for a line perpendicular to each of the lines described in Exercises 39–42 if its direction numbers are $(1, 2, k)$ and it passes through $(1, -4, 2)$.

10.3 THREE-DIMENSIONAL VECTORS

Definition of Three-Dimensional Vector A three-dimensional vector, $\mathbf{v}$, over the set of real numbers R is an ordered triple of real numbers. In symbols,

$$\mathbf{v} = (v_x, v_y, v_z),$$

where the real numbers v_x, v_y, and v_z are called the components of the vector.

Definition of Scalar A scalar is any real number.

Definition of Equal Vectors Two vectors are equal if and only if their corresponding components are equal. In symbols, if

$$\mathbf{a} = (a_x, a_y, a_z) \quad \text{and} \quad \mathbf{b} = (b_x, b_y, b_z),$$

then

$\mathbf{a} = \mathbf{b}$ if and only if $a_x = b_x$, $\quad a_y = b_y$, $\quad$ and $\quad a_z = b_z$.

Definition of Vector Addition If $\mathbf{a} = (a_x, a_y, a_z)$ and $\mathbf{b} = (b_x, b_y, b_z)$, then

$$\mathbf{a} + \mathbf{b} = (a_x + b_x,\ a_y + b_y,\ a_z + b_z).$$

With the definitions above, it may be shown that the closure, commutative, associative, identity, and inverse properties are valid for vector addition, with the zero vector $\mathbf{0} = (0, 0, 0)$ and the addition inverse vector $-\mathbf{a} = (-a_x, -a_y, -a_z)$. The proofs of these statements are left for the student.

Definition of Magnitude The magnitude, $|\mathbf{v}|$, of a vector $\mathbf{v} = (v_x, v_y, v_z)$ is defined as follows:

$$|\mathbf{v}| = \sqrt{v_x^2 + v_y^2 + v_z^2}$$

Definition of Direction The direction cosines of the nonzero vector

$$\mathbf{v} = (v_x, v_y, v_z)$$

are defined as follows:

$$\cos \alpha = \frac{v_x}{|\mathbf{v}|}, \quad \cos \beta = \frac{v_y}{|\mathbf{v}|}, \quad \cos \gamma = \frac{v_z}{|\mathbf{v}|}.$$

The zero vector is not assigned a direction.

Definition of Unit Vector A unit vector is a vector whose magnitude is 1; that is, $\mathbf{v}$ is a unit vector if and only if $v_x^2 + v_y^2 + v_z^2 = 1$.

Geometrically, a three-dimensional vector, $\mathbf{v} = (v_x, v_y, v_z)$, may be interpreted as a directed line segment, $\overrightarrow{PQ}$, in a three-dimensional coordinate space, where any point $P: (x, y, z)$ may be chosen as the initial point and the terminal point is then $Q: (x + v_x, y + v_y, z + v_z)$. The length of the directed line segment is identified as the magnitude of the vector and the direction of the line segment is the direction of the vector. The sum of two vectors is interpreted as either the directed third side of a triangle or as the directed diagonal of a parallelogram, similar to the two-dimensional case.

Definition of Multiplication by a Scalar If k is a scalar and $\mathbf{v} = (v_x, v_y, v_z)$ is a vector, then

$$k\mathbf{v} = (kv_x, kv_y, kv_z).$$

As in the two-dimensional case, multiplication by a positive scalar produces a vector whose magnitude is k times that of the original one but with the same direction. If the scalar is negative, the direction is reversed.

Definition of Dot Product (Inner Product) If $\mathbf{a} = (a_x, a_y, a_z)$ and $\mathbf{b} = (b_x, b_y, b_z)$, then their dot product $\mathbf{a} \cdot \mathbf{b}$ is defined as follows:

$$\mathbf{a} \cdot \mathbf{b} = a_x b_x + a_y b_y + a_z b_z.$$

Theorem (Angle between Two Vectors) If θ is the angle between two nonzero vectors $\mathbf{a}$ and $\mathbf{b}$ where θ is defined as the angle between the half-lines of two corresponding line segments, then

$$\mathbf{a} \cdot \mathbf{b} = |\mathbf{a}| \cdot |\mathbf{b}| \cos \theta.$$

Proof
$$\begin{aligned}
\mathbf{a} \cdot \mathbf{b} &= a_x b_x + a_y b_y + a_z b_z \\
&= |\mathbf{a}| \cdot |\mathbf{b}| \left(\frac{a_x b_x}{|\mathbf{a}||\mathbf{b}|} + \frac{a_y b_y}{|\mathbf{a}||\mathbf{b}|} + \frac{a_z b_z}{|\mathbf{a}||\mathbf{b}|} \right) \\
&= |\mathbf{a}| \cdot |\mathbf{b}| \cos \theta
\end{aligned}$$

using the formula for the angle between two lines.

THREE-DIMENSIONAL VECTORS

As in the two-dimensional case, the dot product is interpreted geometrically as the projection of one directed line segment on a second multiplied by the length of the second.

Definition (Basic Unit Vectors) The three basic unit vectors, **i**, **j**, and **k**, are the vectors with their initial points at the origin and with their terminal points having Cartesian coordinates $(1, 0, 0)$, $(0, 1, 0)$, and $(0, 0, 1)$, respectively. In symbols,

$$\mathbf{i} = (1, 0\ 0), \qquad \mathbf{j} = (0, 1, 0), \qquad \mathbf{k} = (0, 0, 1).$$

Theorem Every vector can be expressed as a sum of scalar multiples of the basic unit vectors. In symbols,

$$(v_x, v_y, v_z) = v_x \mathbf{i} + v_y \mathbf{j} + v_z \mathbf{k}.$$

Proof
$$v_x \mathbf{i} + v_y \mathbf{j} + v_z \mathbf{k} = (v_x, 0, 0) + (0, v_y, 0) + (0, 0, v_z)$$
$$= (v_x, v_y, v_z).$$

Definition of Position Vector The position vector, **R**, from the origin to the point $P: (x, y, z)$ is as follows:

$$\mathbf{R} = \overrightarrow{OP} = x\mathbf{i} + y\mathbf{j} + z\mathbf{k} = (x, y, z)$$

Another useful product because of its many physical and mathematical applications is the cross product, also called the outer product.

Definition of Cross Product (Outer Product) If $\mathbf{a} = (a_x, a_y, a_z)$ and $\mathbf{b} = (b_x, b_y, b_z)$, then their cross product $\mathbf{a} \times \mathbf{b}$ is defined as follows:

$$\mathbf{a} \times \mathbf{b} = (a_y b_z - a_z b_y, a_z b_x - a_x b_z, a_x b_y - a_y b_x).$$

The components of the cross product vector can be memorized more easily by considering the following array:

$$\begin{array}{ccc} v_x & v_y & v_z \\ a_x & a_y & a_z \\ b_x & b_y & b_z \end{array}$$

with

$$v_x = \begin{vmatrix} a_y & a_z \\ b_y & b_z \end{vmatrix} = a_y b_z - a_z b_y$$

$$v_y = -\begin{vmatrix} a_x & a_z \\ b_x & b_z \end{vmatrix} = -a_x b_z + a_z b_x$$

$$v_z = \begin{vmatrix} a_x & a_y \\ b_x & b_y \end{vmatrix} = a_x b_y - a_y b_x.$$

Theorem (Algebra of the Cross Product) If **a**, **b**, and **c** are vectors, then

(1) $\mathbf{a} \times \mathbf{b} = -(\mathbf{b} \times \mathbf{a})$ Anticommutative
(2) $(\mathbf{a} \times \mathbf{b}) \times \mathbf{c} = \mathbf{a} \times (\mathbf{b} \times \mathbf{c})$ Associative
(3) $\mathbf{a} \times (\mathbf{b} + \mathbf{c}) = (\mathbf{a} \times \mathbf{b}) + (\mathbf{a} \times \mathbf{c})$ Distributive
(4) $(\mathbf{a} \times \mathbf{b}) \times \mathbf{c} = (\mathbf{a} \cdot \mathbf{c})\mathbf{b} - (\mathbf{a} \cdot \mathbf{b})\mathbf{c}$
(5) $\mathbf{a} \cdot (\mathbf{b} \times \mathbf{c}) = (\mathbf{a} \times \mathbf{b}) \cdot \mathbf{c}$
(6) $\mathbf{a} \times \mathbf{a} = 0$.

The theorems above are left as exercises for the student.

Theorem (Magnitude of Cross Product) The magnitude of the cross product of **a** and **b** is the product of the magnitudes of **a** and **b** and the sine of the angle θ between **a** and **b**. In symbols,

$$|\mathbf{a} \times \mathbf{b}| = |\mathbf{a}||\mathbf{b}| \sin \theta.$$

Proof

$$\begin{aligned}
|\mathbf{a} \times \mathbf{b}|^2 &= (a_y b_z - a_z b_y)^2 + (a_z b_x - a_x b_z)^2 + (a_x b_y - a_y b_x)^2 \\
&= (a_x^2 + a_y^2 + a_z^2)(b_x^2 + b_y^2 + b_z^2) - (a_x b_x + a_y b_y + a_z b_z)^2 \\
&= |\mathbf{a}|^2 \cdot |\mathbf{b}|^2 - (\mathbf{a} \cdot \mathbf{b})^2 \\
&= |\mathbf{a}|^2 |\mathbf{b}|^2 - |\mathbf{a}|^2 |\mathbf{b}|^2 \cos^2 \theta \\
&= |\mathbf{a}|^2 |\mathbf{b}|^2 \sin^2 \theta
\end{aligned}$$

Since $\sin \theta \geq 0$ for $0 \leq \theta < 180°$,

$$|\mathbf{a} \times \mathbf{b}| = |\mathbf{a}||\mathbf{b}| \sin \theta.$$

Geometrically, the magnitude of the cross product $\mathbf{a} \times \mathbf{b}$ may be interpreted as the area of a parallelogram with **a** and **b** as adjacent sides. See Figure 10.11.

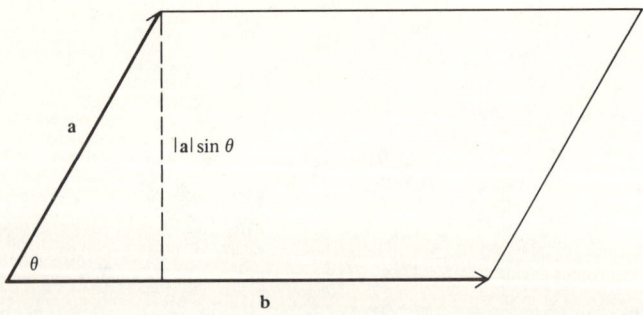

Figure 10.11 Area of Parallelogram $= |\mathbf{a} \times \mathbf{b}| = |\mathbf{a}||\mathbf{b}| \sin \theta$.

THREE-DIMENSIONAL VECTORS

Theorem (Direction of Cross Product) The direction of $\mathbf{a} \times \mathbf{b}$ is perpendicular to the plane of $\mathbf{a}$ and $\mathbf{b}$ and oriented so that the rotation from $\mathbf{a}$ to $\mathbf{b}$ to $\mathbf{a} \times \mathbf{b}$ is counterclockwise.

This theorem is proved by first showing that $\mathbf{a} \times \mathbf{b}$ is perpendicular to $\mathbf{a}$ and to $\mathbf{b}$ by using the dot product to show that the cosine of each of the angles involved is 0. The orientation is verified by selecting $\mathbf{a}$ and $\mathbf{b}$ in an xy-coordinate plane with $\mathbf{a}$ in the positive direction of the x-axis, and then showing that $\mathbf{a} \times \mathbf{b}$ is in the positive z-direction. See Figure 10.12.

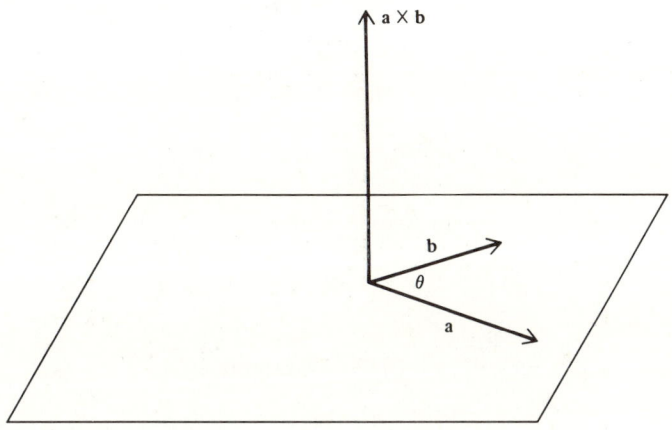

Figure 10.12 The Cross Product Vector, $\mathbf{a} \times \mathbf{b}$

Example 1 Find the magnitude and direction of the vector $(2, -3, 6)$.

Solution $|\mathbf{v}| = \sqrt{4 + 9 + 36} = 7$

$$\cos \alpha = \tfrac{2}{7}, \qquad \cos \beta = -\tfrac{3}{7}, \qquad \cos \gamma = \tfrac{6}{7}$$

Example 2 Let $\mathbf{a} = (2, -3, 6)$ and $\mathbf{b} = (5, 3, 8)$. Find each of the following:

(a) $\mathbf{a} + \mathbf{b}$
(b) $\mathbf{a} - \mathbf{b}$
(c) $3\mathbf{a}$
(d) $-2\mathbf{a}$
(e) $\mathbf{a} \cdot \mathbf{b}$
(f) $\mathbf{a} \times \mathbf{b}$

Solution

(a) $(2 + 5, -3 + 3, 6 + 8) = (7, 0, 14)$
(b) $(2 - 5, -3 - 3, 6 - 8) = (-3, -6, -2)$

(c) $(6, -9, 18)$
(d) $(-4, 6, -12)$
(e) $2(5) + (-3)(3) + 6(8) = 49$
(f) $\begin{vmatrix} \mathbf{i} & \mathbf{j} & \mathbf{k} \\ 2 & -3 & 6 \\ 5 & 3 & 8 \end{vmatrix}$

$(-3(8) - 3(6), -[2(8) - 5(6)], 2(3) - 5(-3)) = (-42, 14, 21)$

Example 3 Find the angle between $2\mathbf{i} - 3\mathbf{j} + 6\mathbf{k}$ and $5\mathbf{i} + 3\mathbf{j} + 8\mathbf{k}$.

Solution $\cos \theta = \dfrac{\mathbf{a} \cdot \mathbf{b}}{|\mathbf{a}||\mathbf{b}|} = \dfrac{10 - 9 + 48}{7(7\sqrt{2})} = \dfrac{1}{\sqrt{2}}$

$\theta = \pi/4$ radians $= 45°$

Example 4 Find a unit vector perpendicular to

$$\mathbf{a} = (2, -3, 6) \quad \text{and} \quad \mathbf{b} = (5, 3, 8).$$

Solution $\mathbf{a} \times \mathbf{b} = (-42, 14, 21)$ [see Example 2(f)]

$|\mathbf{a} \times \mathbf{b}| = 49$

$$\text{Unit vector} = \left(\dfrac{-42}{49}, \dfrac{14}{49}, \dfrac{21}{49} \right)$$

$$= \left(-\dfrac{6}{7}, \dfrac{2}{7}, \dfrac{3}{7} \right)$$

Example 5 Find the projection of $\mathbf{a} = (2, -3, 6)$ on $\mathbf{b} = (5, 3, 8)$.

Solution Projection of $\mathbf{a}$ on $\mathbf{b} = \dfrac{\mathbf{a} \cdot \mathbf{b}}{|\mathbf{b}|} = \dfrac{49}{7\sqrt{2}} = \dfrac{7\sqrt{2}}{2}$

Example 6 Find the area of the parallelogram where $\overrightarrow{OP} = (2, -3, 6)$ and $\overrightarrow{OQ} = (5, 3, 8)$ are two adjacent sides.

Solution Area $= |\overrightarrow{OP} \times \overrightarrow{OQ}| = 49$ (see Example 4)

THREE-DIMENSIONAL VECTORS

EXERCISES 10.3

1–10. Find the magnitude and direction of each of the following vectors.

1. $(10, -2, 11)$
2. $(-9, 6, -2)$
3. $(1, -4, -8)$
4. $(-4, 12, 3)$
5. $(3, 6, -2)$
6. $3\mathbf{i} + 2\mathbf{j} + \mathbf{k}$
7. $-2\mathbf{i} - \mathbf{j} + 2\mathbf{k}$
8. $\mathbf{i} + \mathbf{j} + \mathbf{k}$
9. $4\mathbf{i} + 3\mathbf{j} - 2\mathbf{k}$
10. $x\mathbf{i} + y\mathbf{j} + z\mathbf{k}$

11–14. For each of the following pairs of vectors, find $3\mathbf{r} - 2\mathbf{s}$, $\mathbf{r} \cdot \mathbf{s}$, $\mathbf{r} \times \mathbf{s}$ and illustrate each geometrically.

11. $\mathbf{r} = 2\mathbf{i} + \mathbf{j} + \mathbf{k}$, $\mathbf{s} = \mathbf{i} - \mathbf{j} - 2\mathbf{k}$
12. $\mathbf{r} = -\mathbf{i} + 2\mathbf{j} - 2\mathbf{k}$, $s = 6\mathbf{i} - 3\mathbf{j} + 3\mathbf{k}$
13. $\mathbf{r} = 2\mathbf{i} + \mathbf{j} - \mathbf{k}$, $\mathbf{s} = -3\mathbf{i} - 2\mathbf{j} - \mathbf{k}$
14. $\mathbf{r} = \mathbf{i} + 2\mathbf{j} + 3\mathbf{k}$, $\mathbf{s} = -3\mathbf{i} - 2\mathbf{j} - \mathbf{k}$

15–18. Find the angle between each of the following pairs of vectors.

15. $2\mathbf{i} + 3\mathbf{j} - 6\mathbf{k}$ and $3\mathbf{i} + 2\mathbf{j} + 2\mathbf{k}$
16. $2\mathbf{i} - \mathbf{j} - \mathbf{k}$ and $-2\mathbf{i} + \mathbf{j} + \mathbf{k}$
17. $(4, 1, 1)$ and $(1, 1, 4)$
18. $(2, -1, 1)$ and $(2, -2, 0)$

19–22. Find a unit vector perpendicular to each of the following pairs of vectors.

19. $(1, 2, -1)$ and $(3, -1, 1)$
20. $(2, 3, 4)$ and $(3, 4, 6)$
21. $(-1, 2, 3)$ and $(2, -4, -5)$
22. $(3, -3, 2)$ and $(2, 2, -3)$

23–26. Find the projection of the first on the second for each of the following pairs of vectors.

23. $2\mathbf{i} - \mathbf{j} + 2\mathbf{k}$ and $4\mathbf{i} + 8\mathbf{j} - \mathbf{k}$
24. $2\mathbf{i} + 3\mathbf{j} + 2\mathbf{k}$ and $10\mathbf{i} + 2\mathbf{j} - 11\mathbf{k}$
25. $\mathbf{i} + 3\mathbf{j} - 2\mathbf{k}$ and $9\mathbf{i} - 2\mathbf{j} - 6\mathbf{k}$
26. $2\mathbf{i} - 3\mathbf{j} + 4\mathbf{k}$ and $3\mathbf{i} + 4\mathbf{j} - 12\mathbf{k}$

27–28. Find the area of the parallelogram having as adjacent sides the given pair of vectors.

27. $(-1, 2, 3)$ and $(2, -4, -5)$
28. $(3, -3, 2)$ and $(2, 2, -3)$

29–30. Find the area of the triangle having as adjacent sides the given pairs of vectors in Exercises 27 and 28.

31. A vector with magnitude twelve units makes equal angles with the three coordinate axes. Express the vector in terms of the basic unit vectors $\mathbf{i}$, $\mathbf{j}$, and $\mathbf{k}$.

32. Show that $\dfrac{|\mathbf{a}|\mathbf{b} + |\mathbf{b}|\mathbf{a}}{|\mathbf{a}| + |\mathbf{b}|}$ bisects the angle between $\mathbf{a}$ and $\mathbf{b}$.

33. Show that if neither **a** nor **b** is the zero vector, then $\mathbf{a} \times \mathbf{b} = 0$ if and only if **a** and **b** are parallel.
34. Show that $|\mathbf{a}|\mathbf{b} + |\mathbf{b}|\mathbf{a}$ and $|\mathbf{b}|\mathbf{a} - |\mathbf{a}|\mathbf{b}$ are perpendicular.

35–41. Let $\mathbf{a} = \mathbf{i} + 2\mathbf{j} + \mathbf{k}$, $\mathbf{b} = \mathbf{i} + 2\mathbf{j} - \mathbf{k}$, and $\mathbf{c} = 2\mathbf{i} + 3\mathbf{j} + 2\mathbf{k}$. Find each of the following.

35. $\mathbf{a} \times \mathbf{b}$
36. $\mathbf{b} \times \mathbf{a}$
37. $\mathbf{a} \cdot (\mathbf{b} \times \mathbf{c})$
38. $\mathbf{a} \times (\mathbf{b} \times \mathbf{c})$
39. $\mathbf{a} \times (\mathbf{b} + \mathbf{c})$
40. $(\mathbf{a} + \mathbf{b}) \times \mathbf{c}$
41. $(\mathbf{a} \times \mathbf{c}) \cdot \mathbf{b}$

42. Prove that the area of a parallelogram whose sides are **v** and **u** is $|\mathbf{v} \times \mathbf{u}|$, the magnitude of the cross product of **v** and **u**.

43–48. If **a**, **b**, and **c** are vectors, prove each of the following properties.

43. $\mathbf{a} \times \mathbf{b} = -(\mathbf{b} \times \mathbf{a})$ (anticommutative)
44. $(\mathbf{a} \times \mathbf{b}) \times \mathbf{c} = \mathbf{a} \times (\mathbf{b} \times \mathbf{c})$ (associative)
45. $\mathbf{a} \times (\mathbf{b} + \mathbf{c}) = (\mathbf{a} \times \mathbf{b}) + (\mathbf{a} \times \mathbf{c})$ (distributive)
46. $(\mathbf{a} \times \mathbf{b}) \times \mathbf{c} = (\mathbf{a} \cdot \mathbf{c})\mathbf{b} - (\mathbf{a} \cdot \mathbf{b})\mathbf{c}$
47. $\mathbf{a} \cdot (\mathbf{b} \times \mathbf{c}) = (\mathbf{a} \times \mathbf{b}) \cdot \mathbf{c}$
48. $\mathbf{a} \times \mathbf{a} = 0$

10.4 PLANES IN SPACE

Definition (Graph of an Equation in Three-Space) The graph in three-space of an equation involving any of the three variables x, y, and z is the set of points in space whose coordinates satisfy the equation; that is, $P: (a, b, c)$ is on the graph if and only if the open equation becomes a true statement when x is replaced by a, y by b, and z by c.

Definition (Normal to a Plane) A normal to a plane is any line perpendicular to the plane.

Theorem (Equation of Plane) If, A, B, C, and D are real numbers such that A, B, and C are not all zero, then the graph of $Ax + By + Cz + D = 0$ is a plane. Conversely, the points of every plane in three-space satisfy an equation of this form.

Proof (See Figure 10.13.)

Let $P: (x_1, y_1, z_1)$ be a point on a given plane and let (l, m, n) be direction numbers of the normal to the plane at P. Then a point $Q: (x, y, z)$ is on the plane

PLANES IN SPACE

if and only if line PQ is perpendicular to the normal, or equivalently, the cosine of the angle θ between them is 0.

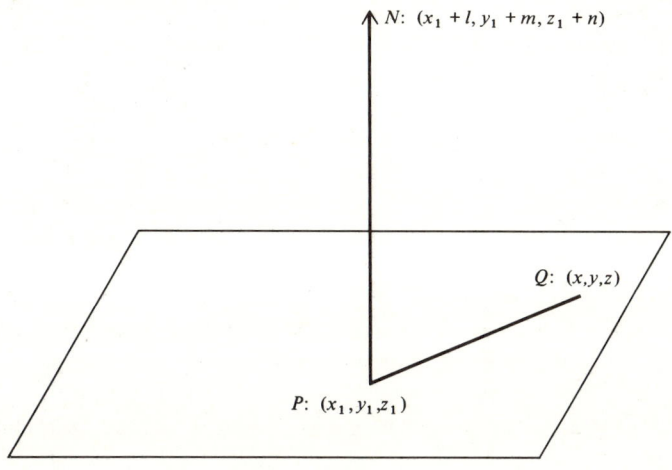

Figure 10.13

Direction numbers for PQ are $(x - x_1, y - y_1, z - z_1)$. Then $\cos \theta = 0$ if and only if

$$l(x - x_1) + m(y - y_1) + n(z - z_1) = 0$$

or

$$Ax + By + Cz + D = 0,$$

where $A = l$, $B = m$, $C = n$, and $D = -lx_1 - my_1 - nz_1$.

Conversely, given the equation $Ax + By + Cz + D = 0$, where at least one of A, B, and C is not 0, suppose $C \neq 0$. Then $A(x - 0) + B(y - 0) + C\left(z - \dfrac{-D}{C}\right) = 0$. This, however, is a plane on the point $\left(0, 0, -\dfrac{D}{C}\right)$ with a normal line having direction numbers (A, B, C).

Example 1 Find the equation of the plane on the point $(3, -1, 5)$ if a normal to the plane has direction numbers $(2, 7, -4)$.

Solution $2(x - 3) + 7(y + 1) - 4(z - 5) = 0$

$2x + 7y - 4z + 21 = 0$

Example 2 Find the direction cosines of a normal to the plane

$$3x + 2y + 6z - 12 = 0.$$

Solution Direction numbers are $(3, 2, 6)$.

$$\sqrt{3^2 + 2^2 + 6^2} = \sqrt{9 + 4 + 36} = \sqrt{49} = 7$$

Therefore, the direction cosines are $(\frac{3}{7}, \frac{2}{7}, \frac{6}{7})$ or $(-\frac{3}{7}, -\frac{2}{7}, -\frac{6}{7})$.

Definition of Intercepts If a graph intersects the x-axis at a point whose coordinates are $(a, 0, 0)$, then a is called an **x-intercept** of the graph.

If a graph intersects the y-axis at a point whose coordinates are $(0, b, 0)$, then b is called a **y-intercept** of the graph.

If a graph intersects the z-axis at a point whose coordinates are $(0, 0, c)$, then c is called a **z-intercept** of the graph.

Example 3 Find the intercepts of $3x + 2y + 6z = 12$ and graph the plane in Octant I. (See Figure 10.14.)

Solution

(1) x-intercept. $3x + 2(0) + 6(0) = 12$, $3x = 12$, $x = 4$. The x-intercept is 4.
(2) y-intercept. $3(0) + 2y + 6(0) = 12$, $2y = 12$, $y = 6$. The y-intercept is 6.
(3) z-intercept. $3(0) + 2(0) + 6z = 12$, $6z = 12$, $z = 2$. The z-intercept is 2.

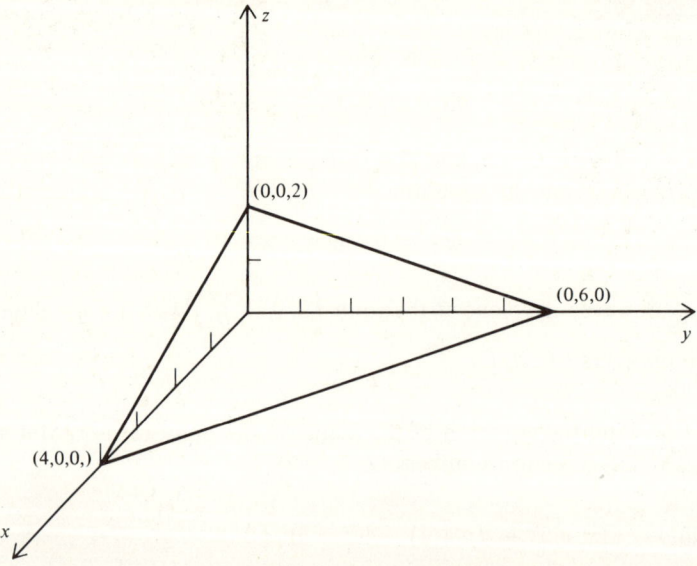

Figure 10.14 Graph of $3x + 2y + 6z = 12$ in Octant I

PLANES IN SPACE

Theorem (Equations of Coordinate Planes)
   The xy-plane has equation $z = 0$.
   The xz-plane has equation $y = 0$.
   The yz-plane has equation $x = 0$.

(Proof left for student.)

Theorem (Equations of Planes Parallel to the Coordinate Planes)
   A plane parallel to the xy-plane has equation $z = k$.
   A plane parallel to the xz-plane has equation $y = k$.
   A plane parallel to the yz-plane has equation $x = k$.

(Proof left for student.)

Example 4 Find the equation of a plane passing through the point $(3, -4, 5)$ and

(a) parallel to the yz-plane,
(b) parallel to the x-axis and the z-axis,
(c) perpendicular to the z-axis.

   Solution

(a) $x = 3$
(b) $y = -4$
(c) $z = 5$

Theorem (Parallel Planes) Two planes, $A_1 x + B_1 y + C_1 z + D_1 = 0$ and $A_2 x + B_2 y + C_2 z + D_2 = 0$ are parallel if and only if there is a real number k so that
$$A_1 = kA_2, \quad B_1 = kB_2, \quad \text{and} \quad C_1 = kC_2.$$

Proof Two planes are parallel if and only if their normals are parallel, and two lines are parallel if and only if their direction numbers are proportional.

Theorem (Perpendicular Planes) Two planes, $A_1 x + B_1 y + C_1 z + D_1 = 0$ and $A_2 x + B_2 y + C_2 z + D_2 = 0$, are perpendicular if and only if
$$A_1 A_2 + B_1 B_2 + C_1 C_2 = 0.$$

Proof Two planes are perpendicular if and only if their normals are perpendicular from which follows the relation for the direction numbers of the normals.

It should be observed that through a given point there is one and only one plane parallel to a given plane, but there are infinitely many planes perpendicular to the given plane.

Example 5 Find the equation of a plane through $(2, -1, 1)$ and

(a) parallel to $4x + 2y - z = 6$
(b) perpendicular to $4x + 2y - z = 6$.

Solution

(a) Direction numbers of the normal are $(4, 2, -1)$.
$$4(x - 2) + 2(y + 1) - (z - 1) = 0$$
$$4x + 2y - z - 5 = 0$$

(b) Let l, m, n be direction numbers of a perpendicular plane. Then
$$4l + 2m - n = 0 \text{ or } n = 4l + 2m, \text{ or } (l, m, n) = (l, m, 4l + 2m).$$
$$l(x - 2) + m(y + 1) + (4l + 2m)(z - 1) = 0.$$

$lx + my + (4l + 2m)z - 6l - m = 0$, where l and m may be selected as any real numbers.

Thus, there are many planes passing through a given point and perpendicular to a given plane. In particular,

$$\begin{array}{ll} x + 4z - 6 = 0 & \text{for } l = 1, \quad m = 0 \\ y + 2z - 1 = 0 & \text{for } l = 0, \quad m = 1 \\ 2x + 3y + 14z - 15 = 0 & \text{for } l = 2, \quad m = 3. \end{array}$$

Since two intersecting planes determine a unique line, a line may be represented as the intersection of two nonparallel planes.

Theorem (Line as Intersection of Two Planes)
The intersection of the planes
$$A_1 x + B_1 y + C_1 z + D_1 = 0$$
$$A_2 x + B_2 y + C_2 z + D_2 = 0$$
is a line if and only if the planes are not parallel; that is, for all real numbers k, $(A_2, B_2, C_2) \neq (kA_1, kB_1, kC_1)$.

Example 6 Find parametric equations of the line that is the intersection of the planes
$$3x - 2y + z - 6 = 0 \quad \text{and} \quad x + 4y - 4z + 8 = 0.$$

PLANES IN SPACE

Solution

(1) First eliminate z.
If (x, y, z) is on both planes, then (x, y, z) is also on
$$4(3x - 2y + z - 6) + (x + 4y - 4z + 8) = 0$$
$$13x - 4y - 16 = 0.$$

(2) Eliminate y.
$$2(3x - 2y + z - 6) + (x + 4y - 4z + 8) = 0$$
$$7x - 2z - 4 = 0$$

(3) Solve each of the two resulting equations for x.
$$x = \frac{4y + 16}{13} = \frac{4(y + 4)}{13}$$
$$x = \frac{2z + 4}{7} = \frac{4(z + 2)}{14}$$

(4)
$$\frac{x}{4} = \frac{y + 4}{13} = \frac{z + 2}{14} = t$$
$$x = 4t, \quad y = -4 + 13t, \quad z = -2 + 14t$$

are parametric equations for the line.

EXERCISES 10.4

1–4. Find the equation of the plane on each of the following points, given the direction numbers of a normal to the plane.

1. $P: (3, 1, 3)$, direction numbers of normal: $(1, 1, -1)$
2. $P: (4, 2, 0)$, direction numbers of normal: $(2, -1, 1)$
3. $P: (2, 1, -1)$, direction numbers of normal: $(3, -4, -1)$
4. $P: (-1, 2, -3)$, direction numbers of normal: $(1, -2, 3)$

5–8. Find the direction cosines of a normal to each of the following planes.

5. $3x - 2y + 6z - 5 = 0$
6. $2x + 2y - z + 2 = 0$
7. $x - y + z - 1 = 0$
8. $2x - 3y - 4z + 5 = 0$

9–13. Find the intercepts of each of the following planes and graph the plane in the first octant.

9. $2x + 3y + z - 6 = 0$
10. $x + 2y + 2z - 4 = 0$
11. $3x - y = 0$
12. $4x + z - 2 = 0$
13. $2x - 3y + 4z - 12 = 0$

14–20. Find the equation of each of the following planes satisfying the stated conditions.

14. Through the point (1, 2, 3) and
    (a) parallel to the yz-plane
    (b) parallel to the xz-plane
    (c) parallel to the xy-plane.

15. Through the point $(-3, -2, -1)$ and perpendicular to
    (a) the x-axis
    (b) the y-axis
    (c) the z-axis.

16. Through the point $(-1, 1, -1)$ and perpendicular to the plane
$$3x + y - z - 4 = 0$$
and
    (a) parallel to the x-axis
    (b) parallel to the y-axis
    (c) parallel to the z-axis
    (d) parallel to the plane $x + y + 2z = 6$.

17. Through the point $(3, -1, 2)$ and perpendicular to the planes
$$2x + 2y - z = 6 \quad \text{and} \quad 2x - y + 2z = 4.$$

18. Through the point (5, 2, 3) and
    (a) perpendicular to the line $x = 2 + 3t, y = 1 - 4t, z = 3 + t$
    (b) parallel to the lines $x = 2 + 3t, \quad y = 1 - 4t, \quad z = 3 + t$
    $\qquad\qquad\qquad\qquad x = 7 + t, \quad y = 6 - t, \quad z = 5 + t.$

19. Through the point $(-3, -2, -1)$ and parallel to the plane
$$3x + y - z - 4 = 0.$$

20. Through the point (2, 3, 1) and parallel to the plane $x - 2y + 3z = 7$.

21–26. Find parametric equations of the line that is the intersection of each of the following pairs of planes.

21. $x + 2y - z - 7 = 0$ and $x - 2y + z - 3 = 0$
22. $2x + 3y - 5z + 3 = 0$ and $3x + 2y + 5z - 8 = 0$
23. $3x + y - z + 1 = 0$ and $x + y + z - 3 = 0$
24. $2x + 3y - 5z = 0$ and $3x - 4y + z = 0$
25. $5x - y - z = 0$ and $4x + y - 2z = 0$
26. $x + y + z = 9$ and $7x - 2y - 2z = 0$

27. A normal to the xy-plane passes through the point (2, 3, 0). Write parametric equations of this line.

28. Find the equation of the plane containing the point P: (3, −6, 2) and perpendicular to the position vector $\overrightarrow{OP}$.

29–34. Prove each of the following theorems.

29. The xy-plane has equation $z = 0$.
30. The xz-plane has equation $y = 0$.
31. The yz-plane has equation $x = 0$.
32. A plane parallel to the xy-plane has equation $z = k$.
33. A plane parallel to the xz-plane has equation $y = k$.
34. A plane parallel to the yz-plane has equation $x = k$.
35. Derive the **normal equation of the plane**

$$(\cos \alpha)x + (\cos \beta)y + (\cos \gamma)z = d$$

from

$$Ax + By + Cz + D = 0, \quad \text{where } (\cos \alpha, \cos \beta, \cos \gamma)$$

are chosen so that $d \geq 0$, d is the distance from the origin to the plane, and $(\cos \alpha, \cos \beta, \cos \gamma)$ are direction cosines of the normal.

36. Find the distance from the origin to the plane

$$3x + 2y + 6z + 12 = 0.$$

37. Show that the distance from a point $P: (x_1, y_1, z_1)$ to a plane

$$Ax + By + Cz + D = 0$$

is

$$\left| \frac{Ax_1 + By_1 + Cz_1 + D}{\sqrt{A^2 + B^2 + C^2}} \right|.$$

38. Find the distance from the point (2, 3, 1) to the plane $2x - 4y - 4z - 9 = 0$.

10.5 SPHERES; CYLINDERS; CONES

In general, the three-dimensional graph of an equation in one or more of the variables x, y, and z is a surface. In the preceding section the plane surface was discussed as the graph of the linear or first degree equation $Ax + By + Cz + D = 0$. In this section will be treated special quadric surfaces, the graphs of certain second degree equations.

Definition of Sphere A *sphere* is the set of points in space equidistant from a fixed point called the **center**. The distance from any point on the sphere to the center is called the **radius** of the sphere.

Theorem (Equation of Sphere) The equation of a sphere with center (a, b, c) and with radius r is
$$(x - a)^2 + (y - b)^2 + (z - c)^2 = r^2.$$

Proof The point (x, y, z) is on the sphere if and only if its distance from the center (a, b, c) is r. Using the distance formula,
$$\sqrt{(x - a)^2 + (y - b)^2 + (z - c)^2} = r.$$
Squaring both sides yields the equation of the sphere.

Example 1 Show that $x^2 + y^2 + z^2 - 6x + 4y - 2z - 11 = 0$ is a sphere and find its center and radius.

Solution Completing the squares for x, y, and z,
$$(x^2 - 6x +) + (y^2 + 4y +) + (z^2 - 2z +) = 11$$
$$(x^2 - 6x + 9) + (y^2 + 4y + 4) + (z^2 - 2z + 1) = 11 + 9 + 4 + 1$$
$$(x - 3)^2 + (y + 2)^2 + (z - 1)^2 = 25.$$

This is the equation of a sphere with center $(3, -2, 1)$ and with radius $r = 5$. (See Figure 10.15.)

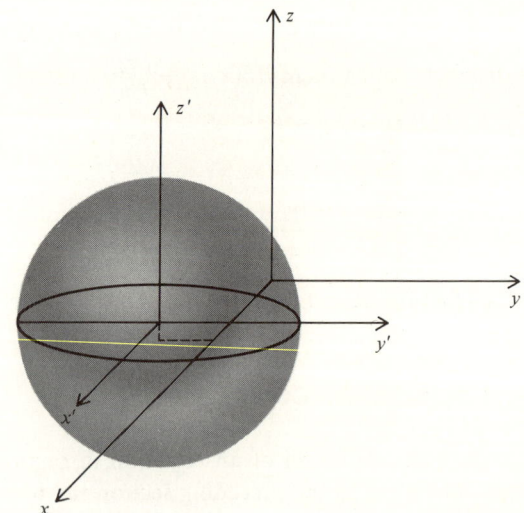

Figure 10.15 Sphere: $(x - 3)^2 + (y + 2)^2 + (z - 1)^2 = 25$

Definition of Cylinder A **cylinder** is the set of points in space that lie on a set of parallel lines (called **generators**) such that each line passes through a given plane curve (called the **directrix**) and no line of the set is in the plane of the curve.

SPHERES; CYLINDERS; CONES

Theorem The graph of any second degree equation with one of the variables missing, but not all, is a cylinder.

Proof Suppose z is the missing variable, and the equation is $Ax^2 + Bxy + Cy^2 + Dx + Ey + F = 0$. Then the intersection of the graph of this equation with the $z = k$ plane is a curve in either the xy-plane (for $z = 0$) or a plane parallel to the xy-plane. Furthermore, any one of these curves is congruent to another.

Therefore, the graph consists of the set of lines parallel to the z-axis and intersecting the curve in the xy-plane. Thus, by definition, the surface is a cylinder. A similar argument holds if x or y is the missing variable.

Example 2 Describe and sketch the graph of $(y - 3)^2 + x^2 = 9$.

Solution Since z is missing and the curve in the xy-plane is a circle, the graph is a circular cylinder. The directrix is the circle in the xy-plane.

Since the intersection of the surface with any plane parallel to the xy-plane produces the same curve, the generators are perpendicular to the xy-plane, and the surface is a right circular cylinder. (See Figure 10.16.)

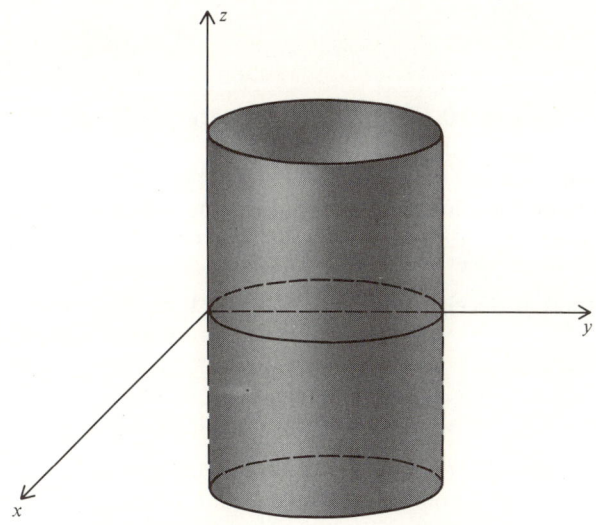

Figure 10.16 Cylinder: $x^2 + (y - 3)^2 = 9$

Example 3 Describe and sketch the graph of $y = 4 - x^2$.

Solution Since the z terms are missing and since the curve in the xy-plane is a parabola, the graph is a parabolic cylinder with directrix a parabola in the xy-plane. (See Figure 10.17.)

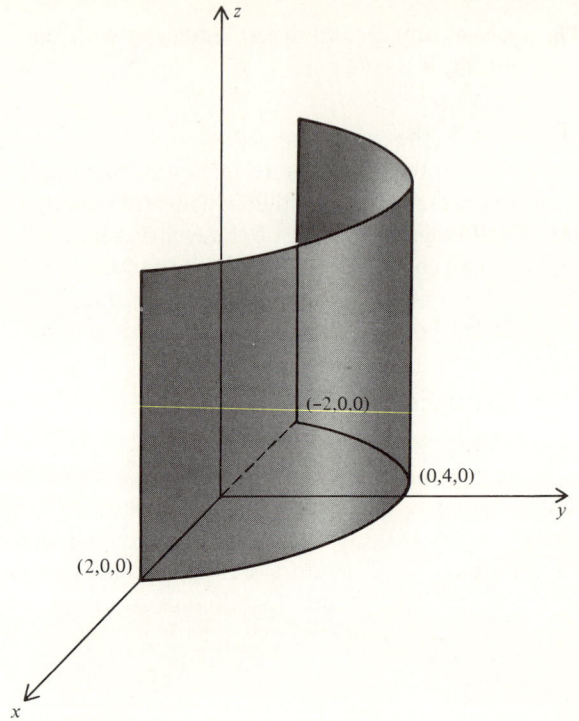

Figure 10.17 Parabolic Cylinder: $y = 4 - x^2$

Definition of Cone A **cone** is the set of points in space that lie on a set of lines (called **generators**) passing through a fixed point (called the **vertex**) and also through a point on a plane curve.

Theorem (Equations of Special Cones) If $abc \neq 0$, then the graph of

$$\frac{x^2}{a^2} + \frac{y^2}{b^2} - \frac{z^2}{c^2} = 0$$

is a cone.

Proof Note that the origin $O: (0, 0, 0)$ is on the graph.

Let $P: (x_1, y_1, c)$ be a point on the intersection of the graph and the $z = c$ plane.

Then the line through O and P has parametric equations

$$x = x_1 t, \quad y = y_1 t, \quad z = ct.$$

Now, any point on line OP is also on the graph since

$$\frac{x_1^2 t^2}{a^2} + \frac{y_1^2 t^2}{b^2} - \frac{c^2 t^2}{c^2} = t^2 \left(\frac{x_1^2}{a^2} + \frac{y_1^2}{b^2} - \frac{c^2}{c^2} \right) = 0.$$

Thus, by definition, the graph is a cone.

SPHERES; CYLINDERS; CONES

Corollary The graphs of $x^2/a^2 + z^2/b^2 = y^2/c^2$ and $y^2/a^2 + z^2/b^2 = x^2/c^2$ are cones.

Example 4 Describe and sketch the graph of $x^2 + z^2 = y^2$.

Solution The graph is a right circular cone with vertex at the origin and generators having parametric equations

$$x = x_1 t, \quad z = z_1 t, \quad y = ct, \quad \text{where } x_1^2 + z_1^2 = c^2.$$

(See Figure 10.18.)

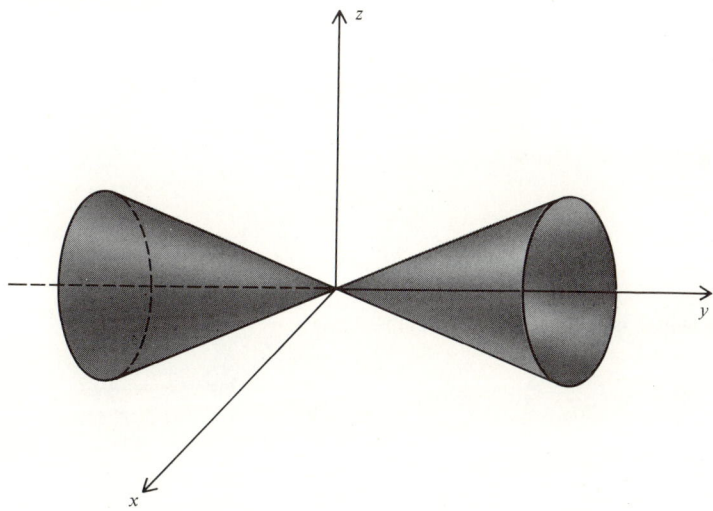

Figure 10.18 Right Circular Cone: $x^2 + z^2 = y^2$

EXERCISES 10.5

1–4. Find the equation of each of the following spheres, and sketch the graph of each.

1. Center at (1, 2, 1) and radius 4
2. Center at (2, 3, 1) and radius 3
3. Center at (−1, −1, 1) and passing through the point (1, 2, 7)
4. Center at (3, 0, 2) and passing through the point (2, 1, 4)

5–8. Show that each of the following is an equation of a sphere, and find the center and radius of the sphere

5. $x^2 + y^2 + z^2 + 6x - 2y - 8z - 10 = 0$
6. $x^2 + y^2 + z^2 - 12x + 14y - 8z + 1 = 0$
7. $x^2 + y^2 + z^2 + 2x - 6y + 22z + 122 = 0$

8. $2x^2 + 2y^2 + 2z^2 - 9z = 18$

9–20. Describe and sketch each of the following graphs.

9. $(x-2)^2 + y^2 - 4 = 0$
10. $z = 9 - x^2$
11. $z^2 = 4y$
12. $(y-2)^2 + (z-3)^2 - 1 = 0$
13. $x^2 + y^2 - 8x + 2y + 13 = 0$
14. $x^2 + 9y^2 - z^2 = 0$
15. $2x^2 + 3z^2 - 6(y-2)^2 = 0$
16. $x^2 - 2x + 1 - y^2 - z^2 = 0$
17. $x^2 = 4y^2 + 4z^2$
18. $y^2 = 4x^2 + z^2$
19. $4x^2 + z^2 = 4$
20. $4x^2 - z^2 = 4$

CHAPTER SUMMARY

Distance Formula $|P_1P_2| = \sqrt{(x_2 - x_1)^2 + (y_2 - y_1)^2 + (z_2 - z_1)^2}$

Midpoint Formula $M = \left(\dfrac{x_1 + x_2}{2}, \dfrac{y_1 + y_2}{2}, \dfrac{z_1 + z_2}{2}\right)$

Direction Cosines of Directed Line Segment $\overrightarrow{P_1P_2}$

$$\cos \alpha = \frac{x_2 - x_1}{d}, \quad \cos \beta = \frac{y_2 - y_1}{d}, \quad \cos \gamma = \frac{z_2 - z_1}{d},$$

where $d = |P_1P_2|$.

Direction Cosine Dependence $\cos^2 \alpha + \cos^2 \beta + \cos^2 \gamma = 1$

Direction Numbers of Line $(l, m, n) = (k \cos \alpha, k \cos \beta, k \cos \gamma)$ where k is any real number and

$$\cos \alpha = \frac{l}{\sqrt{l^2 + m^2 + n^2}}, \quad \cos \beta = \frac{m}{\sqrt{l^2 + m^2 + n^2}}, \quad \cos \gamma = \frac{n}{\sqrt{l^2 + m^2 + n^2}}.$$

Direction Numbers of Line on P_1 and P_2 $(x_2 - x_1, y_2 - y_1, z_2 - z_1)$

Parametric Equations of Line on P_1 and P_2

$$x = x_1 + (x_2 - x_1)t$$
$$y = y_1 + (y_2 - y_1)t$$
$$z = z_1 + (z_2 - z_1)t$$

Angle between Two Lines $\cos \theta = a_1a_2 + b_1b_2 + c_1c_2$, where (a_1, b_1, c_1) and (a_2, b_2, c_2) are direction cosines for the two lines.

Parallel Lines Two lines with direction numbers (l_1, m_1, n_1) and (l_2, m_2, n_2), respectively, are parallel if and only if there is a real number k so that

$$l_2 = kl_1, \quad m_2 = km_1, \quad \text{and} \quad n_2 = kn_1.$$

Perpendicular Lines Two lines with direction numbers (l_1, m_1, n_1) and (l_2, m_2, n_2) are perpendicular if and only if

$$l_1 l_2 + m_1 m_2 + n_1 n_2 = 0.$$

Normal to Plane A normal to a plane is any line perpendicular to the plane.

Equation of Plane The equation of a plane on point (x_1, y_1, z_1) and with direction numbers of a normal (l, m, n) is

$$l(x - x_1) + m(y - y_1) + n(z - z_1) = 0.$$

Parallel Planes Planes $A_1 x + B_1 y + C_1 z + D_1 = 0$ and $A_2 x + B_2 y + C_2 z + D_2 = 0$ are parallel if and only if there is a real number k so that $A_2 = kA_1$, $B_2 = kB_1$, and $C_2 = kC_1$.

Perpendicular Planes Planes $A_1 x + B_1 y + C_1 z + D_1 = 0$ and $A_2 x + B_2 y + C_2 z + D_2 = 0$ are perpendicular if and only if $A_1 A_2 + B_1 B_2 + C_1 C_2 = 0$.

Line as Intersection of Two Planes The set of equations

$$A_1 x + B_1 y + C_1 z + D_1 = 0$$
$$A_2 x + B_2 y + C_2 z + D_2 = 0$$

represent a line if and only if the planes are not parallel.

Sphere $(x - a)^2 + (y - b)^2 + (z - c)^2 = r^2$. Center: (a, b, c); radius: r

Cylinder (Special Cases) Any equation in which one of the variables is missing.

Cone (Special Cases)

$$\frac{x^2}{a^2} + \frac{y^2}{b^2} = \frac{z^2}{c^2}, \quad \frac{x^2}{a^2} + \frac{z^2}{b^2} = \frac{y^2}{c^2}, \quad \frac{y^2}{a^2} + \frac{z^2}{b^2} = \frac{x^2}{c^2}$$

REVIEW EXERCISES

1-4. Find the distance between each of the following points and the coordinates of the midpoint of the line segment joining these points.

1. $(2, 1, -1)$ and $(3, 5, 1)$
2. $(-1, 0, 2)$ and $(4, 5, 6)$
3. $(2, 4, 1)$ and $(-2, -4, -1)$
4. $(a, b, 0)$ and $(0, -b, 1)$

5–9. Find the direction cosines of the position vector $\overrightarrow{OP}$ for each of the following points.

5. $P: (1, 2, 3)$
6. $P: (-1, -2, -3)$
7. $P: (2, 0, 4)$
8. $P: (1, -2, 2)$
9. $P: (-2, 11, -10)$

10. Show that the points $(1, 6, -3)$, $(-2, 0, 3)$, and $(3, 10, -7)$ are collinear.
11. Find a set of parametric equations for the line through the point $(2, -3, 2)$ if its direction angles are $\alpha = \pi/3$, $\beta = 2\pi/3$, $\gamma = 3\pi/4$ (radians).
12. Determine whether the lines below are parallel, perpendicular, or neither parallel nor perpendicular to the line $x = 4 - t$, $y = -3 + 2t$, $z = 2 - 3t$.
    (a) $x = 3 + 2t$, $y = 5 - 4t$, $z = 4 + 6t$
    (b) $x = 4 + t$, $y = -3 + 2t$, $z = 2 + 2t$
    (c) $x = 4 - t$, $y = -3 - 2t$, $z = 2 - 3t$
    (d) $x = -1$, $y = 2 - 3t$, $z = -2 + 2t$
    (e) $x = 4 - 2t$, $y = -3 + 4t$, $z = 2 - 6t$
13. Determine whether the planes below are parallel, perpendicular, or neither parallel nor perpendicular to the plane $2x - 4y + 6z = 1$.
    (a) $3x - 6y - 5z = 3$
    (b) $3x - 6y + 9z = 5$
    (c) $2x + 4y - 6z = 1$

14–23. Let $\mathbf{v} = 2\mathbf{i} + \mathbf{j} - 3\mathbf{k}$, $\mathbf{u} = -2\mathbf{i} + 3\mathbf{j} + \mathbf{k}$, and $\mathbf{r} = \mathbf{i} - \mathbf{j} + 2\mathbf{k}$. Find each of the following.

14. $|\mathbf{v}|$
15. $\mathbf{v} \cdot \mathbf{u}$
16. $\mathbf{v} \times \mathbf{u}$
17. $\mathbf{u} \times \mathbf{v}$
18. $\mathbf{u} \times (\mathbf{v} + \mathbf{r})$
19. $(\mathbf{u} + \mathbf{v}) \times \mathbf{r}$
20. $(\mathbf{v} \times \mathbf{u}) \cdot \mathbf{r}$
21. $|\mathbf{v} \times \mathbf{u}|$
22. The angle θ between $\mathbf{u}$ and $\mathbf{v}$, where $0 < \theta \le 90°$.
23. The angle θ between $\mathbf{v}$ and $\mathbf{r}$, where $0 < \theta \le 90°$.

24–25. Find the equation of the plane on each of the following points given the direction numbers of the normal to the plane.

24. $P: (1, 2, 3)$, direction numbers of normal $(2, 1, 1)$
25. $P: (-1, 2, -3)$, direction numbers of normal $(-2, 1, 3)$
26. Find the direction cosines of the normal to the plane $9x + 2y + 6z + 4 = 0$.

27–30. Find the intercepts of each of the following planes and sketch the plane in the first octant.

27. $x + 3y + 2z - 12 = 0$
28. $2x - y - 2z + 4 = 0$
29. $2y + 5z = 10$
30. $x = 3$

31. Find the equation of the plane through the point (3, 2, 1) and perpendicular to the planes $x + y + z - 2 = 0$ and $2x - 2y - z - 1 = 0$.
32. Find parametric equations of the line that is the intersection of the planes $2x + y - z + 4 = 0$ and $x - 4y + 3z + 1 = 0$.
33. Find parametric equations of the line on the points $P: (2, 1, 5)$ and $Q: (3, -2, -1)$.
34. Find the normal equation of the plane $x + 5y - z - 9 = 0$. (See Exercises 10.4, Problem 35.)
35. Find the equation of the sphere with center at (1, 3, 2) and radius 5.
36. Find the equation of the sphere with center at (1, -1, 1) and passing through the point (3, 1, 2).
37. Show that $x^2 + y^2 + z^2 - 4x + 6y - 4z - 32 = 0$ is the equation of a sphere and find its center and radius.

38–41. Describe and sketch the graph of each of the following.

38. $x^2 + y^2 + z^2 = 16$
39. $x^2 + y^2 = 16$
40. $4y^2 + 9z^2 - 36 = 0$
41. $\dfrac{x^2}{9} + \dfrac{y^2}{9} - z^2 = 0$

42. Sketch the line through the origin if its direction angles are $\alpha = \pi/4$, $\beta = \pi/3$, $\gamma = 2\pi/3$ (radians).

11
SYSTEMS OF EQUATIONS

11.1 LINEAR SYSTEMS

11.1.1 Two Variables

Definition *A linear equation in two variables, x and y, is an equation of the form $ax + by = c$, where a, b, and c are constants, a and b not both zero.*

Definition *A solution of an equation in two variables* is an ordered pair of constants (p, q) such that the open equation becomes true when x is replaced by the first member of the ordered pair and y is replaced by the second member of the ordered pair.

Definition *The solution set of an equation in two variables* is the set of all solutions of the equation.

Definition *The solution set of a system of two linear equations in two variables* is the intersection of the solution sets of each of the equations in the system. In symbols, the solution set of the system

$$a_1 x + b_1 y = c_1$$
$$a_2 x + b_2 y = c_2$$

LINEAR SYSTEMS

is the set

$$\{(x, y) | a_1 x + b_1 y = c_1\} \cap \{(x, y) | a_2 x + b_2 y = c_2\}.$$

One method for obtaining this solution is based on the following theorem:

Theorem (Linear Combinations of Equations) If (p, q) is a solution of both $a_1 x + b_1 y = c_1$ and $a_2 x + b_2 y = c_2$, then (p, q) is a solution of $A(a_1 x + b_1 y - c_1) + B(a_2 x + b_2 y - c_2) = 0$, where A and B are any constants.

This theorem is valid because the replacement of x by p and y by q results in the statement $A \cdot 0 + B \cdot 0 = 0$ which is true for all constants A and B.

Example 1 Solve the system $3x - y = 9$, $x + 2y = 10$.

Solution

$$\begin{array}{l} 2(3x - y = 9) \\ 1(x + 2y = 10) \end{array} \rightarrow \begin{array}{l} 6x - 2y = 18 \\ \underline{x + 2y = 10} \\ 7x \qquad\quad = 28 \quad \text{and} \quad x = 4 \end{array}$$

$$\begin{array}{l} 1(3x - y = 9) \\ -3(x + 2y = 10) \end{array} \rightarrow \begin{array}{l} 3x - y = 9 \\ \underline{-3x - 6y = -30} \\ -7y = -21 \quad \text{and} \quad y = 3 \end{array}$$

Thus, the solution set is $\{(4, 3)\}$, which is verified by substituting these values into the two equations of the given system.

There are three possibilities for the solution set of a system of two linear equations in two variables:

(1) There is exactly one solution.
(2) The solution set is the empty set.
(3) There are infinitely many solutions.

The geometric interpretations for these three possibilities are, respectively:

(1) The lines intersect in exactly one point,
(2) The lines are parallel,
(3) The lines coincide.

11.1.2 Three Variables

Definition *A linear equation in three variables, x, y, and z, is an equation of the form $ax + by + cz = d$, where a, b, c, and d are constants, a, b, and c not all zero.*

Definition *A solution of an equation in three variables* is an ordered triple of numbers, (x, y, z), such that the open equation becomes true when x is replaced by the first member of the ordered triple, y by the second member, and z by the third member.

Definition The *solution set of an equation in three variables* is the set of all solutions of the equation.

As is the case with an equation in two variables, there are infinitely many solutions in the solution set.

Definition The *solution set of a system of three linear equations in three variables* is the intersection of the solution sets of each of the three equations in the system. In symbols, the solution set of the system

$$a_1 x + b_1 y + c_1 z = d_1$$
$$a_2 x + b_2 y + c_2 z = d_2$$
$$a_3 x + b_3 y + c_3 z = d_3$$

is $S_1 \cap S_2 \cap S_3$ where

$$S_1 = \{(x, y, z) | a_1 x + b_1 y + c_1 z = d_1\}$$
$$S_2 = \{(x, y, z) | a_2 x + b_2 y + c_2 z = d_2\}$$
$$S_3 = \{(x, y, z) | a_3 x + b_3 y + c_3 z = d_3\}.$$

Theorem (Linear Combinations of Equations) If (p, q, r) is a solution of both $a_1 x + b_1 y + c_1 z = d_1$ and $a_2 x + b_2 y + c_2 z = d_2$, then (p, q, r) is a solution of

$$A(a_1 x + b_1 y + c_1 z - d_1) + B(a_2 x + b_2 y + c_2 z - d_2) = 0,$$

where A and B are any constants.

This theorem is valid because the replacement of x by p, y by q, and z by r results in the statement $A \cdot 0 + B \cdot 0 = 0$ which is true for all constants A and B.

The solution set of a system of three linear equations in three variables may be obtained by a method similar to the one used for systems of two linear equations in two variables. By adding constant multiples of two equations of the system, one of the variables may be eliminated. Repeating this operation with a different pair of equations, the same variable may be eliminated again. There then result two linear equations in two variables which can be solved by the method discussed in the previous section.

LINEAR SYSTEMS

Again, there are three possibilities.

(1) There is exactly one solution.
(2) The solution set is the empty set.
(3) There are infinitely many solutions.

Since a linear equation in three variables represents a plane, the solution set of a system of three equations in three variables is the geometric intersection of the three planes. The possibilities are summarized below.

ALGEBRAIC STATEMENT	GEOMETRIC STATEMENT	ILLUSTRATION
There is exactly one solution.	The three planes intersect in exactly one point, P.	(a)
The solution set is empty.	The three planes do not have a point in common.	(b)
There are infinitely many solutions.	The three planes intersect in a line, or the three planes are coincident.	(c)

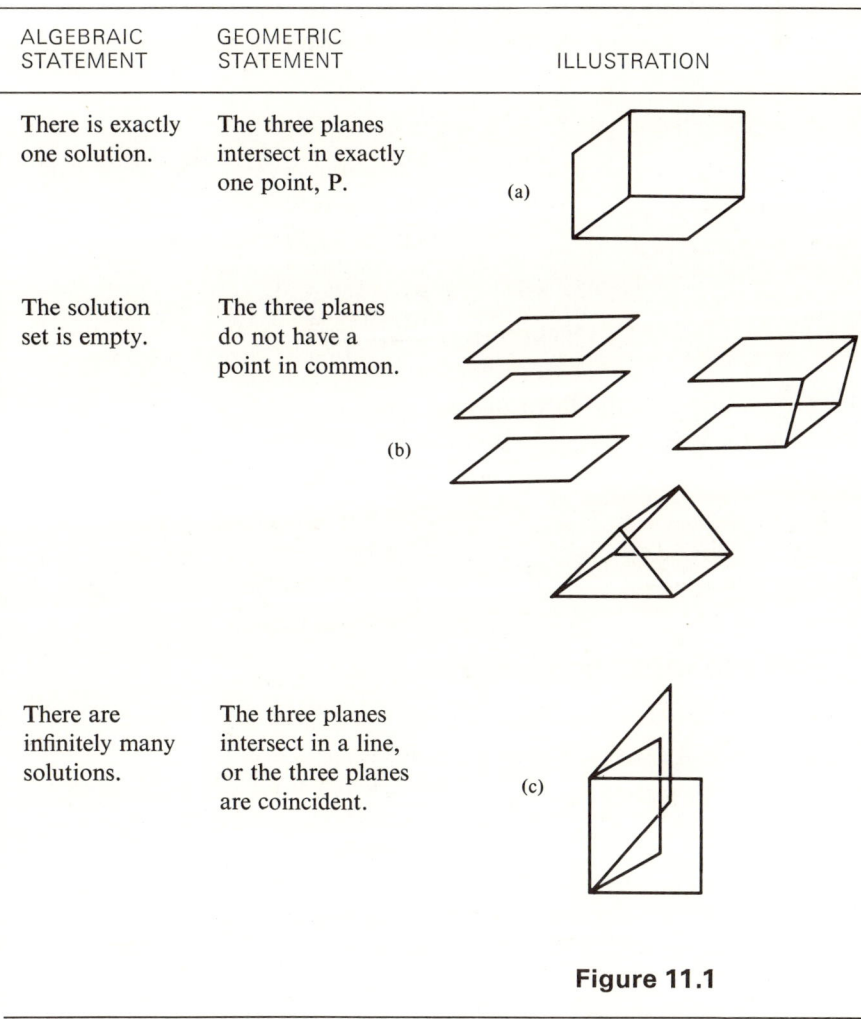

Figure 11.1

Example 2 Solve the system $x + 2y - 3z = -11$
$x - y - z = 2$
$x + 3y + 2z = -4.$

Solution

(1) Eliminate z by using the first and second equations:

$$-1(x + 2y - 3z = -11) \qquad -x - 2y + 3z = 11$$
$$3(x - y - z = 2) \qquad \underline{3x - 3y - 3z = 6}$$
$$2x - 5y \quad\quad = 17$$

(2) Eliminate z by using the second and third equations:

$$2(x - y - z = 2) \qquad 2x - 2y - 2z = 4$$
$$1(x + 3y + 2z = -4) \qquad \underline{x + 3y + 2z = -4}$$
$$3x + y \quad\quad = 0$$

(3) Now solve the system $2x - 5y = 17$, $3x + y = 0$.

$$1(2x - 5y = 17) \qquad 2x - 5y = 17$$
$$5(3x + y = 0) \qquad \underline{15x + 5y = 0}$$
$$17x \quad\quad = 17 \text{ or } x = 1$$

$$3(2x - 5y = 17) \qquad 6x - 15y = 51$$
$$-2(3x + y = 0) \qquad \underline{-6x - 2y = 0}$$
$$-17y = 51 \text{ or } y = -3$$

(4) Use any one of the three original equations to find z. Using the third equation, $x + 3y + 2z = -4$, with $x = 1$, and $y = -3$,

$$1 + 3(-3) + 2z = -4$$
$$-8 + 2z = -4$$
$$2z = 4 \text{ or } z = 2.$$

(5) State the solution set.

The solution set is $\{(1, -3, 2)\}$.

(6) Check the solution in each equation.

$x + 2y - 3z = 1 + 2(-3) - 3(2) = 1 - 6 - 6 = 1 - 12 = -11$
$x - y - z = 1 - (-3) - 2 = 1 + 3 - 2 = 4 - 2 = 2$
$x + 3y + 2z = 1 + 3(-3) + 2(2) = 1 - 9 + 4 = -8 + 4 = -4$

Thus, if

$$A = \{(x, y, z) | x + 2y - 3z = -11\}$$
$$B = \{(x, y, z) | x - y - z = 2\}$$
$$C = \{(x, y, z) | x + 3y + 2z = -4\},$$

then the solution set of the system is $A \cap B \cap C = \{(1, -3, 2)\}$.

LINEAR SYSTEMS

Example 3 Solve the system $2x + 2y - 2z = 1$.
$$5x - 2y + z = 2.$$
$$3x + 3y - 3z = 5.$$

Solution

(1) $\begin{array}{l} 1(2x + 2y - 2z = 1) \\ 2(5x - 2y + z = 2) \end{array}$ $\quad \begin{array}{l} 2x + 2y - 2z = 1 \\ 10x - 4y + 2z = 4 \\ \hline 12x - 2y \quad\quad = 5 \end{array}$

(2) $\begin{array}{l} 3(5x - 2y + z = 2) \\ 1(3x + 3y - 3z = 5) \end{array}$ $\quad \begin{array}{l} 15x - 6y + 3z = 6 \\ 3x + 3y - 3z = 5 \\ \hline 18x - 3y \quad\quad = 11 \end{array}$

(3) $\begin{array}{l} 3(12x - 2y = 5) \\ -2(18x - 3y = 11) \end{array}$ $\quad \begin{array}{l} 36x - 6y = 15 \\ -36x + 6y = -22 \\ \hline 0 = -7, \quad \text{a false statement} \end{array}$

(4) The solution set is the empty set, ∅.

Example 4 Solve the system $x - 2y + 3z = 4$
$$2x + y - z = 1$$
$$3x - y + 2z = 5.$$

Solution

(1) $\begin{array}{l} 1(x - 2y + 3z = 4) \\ 3(2x + y - z = 1) \end{array}$ $\quad \begin{array}{l} x - 2y + 3z = 4 \\ 6x + 3y - 3z = 3 \\ \hline 7x + y \quad\quad = 7 \end{array}$

(2) $\begin{array}{l} 2(2x + y - z = 1) \\ 1(3x - y + 2z = 5) \end{array}$ $\quad \begin{array}{l} 4x + 2y - 2z = 2 \\ 3x - y + 2z = 5 \\ \hline 7x + y \quad\quad = 7 \end{array}$

Since the same equation was obtained in both cases, the solution set is infinite. Solving this equation for y, $y = 7 - 7x$.

Now, replacing y by $7 - 7x$ in the second equation,
$$2x + (7 - 7x) - z = 1$$
$$-5x - z = -6$$
$$z = 6 - 5x.$$

The solution set is $\{(x, 7 - 7x, 6 - 5x) | x \text{ is any real number}\}$.

CHECK
$$x - 2y + 3z = x - 2(7 - 7x) + 3(6 - 5x)$$
$$= x - 14 + 14x + 18 - 15x = 4$$
$$2x + y - z = 2x + (7 - 7x) - (6 - 5x)$$
$$= 2x + 7 - 7x - 6 + 5x = 1$$
$$3x - y + 2z = 3x - (7 - 7x) + 2(6 - 5x)$$
$$= 3x - 7 + 7x + 12 - 10x = 5$$

A system of four linear equations in four variables may be solved by eliminating the same variable using three different pairs of equations. Each of the four equations should be represented in at least one pair. A system of three linear equations in three variables is thus obtained and this may be solved by the preceding technique.

EXERCISES 11.1

1–16. Solve and check.

1. $x - 2y - 3z = 3$
   $x + y - z = 2$
   $2x - 3y - 5z = 5$

2. $x + y = 3$
   $z - y = 3$
   $2x + y = 8$

3. $x + y + 3z - 3 = 0$
   $3x - y + 2z - 1 = 0$
   $4x - 4y + z + 2 = 0$

4. $2r + s - t = 3$
   $4r + 2s + 3t = 1$
   $6r + 3s - 2t = 2$

5. $a + b + c = 5$
   $2a - b - 2c = 3$
   $3a - 3b - 5c = 1$

6. $x + y = z + 2$
   $4x + 3z = y + 1$
   $2x - 3y = 4 - z$

7. $x + y - z = 0$
   $x - 2y + 3z = 0$
   $2x - y - 2z = 0$

8. $x + 2y + z = 3$
   $2x - y - 2z = 1$
   $3x - 3y + 3z = 0$

9. $4a - b + 2c = 3$
   $a + 3b - 4c = 2$
   $10a - 9b + 14c = 5$

10. $3x - y = z - 2$
    $5x + 2y = 22 - z$
    $2x + 4y - 5z = 0$

11. $x - y - z = 4$
    $6x + y - 2z = 5$
    $-3x + 3y + 3z = -7$

12. $x - y + 2z = 0$
    $2x + y - z = 0$
    $x + 5y - 8z = 0$

13. $x + 2y - 3z + 2t = 6$
    $2x + 3y + 4z + t = -5$
    $3x - y + 2z - t = 0$
    $2x + y + z - t = -4$

14. $a + 2c = 11$
    $a + 3b - 2d - 13 = 0$
    $2a - 3c - d + 4 = 0$
    $3a - 3b - 2c + 2d - 1 = 0$

15. $x + y + z + t = 1$
    $x + 2y - t = 10$
    $2y - z + t = 1$
    $x + 2z - 2t = 7$

16. $a - b + c = 5$
    $2a - 3c + d = 2$
    $a - 3d = 6$
    $2b - c + 2d = -3$

LINEAR SYSTEMS

17. Find the solution set of the system

$$x - 3y + z = 1$$
$$2x + y - 2z = -2.$$

[*Hint:* Express y and z in terms of x.]

18. Find the solution set of the system

$$2x + 3y - z = 2$$
$$x + y + 3z = 5.$$

19. Find the solution set of the system

$$x + y - z + 2w = 4$$
$$2x - y + 2z - w = 2$$
$$x - y + 3z + w = 1.$$

20. Find the solution set of the system

$$x + y - 2z - w = 1$$
$$2x - y + z + 2w = 3.$$

21. Find the solution set of the system

$$x + y + z = 4$$
$$2x - y + 3z = 14$$
$$3x - y - 2z = 1$$
$$6x - y + 2z = 19.$$

[*Hint:* Find the solution set for the system consisting of the first three equations. Show that the solution also satisfies the fourth equation.]

22. Find the solution set of the system

$$x - 2y = 5$$
$$2x + 3y = 7$$
$$x + 5y = 2.$$

23. Find the solution set of the system

$$x + y + z = 6$$
$$2x - y + z = 3$$
$$5x + 2y - 2z = 12$$
$$3x - y - z = 2.$$

24. Find the solution set of the system

$$x + 2y = 3$$
$$2x + y = 1$$
$$x - 3y = 2$$
$$3x - 2y = 5.$$

11.2 MATRICES

11.2.1 Motivation

The *general solution* of a system of two linear equations in two variables is obtained as follows:

$$\begin{aligned} b_2(a_1x + b_1y &= c_1) \\ -b_1(a_2x + b_2y &= c_2) \end{aligned} \qquad \begin{aligned} a_1b_2x + b_1b_2y &= b_2c_1 \\ -a_2b_1x - b_1b_2y &= -b_1c_2 \\ \hline (a_1b_2 - a_2b_1)x &= b_2c_1 - b_1c_2 \end{aligned}$$

$$\begin{aligned} -a_2(a_1x + b_1y &= c_1) \\ a_1(a_2x + b_2y &= c_2) \end{aligned} \qquad \begin{aligned} -a_1a_2x - a_2b_1y &= -a_2c_1 \\ a_1a_2x + a_1b_2y &= a_1c_2 \\ \hline (a_1b_2 - a_2b_1)y &= a_1c_2 - a_2c_1 \end{aligned}$$

Now, if $a_1b_2 - a_2b_1 \neq 0$, then

$$x = \frac{b_2c_1 - b_1c_2}{a_1b_2 - a_2b_1} \quad \text{and} \quad y = \frac{a_1c_2 - a_2c_1}{a_1b_2 - a_2b_1}.$$

Although these equations provide a formula for the solution of the system, they are not easy to remember. Introducing the symbol $\begin{vmatrix} a & b \\ c & d \end{vmatrix}$ to designate the number $ad - bc$, then the equations become

$$x = \frac{\begin{vmatrix} c_1 & b_1 \\ c_2 & b_2 \end{vmatrix}}{\begin{vmatrix} a_1 & b_1 \\ a_2 & b_2 \end{vmatrix}} \quad \text{and} \quad y = \frac{\begin{vmatrix} a_1 & c_1 \\ a_2 & c_2 \end{vmatrix}}{\begin{vmatrix} a_1 & b_1 \\ a_2 & b_2 \end{vmatrix}}.$$

These arrays are more easily memorized, and moreover, they can be generalized to indicate the solution of a consistent system of three linear equations in three variables, or of four linear equations in four variables, or of n linear equations in n variables. The next sections are devoted to arrays of numbers, their properties, and their application to the solution of systems of equations.

11.2.2 Terminology

Definition A **matrix** is a rectangular (or square) array of numbers (or elements) arranged in rows and columns enclosed by brackets, double vertical lines, or by large parentheses.

Some examples of matrices are:

$$\begin{bmatrix} 1 & 0 & 2 \\ 3 & -1 & 5 \\ 4 & 2 & 6 \end{bmatrix}, \quad \begin{Vmatrix} 5 & 1 & 2 \\ 3 & 2 & -7 \end{Vmatrix}, \quad \begin{pmatrix} 2 & 0 \\ 1 & 3 \end{pmatrix}, \quad \begin{bmatrix} 2 \\ 5 \end{bmatrix}, \quad [3 \quad 1].$$

Definition The **dimensions** of a matrix are $n \times k$ (read "n by k") if the matrix is an array of n rows and k columns.

MATRICES

If the number of rows is equal to the number of columns, then the matrix is called a **square** matrix.

If a matrix has only one row or one column, then it is called a **row matrix** or a **column matrix** (or a **row vector** or a **column vector**).

In the examples above, the first matrix is a 3 × 3 square matrix, the second is a 2 × 3 matrix, the third is a 2 × 2 square matrix, the fourth is a column matrix or column vector, and the fifth is a row matrix or a row vector.

Matrix algebra was originated by the English mathematician Arthur Cayley (1821–1895) in his work *A Memoir on the Theory of Matrices* published in 1858. The term "matrix" was first used in 1850 by the English mathematician James Joseph Sylvester (1814–1897), a life-long friend and collaborator of Cayley, both in mathematics and in the legal profession.

The nineteenth century witnessed the development of the logical foundations of elementary algebra and the creation of modern abstract algebra. The discovery of noncommutative algebras such as Cayley's matrix algebra and Hamilton's quaternion algebra were events of major significance.

11.2.3 Matrix Algebra

Definition of Equal Matrices Two matrices are equal if and only if they have the same dimensions and their corresponding elements are equal.

For example,

$$\begin{bmatrix} 2 & 1 \\ 3 & 0 \end{bmatrix} = \begin{bmatrix} \frac{6}{3} & \frac{3}{3} \\ \frac{9}{3} & 0 \end{bmatrix} \quad \text{and} \quad \begin{bmatrix} 4 & -1.5 & 2.5 \\ 1.5 & 3 & 5 \end{bmatrix} = \begin{bmatrix} 4 & -\frac{3}{2} & \frac{5}{2} \\ \frac{3}{2} & 3 & 5 \end{bmatrix}$$

but

$$\begin{bmatrix} 1 & 4 \\ 3 & 2 \end{bmatrix} \neq \begin{bmatrix} 1 & 3 \\ 4 & 2 \end{bmatrix} \quad \text{and} \quad \begin{bmatrix} 0 & 0 & 0 \\ 0 & 0 & 0 \end{bmatrix} \neq \begin{bmatrix} 0 & 0 \\ 0 & 0 \end{bmatrix}.$$

Definition of Sum of Two Matrices The sum of two matrices having the same dimensions is the matrix whose elements are the sums of the corresponding elements of the matrices being added.

Examples

$$\begin{bmatrix} a & b \\ c & d \end{bmatrix} + \begin{bmatrix} 1 & 2 \\ 3 & 4 \end{bmatrix} = \begin{bmatrix} a+1 & b+2 \\ c+3 & d+4 \end{bmatrix}$$

$$\begin{bmatrix} 1 & 0 & 2 \\ 2 & 1 & -3 \end{bmatrix} + \begin{bmatrix} 2 & 4 & -1 \\ 0 & 2 & 1 \end{bmatrix} = \begin{bmatrix} (1+2) & (0+4) & (2-1) \\ (2+0) & (1+2) & (-3+1) \end{bmatrix}$$

$$= \begin{bmatrix} 3 & 4 & 1 \\ 2 & 3 & -2 \end{bmatrix}.$$

It should be noted that addition is defined only for matrices having the same dimensions. It is left for the student to verify that matrix addition is commutative and associative.

An element of a matrix is also called a **scalar**.

If the elements of a matrix are members of the set of real numbers, then a **scalar** is any real number.

Definition of Scalar Multiplication The product of a scalar and a matrix is the matrix each of whose elements is the product of the scalar and the corresponding element of the matrix being multiplied. For example,

$$k \begin{bmatrix} 2 & 1 \\ 3 & 0 \end{bmatrix} = \begin{bmatrix} 2k & k \\ 3k & 0 \end{bmatrix} \text{ and } 2 \begin{bmatrix} 1 & -1 & 2 \\ 3 & 0 & 4 \end{bmatrix} = \begin{bmatrix} 2 & -2 & 4 \\ 6 & 0 & 8 \end{bmatrix}.$$

Scalar multiplication may be shown to be commutative and associative.

Definition of Matrix Multiplication The product of an $n \times k$ matrix, A, and a $k \times m$ matrix, B, is an $n \times m$ matrix, AB, in which the element in the ath row and bth column is the sum of the products of the elements of the ath row of A multiplied by the corresponding elements of the bth row of B.

Example 1 If $A = \begin{bmatrix} 2 & 3 \\ 4 & 5 \end{bmatrix}$ and $B = \begin{bmatrix} a & b & c \\ x & y & z \end{bmatrix}$, find the matrix product AB.

Solution

$$AB = \begin{bmatrix} 2a + 3x & 2b + 3y & 2c + 3z \\ 4a + 5x & 4b + 5y & 4c + 5z \end{bmatrix}$$

Example 2 Multiply

$$\begin{bmatrix} 2 & -1 & 3 \\ 3 & 4 & 1 \end{bmatrix} \text{ by } \begin{bmatrix} x \\ y \\ z \end{bmatrix}.$$

Solution

$$\begin{bmatrix} 2 & -1 & 3 \\ 3 & 4 & 1 \end{bmatrix} \begin{bmatrix} x \\ y \\ z \end{bmatrix} = \begin{bmatrix} 2x - y + 3z \\ 3x + 4y + z \end{bmatrix}$$

Example 3 If $A = \begin{bmatrix} 2 & 0 \\ 1 & 3 \end{bmatrix}$ and $B = \begin{bmatrix} 1 & 5 \\ 3 & 2 \end{bmatrix}$, find AB and BA.

Solution

$$AB = \begin{bmatrix} 2 & 0 \\ 1 & 3 \end{bmatrix} \begin{bmatrix} 1 & 5 \\ 3 & 2 \end{bmatrix} = \begin{bmatrix} 2 \cdot 1 + 0 \cdot 3 & 2 \cdot 5 + 0 \cdot 2 \\ 1 \cdot 1 + 3 \cdot 3 & 1 \cdot 5 + 3 \cdot 2 \end{bmatrix} = \begin{bmatrix} 2 & 10 \\ 10 & 11 \end{bmatrix}$$

$$BA = \begin{bmatrix} 1 & 5 \\ 3 & 2 \end{bmatrix} \begin{bmatrix} 2 & 0 \\ 1 & 3 \end{bmatrix} = \begin{bmatrix} 2 + 5 & 0 + 15 \\ 6 + 2 & 0 + 6 \end{bmatrix} = \begin{bmatrix} 7 & 15 \\ 8 & 6 \end{bmatrix}$$

MATRICES

Example 3 illustrates that matrix multiplication is *not commutative*. Note that it is not necessary to compute all entries in BA to show that $AB \neq BA$. Only two corresponding entries need to be shown unequal; for example, the entries in row 1, column 1 are not equal, $2 \neq 7$ and thus $AB \neq BA$.

On the other hand, matrix multiplication *is* associative and distributive from both the right and the left; that is,

$$(AB)C = A(BC)$$
$$A(B + C) = AB + AC$$
$$(B + C)A = BA + CA.$$

It should be noted that the product of two matrices is defined *only* for the case where the number of columns of the left matrix is equal to the number of rows of the right matrix.

It may seem strange that matrix multiplication is defined as it is and not some other way; for example, as the matrix whose elements are the products of the corresponding elements. There are very few applications of such a definition whereas there is a wide range of applications for the definition as stated, particularly in the solution of systems of equations. This will be shown in a later section.

Definition The **main diagonal** (or **principal diagonal**) of a square matrix is the diagonal from the upper left element to the lower right element.

For example, the main diagonal of the matrix $\begin{bmatrix} 1 & 2 & 3 \\ 4 & 5 & 6 \\ 7 & 8 & 9 \end{bmatrix}$ contains the elements 1, 5, and 9.

Definition An **identity matrix**, I, is a square matrix whose elements along the main diagonal are ones and the rest of whose elements are zeros.

For example,

$$\begin{bmatrix} 1 & 0 \\ 0 & 1 \end{bmatrix}, \quad \begin{bmatrix} 1 & 0 & 0 \\ 0 & 1 & 0 \\ 0 & 0 & 1 \end{bmatrix}, \quad \text{and} \quad \begin{bmatrix} 1 & 0 & 0 & 0 \\ 0 & 1 & 0 & 0 \\ 0 & 0 & 1 & 0 \\ 0 & 0 & 0 & 1 \end{bmatrix}.$$

are identity matrices.

Theorem If I is an identity matrix and if A is any matrix having the same dimensions as I, then $IA = AI = A$.

For example,

$$\begin{bmatrix} 1 & 0 \\ 0 & 1 \end{bmatrix} \begin{bmatrix} a & b \\ c & d \end{bmatrix} = \begin{bmatrix} a & b \\ c & d \end{bmatrix} = \begin{bmatrix} a & b \\ c & d \end{bmatrix} \begin{bmatrix} 1 & 0 \\ 0 & 1 \end{bmatrix}.$$

Definition An **inverse matrix**, A^{-1}, of a square matrix A is a square matrix having the same dimensions as A, such that $AA^{-1} = I$.

For example, to find A^{-1}, if $A = \begin{bmatrix} 4 & 2 \\ 1 & 1 \end{bmatrix}$, let

$$A^{-1} = \begin{bmatrix} a & b \\ c & d \end{bmatrix}.$$

Then

$$AA^{-1} = \begin{bmatrix} 4 & 2 \\ 1 & 1 \end{bmatrix} \cdot \begin{bmatrix} a & b \\ c & d \end{bmatrix}$$

$$= \begin{bmatrix} 4a + 2c & 4b + 2d \\ a + c & b + d \end{bmatrix} = \begin{bmatrix} 1 & 0 \\ 0 & 1 \end{bmatrix} = I.$$

$$4a + 2c = 1 \qquad 4b + 2d = 0$$
$$a + c = 0 \qquad b + d = 1.$$

Solving,

$$a = \tfrac{1}{2}, \qquad b = -1, \qquad c = -\tfrac{1}{2}, \qquad d = 2$$

$$A^{-1} = \begin{bmatrix} \tfrac{1}{2} & -1 \\ -\tfrac{1}{2} & 2 \end{bmatrix}.$$

EXERCISES 11.2

1–23. Perform the indicated operations, if possible. If not possible, state why not.

1. $\begin{bmatrix} 3 & 5 \\ -2 & 4 \end{bmatrix} + \begin{bmatrix} 1 & 0 \\ -5 & -5 \end{bmatrix}$

2. $\begin{bmatrix} 2 & 4 & -3 \\ -1 & 0 & 1 \end{bmatrix} + \begin{bmatrix} 3 & 4 & 5 \\ 6 & 7 & 8 \end{bmatrix}$

3. $\begin{bmatrix} 4 & 2 & 9 \\ 2 & -1 & 0 \end{bmatrix} + \begin{bmatrix} 3 & 5 \\ 1 & -7 \end{bmatrix}$

4. $\begin{bmatrix} 5 & -2 & 7 \end{bmatrix} + \begin{bmatrix} 3 \\ 8 \\ 2 \end{bmatrix}$

5. $\begin{bmatrix} 3 & -2 & 4 \\ 1 & 0 & 2 \\ 1 & 0 & 1 \end{bmatrix} + \begin{bmatrix} 2 & 5 & -8 \\ 1 & -3 & -2 \\ 3 & -2 & -2 \end{bmatrix}$

6. $\begin{bmatrix} 4 & 6 & -3 & 5 \\ 1 & -4 & 2 & 1 \\ 3 & 0 & 0 & 1 \end{bmatrix} + \begin{bmatrix} 7 & 0 & -5 \\ 2 & -1 & 0 \end{bmatrix}$

7. $[4 \quad 7 \quad 9] + [2 \quad 8 \quad 5]$

8. $-2 \begin{bmatrix} \frac{1}{2} & 3 & -\frac{5}{2} \\ 1 & 0 & -1 \\ 3 & -2 & 1 \end{bmatrix}$

9. $3[2 \quad -3 \quad 4]$.

10. $5 \begin{bmatrix} 2 & -3 & 5 \\ 1 & 4 & -2 \end{bmatrix}$

11. $\begin{bmatrix} 3 & 1 \\ 2 & 5 \end{bmatrix} \begin{bmatrix} -2 & 3 \\ 4 & 1 \end{bmatrix}$

12. $\begin{bmatrix} 2 & 3 \\ 1 & -2 \\ 4 & 3 \end{bmatrix} \begin{bmatrix} a & b & c \\ 4 & 2 & 3 \end{bmatrix}$

13. $\begin{bmatrix} 2 & 1 & 4 \\ 3 & -2 & 3 \end{bmatrix} \begin{bmatrix} a & b & c \\ 4 & 2 & 3 \end{bmatrix}$

14. $\begin{bmatrix} 1 & -1 & 1 \\ 2 & 1 & -3 \\ 3 & 2 & -1 \end{bmatrix} \begin{bmatrix} x \\ y \\ z \end{bmatrix}$

15. $\begin{bmatrix} 1 & -2 & 1 & 3 \\ 2 & 0 & 5 & 4 \end{bmatrix} \begin{bmatrix} a & 2 \\ b & -3 \\ c & -1 \\ d & 0 \end{bmatrix}$

16. $\begin{bmatrix} 1 & 3 & -4 \\ 5 & -1 & 2 \end{bmatrix} \begin{bmatrix} x \\ y \\ z \end{bmatrix}$

17. $\begin{bmatrix} 1 & 0 & 0 \\ 0 & 1 & 0 \\ 0 & 0 & 1 \end{bmatrix} \begin{bmatrix} a & b & c \\ d & e & f \\ g & h & i \end{bmatrix}$

18. $\begin{bmatrix} 1 & 2 & 1 & 3 \\ 1 & 1 & -2 & 1 \end{bmatrix} \begin{bmatrix} a & 0 \\ b & 1 \\ c & -1 \\ d & 0 \end{bmatrix}$

19. $\begin{bmatrix} 1 & 2 & 1 & 3 \\ 0 & 1 & -2 & 1 \end{bmatrix} + \begin{bmatrix} a & 0 \\ b & 1 \\ c & -1 \\ d & 0 \end{bmatrix}$

20. $\begin{bmatrix} 4 & 2 & 4 & -1 \\ 3 & -1 & 0 & 2 \end{bmatrix} \begin{bmatrix} 2 & x & 0 & 1 \\ y & 1 & -1 & 0 \end{bmatrix}$

21. $\begin{bmatrix} 4 & 2 & 4 & -1 \\ 3 & -1 & 0 & 2 \end{bmatrix} + \begin{bmatrix} 2 & x & 0 & 1 \\ y & 1 & -1 & 0 \end{bmatrix}$

22. $\begin{bmatrix} 1 & 2 \\ 2 & 4 \end{bmatrix} \begin{bmatrix} 6 & -2 \\ -3 & 1 \end{bmatrix}$

23. $\begin{bmatrix} 1 & 0 & 0 & 0 \\ 0 & 1 & 0 & 0 \\ 0 & 0 & 1 & 0 \\ 0 & 0 & 0 & 1 \end{bmatrix} \begin{bmatrix} w \\ x \\ y \\ z \end{bmatrix} = \begin{bmatrix} -2 \\ 3 \\ 1 \\ 5 \end{bmatrix}$

24–28. Solve for the variables.

24. $\begin{bmatrix} 1 & x & 3 \\ y & 2 & z \end{bmatrix} = \begin{bmatrix} 1 & 0 & 3 \\ 6 & 2 & 9 \end{bmatrix}$

25. $5\begin{bmatrix} x & y \\ 3 & -2 \end{bmatrix} + \begin{bmatrix} 3 & -2 \\ -1 & 4 \end{bmatrix} = 2\begin{bmatrix} -6 & 5 \\ 7 & -3 \end{bmatrix}$

26. $\begin{bmatrix} 1 & 0 & 0 \\ 0 & -1 & 0 \\ 0 & 0 & 2 \end{bmatrix} \begin{bmatrix} x \\ y \\ z \end{bmatrix} = \begin{bmatrix} 2 \\ 3 \\ 8 \end{bmatrix}$

27. $\begin{bmatrix} 2 & 0 & 0 \\ 0 & 3 & 0 \\ 0 & 0 & -4 \end{bmatrix} \begin{bmatrix} x & -1 \\ y & 2 \\ z & -3 \end{bmatrix} = 3\begin{bmatrix} 3 & -\frac{2}{3} \\ -2 & 2 \\ 5 & 4 \end{bmatrix}$

28. $\begin{bmatrix} 1 & 2 & -1 \\ 0 & 1 & 1 \\ 0 & 0 & 1 \end{bmatrix} \begin{bmatrix} x \\ y \\ z \end{bmatrix} = \begin{bmatrix} -2 \\ 5 \\ 3 \end{bmatrix}$

29–37. Let $A = \begin{bmatrix} 1 & 2 \\ -1 & 1 \end{bmatrix}$ and $B = \begin{bmatrix} 1 & -1 \\ 2 & 1 \end{bmatrix}$. Find each of the following.

29. $-B$
30. $-BA$
31. AB
32. BA.
33. $A(-B)$
34. B^{-1}
35. $(AB)^{-1}$
36. $B^{-1}A^{-1}$
37. $-A(B)^{-1}$

38–40. Determine the relationship between each of the following:

38. AB and BA
39. $A(-B)$ and $(-B)A$
40. $(AB)^{-1}$ and $B^{-1}A^{-1}$

41. Let $A = \begin{bmatrix} x & y \\ 2x & 2y \end{bmatrix}$ and $B = \begin{bmatrix} y & 3y \\ -x & -3x \end{bmatrix}$. Find AB. If $AB = 0$, does it follow that $A = 0$ or $B = 0$?

42–52. Let $A = \begin{bmatrix} a & b \\ c & d \end{bmatrix}$, $B = \begin{bmatrix} 1 & 1 \\ -1 & 1 \end{bmatrix}$, $C = \begin{bmatrix} 1 & 1 \\ 1 & -1 \end{bmatrix}$, and let $S =$ the set of 2×2 square matrices whose elements are real numbers.

42. Show that $A + B \in S$ (closure, addition), and that $AB \in S$ (closure, multiplication).

43. Show that $A + B = B + A$ (commutativity, addition).

MATRIX SOLUTION OF LINEAR SYSTEMS

44. Show that $BC \neq CB$ (multiplication is not commutative).
45. Show that $(A + B) + C = A + (B + C)$ and $(AB)C = A(BC)$ (associativity, addition and multiplication).
46. Find 0 so that $0 + A = A = A + 0$ (addition identity).
47. Find I so that $IA = A = AI$ (multiplication identity).
48. Find $-A$ so that $A + (-A) = 0$ (addition inverse).
49. Assuming that $ad - bc \neq 0$, find A^{-1} so that $AA^{-1} = I$ (multiplication inverse). Does $A^{-1}A = AA^{-1}$? Justify your answer.
50. Find X so that $A = B + X$. Define subtraction.
51. Find Y so that $AY = B$. Define division.
52. Assuming $ad - bc \neq 0$, if $AX = B$, then which of the following is correct: $X = BA^{-1}$ or $X = A^{-1}B$? Justify your answer.

53–56. Solve for the variables.

53. $\begin{bmatrix} x & y^2 + 9 \\ \log_2 z & 10^w \end{bmatrix} = \begin{bmatrix} \operatorname{Sin}^{-1} - \tfrac{1}{2} & 6y \\ 5 & 0.001 \end{bmatrix}$

54. $\begin{bmatrix} 2 & 3 \\ 3 & -4 \end{bmatrix} \begin{bmatrix} x \\ y \end{bmatrix} = \begin{bmatrix} 16 \\ 7 \end{bmatrix}$

55. $\begin{bmatrix} x & y \\ x & -y \end{bmatrix} \begin{bmatrix} x \\ y \end{bmatrix} = \begin{bmatrix} 13 \\ 5 \end{bmatrix}$

56. $\begin{bmatrix} \sin^2 x & \cos x \\ 2y^3 & y \end{bmatrix} \begin{bmatrix} 2 \\ -1 \end{bmatrix} = \begin{bmatrix} 2 \\ 12 \end{bmatrix}$

11.3 MATRIX SOLUTION OF LINEAR SYSTEMS

11.3.1 Transformation to a Matrix Equation

The solution of linear systems by using matrices may be considered as an abbreviated form of the solution by the addition method previously discussed.

By applying the definition of equal matrices and the definition of matrix multiplication, it follows that the linear system

$$a_1 x + b_1 y = c_1$$
$$a_2 x + b_2 y = c_2$$

is equivalent to

$$\begin{bmatrix} a_1 & b_1 \\ a_2 & b_2 \end{bmatrix} \begin{bmatrix} x \\ y \end{bmatrix} = \begin{bmatrix} c_1 \\ c_2 \end{bmatrix}$$

and the system

$$a_1 x + b_1 y + c_1 z = d_1$$
$$a_2 x + b_2 y + c_2 z = d_2$$
$$a_3 x + b_3 y + c_3 z = d_3$$

is equivalent to

$$\begin{bmatrix} a_1 & b_1 & c_1 \\ a_2 & b_2 & c_2 \\ a_3 & b_3 & c_3 \end{bmatrix} \begin{bmatrix} x \\ y \\ z \end{bmatrix} = \begin{bmatrix} d_1 \\ d_2 \\ d_3 \end{bmatrix}.$$

Similar statements are valid for the cases involving four or more variables. Each of the matrices,

$$\begin{bmatrix} a_1 & b_1 \\ b_2 & b_2 \end{bmatrix} \quad \text{and} \quad \begin{bmatrix} a_1 & b_1 & c_1 \\ a_2 & b_2 & c_2 \\ a_3 & b_3 & c_3 \end{bmatrix},$$

is called the **matrix of the coefficients**.

In solving a linear system by using matrices, it is convenient to work with the following matrices which are called the **augmented matrices** of the linear system

$$\begin{bmatrix} a_1 & b_1 & c_1 \\ a_2 & b_2 & c_2 \end{bmatrix}$$

for two variables and

$$\begin{bmatrix} a_1 & b_1 & c_1 & d_1 \\ a_2 & b_2 & c_2 & d_2 \\ a_3 & b_3 & c_3 & d_3 \end{bmatrix}$$

for three variables.

11.3.2 Equivalent Transformations of Matrices

Since the linear system

$$\begin{aligned} x &= p \\ y &= q \\ z &= r \end{aligned}$$

is equivalent to the open matrix equation

$$\begin{bmatrix} 1 & 0 & 0 \\ 0 & 1 & 0 \\ 0 & 0 & 1 \end{bmatrix} \begin{bmatrix} x \\ y \\ z \end{bmatrix} = \begin{bmatrix} p \\ q \\ r \end{bmatrix},$$

the solution set can be recognized immediately in either of these forms as $\{(p, q, r)\}$. Thus, the objective is to transform the augmented matrix into one having zeros below and above the main diagonal of the matrix of the coefficients, if this is possible. Also it is necessary that the linear system of the transformed matrix should have the same solution set as that of the original matrix.

Definition Two **augmented matrices of linear systems are equivalent**, $A \sim B$, if and only if the associated linear systems of the two matrices are equivalent (that is, have the same solution set).

MATRIX SOLUTION OF LINEAR SYSTEMS

There are three elementary transformations which produce equivalent augmented matrices; interchanging rows, multiplying a row by a constant, adding a constant multiple of a row to another row.

Theorem 1 Two augmented matrices are equivalent if one is obtained from the other by **interchanging any two rows**.

Theorem 2 Two augmented matrices are equivalent if one is obtained from the other by **multiplying each element of a row by a nonzero constant**.

Theorem 3 Two augmented matrices are equivalent if one is obtained from the other by **adding a nonzero constant multiple of the elements of one row to the corresponding elements of another row**.

Theorem 1 states, for example, that

$$\begin{bmatrix} a_1 & b_1 & c_1 & d_1 \\ a_2 & b_2 & c_2 & d_2 \\ a_3 & b_3 & c_3 & d_3 \end{bmatrix} \sim \begin{bmatrix} a_3 & b_3 & c_3 & d_3 \\ a_2 & b_2 & c_2 & d_2 \\ a_1 & b_1 & c_1 & d_1 \end{bmatrix}$$

or that the associated linear systems

$$\begin{aligned} a_1 x + b_1 y + c_1 z &= d_1 & & & a_3 x + b_3 y + c_3 z &= d_3 \\ a_2 x + b_2 y + c_2 z &= d_2 & \text{and} & & a_2 x + b_2 y + c_2 z &= d_2 \\ a_3 x + b_3 y + c_3 z &= d_3 & & & a_1 x + b_1 y + c_1 z &= d_1 \end{aligned}$$

are equivalent. Since the equations of a linear system may be written in any order without changing the solution set of the system, the augmented matrices are equivalent. A similar argument holds for the other possibilities.

Theorem 2 implies that an equivalent system is obtained if any equation of the system is multiplied by a constant. Thus

$$\begin{aligned} a_1 x + b_1 y + c_1 z &= d_1 & & & k a_1 x + k b_1 y + k c_1 z &= k d_1 \\ a_2 x + b_2 y + c_2 z &= d_2 & \text{and} & & l a_2 x + l b_2 y + l c_2 z &= l d_2 \\ a_3 x + b_3 y + c_3 z &= d_3 & & & m a_3 x + m b_3 y + m c_3 z &= m d_3 \end{aligned}$$

are equivalent, for $klm \neq 0$, as this has been established previously. Thus the augmented matrices are equivalent.

Theorem 3 implies that any equation of a system may be replaced by a linear combination of two equations of the system. Since this equivalence has also been established, it follows that the augmented matrices of the systems

$$\begin{aligned} a_1 x + b_1 y + c_1 z &= d_1 & & & (a_1 + k a_2)x + (b_1 + k b_2)y & \\ a_2 x + b_2 y + c_2 z &= d_2 & \text{and} & & + (c_1 + k c_2)z &= d_1 + k d_2 \\ a_3 x + b_3 y + c_3 z &= d_3 & & & a_2 x + b_2 y + c_2 z &= d_2 \\ & & & & a_3 x + b_3 y + c_3 z &= d_3 \end{aligned}$$

are equivalent.

11.3.3 Gauss' Reduction Method

The method of solution illustrated in this section is known as Gauss' reduction method in honor of the German mathematician Carl Friedrich Gauss (1777–1855) who is generally acknowledged as one of the greatest mathematicians of all times.

The objective of this method is to obtain zeros below the main diagonal of the matrix of the coefficients and ones along the main diagonal.

Example 1 By using matrices, solve the system
$$3x + 2y - z = 5$$
$$x - 3y + z = 2$$
$$2x - y - 2z = 1.$$

Solution

(1) Form the augmented matrix.
$$\begin{bmatrix} 3 & 2 & -1 & 5 \\ 1 & -3 & 1 & 2 \\ 2 & -1 & -2 & 1 \end{bmatrix}$$

(2) Interchange rows to have a 1 in Row 1, Column 1.
$$\begin{bmatrix} 1 & -3 & 1 & 2 \\ 2 & -1 & -2 & 1 \\ 3 & 2 & -1 & 5 \end{bmatrix}$$

(3) Multiply Row 1 by -2, add to Row 2. Multiply Row 1 by -3, add to Row 3.
$$\begin{bmatrix} 1 & -3 & 1 & 2 \\ 0 & 5 & -4 & -3 \\ 0 & 11 & -4 & -1 \end{bmatrix}$$

(4) Multiply Row 2 by $\frac{1}{5}$.
$$\begin{bmatrix} 1 & -3 & 1 & 2 \\ 0 & 1 & -\frac{4}{5} & -\frac{3}{5} \\ 0 & 11 & -4 & -1 \end{bmatrix}$$

(5) Multiply Row 2 by -11, add to Row 3.
$$\begin{bmatrix} 1 & -3 & 1 & 2 \\ 0 & 1 & -\frac{4}{5} & -\frac{3}{5} \\ 0 & 0 & \frac{24}{5} & \frac{28}{5} \end{bmatrix}$$

(6) Multiply row 3 by $\frac{5}{24}$.
$$\begin{bmatrix} 1 & -3 & 1 & 2 \\ 0 & 1 & -\frac{4}{5} & -\frac{3}{5} \\ 0 & 0 & 1 & \frac{7}{6} \end{bmatrix}$$

(7) Write the associated linear system.
$$x - 3y + z = 2$$
$$y - \tfrac{4}{5}z = -\tfrac{3}{5}$$
$$z = \tfrac{7}{6}$$

(8) Solve the system.
$$y = \tfrac{4}{5}(\tfrac{7}{6}) - \tfrac{3}{5} = \tfrac{14}{15} - \tfrac{9}{15} = \tfrac{1}{3}$$
$$x = 3(\tfrac{1}{3}) - \tfrac{7}{6} + 2 = \tfrac{18}{6} - \tfrac{7}{6}$$
$$= \tfrac{11}{6}.$$

(9) State the solution. $(\tfrac{11}{6}, \tfrac{1}{3}, \tfrac{7}{6})$

(10) Check the solution. Check left for student.

MATRIX SOLUTION OF LINEAR SYSTEMS

11.3.4 Extension of Gauss' Method

It is sometimes possible to introduce zeros above the main diagonal of the matrix of the coefficients as the following example illustrates.

Example 2 By using matrices, solve the system
$$x - 2y + 3z = 1$$
$$2x + y - 2z = 13$$
$$x + 3y - z = 4.$$

Solution

$$\begin{bmatrix} 1 & -2 & 3 & 1 \\ 2 & 1 & -2 & 13 \\ 1 & 3 & -1 & 4 \end{bmatrix} \sim \begin{bmatrix} 1 & -2 & 3 & 1 \\ 0 & 5 & -8 & 11 \\ 0 & 5 & -4 & 3 \end{bmatrix}$$ Multiply Row 1 by -2, add to Row 2. Multiply Row 1 by -1, add to Row 3.

$$\sim \begin{bmatrix} 1 & -2 & 3 & 1 \\ 0 & 5 & -8 & 11 \\ 0 & 0 & 4 & -8 \end{bmatrix}$$ Multiply Row 2 by -1, add to Row 3.

$$\sim \begin{bmatrix} 1 & -2 & 3 & 1 \\ 0 & 5 & 0 & -5 \\ 0 & 0 & 1 & -2 \end{bmatrix}$$ Multiply Row 3 by 2, add to Row 2. Multiply Row 3 by $\frac{1}{4}$.

$$\sim \begin{bmatrix} 1 & -2 & 3 & 1 \\ 0 & 1 & 0 & -1 \\ 0 & 0 & 1 & -2 \end{bmatrix}$$ Multiply Row 2 by $\frac{1}{5}$.

$$\sim \begin{bmatrix} 1 & 0 & 3 & -1 \\ 0 & 1 & 0 & -1 \\ 0 & 0 & 1 & -2 \end{bmatrix}$$ Multiply Row 2 by 2, add to Row 1.

$$\sim \begin{bmatrix} 1 & 0 & 0 & 5 \\ 0 & 1 & 0 & -1 \\ 0 & 0 & 1 & -2 \end{bmatrix}$$ Multiply Row 3 by -3, add to Row 1.

The associated linear system is $x = 5$, $y = -1$, $z = -2$. Thus, the solution set is $\{(5, -1, -2)\}$. The solution should be checked in each of the three equations of the original system.

This technique, using matrices, is readily adaptable to the electronic computer. Moreover, it becomes increasingly effective as the number of equations and the number of variables increase.

EXERCISES 11.3

1–20. Solve by using matrices.

1. $2x - 3y = 16$
   $5x - 2y = -4$

2. $5x + 4y = 12$
   $3x - 2y = 16$

3. $x - 2y = 7$
   $3x + 2y = 5$

4. $5x + 6y = 8$
   $3x - 4y = -18$

5. $2x - y = 0$
   $2y - z = 0$
   $x + 2y - z = 3$

6. $4x + 5y - 6z = 15$
   $3x - 7y - 4z = -19$
   $6x + 2y + z = 46$

7. $y - x = 1$
   $z - x = 2$
   $74x - 16y = 25z$

8. $x + y - z = 0$
   $4x + 4y + 2z = 3$
   $2x + 5y - z = 1$

9. $x + y + 2z - 7 = 0$
   $4x - 2y + 3z - 1 = 0$
   $9x - 3y + 8z - 4 = 0$

10. $2x + y - 3z - 2 = 0$
    $3x - y - 2z - 3 = 0$
    $x - z - 1 = 0$

11. $2a + b + c = 2$
    $3a - b + c = 2$
    $7a - 5b - 3c = 0$

12. $x + 5y - 2z = 8$
    $2x - 4y + 2z = 3$
    $3x - 6y + 3z = 4$

13. $x - 2y + z = 0.5$
    $-2x + 4y - 2z = 1$
    $3x - 6y + 3z = 1.5$

14. $x + y + z = 0$
    $x + y - 4z = 2$
    $x - 2y + z = 1$
    $2x - y - 3z = 3$

15. $x - y + z = 6$
    $2x - y - 2z = 2$
    $3x - 2y - z = 8$

16. $4r - 3s - t = 2$
    $-r + 2s + 3t = 1$
    $5r + s - 2t = 3$

17. $x + y + z + t = 0$
    $2x - y + t = 7$
    $x - 2z - t = -1$
    $2y - z - 2t = -3$

18. $x - y + 2z - t = 0$
    $2y - 3z + 2t = 5$
    $x - 2z - 3t = 2$
    $2x + y - 3z - 2t = 7$

19. $x + 2y - t = 0$
    $3x + y + 5z + 2t = 5$
    $x + 2z + 2t = -1$
    $2x - 3y - z + 5t = 1$

20. $2a + b + c + d = -3$
    $a - b + c - d = 2$
    $3a + 2b + 2c + d = 3$
    $4a + b - c - d = -5$

21. Show that $\begin{bmatrix} a_1 & b_1 & c_1 & d_1 \\ a_2 & b_2 & c_2 & d_2 \\ a_3 & b_3 & c_3 & d_3 \\ a_4 & b_4 & c_4 & d_4 \end{bmatrix} \begin{bmatrix} x \\ y \\ z \\ t \end{bmatrix} = \begin{bmatrix} e_1 \\ e_2 \\ e_3 \\ e_4 \end{bmatrix}.$

DETERMINANTS

is equivalent to the linear system

$$a_1 x + b_1 y + c_1 z + d_1 t = e_1$$
$$a_2 x + b_2 y + c_2 z + d_2 t = e_2$$
$$a_3 x + b_3 y + c_3 z + d_3 t = e_3$$
$$a_4 x + b_4 y + c_4 z + d_4 t = e_4.$$

22. A problem in genetics is to determine the probable genetic structure of a future generation if the offspring are repeatedly bred with hybrids. The solution is obtained by determining a unique fixed point probability vector; that is, by solving the matrix equation below.

$$\begin{bmatrix} \tfrac{1}{2} & \tfrac{1}{4} & 0 \\ \tfrac{1}{2} & \tfrac{1}{2} & \tfrac{1}{2} \\ 0 & \tfrac{1}{4} & \tfrac{1}{2} \end{bmatrix} \begin{bmatrix} d \\ h \\ r \end{bmatrix} = \begin{bmatrix} d \\ h \\ r \end{bmatrix}.$$

Solve this equation for (d, h, r) if $d + h + r = 1$.

11.4 DETERMINANTS

A real number can be assigned to every square matrix whose elements are real numbers. The real number is called the determinant of the matrix.

Definition A **determinant of the second order**, denoted by the symbol $\begin{vmatrix} a_1 & b_1 \\ a_2 & b_2 \end{vmatrix}$, where a_1, b_1, a_2, and b_2 are real numbers, is the real number, $a_1 b_2 - a_2 b_1$; that is,

$$\begin{vmatrix} a_1 & b_1 \\ a_2 & b_2 \end{vmatrix} = a_1 b_2 - a_2 b_1.$$

Definition A **determinant of the third order**, denoted by the symbol

$$\begin{vmatrix} a_1 & b_1 & c_1 \\ a_2 & b_2 & c_2 \\ a_3 & b_3 & c_3 \end{vmatrix},$$

where the elements are real numbers, is the real number

$$a_1 \begin{vmatrix} b_2 & c_2 \\ b_3 & c_3 \end{vmatrix} - a_2 \begin{vmatrix} b_1 & c_1 \\ b_3 & c_3 \end{vmatrix} + a_3 \begin{vmatrix} b_1 & c_1 \\ b_2 & c_2 \end{vmatrix};$$

that is,

$$\begin{vmatrix} a_1 & b_1 & c_1 \\ a_2 & b_2 & c_2 \\ a_3 & b_3 & c_3 \end{vmatrix} = a_1 \begin{vmatrix} b_2 & c_2 \\ b_3 & c_3 \end{vmatrix} - a_2 \begin{vmatrix} b_1 & c_1 \\ b_3 & c_3 \end{vmatrix} + a_3 \begin{vmatrix} b_1 & c_1 \\ b_2 & c_2 \end{vmatrix}.$$

Thus, it may be observed that the association of a real number, the determinant, to a square matrix, M, defines a function, $d(M)$. The domain is the set of square matrices, and the range is the set of real numbers. The function is the set of ordered pairs $\{(M, d(M))\}$.

The **order** of a determinant is the number of rows (or columns) in the square array of the matrix associated with the determinant.

When the determinant is being indicated, the array of numbers is enclosed by **vertical lines**, in contrast to the use of brackets, parentheses, or double lines used to designate the matrix.

The **expansion** of a determinant is the indicated calculation in the definition of the determinant.

The **minor** of an element of a determinant is the determinant obtained by deleting the row and column in which the element lies.

For example, for a determinant of order three,

$$A_1 = \begin{vmatrix} b_2 & c_2 \\ b_3 & c_3 \end{vmatrix} \text{ is the minor of } a_1,$$

$$A_2 = \begin{vmatrix} b_1 & c_1 \\ b_3 & c_3 \end{vmatrix} \text{ is the minor of } a_2,$$

$$A_3 = \begin{vmatrix} b_1 & c_1 \\ b_2 & c_2 \end{vmatrix} \text{ is the minor of } a_3,$$

$$B_2 = \begin{vmatrix} a_1 & c_1 \\ a_3 & c_3 \end{vmatrix} \text{ is the minor of } b_2, \text{ and so on.}$$

Thus,

$$D = \begin{vmatrix} a_1 & b_1 & c_1 \\ a_2 & b_2 & c_2 \\ a_3 & b_3 & c_3 \end{vmatrix} = a_1 A_1 - a_2 A_2 + a_3 A_3$$

is called an expansion of D by the minors of the elements of the first column.

Theorem 1 Any row or column may be used to expand a determinant of the third order; that is,

$$\begin{aligned} D &= a_1 A_1 - a_2 A_2 + a_3 A_3 \\ &= -b_1 B_1 + b_2 B_2 - b_3 B_3 \\ &= c_1 C_1 - c_2 C_2 + c_3 C_3 \\ &= a_1 A_1 - b_1 B_1 + c_1 C_1 \\ &= -a_2 A_2 + b_2 B_2 - c_2 C_2 \\ &= a_3 A_3 - b_3 B_3 + c_3 C_3. \end{aligned}$$

DETERMINANTS

Whether a product in the expansion is to be multiplied by $+1$ or -1 is determined by the following array, called the "checkerboard of signs."

$$\begin{vmatrix} + & - & + \\ - & + & - \\ + & - & + \end{vmatrix}$$

The proof given below shows that $a_1 A_1 - a_2 A_2 + a_3 A_3 = -b_1 B_1 + b_2 B_2 - b_3 B_3$. The other cases are proved similarly and are left for the student to verify.

$$\begin{aligned} a_1 A_1 - a_2 A_2 + a_3 A_3 &= a_1 \begin{vmatrix} b_2 & c_2 \\ b_3 & c_3 \end{vmatrix} - a_2 \begin{vmatrix} b_1 & c_1 \\ b_3 & c_3 \end{vmatrix} + a_3 \begin{vmatrix} b_1 & c_1 \\ b_2 & c_2 \end{vmatrix} \\ &= a_1(b_2 c_3 - b_3 c_2) - a_2(b_1 c_3 - b_3 c_1) + a_3(b_1 c_2 - b_2 c_1) \\ &= a_1 b_2 c_3 + a_2 b_3 c_1 + a_3 b_1 c_2 - a_1 b_3 c_2 - a_2 b_1 c_3 \\ &\quad - a_3 b_2 c_1 \\ &= -b_1(a_2 c_3 - a_3 c_2) + b_2(a_1 c_3 - a_3 c_1) - b_3(a_1 c_2 - a_2 c_1) \\ &= -b_1 \begin{vmatrix} a_2 & a_3 \\ c_2 & c_3 \end{vmatrix} + b_2 \begin{vmatrix} a_1 & a_3 \\ c_1 & c_3 \end{vmatrix} - b_3 \begin{vmatrix} a_1 & a_2 \\ c_1 & c_2 \end{vmatrix} \\ &= -b_1 B_1 + b_2 B_2 - b_3 B_3 \end{aligned}$$

Example 1 Expand $\begin{vmatrix} 2 & 3 & -1 \\ 1 & -2 & 2 \\ -4 & -1 & 5 \end{vmatrix}$ by Row 1.

Solution

$$\begin{aligned} D &= 2 \begin{vmatrix} -2 & 2 \\ -1 & 5 \end{vmatrix} - 3 \begin{vmatrix} 1 & 2 \\ -4 & 5 \end{vmatrix} + (-1) \begin{vmatrix} 1 & -2 \\ -4 & -1 \end{vmatrix} \\ &= 2(-10 - (-2)) - 3(5 - (-8)) - (-1 - 8) \\ &= 2(-8) - 3(13) - (-9) \\ &= -16 - 39 + 9 = -46 \end{aligned}$$

Example 2 Expand $\begin{vmatrix} 3 & 2 & -5 \\ 2 & 0 & 1 \\ 1 & 0 & 4 \end{vmatrix}$ by Column 2.

Solution

$$\begin{aligned} D &= -2 \begin{vmatrix} 2 & 1 \\ 1 & 4 \end{vmatrix} + 0 \begin{vmatrix} 3 & -5 \\ 1 & 4 \end{vmatrix} - 0 \begin{vmatrix} 3 & -5 \\ 2 & 1 \end{vmatrix} \\ &= -2(8 - 1) + 0 - 0 \\ &= -14 \end{aligned}$$

The two preceding examples illustrate that the determinant is much easier to evaluate when there are zeros in a row or column. Thus, it is desirable to establish some theorems indicating the transformations that can be performed on determinants in order to obtain an equal determinant with zeros as some of its elements.

Theorem 2 *Two determinants of the same order are equal if one is obtained from the other by* interchanging the rows and columns. *In symbols,*

$$\begin{vmatrix} a_1 & b_1 \\ a_2 & b_2 \end{vmatrix} = \begin{vmatrix} a_1 & a_2 \\ b_1 & b_2 \end{vmatrix} \quad \text{and} \quad \begin{vmatrix} a_1 & b_1 & c_1 \\ a_2 & b_2 & c_2 \\ a_3 & b_3 & c_3 \end{vmatrix} = \begin{vmatrix} a_1 & a_2 & a_3 \\ b_1 & b_2 & b_3 \\ c_1 & c_2 & c_3 \end{vmatrix}.$$

This is proved by applying the definition, regrouping the terms, and applying the definition again.

Theorem 3 *If two rows (or two columns) of a determinant are interchanged, then the resulting determinant is the negative of the original one.*

For example, let $D = \begin{vmatrix} a_1 & b_1 & c_1 \\ a_2 & b_2 & c_2 \\ a_3 & b_3 & c_3 \end{vmatrix}$ and let $E = \begin{vmatrix} c_1 & b_1 & a_1 \\ c_2 & b_2 & a_2 \\ c_3 & b_3 & a_3 \end{vmatrix}.$

Then, $E = -D$.

This is proved by expanding E, regrouping the terms, and applying the definition.

Theorem 4 *If each element of a row (or column) is multiplied by a constant, then the determinant is multiplied by a constant.*

For example,

$$\begin{vmatrix} ka_1 & kb_1 & kc_1 \\ a_2 & b_2 & c_2 \\ a_3 & b_3 & c_3 \end{vmatrix} = k \begin{vmatrix} a_1 & b_1 & c_1 \\ a_2 & b_2 & c_2 \\ a_3 & b_3 & c_3 \end{vmatrix}$$

and

$$\begin{vmatrix} 6 & 9 & -12 \\ 2 & 1 & 5 \\ 3 & -1 & 2 \end{vmatrix} = 3 \begin{vmatrix} 2 & 3 & -4 \\ 2 & 1 & 5 \\ 3 & -1 & 2 \end{vmatrix}.$$

The proof is left for the student.

Theorem 5 *If the* corresponding elements of two rows (or columns) are equal, *then the value of the determinant is 0.*

DETERMINANTS

For example,

$$\begin{vmatrix} 1 & 4 & 1 \\ 3 & 1 & 3 \\ -2 & 5 & -2 \end{vmatrix} = 0 \quad \text{and} \quad \begin{vmatrix} 2 & 1 & 5 \\ 2 & 1 & 5 \\ 3 & 7 & 9 \end{vmatrix} = 0.$$

The proof is left for the student.

Theorem 6 If each element of a row (or column) is multiplied by a constant and then added to the corresponding element of another row (or column), then the resulting determinant is equal to the original determinant.

For example,

$$\begin{vmatrix} a_1 + ka_3 & b_1 + kb_3 & c_1 + kc_3 \\ a_2 & b_2 & c_2 \\ a_3 & b_3 & c_3 \end{vmatrix} = \begin{vmatrix} a_1 & b_1 & c_1 \\ a_2 & b_2 & c_2 \\ a_3 & b_3 & c_3 \end{vmatrix}$$

and

$$\begin{vmatrix} a_1 & b_1 + kc_1 & c_1 \\ a_2 & b_2 + kc_2 & c_2 \\ a_3 & b_3 + kc_3 & c_3 \end{vmatrix} = \begin{vmatrix} a_1 & b_1 & c_1 \\ a_2 & b_2 & c_2 \\ a_3 & b_3 & c_3 \end{vmatrix}.$$

This theorem may be proved by expanding the determinant by the altered row (or column) and expressing the expansion as the sum of two determinants, one of which is zero.

For example,

$$\begin{vmatrix} a_1 + ka_3 & b_1 + kb_3 & c_1 + kc_3 \\ a_2 & b_2 & c_2 \\ a_3 & b_3 & c_3 \end{vmatrix} = (a_1 + ka_3)A_1 - (b_1 + kb_3)B_1 + (c_1 + kc_3)C_1$$

$$= a_1 A_1 - b_1 B_1 + c_1 C_1$$
$$\quad + k(a_3 A_1 - b_3 B_1 + c_3 C_1)$$

$$= \begin{vmatrix} a_1 & b_1 & c_1 \\ a_2 & b_2 & c_2 \\ a_3 & b_3 & c_3 \end{vmatrix} + k \begin{vmatrix} a_3 & b_3 & c_3 \\ a_2 & b_2 & c_2 \\ a_3 & b_3 & c_3 \end{vmatrix}$$

$$= \begin{vmatrix} a_1 & b_1 & c_1 \\ a_2 & b_2 & c_2 \\ a_3 & b_3 & c_3 \end{vmatrix} + k(0) \quad \text{(by Theorem 5)}$$

$$= \begin{vmatrix} a_1 & b_1 & c_1 \\ a_2 & b_2 & c_2 \\ a_3 & b_3 & c_3 \end{vmatrix}.$$

Example 3 Find a determinant equal to $\begin{vmatrix} -2 & 3 & 1 \\ 1 & 2 & 3 \\ 2 & 3 & 3 \end{vmatrix}$ having zeros everywhere in Column 3 except the first row. Expand the resulting determinant.

Solution

$\begin{vmatrix} -2 & 3 & 1 \\ 1 & 2 & 3 \\ 2 & 3 & 3 \end{vmatrix} = \begin{vmatrix} -2 & 3 & 1 \\ 7 & -7 & 0 \\ 2 & 3 & 3 \end{vmatrix}$ Multiply Row 1 by -3 and add to Row 2.

$= \begin{vmatrix} -2 & 3 & 1 \\ 7 & -7 & 0 \\ 8 & -6 & 0 \end{vmatrix}$ Multiply Row 1 by -3 and add to Row 3.

$= 1 \begin{vmatrix} 7 & -7 \\ 8 & -6 \end{vmatrix} = 7(-6) - 8(-7)$

$= -42 + 56 = 14$

Example 4 Expand the determinant having zeros everywhere in Row 2 except Column 2 and equal to $\begin{vmatrix} 2 & -1 & 4 \\ 5 & 1 & -2 \\ 3 & -3 & -4 \end{vmatrix}$.

Solution

$\begin{vmatrix} 2 & -1 & 4 \\ 5 & 1 & -2 \\ 3 & -3 & -4 \end{vmatrix} = \begin{vmatrix} 7 & -1 & 4 \\ 0 & 1 & -2 \\ 18 & -3 & -4 \end{vmatrix}$ Multiply Column 2 by -5 and add to Column 1.

$= \begin{vmatrix} 7 & -1 & 2 \\ 0 & 1 & 0 \\ 18 & -3 & -10 \end{vmatrix}$ Multiply Column 2 by 2 and add to Column 3.

$= 1 \begin{vmatrix} 7 & 2 \\ 18 & -10 \end{vmatrix}$

$= 7(-10) - 18(2)$

$= -70 - 36$

$= -106$

Definition A **determinant of order n**, denoted by the symbol

$$\begin{vmatrix} a_{11} & a_{12} & a_{13} & \cdots & a_{1n} \\ a_{21} & a_{22} & a_{23} & \cdots & a_{2n} \\ a_{31} & a_{32} & a_{33} & \cdots & a_{3n} \\ \vdots & & & & \vdots \\ a_{n1} & a_{n2} & a_{n3} & \cdots & a_{nn} \end{vmatrix}$$

DETERMINANTS

where each element a_{ij} is a real number (the double subscript notation is used with the first subscript i indicating the row in which the element lies and the second subscript j indicating the column in which the element lies), is the real number

$$a_{11}A_{11} - a_{21}A_{21} + a_{31}A_{31} - \ldots + (-1)^{1+n}a_{n1}A_{n1},$$

where A_{ij} is the *minor* of a_{ij}; that is, the determinant obtained by deleting the row and column in which a_{ij} lies.

It may be shown that Theorems 1–6 remain valid for determinants of order n. In particular, any row or column may be used to expand a determinant of order n, by the following procedure:

(1) Select any row (or column);
(2) Multiply each element in this row (or column) by its minor and by $(-1)^{\text{row number + column number of the element}}$;
(3) Add the products thus obtained.

EXERCISES 11.4

1–4. Expand the given determinant about the indicated row or column.

1. $\begin{vmatrix} 3 & 5 \\ 4 & 9 \end{vmatrix}$
Column 1

2. $\begin{vmatrix} 2 & 3 \\ 4 & -5 \end{vmatrix}$
Row 2

3. $\begin{vmatrix} 2 & 3 & 5 \\ 1 & 4 & 2 \\ 3 & 1 & 1 \end{vmatrix}$
Row 1

4. $\begin{vmatrix} 7 & -3 & 5 \\ 1 & 2 & -3 \\ 2 & -1 & 0 \end{vmatrix}$
Column 3

5–8. Find a determinant equal to the given one and satisfying the stated conditions. Expand the determinant.

5. $\begin{vmatrix} -2 & 3 & 1 \\ 1 & 2 & 3 \\ 2 & 3 & 3 \end{vmatrix}$ zeros everywhere in Row 2 except Column 1.

6. $\begin{vmatrix} 1 & 1 & 1 \\ 1 & 4 & 9 \\ 1 & 8 & 27 \end{vmatrix}$ zeros everywhere in Column 1 except Row 1.

7. $\begin{vmatrix} -1 & 4 & -4 \\ 2 & 3 & 2 \\ 1 & -1 & 2 \end{vmatrix}$ zeros everywhere in Column 3 except Row 3.

8. $\begin{vmatrix} 5 & -2 & 3 \\ -12 & 3 & 9 \\ 4 & 1 & -2 \end{vmatrix}$ zeros everywhere in Row 2 except Column 2.

9–10. Find an equal determinant having zeros everywhere below the main diagonal.

9. $\begin{vmatrix} 1 & 2 & 3 \\ 1 & 4 & 9 \\ 1 & 8 & 27 \end{vmatrix}$

10. $\begin{vmatrix} 2 & 3 & 4 \\ 1 & -5 & 6 \\ 4 & -7 & -1 \end{vmatrix}$

11–14. Solve for x:

11. $\begin{vmatrix} 2 & 0 & 1 \\ 3 & 0 & 4 \\ 1 & x & 2 \end{vmatrix} = 15$

12. $\begin{vmatrix} 1 & 2 & 3 \\ 5 & 3 & 4 \\ 1 & 2 & x \end{vmatrix} = 0$

13. $\begin{vmatrix} 1 & 1 & 1 \\ 1 & 2 & 3 \\ 1 & 4 & 9 \end{vmatrix} = \begin{vmatrix} 1 & 1 & 1 \\ 0 & 1 & 2 \\ 0 & 3 & x \end{vmatrix}$

14. $\frac{1}{2}\begin{vmatrix} 0 & 3 & 1 \\ 0 & -2 & 1 \\ x & 5 & 1 \end{vmatrix} = 30$

15. Without expanding, show that

$$\begin{vmatrix} 1 & 1 & 2x - 2y \\ 0 & x & x^2 - xy \\ x & y & x^2 - y^2 \end{vmatrix} = 0$$

State the reason for each step.

16. Find an equal determinant having zeros everywhere in Column 1 except Row 1. Then expand the determinant and find its value.

$$\begin{vmatrix} 1 & 3 & 2 & -2 \\ 2 & 5 & -1 & 1 \\ -3 & 2 & 1 & 6 \\ -1 & 4 & 5 & 2 \end{vmatrix}$$

17. Find an equal determinant having zeros everywhere below the main diagonal. Then expand the determinant and find its value.

$$\begin{vmatrix} 1 & 1 & 1 & 1 \\ 1 & 2 & 3 & 4 \\ 1 & 4 & 9 & 16 \\ 1 & 8 & 27 & 64 \end{vmatrix}$$

18–20. Expand each of the following determinants, and find the value of each.

18. $\begin{vmatrix} 2 & 4 & 2 & 2 \\ 3 & 1 & 0 & 1 \\ 2 & 1 & 0 & -1 \\ 4 & 2 & 1 & 1 \end{vmatrix}$

19. $\begin{vmatrix} 2 & -1 & 0 & 3 \\ 1 & 2 & -1 & -3 \\ 3 & 0 & 2 & 0 \\ 4 & 3 & 1 & 2 \end{vmatrix}$

20. $\begin{vmatrix} 3 & 1 & 0 & 6 & 1 \\ 1 & 2 & 1 & 3 & 1 \\ 1 & -1 & 1 & 1 & 1 \\ -1 & 1 & 1 & 1 & 1 \\ 0 & 0 & 1 & 1 & 1 \end{vmatrix}$

DETERMINANT SOLUTION OF LINEAR SYSTEMS   459

21. Without expanding, show that

$$\begin{vmatrix} 1 & 1 & 1 & 1 \\ 1 & a & a^2 & a^3 \\ 1 & b & b^2 & b^3 \\ 1 & c & c^2 & c^3 \end{vmatrix} = (a-1)(b-1)(c-1)(b-a)(c-a)(c-b).$$

22. (a) Express $f(x)$ as a product of linear factors.

$$f(x) = \begin{vmatrix} 1 & 1 & 1 & 1 \\ x^2 & 2^2 & 3^2 & 4^2 \\ x^3 & 2^3 & 3^3 & 4^3 \\ x^4 & 2^4 & 3^4 & 4^4 \end{vmatrix}$$

(b) By noting that $f(1) = D$, evaluate

$$D = \begin{vmatrix} 1 & 1 & 1 & 1 \\ 1 & 2^2 & 3^2 & 4^2 \\ 1 & 2^3 & 3^3 & 4^3 \\ 1 & 2^4 & 3^4 & 4^4 \end{vmatrix}$$

23. Evaluate

$$\begin{vmatrix} 1 & 1 & 1 & 1 \\ 1 & 2 & 2^2 & 2^3 \\ 1 & 3 & 3^2 & 3^3 \\ 1 & 4 & 4^2 & 4^3 \end{vmatrix}$$

by (a) replacing the first row by $1, x, x^2, x^3$ and calling the resulting determinant $f(x)$,
(b) factoring $f(x)$, and
(c) evaluating $f(1)$ using the factored form of $f(x)$.
[*Note:* This example illustrates a useful mathematical technique for solving problems; that is, first generalize the problem and later specialize.]

24–29. Prove Theorems 1–6 for determinants of order four.

30–35. Show that Theorems 1–6 remain valid for determinants of order n.

11.5 DETERMINANT SOLUTION OF LINEAR SYSTEMS

CASE 1 Two equations, two variables.
In solving the system

$$a_1 x + b_1 y = k_1$$
$$a_2 x + b_2 y = k_2$$

by the addition method, the following equations were obtained:

$$(a_1 b_2 - a_2 b_1)x = b_2 k_1 - b_1 k_2$$
$$(a_1 b_2 - a_2 b_1)y = a_1 k_2 - a_2 k_1.$$

Now, letting

$$D = \begin{vmatrix} a_1 & b_1 \\ a_2 & b_2 \end{vmatrix}, \quad X = \begin{vmatrix} k_1 & b_1 \\ k_2 & b_2 \end{vmatrix}, \quad \text{and} \quad Y = \begin{vmatrix} a_1 & k_1 \\ a_2 & k_2 \end{vmatrix},$$

these equations become $Dx = X$, $Dy = Y$.

The determinant, D, is called the **determinant of the coefficients**.

If $D \neq 0$, then there is a unique solution, $(X/D, Y/D)$. (The case of two intersecting lines.)

If $D = 0$ and $X = Y = 0$, then the solution set is infinite. (The case of coincident lines.)

If $D = 0$ and either $X \neq 0$ or $Y \neq 0$, then the solution set is empty. (The case of parallel lines.)

CASE 2 *Three equations, three variables.* If in the system

$$a_1 x + b_1 y + c_1 z = k_1$$
$$a_2 x + b_2 y + c_2 z = k_2$$
$$a_3 x + b_3 y + c_3 z = k_3,$$

the first equation is multiplied by A_1, the minor of a_1 for the determinant of the coefficients, and the second equation is multiplied by $-A_2$, and the third equation is multiplied by A_3, and the resulting three equations are added, then

$$(a_1 A_1 - a_2 A_2 + a_3 A_3)x + (b_1 A_1 - b_2 A_2 + b_3 A_3)y$$
$$+ (c_1 A_1 - c_2 A_2 + c_3 A_3)z = k_1 A_1 - k_2 A_2 + k_3 A_3.$$

The coefficients of the variables and the constant term can be recognized as the expansions of determinants as follows:

$$\begin{vmatrix} a_1 & b_1 & c_1 \\ a_2 & b_2 & c_2 \\ a_3 & b_3 & c_3 \end{vmatrix} x + \begin{vmatrix} b_1 & b_1 & c_1 \\ b_2 & b_2 & c_2 \\ b_3 & b_3 & c_3 \end{vmatrix} y + \begin{vmatrix} c_1 & b_1 & c_1 \\ c_2 & b_2 & c_2 \\ c_3 & b_3 & c_3 \end{vmatrix} z = \begin{vmatrix} k_1 & b_1 & c_1 \\ k_2 & b_2 & c_2 \\ k_3 & b_3 & c_3 \end{vmatrix}.$$

Thus,

$$\begin{vmatrix} a_1 & b_1 & c_1 \\ a_2 & b_2 & c_2 \\ a_3 & b_3 & c_3 \end{vmatrix} x + 0 \cdot y + 0 \cdot z = \begin{vmatrix} k_1 & b_1 & c_1 \\ k_2 & b_2 & c_2 \\ k_3 & b_3 & c_3 \end{vmatrix}$$

since the determinants which are the coefficients of y and z each contain two identical columns and thus equal 0.

Similar equations may be obtained where x and y or x and z are eliminated. Finally,

$$\begin{vmatrix} a_1 & b_1 & c_1 \\ a_2 & b_2 & c_2 \\ a_3 & b_3 & c_3 \end{vmatrix} x = \begin{vmatrix} k_1 & b_1 & c_1 \\ k_2 & b_2 & c_2 \\ k_3 & b_3 & c_3 \end{vmatrix}$$

$$\begin{vmatrix} a_1 & b_1 & c_1 \\ a_2 & b_2 & c_2 \\ a_3 & b_3 & c_3 \end{vmatrix} y = \begin{vmatrix} a_1 & k_1 & c_1 \\ a_2 & k_2 & c_2 \\ a_3 & k_3 & c_3 \end{vmatrix}$$

DETERMINANT SOLUTION OF LINEAR SYSTEMS

$$\begin{vmatrix} a_1 & b_1 & c_1 \\ a_2 & b_2 & c_2 \\ a_3 & b_3 & c_3 \end{vmatrix} z = \begin{vmatrix} a_1 & b_1 & k_1 \\ a_2 & b_2 & k_2 \\ a_3 & b_3 & k_3 \end{vmatrix}.$$

The determinant of the coefficients, D, may be readily obtained from the equations when they are expressed in the form stated at the beginning of this section. The determinants on the right of each equation above can be obtained from the determinant of the coefficients by replacing the column containing the coefficients of the variable in the equation by the constant terms of the system.

Now, designating the determinants on the right of each equation by X, Y, and Z, respectively, these equations can now be expressed as follows:

$$Dx = X$$
$$Dy = Y$$
$$Dz = Z.$$

If $D \neq 0$, then there is exactly one solution: $(X/D, Y/D, Z/D)$.
If $D = 0$ and $X = Y = Z = 0$, then the solution set is infinite.
If $D = 0$ and either $X \neq 0$ or $Y \neq 0$ or $Z \neq 0$, then the solution set is empty.
This solution of linear systems using determinants is known as **Cramer's rule** in honor of the Swiss mathematician, Gabriel Cramer (1704–1752).

Example 1 Solve by using determinants. Check.

$$x + y + z = 4$$
$$2x - y - 2z = -1$$
$$x - 2y - z = 1.$$

Solution

$$D = \begin{vmatrix} 1 & 1 & 1 \\ 2 & -1 & -2 \\ 1 & -2 & -1 \end{vmatrix} = \begin{vmatrix} 1 & 1 & 2 \\ 2 & -1 & 0 \\ 1 & -2 & 0 \end{vmatrix} = 2 \begin{vmatrix} 2 & -1 \\ 1 & -2 \end{vmatrix} = 2(-4 + 1) = -6$$

$$X = \begin{vmatrix} 4 & 1 & 1 \\ -1 & -1 & -2 \\ 1 & -2 & -1 \end{vmatrix} = \begin{vmatrix} 4 & -3 & -7 \\ -1 & 0 & 0 \\ 1 & -3 & -3 \end{vmatrix} = \begin{vmatrix} -3 & -7 \\ -3 & -3 \end{vmatrix} = \begin{vmatrix} 3 & 7 \\ 3 & 3 \end{vmatrix}$$

$$= 9 - 21 = -12$$

$$Y = \begin{vmatrix} 1 & 4 & 1 \\ 2 & -1 & -2 \\ 1 & 1 & -1 \end{vmatrix} = \begin{vmatrix} 1 & 4 & 1 \\ 0 & -3 & 0 \\ 1 & 1 & -1 \end{vmatrix} = -3 \begin{vmatrix} 1 & 1 \\ 1 & -1 \end{vmatrix} = -3(-1 - 1) = 6$$

$$Z = \begin{vmatrix} 1 & 1 & 4 \\ 2 & -1 & -1 \\ 1 & -2 & 1 \end{vmatrix} = \begin{vmatrix} 1 & 1 & 4 \\ 2 & -1 & -1 \\ 3 & -3 & 0 \end{vmatrix} = \begin{vmatrix} 1 & 2 & 4 \\ 2 & 1 & -1 \\ 3 & 0 & 0 \end{vmatrix} = 3 \begin{vmatrix} 2 & 4 \\ 1 & -1 \end{vmatrix}$$

$$= 3(-2 - 4) = -18$$

Thus,

$$x = \frac{X}{D} = \frac{-12}{-6} = 2$$

$$y = \frac{Y}{D} = \frac{6}{-6} = -1$$

$$z = \frac{Z}{D} = \frac{-18}{-6} = 3.$$

The solution is $(2, -1, 3)$.

CHECK $x + y + z = 2 - 1 + 3 = 4.$
$2z - y - 2z = 2(2) - (-1) - 2(3) = 4 + 1 - 6 = -1.$
$x - 2y - z = 2 - 2(-1) - 3 = 2 + 2 - 3 = 1.$

Example 2 Solve by using determinants.

$$2x - 4y + 2z = 3$$
$$x + y - z = 2$$
$$3x - 6y + 3z = 2$$

Solution

$$D = \begin{vmatrix} 2 & -4 & 2 \\ 1 & 1 & -1 \\ 3 & -6 & 3 \end{vmatrix} = \begin{vmatrix} 4 & -2 & 2 \\ 0 & 0 & -1 \\ 6 & -3 & 3 \end{vmatrix} = -(-1)\begin{vmatrix} 4 & -2 \\ 6 & -3 \end{vmatrix} = -12 + 12 = 0$$

$$X = \begin{vmatrix} 3 & -4 & 2 \\ 2 & 1 & -1 \\ 2 & -6 & 3 \end{vmatrix} = \begin{vmatrix} 7 & -2 & 2 \\ 0 & 0 & -1 \\ 8 & -3 & 3 \end{vmatrix} = -(-1)\begin{vmatrix} 7 & -2 \\ 8 & -3 \end{vmatrix} = -21 + 16 = -5$$

Since $D = 0$ and $X \neq 0$, the solution set is the empty set, $\emptyset$.

Example 3 Solve by using determinants and check.

$$x + 2y - 3z = 4$$
$$2x - y + z = 1$$
$$3x + y - 2z = 5$$

Solution

$$D = \begin{vmatrix} 1 & 2 & -3 \\ 2 & -1 & 1 \\ 3 & 1 & -2 \end{vmatrix} = \begin{vmatrix} 1 & 0 & 0 \\ 2 & -5 & 7 \\ 3 & -5 & 7 \end{vmatrix} = 0$$

$$X = \begin{vmatrix} 4 & 2 & -3 \\ 1 & -1 & 1 \\ 5 & 1 & -2 \end{vmatrix} = \begin{vmatrix} 6 & 2 & -1 \\ 0 & -1 & 0 \\ 6 & 1 & -1 \end{vmatrix} = -1\begin{vmatrix} 6 & -1 \\ 6 & -1 \end{vmatrix} = 0$$

DETERMINANT SOLUTION OF LINEAR SYSTEMS

$$Y = \begin{vmatrix} 1 & 4 & -3 \\ 2 & 1 & 1 \\ 3 & 5 & -2 \end{vmatrix} = \begin{vmatrix} 1 & 1 & -3 \\ 2 & 2 & 1 \\ 3 & 3 & -2 \end{vmatrix} = 0$$

$$Z = \begin{vmatrix} 1 & 2 & 4 \\ 2 & -1 & 1 \\ 3 & 1 & 5 \end{vmatrix} = \begin{vmatrix} 1 & 2 & 4 \\ 3 & 1 & 5 \\ 3 & 1 & 5 \end{vmatrix} = 0$$

Thus, the solution set is infinite. Now, try to solve two equations for which the determinant of the coefficients of two of the variables, say x and y, is *not* zero. Using the first and second equations,

$$x + 2y = 4 + 3z$$
$$2x - y = 1 - z.$$

$$D = \begin{vmatrix} 1 & 2 \\ 2 & -1 \end{vmatrix} = -1 - 4 = -5, \quad X = \begin{vmatrix} 4+3z & 2 \\ 1-z & -1 \end{vmatrix} = -6 - z,$$

$$Y = \begin{vmatrix} 1 & 4+3z \\ 2 & 1-z \end{vmatrix} = -7 - 7z.$$

Thus,

$$x = \frac{X}{D} = \frac{-6-z}{-5} = \frac{6+z}{5} \quad \text{and} \quad y = \frac{Y}{D} = \frac{-7-7z}{-5} = \frac{7+7z}{5}.$$

Thus, the solution set is

$$\left\{ \left(\frac{6+z}{5}, \frac{7+7z}{5}, z\right) \;\middle|\; z \text{ is any real number.} \right\}$$

CHECK $\quad x + 2y - 3z = \dfrac{6+z}{5} + 2\left(\dfrac{7+7z}{5}\right) - 3z$

$$= \frac{6 + 14 + z + 14z - 15z}{5} = 4$$

$$2x - y + z = 2\left(\frac{6+z}{5}\right) - \frac{7+7z}{5} + z$$

$$= \frac{12 - 7 + 2z - 7z + 5z}{5} = 1$$

$$3x + y - 2z = 3\left(\frac{6+z}{5}\right) + \frac{7+7z}{5} - 2z$$

$$= \frac{18 + 7 + 3z + 7z - 10z}{5} = 5$$

Cramer's Rule for n Linear Equations in n Variables Let the following be a system of n linear equations in the variables $x_1, x_2, \ldots, x_n$:
$$a_{11}x_1 + a_{12}x_2 + \ldots + a_{1n}x_n = c_1$$
$$a_{21}x_2 + a_{22}x_2 + \ldots + a_{2n}x_n = c_2$$
$$\vdots \qquad\qquad\qquad\qquad\qquad \vdots$$
$$a_{n1}x_1 + a_{n2}x_2 + \ldots + a_{nn}x_n = c_n$$

Let D be the determinant of the coefficients:
$$D = \begin{vmatrix} a_{11} & a_{12} & \ldots & a_{1n} \\ a_{21} & a_{22} & \ldots & a_{2n} \\ \ldots & \ldots & \ldots & \ldots \\ a_{n1} & a_{n2} & \ldots & a_{nn} \end{vmatrix}$$

Let $N_1, N_2, \ldots, N_n$ be the determinants obtained by replacing the 1st, 2nd, ..., nth columns, respectively, of D by the column of constant terms, $c_1, c_2, \ldots, c_n$.

Then the given linear system is equivalent to the system
$$Dx_1 = N_1, \qquad Dx_2 = N_2, \ldots, \qquad Dx_n = N_n.$$

If $D \neq 0$, then there is exactly one solution, an ordered n-tuple.

If $D = 0$ and $N_1 = N_2 = \ldots = N_n = 0$, then there are infinitely many solutions.

If $D = 0$ and $N_j \neq 0$ for some j, then the solution set is empty.

The proof of this theorem is similar to the proof for the case of three linear equations in three variables. Although it is simple and direct, it is omitted because it is rather tedious to write.

Definition (Homogeneous Linear Systems) A *homogeneous linear system* of n linear equations in n variables is a linear system in which all the constant terms are 0; that is, $c_1 = c_2 = \ldots = c_n = 0$.

Theorem on Homogeneous Linear Systems A homogeneous linear system of n linear equations in n variables has a solution different from the trivial solution $(0, 0, \ldots, 0)$ (which is a solution of all homogeneous linear systems) if and only if the determinant of the coefficients, $D = 0$.

The proof of this theorem is left for the student.

Example 4 Solve the system
$$x + 2y - z = 0$$
$$2x - y + 2z = 0$$
$$2x + 9y - 6z = 0.$$

Solution The determinant of the coefficients,
$$D = \begin{vmatrix} 1 & 2 & -1 \\ 2 & -1 & 2 \\ 2 & 9 & -6 \end{vmatrix} = 0.$$

DETERMINANT SOLUTION OF LINEAR SYSTEMS

Now solve the system of the first two equations as follows:

$$\begin{cases} x + 2y = z \\ 2x - y = -2z \end{cases} \quad \text{and} \quad x = \frac{-z + 4z}{-1 - 4} = \frac{-3z}{5}, \quad y = \frac{-2z - 2z}{-5} = \frac{4z}{5}.$$

Checking in the third equation,

$$2x + 9y - 6z = 2\left(\frac{-3z}{5}\right) + 9\left(\frac{4z}{5}\right) - 6z = \frac{30z}{5} - 6z = 0.$$

The solution set is

$$\left\{ \left(\frac{-3z}{5}, \frac{4z}{5}, z \right) \right\}.$$

EXERCISES 11.5

1–16. Solve each of the following systems by determinants.

1. $x - 2y - 3z = -20$
   $2x + 4y - 5z = 11$
   $3x + 7y - 4z = 33$

2. $4x + y - 3z = 0$
   $x - y + z = -7$
   $3x + 2y - z = -5$

3. $8x + 2y = 5$
   $4y - 3z = 0$
   $4x + 6z = -1$

4. $x + y = z$
   $2x + 2z = 3 - 2y$
   $5x + 2y = z + 2$

5. $3x + 2y - 4 = 0$
   $4y - z + 2 = 0$
   $5x + 2z - 20 = 0$

6. $6x + 3y - 3z = 2$
   $2x + 2y - 2z = 5$
   $3x - 3y + 3z = 7$

7. $2x + y - z + t = 1$
   $x - y + 2z - t = 2$
   $-x - y + z - t = -1$
   $3x + y - z + 2t = 0$

8. $x - 2z = 4$
   $y + z + t = 6$
   $x + 3t = 6$
   $x + y + z + t = 0$

9. $2x + y + 3z = 15$
   $2x + 7z = 25$
   $3x + 2y + 6z = 35$

10. $x + 2y = 2 + z$
    $x - 4y = 5z - 7$
    $x + 3y + 4z = 5$

11. $x + 2y = z + 7$
    $4x + 3y = 1 - 2z$
    $9x + 8y = 4 - 3z$

12. $5x + 2y - z + 1 = 0$
    $2x - y + z = 0$
    $3x + z - 1 = 0$

13. $2x - 4y + 7z = 5$
    $3x + 2y - z = 2$
    $x - 10y + 15z = 8$

14. $x - 2y + 3 = 0$
    $4x + z - 2 = 0$
    $3x + 2y - 2z = 5$

15. $x + y + z + t = 2$
    $2x + 2y - z = -3$
    $y - 2z - 2t = 1$
    $x + y - t = 5$

16. $x + y + z = 14$
    $x + z + t = 13$
    $y + z + t = 10$
    $x - y + z - t = 9$

17. There are 47 coins in a collection of nickels, dimes, and quarters. The total value of the collection is $5. The total number of nickels and quarters is three less than the number of dimes. How many coins of each kind are there?

18. Three machines, working together, require 20 min to complete a certain job. If the first two machines require 30 min to complete the job when working together, and if the first and the third require 36 min, find the time it would take for each machine alone.

19. Applying Kirchhoff's laws to the electrical circuit shown in Figure 11.2, the following equations are obtained:

$$I_1 - I_2 - I_3 = 0$$
$$5I_1 + 20I_3 = 100$$
$$15I_2 - 20I_3 = 15$$
$$5I_1 + 15I_2 = 115$$

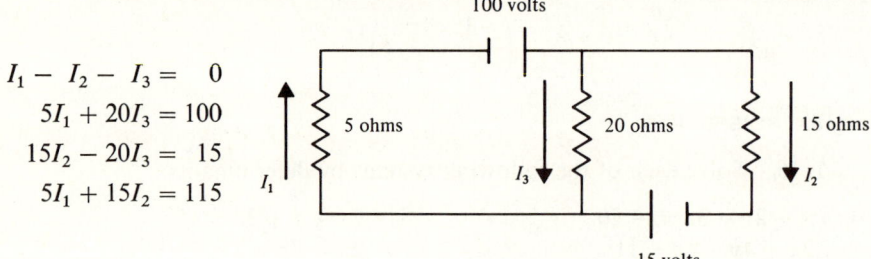

Figure 11.2

Solve this system by showing that the common solution of the first three equations is also a solution of the fourth equation.

20. A man invested $40,000 in three different investments. The first yielded 6 percent, the second 5 percent, and the third 10 percent. The total income was $2660. If the last two amounts had been interchanged, his income would have been $2860. Find the amount of money invested at each rate.

21. A factory has 2400 lb of A, 310 lb of B, and 28 lb of C. Product P requires 25 lb of A and 5 lb of B. Product Q requires 20 lb of A, 2 lb of B, and 1 lb of C. Product R requires 150 lb of A, 10 lb of B, and $\frac{1}{2}$ lb of C. How many items of each product should be produced in order to use all of the raw material?

22. The copper content of U.S. coins is 95 percent for the cent, 75 percent for the nickel, and 10 percent for the dime. For 82 lb of coins, all of the nickels and one third of the dimes were made from 50 lb of ore containing 50 percent copper and the rest of the dimes and all of the cents were made from 50 lb of ore containing 42 percent copper. How many pounds of each type were there?

23. A man went to a bus station in a cab averaging 25 mph, and took a bus averaging 40 mph to an airport. At the airport he boarded a plane, which flew him to his destination. The plane averaged 600 mph. The entire trip of 1469 miles required 3 hr and 12 min. The plane trip was three times as long as the other two trips combined. How much time was spent in each type of travel?

24. Find the equation of the plane $Ax + By + Cz = D$ determined by the three points $P(2, -1, 0)$, $Q(0, 3, 4)$, $R(-3, 0, 5)$.
25. Find all possible amounts of the following four foods that will provide precisely the amounts of nutrients indicated in the last column of the array if each contains the amounts per unit as indicated.

NUTRIENT	FOOD				
	I	II	III	IV	TOTAL
A	1	3	3	0	12
B	2	2	0	3	26
C	3	1	5	4	44
D	3	9	1	2	32

If food I costs 40 cents per unit, food II 40 cents per unit, food III 10 cents per unit, and food IV 20 cents per unit, is there a solution costing exactly $1? Exactly $3?

26. Prove that a homogeneous linear system of n linear equations in n variables has a solution different from the trivial solution $(0, 0, \ldots, 0)$ if and only if the determinant of the coefficients, $D = 0$.
27. Show that the system
$$ax + by = 0$$
$$cx + dy = 0$$
has more than one solution if and only if $ad - bc = 0$.
28. (a) Find k so that the system
$$x + y - kz = 0$$
$$x - y + 3z = 0$$
$$x + 2y - 2z = 0$$
has a solution different from $(0, 0, 0)$.
   (b) Solve the system for the value of k obtained in part (a).

29–30. State the solution set of each of the following systems.

29. $x - 2y + z = 0$
    $2x + y - z = 0$
    $4x - 3y + z = 0$

30. $2x - y + z = 0$
    $x + y + z = 0$
    $3x - 2y - z = 0$

31–32. Find k so that each of the following systems has a solution for which $z \neq 0$.

31. $3x + 2y + 3z = 0$
    $2x + 5y - 2z = 0$
    $4x - y + kz = 0$

32. $2x - y + 5z = 0$
    $x + 3y - 4z = 0$
    $5x + y + kz = 0$

11.6 QUADRATIC SYSTEMS

A quadratic system of two equations in two variables, x and y, is a set of two quadratic equations having the form

$$Ax^2 + Bxy + Cy^2 + Dx + Ey + F = 0, \qquad A^2 + B^2 + C^2 \neq 0$$

or a set of one quadratic equation having the above form and one linear equation having the form $Ax + By + C = 0$; $A^2 + B^2 \neq 0$.

The solution set of a quadratic system is the set of ordered pairs that is the intersection of the solutions sets of the equations of the system.

Since a real solution (a, b) of a system of two equations in two variables x and y is also the name of a geometric point of intersection of the graphs of the two equations, the graphs can be used to locate or to approximate the *real* solutions of the quadratic system.

A quadratic system consisting of one linear equation and one quadratic equation has zero, one, or two real solutions. A quadratic system consisting of two quadratic equations has zero, one, two, three, or four real solutions. The different possibilities are illustrated in Figure 11.3.

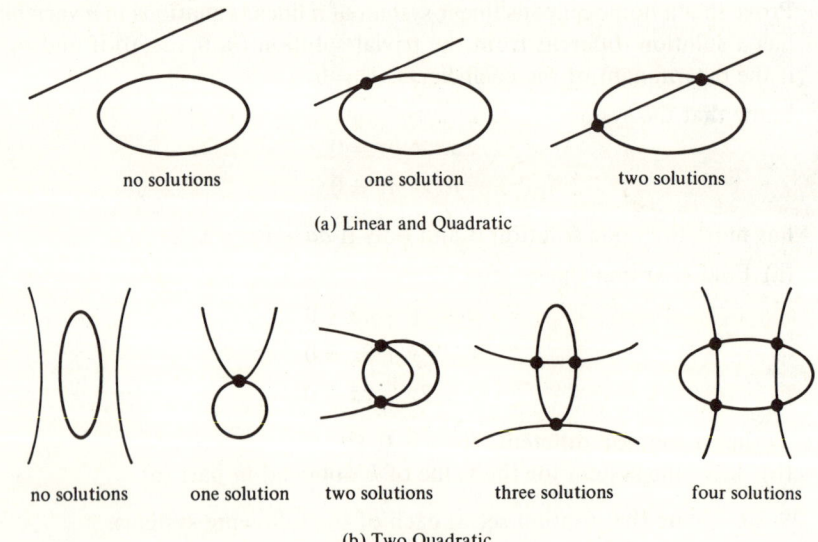

Figure 11.3

11.6.1 Graphic Solution

The graphic solution of a quadratic system yields the real solutions *only*. Moreover, due to the limitations imposed by eyesight and drawing techniques, these real solutions, as a general rule, can only be approximated. Thus, it is essential to check a proposed solution obtained by the method of graphs.

QUADRATIC SYSTEMS

Example 1 Estimate graphically the real solutions of the quadratic system:

$$x^2 + 16y^2 = 169$$
$$x^2 - y^2 = 16.$$

Solution

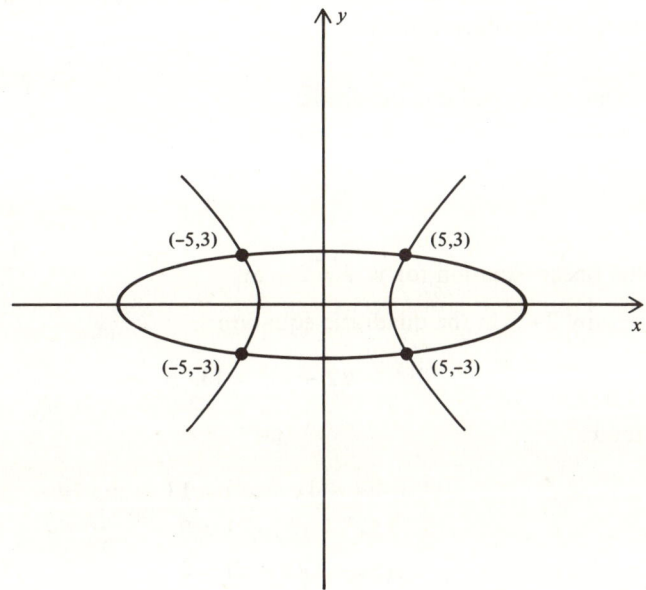

Figure 11.4

(1) Graph both equations on the same axes.

(2) Determine the coordinates of the points of intersection $(5, 3)$, $(-5, 3)$, $(-5, -3)$, $(5, -3)$.

(3) Check each of these proposed solutions.
The check is illustrated for $(5, 3)$. The other checks are left to the student.
If $x = 5$ and $y = 3$, then

$$x^2 + 16y^2 = 5^2 + 16(3^2) = 25 + 144 = 169$$

and

$$x^2 - y^2 = 5^2 - 3^2 = 25 - 9 = 16.$$

Thus $(5, 3)$ is a solution.

(4) List the solution set.

$$\{(5, 3), (-5, 3), (-5, -3), (5, -3)\}$$

11.6.2 Algebraic Solution, Substitution Method

If one of the equations in a quadratic system can be solved for one variable, expressed explicitly as a function of the other variable, then the expression obtained may be *substituted* for the variable in the other equation. This replacement yields an equation in one variable. The solution is completed by solving this equation for the one variable, and then by using the function to obtain the corresponding values of the other variable.

Example 2 One linear and one quadratic.
Solve and check: $x^2 + 4y^2 = 13$, $x + y = 2$.

Solution

(1) Solve the linear equation for y: $y = 2 - x$.

(2) Replace y by $2 - x$ in the quadratic equation:

$$x^2 + 4(2 - x)^2 = 13.$$

(3) Solve for x:

$$x^2 + 4(4 - 4x + x^2) = 13$$
$$5x^2 - 16x + 3 = 0$$
$$(5x - 1)(x - 3) = 0$$
$$5x - 1 = 0 \quad \text{or} \quad x - 3 = 0$$
$$x = \frac{1}{5} \quad \text{or} \quad x = 3$$

(4) Solve for y, using the linear equation, $y = 2 - x$:

$$\text{If } x = \frac{1}{5}, \text{ then } y = 2 - \frac{1}{5} = \frac{9}{5}.$$

$$\text{If } x = 3, \text{ then } y = 2 - 3 = -1.$$

(5) Check each solution in both equations.

$$\left(\frac{1}{5}, \frac{9}{5}\right): x^2 + 4y^2 = \frac{1}{25} + 4\left(\frac{81}{25}\right) = \frac{325}{25} = 13$$

$$x + y = \frac{1}{5} + \frac{9}{5} = \frac{10}{5} = 2$$

$(3, -1): x^2 + 4y^2 = 9 + 4(1) = 13$, $x + y = 3 + (-1) = 2$

(6) State the solution set:

$$\left\{\left(\frac{1}{5}, \frac{9}{5}\right), (3, -1)\right\}$$

Example 3 Two quadratic equations.
Solve and check: $4x^2 + y^2 = 16$, $x^2 - y = 4$.

Solution

(1) Solve $x^2 - y = 4$ for x^2: $x^2 = 4 + y$.

(2) Replace x^2 by $4 + y$ in the other equation: $4(4 + y) + y^2 = 16$.

(3) Solve for y:

$$y^2 + 4y = 0$$

$$y(y + 4) = 0$$

Thus, $y = 0$ or $y = -4$.

(4) Solve for x in the first equation used: $(x^2 = 4 + y)$.
If $y = 0$, then $x^2 = 4 + 0 = 4$ and $x = 2$ or $x = -2$.
If $y = -4$, then $x^2 = 4 - 4 = 0$.
Thus, $x = 0$.

(5) Check each solution in both equations.

$(0, -4)$: $4x^2 + y^2 = 4(0) + (-4)^2 = 0 + 16 = 16$
$x^2 - y = 0^2 - (-4) = 4$

$(2, 0)$: $4x^2 + y^2 = 4(2^2) + 0^2 = 4(4) = 16$
$x^2 - y = 2^2 - 0 = 4$

$(-2, 0)$: $4x^2 + y^2 = 4(-2)^2 + 0^2 = 4(4) = 16$
$x^2 - y = (-2)^2 - 0 = 4$

(6) State the solution set:

$$\{(0, -4), (2, 0), (-2, 0)\}$$

Example 4 Two quadratic equations.
Solve and check: $4x^2 - y^2 = 5$, $xy = 3$.

Solution

(1) If $xy = 3$, then $y = \dfrac{3}{x}$ (for $x \neq 0$).

(2) $4x^2 - \left(\dfrac{3}{x}\right)^2 = 5$

$4x^2 - \dfrac{9}{x^2} = 5$

$4x^4 - 9 = 5x^2$

$4x^4 - 5x^2 - 9 = 0$

$(4x^2 - 9)(x^2 + 1) = 0$

$4x^2 - 9 = 0$ or $x^2 + 1 = 0$

$x^2 = \dfrac{9}{4}$ or $x^2 = -1$

Thus, $x = \dfrac{3}{2}$, $x = \dfrac{-3}{2}$, $x = i$, or $x = -i$.

(3) Using $y = \dfrac{3}{x}$ to find y:

If $x = \dfrac{3}{2}$, then $y = \dfrac{3}{\frac{3}{2}} = 2$.

If $x = -\dfrac{3}{2}$, then $y = -2$.

If $x = i$, then $y = \dfrac{3}{i} = \dfrac{3i}{ii} = -3i$.

If $x = -i$, then $y = 3i$.

(4) Check each solution in both equations. The check is left to the student.

(5) The solution set is

$$\left\{\left(\dfrac{3}{2}, 2\right), \left(-\dfrac{3}{2}, -2\right), (i, -3i), (-i, 3i)\right\}.$$

11.6.3 Algebraic Solution, Addition Method

ELIMINATING ONE OF THE VARIABLES

If one equation of a quadratic system is linear, then the solution is easily obtained by the substitution method. However, when both equations are

QUADRATIC SYSTEMS

quadratic, the substitution method generally yields a complicated equation involving radicals, and, upon simplification, a fourth degree equation which is not always easy to solve.

For certain special cases, it is possible to eliminate one of the variables by the *addition method* which is similar to that used for linear systems. Each equation of the system is multiplied by an appropriate constant and the resulting equations are added. The procedure is justified by reasoning similar to that used for the linear systems.

If (p, q) is a solution of both

$$A_1 x^2 + B_1 xy + C_1 y^2 + D_1 x + E_1 y + F_1 = 0$$

and

$$A_2 x^2 + B_2 xy + C_2 y^2 + D_2 x + E_2 y + F_2 = 0,$$

then it is also a solution of any linear combination of the two quadratic equations; that is, (p, q) is a solution of

$$a(A_1 x^2 + B_1 xy + C_1 y^2 + D_1 x + E_1 y + F_1) \\ + b(A_2 x^2 + B_2 xy + C_2 y^2 + D_2 x + E_2 y + F_2) = 0$$

since

$$a(A_1 p^2 + B_1 pq + C_1 q^2 + D_1 p + E_1 q + F_1) \\ + b(A_2 p^2 + B_2 pq + C_2 q^2 + D_2 p + E_2 q + F_2) = 0,$$

or

$$a(0) + b(0) = 0.$$

Example 5 Using the addition method, solve the system:

$$3x^2 + 2y^2 = 23$$
$$x^2 - 3y^2 = -7$$

Solution

(1) Multiply first equation by 3. $\quad 9x^2 + 6y^2 = 69$
    Multiply second equation by 2. $\quad 2x^2 - 6y^2 = -14$
(2) Add the new equations. $\quad 11x^2 = 55$
(3) Solve for x. $\quad x^2 = 5$. Thus $x = \pm\sqrt{5}$.
(4) Multiply first equation by 1. $\quad 3x^2 + 2y^2 = 23$
    Multiply second equation by -3. $\quad -3x^2 + 9y^2 = 21$
(5) Add the new equations. $\quad 11y^2 = 44$
(6) Solve for y. $\quad y^2 = 4$. Thus, $y = \pm 2$.

(7) List the solution set: $\{(\sqrt{5}, 2), (\sqrt{5}, -2), (-\sqrt{5}, 2), (-\sqrt{5}, -2)\}$
(8) Check each solution in both equations of the original system. (The check is left for the student.)

ELIMINATING THE SECOND DEGREE TERMS

When it is not possible to eliminate one of the variables by adding multiples of the two equations, it may be possible to eliminate the second degree terms and obtain a linear equation. Since each solution of the quadratic system must also be a solution of the linear equation, the linear equation may be solved simultaneously with either equation of the quadratic system.

Example 6 Solve $x^2 + y^2 - 2x = 9$
$x^2 + y^2 - 4y = 1$.

Solution

(1) Eliminate the second degree terms:

$(x^2 + y^2 - 2x = 9)(-1)$ $-x^2 - y^2 + 2x = -9$
$(x^2 + y^2 - 4y = 1)(1)$ $x^2 + y^2 - 4y = 1$
Adding, $2x - 4y = -8$
$x - 2y = -4$

(2) Now solve the equivalent system:

$$x^2 + y^2 - 4y = 1$$
$$x - 2y = -4$$

Using the substitution method,

$$x = 2y - 4$$
$$(2y - 4)^2 + y^2 - 4y = 1$$
$$5y^2 - 20y + 15 = 0$$
$$y^2 - 4y + 3 = 0$$
$$(y - 3)(y - 1) = 0.$$

Thus, $y = 3$ or $y = 1$.

If $y = 3$, then $x = 2y - 4 = 2(3) - 4 = 2$.
If $y = 1$, then $x = 2y - 4 = 2(1) - 4 = -2$.

Thus, the solution set is $\{(2, 3), (-2, 1)\}$.

ELIMINATING THE CONSTANTS

Sometimes the solution set of a quadratic system may be easily obtained by using the addition method to eliminate the constants.

QUADRATIC SYSTEMS

Example 7 Solve $x^2 + xy = 4$
$y^2 - xy = 6.$

Solution Eliminating the constants,

$$(x^2 + xy = 4)(3) \qquad 3x^2 + 3xy = 12$$
$$(y^2 - xy = 6)(-2) \qquad -2y^2 + 2xy = -12.$$

Adding, $3x^2 + 5xy - 2y^2 = 0.$
Factoring, $(3x - y)(x + 2y) = 0$

$$3x - y = 0 \quad \text{or} \quad x + 2y = 0$$
$$y = 3x \quad \text{or} \quad x = -2y.$$

Now, either equation of the original system can be replaced by this set of two linear equations. Thus, the original system is equivalent to the *union* of the two systems

$$x^2 + xy = 4 \quad \text{and} \quad x^2 + xy = 4$$
$$y = 3x \qquad\qquad x = -2y.$$

Each of these systems can now be solved by the substitution method.

$$x^2 + x(3x) = 4 \qquad (-2y)^2 + (-2y)y = 4$$
$$4x^2 = 4 \qquad\qquad 2y^2 = 4$$
$$x^2 = 1 \qquad\qquad y^2 = 2$$
$$x = \pm 1 \qquad\qquad y = \pm\sqrt{2}$$

If $x = 1$, then $y = 3x = 3$.   If $y = \sqrt{2}$, then $x = -2y = -2\sqrt{2}.$
If $x = -1$, then $y = -3$.   If $y = -\sqrt{2}$, then $x = 2\sqrt{2}.$

Thus, the solution set is $\{(1, 3), (-1, -3), (-2\sqrt{2}, \sqrt{2}), (2\sqrt{2}, -\sqrt{2})\}$. Each of these solutions should be checked in both equations of the original system.

EXERCISES 11.6

1–12. Estimate graphically, to the nearest tenth, the real solutions of each of the following quadratic systems.

1. $x^2 - 2x - y = 5$
   $x + y = 1$
2. $9x^2 + y^2 = 9$
   $y = 4x + 5$
3. $x^2 - y^2 = 16$
   $x - 3y = 0$
4. $y^2 - 2x^2 = 4$
   $y^2 = 9x$
5. $xy = 6$
   $9x^2 + 16y^2 = 144$
6. $4x + y^2 = 36$
   $x^2 + y^2 = 81$

7. $x^2 + y^2 - 4x - 10y + 4 = 0$
   $3x + y = 6$
8. $xy + 5 = 0$
   $4y - 5x = 20$
9. $y = 9 - x^2$
   $y = x + 10$
10. $x^2 + 3y^2 = 12$
    $x^2 + y + 2 = 0$
11. $x^2 - y^2 = 5$
    $3y^2 - x^2 = 3$
12. $x^2 - 2x - y + 1 = 0$
    $x^2 + y^2 - 2x - 4y + 1 = 0$

13-28. Solve each of the following systems by the substitution method. Check. Leave irrational answers in simplified radical form. Express imaginary answers in the $a + bi$ form.

13. $x^2 + y^2 - 2x = 9$
    $x + 2y + 4 = 0$
14. $x^2 + 2x + 4 = y$
    $x = 4y - 16$
15. $x^2 - y^2 - 16 = 0$
    $3x + y = 8$
16. $4x^2 + y^2 = 25$
    $2x - y = 1$
17. $xy + 12 = 0$
    $2x + 3y = 6$
18. $y^2 = 4x$
    $x^2 + 3 = y^2$
19. $4a^2 + b^2 = 16$
    $b + 2 = a^2$
20. $c^2 - 5d^2 = 1$
    $cd = 2$
21. $3x - y = 6$
    $y^2 + 3x - 8y = 0$
22. $xy = 4$
    $x + 3y = 10$
23. $2x^2 + y^2 = 12$
    $x - 2y = 2$
24. $2x^2 - 2xy + y^2 = 10$
    $2x = y - 2$
25. $9x^2 - 4y^2 = 0$
    $y^2 = 4x + 1$
26. $xy + y = 1$
    $xy - x = 4$
27. $y^2 - x = 9$
    $4y^2 + x - y = 67$
28. $y^2 - x^2 = 16$
    $2y + 2x^2 + 32 = 0$

29-32. Using the addition method, solve by eliminating one of the variables. Check each solution.

29. $2x^2 + y^2 = 13$
    $3x^2 + y^2 = 17$
30. $x^2 - y^2 + 7 = 0$
    $3x^2 + 2y^2 = 24$
31. $x^2 - 5y^2 + 3 = 0$
    $2x^2 + y^2 - 5 = 0$
32. $3x^2 - 2y^2 - 6 = 0$
    $5x^2 - 3y^2 - 7 = 0$

33-37. Solve by eliminating either the second degree terms or the constants. Check each solution.

33. $y^2 + xy = 12$
    $2x^2 - xy = 24$
34. $3xy + y^2 = 10$
    $xy = 3$
35. $x^2 - y^2 + 2x - y = -9$
    $x^2 - y^2 + 4x - 2y = -9$
36. $4x^2 - xy = 4$
    $3y^2 + xy = 3$
37. $xy = 120$
    $xy + x - 4y = 124$

THE RESULTANT (OPTIONAL)

38–42. Solve and check each of the following systems.

38. $2xy + 13x + 13y + 56 = 0$
    $3x^2 + 2xy + 3y^2 - 8x - 8y = 88$

39. $x^2 + y^2 = 4$
    $xy = 10 - x - y$

40. $x^2 + y^2 = x + y$
    $xy = 3$

41. $3x^2 - 4xy + 3y^2 = 22$
    $x^2 + y^2 - 3xy + 3x + 3y = 3$

42. $(x + y)^2 + 9(x + y) = 10$
    $(x - y)^2 = 20$

11.7 THE RESULTANT (Optional)

It was observed in the preceding sections that a system of n linear equations in n variables has exactly one solution if the determinant of its coefficients is not 0, and it has infinitely many solutions if all nth order determinants formed from the associated augmented matrix are 0. There is a determinant R called the **resultant** that provides a method for determining whether equations of higher degree have a common solution.

Definition of Resultant If $f(x) = a_n x^n + a_{n-1} x^{n-1} + \ldots + a_0$ and $g(x) = b_m x^m + b_{m-1} x^{m-1} + \ldots + b_0$ are polynomials in the variable x of degrees n and m, respectively, where $n \geq 1$ and $m \geq 1$, then their **resultant** R is defined as follows:

$$R = \begin{vmatrix} a_n & a_{n-1} & \cdots & & & a_0 & & & \\ & a_n & a_{n-1} & \cdots & & a_1 & a_0 & & \\ & & & & a_n & a_{n-1} & \cdots & a_0 & \\ b_m & b_{m-1} & \cdots & & b_0 & & & & \\ & b_m & b_{m-1} & \cdots & & b_0 & & & \\ & & & b_m & b_{m-1} & \cdots & & b_0 & \end{vmatrix} \begin{matrix} \Big\} m \text{ rows} \\ \\ \Big\} n \text{ rows} \end{matrix}$$

$m + n$ columns

where it is understood that 0's occupy the entries in the rows and columns where nothing is written.

Theorem Polynomials $f(x)$ and $g(x)$ each of degree ≥ 1 have a common factor of degree ≥ 1 if and only if their resultant $R = 0$.

Proof

(1) First, it is established that $f(x)$ and $g(x)$ have a common factor of degree ≥ 1 if and only if there exist polynomials $r(x)$ of degree $\leq m - 1$ and $s(x)$ of degree $\leq n - 1$ (not both the zero polynomial) so that for all x,

$$f(x)r(x) + g(x)s(x) = 0.$$

If $f(x)r(x) + g(x)s(x) = 0$, then $f(x)$ has n linear factors, and only $n - 1$ of these at most can be a factor of $s(x)$ since the degree of $s(x)$ is at most $n - 1$. Therefore, one of these factors must be a factor of $g(x)$.

Now, if $f(x)$ and $g(x)$ have a common factor, say $P(x)$, then select $r(x) = g(x)/P(x)$ and $s(x) = -f(x)/P(x)$ and

$$f(x)r(x) + g(x)s(x) = f(x)\{g(x)/P(x)\} + g(x)\{-f(x)/P(x)\} = 0.$$

(2) Now let $r(x) = r_{m-1}x^{m-1} + r_{m-2}x^{m-2} + \ldots + r_0$ and

$$s(x) = s_{n-1}x^{n-1} + s_{n-2}x^{n-2} + \ldots + s_0.$$

Then, $f(x)r(x) + g(x)s(x) = 0$ becomes

$$(a_n x^n + \ldots + a_0)(r_{m-1}x^{m-1} + \ldots + r_0)$$
$$+ (b_m x^m + \ldots + b_0)(s_{n-1}x^{n-1} + \ldots + s_0) = 0.$$

Rearranging the terms in descending powers of x,

$$(a_n r_{m-1} + b_m s_{n-1})x^{n+m-1}$$
$$+ (a_{n-1}r_{m-1} + a_n r_{m-2} + b_{m-1}s_{n-1} + b_m s_{n-2})x^{n+m-2}$$
$$+ \ldots + (a_1 r_0 + a_0 r_1 + b_1 s_0 + b_0 s_1)x + (a_0 r_0 + b_0 s_0) = 0.$$

Since this equation is valid for all x, the polynomial on the left is the zero polynomial and thus each of its coefficients is 0. Therefore,

$$a_n r_{m-1} + b_m s_{n-1} = 0$$
$$a_{n-1}r_{m-1} + a_n r_{m-2} + b_{m-1}s_{n-1} + b_m s_{n-2} = 0$$
$$\ldots$$
$$\ldots$$
$$a_1 r_0 + a_0 r_1 + b_1 s_0 + b_0 s_1 = 0$$
$$a_0 r_0 + b_0 s_0 = 0.$$

The above system of equations resulting from setting the coefficients equal to 0 is a system of $m + n$ linear equations in the $m + n$ variables $r_{m-1}, r_{m-2}, \ldots, r_0, s_{n-1}, s_{n-2}, \ldots, s_0$. Since the constant term of each equation is 0, this system has the obvious solution $(0, 0, \ldots, 0)$ called the trivial solution. The system will have a solution different from the trivial solution if and only if the determinant of its coefficients is 0; that is,

$$\begin{vmatrix} a_n & 0 & \cdots & 0 & b_m & 0 & \cdots & 0 \\ a_{n-1} & a_n & 0 & 0 & b_{m-1} & b_m & 0 & 0 \\ \vdots & \vdots & & \vdots & \vdots & & & \vdots \\ a_0 & a_1 & \cdots & a_n & b_1 & \cdots & & b_m \\ 0 & a_0 & \cdots & a_{n-1} & b_0 & \cdots & & b_{m-1} \\ \vdots & & & \vdots & 0 & & & \vdots \\ 0 & & 0 & a_0 & 0 & & 0 & b_0 \end{vmatrix} = 0.$$

THE RESULTANT (OPTIONAL)

Interchanging the rows and columns of this determinant, it can be identified as the resultant R.

Example 1 By using the resultant, show that $x^3 - 8 = 0$ and $2x^2 - 3x - 2 = 0$ have a common solution.

Solution Since the sum of the degrees $m + n = 2 + 3 = 5$, the resultant has five rows and five columns. The resultant is

$$R = \begin{vmatrix} 1 & 0 & 0 & -8 & 0 \\ 0 & 1 & 0 & 0 & -8 \\ 2 & -3 & -2 & 0 & 0 \\ 0 & 2 & -3 & -2 & 0 \\ 0 & 0 & 2 & -3 & -2 \end{vmatrix}.$$

Evalulating,

$$R = \begin{vmatrix} 1 & 0 & 0 & -8 & 0 \\ 0 & 1 & 0 & 0 & -8 \\ 0 & -3 & -2 & 16 & 0 \\ 0 & 2 & -3 & -2 & 0 \\ 0 & 0 & 2 & -3 & -2 \end{vmatrix} = \begin{vmatrix} 1 & 0 & 0 & -8 \\ -3 & -2 & 16 & 0 \\ 2 & -3 & -2 & 0 \\ 0 & 2 & -3 & -2 \end{vmatrix}$$

$$= \begin{vmatrix} 1 & -8 & 12 & 0 \\ -3 & -2 & 16 & 0 \\ 2 & -3 & -2 & 0 \\ 0 & 2 & -3 & -2 \end{vmatrix} = (-2) \begin{vmatrix} 1 & -8 & 12 \\ -3 & -2 & 16 \\ 2 & -3 & -2 \end{vmatrix}$$

$$= (-2) \begin{vmatrix} 1 & -8 & 12 \\ 0 & -26 & 52 \\ 0 & 13 & -26 \end{vmatrix} = 0.$$

Therefore $x^3 - 8$ and $2x^2 - 3x - 2$ have a common factor of at least degree 1, and thus $x^3 - 8 = 0$ and $2x^2 - 3x - 2 = 0$ have a common solution. By factoring, it may be seen that the common factor is $x - 2$ and the common solution is $x = 2$.

Example 2 Using the resultant, solve the system

$$x^2 + y^2 - 2y - 4 = 0, \qquad x^2 + y^2 - 4x - 6 = 0.$$

Solution Express each of the equations as polynomial equations in the variable x:

$$x^2 + 0 \cdot x + (y^2 - 2y - 4) = 0$$
$$x^2 - 4x + (y^2 - 6) = 0$$

Now, the polynomial equations will have a common solution for x if and only if their resultant is 0. The resultant is

$$R = \begin{vmatrix} 1 & 0 & y^2 - 2y - 4 & 0 \\ 0 & 1 & 0 & y^2 - 2y - 4 \\ 1 & -4 & y^2 - 6 & 0 \\ 0 & 1 & -4 & y^2 - 6 \end{vmatrix}.$$

Evaluating,

$$R = 20(y^2 - 2y - 3) = 0$$
$$(y - 3)(y + 1) = 0$$

and

$$y = 3 \quad \text{or} \quad y = -1.$$

For $y = 3$, the equations of the system become

$$x^2 - 1 = 0 \quad \text{or} \quad (x + 1)(x - 1) = 0$$

and

$$x^2 - 4x + 3 = 0 \quad \text{or} \quad (x - 3)(x - 1) = 0.$$

These have the common solution $x = 1$, and thus $(1, 3)$ is a solution of the system.

For $y = -1$, the equations of the system become

$$x^2 - 1 = 0 \quad \text{or} \quad (x + 1)(x - 1) = 0$$

and

$$x^2 - 4x - 5 = 0 \quad \text{or} \quad (x - 5)(x + 1) = 0.$$

These have the common solution $x = -1$, and thus $(-1, -1)$ is a solution of the system.

The solution set of the system is $\{(1, 3), (-1, -1)\}$.

EXERCISES 11.7

Use the resultant to find the common real solutions of each of the following:

1. $2x + 5y = 0$
   $2x^2 + 7x + 5y = 0$
2. $x^2 + xy - 6 = 0$
   $2x^2 + xy - 10 = 0$
3. $x^3 - 8y = 0$
   $x^2 + 2x - 8y = 0$
4. $y^2 - x^3 = 0$
   $y^2 - 6y - x^2 = 0$
5. $x^2 - 4y^3 = 0$
   $x^2 - 4x + 4y^2 = 0$
6. $y^2 = x^2 + x^3$
   $y = x^2 + x$
7. $x^2 + y^2 = 1$
   $x^2 + y^2 - 2x = 1$
8. $y^3 = x^2 - 1$
   $y^2 = x^2 - 5$
9. $y^3 - x - 9 = 0$
   $y - x - 3 = 0$
10. $y^4 = x^2 + 7$
    $y^2 = x^2 - 5$

CHAPTER SUMMARY

A solution of an equation in two variables x *and* y *is an ordered pair* (x, y).
A solution of an equation in three variables x, y, *and* z *is an ordered triple* (x, y, z).
A solution of an equation in n variables $x_1, x_2, \ldots, x_n$ *is an ordered n-tuple* $(x_1, x_2, \ldots, x_n)$.
The solution set of a system of equations is the intersection of the solution sets of each of the equations in the system.
Methods for solving a system of linear equations are

(1) elimination of a variable by adding linear combinations of pairs of equations,
(2) transformation to a matrix equation, and
(3) application of determinants.

Matrix transformations producing equivalent associated linear systems are

(1) interchanging rows,
(2) multiplying each element of a row by a constant,
(3) adding a constant multiple of one row to another row.

The expansion of a determinant can be obtained by using any row or any column. For example,

$$\begin{vmatrix} a_{11} & a_{12} & a_{13} \\ a_{21} & a_{22} & a_{23} \\ a_{31} & a_{32} & a_{33} \end{vmatrix} = (-1)^{i+1}a_{i1}A_{i1} + (-1)^{i+2}a_{i2}A_{i2} + (-1)^{i+3}a_{i3}A_{i3}$$

$$= (-1)^{1+j}a_{1j}A_{1j} + (-1)^{2+j}a_{2j}A_{2j} + (-1)^{3+j}a_{3j}A_{3j},$$

where $i = 1, 2,$ or 3, $j = 1, 2,$ or 3, and A_{ij} is the *minor* of a_{ij}; that is, the determinant obtained by deleting the row and column in which a_{ij} lies.

Theorem If a constant multiple of any row (or column) of a determinant is added to another row (or column), then the value of the determinant is unchanged.

Cramer's Rule, $n = 3$:
If

$$D = \begin{vmatrix} a_1 & b_1 & c_1 \\ a_2 & b_2 & c_2 \\ a_3 & b_3 & c_3 \end{vmatrix}, \quad N_1 = \begin{vmatrix} k_1 & b_1 & c_1 \\ k_2 & b_2 & c_2 \\ k_3 & b_3 & c_3 \end{vmatrix},$$

$$N_2 = \begin{vmatrix} a_1 & k_1 & c_1 \\ a_2 & k_2 & c_2 \\ a_3 & k_3 & c_3 \end{vmatrix}, \quad N_3 = \begin{vmatrix} a_1 & b_1 & k_1 \\ a_2 & b_2 & k_2 \\ a_3 & b_3 & k_3 \end{vmatrix},$$

then

$$a_1 x + b_1 y + c_1 z = k_1$$
$$a_2 x + b_2 y + c_2 z = k_2 \quad \text{is equivalent to}$$
$$a_3 x + b_3 y + c_3 z = k_3$$

$$Dx = N_1$$
$$Dy = N_2$$
$$Dz = N_3.$$

A quadratic system of two equations in two variables is a set of two quadratic equations or a set of one linear equation and one quadratic equation.

The graphic solution of a linear or quadratic system of equations in two variables is obtained by graphing both equations on the same set of coordinate axes and then determining the points of intersection. This method yields estimates of the real solutions.

Algebraic Solution by the Substitution Method

(1) Solve one of the equations for one of the variables.
(2) In the other equation, replace this variable by the expression obtained.

Algebraic Solution by the Addition Method Constant multiples of two equations are added in order to

(1) eliminate one of the variables,
(2) eliminate the second degree terms, or
(3) eliminate the constant term.

THE RESULTANT

Polynomials $f(x) = a_n x^n + a_{n-1} x^{n-1} + \ldots + a_0$ and

$$g(x) = b_m x^m + b_{m-1} x^{m-1} + \ldots + b_0$$

of degrees ≥ 1 have a common factor of degree ≥ 1 if and only if their resultant $R = 0$ where

$$R = \begin{vmatrix} a_n & a_{n-1} & \cdots & \cdots & a_0 & & & \\ & a_n & \cdots & \cdots & a_1 & a_0 & & \\ & & \ddots & & & & \ddots & \\ & & & a_n & a_{n-1} & \cdots & \cdots & a_0 \\ b_m & b_{m-1} & \cdots & \cdots & b_0 & & & \\ & b_m & \cdots & \cdots & b_0 & & & \\ & & \ddots & & & & \ddots & \\ & & & b_m & \cdots & \cdots & \cdots & b_0 \end{vmatrix} \begin{matrix} \left.\vphantom{\begin{matrix}1\\1\\1\\1\end{matrix}}\right\} m \text{ rows} \\ \\ \left.\vphantom{\begin{matrix}1\\1\\1\\1\end{matrix}}\right\} n \text{ rows} \end{matrix}$$

$$\underbrace{}_{m + n \text{ columns}}$$

REVIEW EXERCISES

1. Add:

$$\begin{bmatrix} 2 & -4 & 6 \\ 4 & 5 & -4 \\ -1 & 2 & 7 \end{bmatrix} + \begin{bmatrix} 4 & 2 & -3 \\ 6 & -8 & -3 \\ 1 & 6 & 3 \end{bmatrix}$$

2. Multiply:

$$\begin{bmatrix} 2 & 5 & 3 & 0 \\ 3 & 1 & 2 & -1 \\ 1 & 7 & 0 & -2 \end{bmatrix} \begin{bmatrix} 2 & x \\ 3 & y \\ -1 & z \\ 1 & t \end{bmatrix}$$

3. By using elementary equivalence transformations, transform the matrix so that the entries below the main diagonal are zeros.

$$\begin{bmatrix} 2 & -2 & 4 & -2 \\ 1 & -3 & 2 & -5 \\ -3 & 6 & 1 & 2 \end{bmatrix}$$

4. Find an equal determinant having zeros below the main diagonal.

$$\begin{vmatrix} 2 & -4 & 6 \\ -3 & 1 & 1 \\ 5 & 0 & -3 \end{vmatrix}$$

5. Solve, by using matrices:

$$x + y + z = 1$$
$$-x - 2y + 3z = 9$$
$$3x + 4y - 2z = -10$$

6. Solve, using Cramer's rule:

$$2x - 4y - 2z = 5$$
$$3x + 2y - z = 3$$
$$4x - 3y - 2z = 1.$$

7. Solve:

$$\begin{bmatrix} 5 & 1 & 2 \\ 3 & -2 & 6 \\ 2 & 3 & -4 \end{bmatrix} \begin{bmatrix} x \\ y \\ z \end{bmatrix} = \begin{bmatrix} 0 \\ 0 \\ 0 \end{bmatrix}$$

8. The amounts per unit and the total amounts on hand of three different chemicals needed to manufacture three different products is given in the following table. How many units of each product should be made if all of the chemicals are used?

CHEMICAL	PRODUCT			TOTAL
	P	Q	R	
A	3	4	1	330
B	5	0	2	330
C	1	2	3	240

9–12. Solve each of the following systems:

9. $x^2 + 4y^2 = 16$
   $4x^2 - y^2 = 36$

10. $x^2 + y^2 = 25$
    $x - y^2 + 5 = 0$

11. $4y^2 - 4x^2 = 25$
    $x^2 + 4y = 2$

12. $x^2 - 3xy - y^2 = 9$
    $2x^2 + 2xy + 3y^2 = 7$

13. A train traveling at 8 miles per hour less than its usual rate arrived at its destination 5 hours late. The destination was 800 miles from the starting point. What was the usual rate of the train? What was the usual time of the train?

14. A rectangular piece of metal has an area of 200 square inches. A square whose side is 3 inches is cut from each of the four corners. The flaps are then turned up to form an open box whose volume is 168 cubic inches. Find the original dimensions of the sheet of metal.

15. Find two numbers whose product is 2 and the sum of whose reciprocals is 19/6.

16. Find the dimensions of a rectangle if its diagonal is 17 feet and its area is 120 square feet.

17. Determine the values of k so that the line $y = 2x + k$ will be tangent to the circle $x^2 + y^2 = 20$.

12

SEQUENCES AND SERIES

12.1 GENERAL DISCUSSION

12.1.1 Sequences

There are many applications of sets of numbers that are arranged in order. One familiar example is the ordered set of natural numbers, 1, 2, 3, 4, 5, ..., used for counting. An ordered set of numbers is called a **sequence**.

Definition A **finite sequence of** n **terms** is a function c whose domain is the ordered set of natural numbers, 1, 2, 3, ..., n and whose range is the ordered set, $c(1), c(2), c(3), \ldots, c(n)$, also written as $c_1, c_2, c_3, \ldots, c_n$.

An **infinite sequence** is a function c whose domain is the ordered set of all natural numbers, 1, 2, 3, ..., and whose range is $c(1), c(2), c(3), \ldots, c(n)$, ..., also written as $c_1, c_2, c_3, \ldots, c_n, \ldots$.

The **terms** of the sequence are the elements of the range; that is, the values of $c(n)$. The first term is c_1, the second term is c_2, the third term is c_3, and the nth term or general term is c_n.

Examples of Sequences

(a) 3, 7, 11, 15, ... $c(n) = 4n - 1$
(b) 1, 3, 9, 27, 81, 243. $c(n) = 3^{n-1}$

(c) 12, 9, 6, 3, 0. $\qquad c(n) = 15 - 3n$

(d) $5, -1, \dfrac{1}{5}, -\dfrac{1}{25}, \ldots \qquad c(n) = \left(-\dfrac{1}{5}\right)^{n-1}$

(e) $1, \dfrac{1}{2}, \dfrac{1}{3}, \dfrac{1}{4}, \dfrac{1}{5}, \ldots \qquad c(n) = \dfrac{1}{n}$

Examples (b) and (c) are finite sequences and (a), (d), and (e) are infinite.

When the rule is stated, other terms may be found by using it. In Example (a) above, where $c(n) = 4n - 1$, the eighth term is $c(8) = 4(8) - 1 = 32 - 1 = 31$.

In applications the rule of a sequence function may be stated specifically or it may have to be inferred from given data. In general, if the rule is guessed from the first few terms of a sequence, there is no guarantee that the rule guessed is the correct one. For example, let 1, 2, 4, ... be the first three terms of a sequence. Suppose one guesses that the rule is $c(n) = 2^{n-1}$. Then the fourth term would be $c(4) = 2^{4-1} = 2^3 = 8$. However, it is possible that the rule might be $c(n) = (n^2 - n + 2)/2$. Then $c(4) = (16 - 4 + 2)/2 = 7$. Thus there is no unique (exactly one) solution to the problem of finding the rule from the first few terms of a sequence.

12.1.2 Series

For each sequence, there is an associated *series* which is obtained by replacing the commas in the ordered set of terms by plus symbols.

Definition A *series* is the indicated sum of the terms of a sequence.

The *finite series* $c_1 + c_2 + \ldots + c_n$ is obtained from the finite sequence $c_1, c_2, \ldots, c_n$.

The *infinite series* $c_1 + c_2 + \ldots + c_n + \ldots$ is obtained from the infinite sequence $c_1, c_2, \ldots, c_n, \ldots$.

The nth term of a series is the nth term of its corresponding sequence.

For a finite series, there is always a finite number which is the sum obtained by adding the terms of the series. Thus the sum of the series $1 + 3 + 9 + 27 + 81 + 243$ is 364. On the other hand, it is important that this sum, a number, is not confused with the series which indicates the addition. This is especially important for infinite series. For example, there is no finite number that can be designated as the sum of the infinite series $3 + 7 + 11 + 15 + 19 + \ldots$. However, there are some infinite series which can be assigned a number, called the *sum*. This shall be seen later.

12.1.3 Sigma Notation

The Greek letter $\sum$ (sigma) is used in many branches of mathematics in designating a series. This sigma notation greatly reduces the amount of writing

GENERAL DISCUSSION

that is necessary. For example, the finite series $3 + 7 + 11 + 15 + 19$ can be written as

$$\sum_{k=1}^{5} (4k - 1),$$

where it is understood that k is to be replaced by each of the numbers 1, 2, 3, 4, and 5 in the expression $4k - 1$, and then the sum of the resulting values is to be indicated.

$$\begin{aligned}
\text{If } k = 1, & \quad \text{then } 4k - 1 = 3 \\
k = 2 & \quad 4k - 1 = 7 \\
k = 3 & \quad 4k - 1 = 11 \\
k = 4 & \quad 4k - 1 = 15 \\
k = 5 & \quad 4k - 1 = 19.
\end{aligned}$$

Thus,

$$\sum_{k=1}^{5} (4k - 1) = 3 + 7 + 11 + 15 + 19.$$

In general, if the kth term of the series is $c(k)$, then

$$\sum_{k=1}^{n} c(k) = c(1) + c(2) + c(3) + \ldots + c(n)$$

The symbol $\sum$ is called the **summation symbol**, the letter k is called the **index of summation**, and the replacement set of integers for k is called the **range of summation**. The letter k is often referred to as a **dummy variable** since any other letter could be used and still the same sum would be indicated. Thus,

$$\sum_{k=1}^{5} (4k - 1) = \sum_{i=1}^{5} (4i - 1) = 3 + 7 + 11 + 15 + 19.$$

An infinite series is designated by the notation

$$\sum_{k=1}^{\infty} c(k).$$

For example,

$$\sum_{k=1}^{\infty} \frac{1}{k} = 1 + \frac{1}{2} + \frac{1}{3} + \ldots.$$

EXERCISES 12.1

1–12. Write the first five terms of the sequence whose nth term is given.

1. $c_n = 2n - 1$
2. $c_n = n^2$
3. $c_n = 5(10)^{-n}$
4. $c_n = 4n - 15$
5. $c_n = 5(2^n)$
6. $c_n = \dfrac{(-1)^n}{2n+1}$
7. $c(n) = n$
8. $c(n) = \dfrac{(-1)^n}{n^2}$
9. $c(n) = 3(-\tfrac{1}{2})^n$
10. $c(n) = \dfrac{n-1}{n+1}$
11. $c(n) = 43(0.01)^n$
12. $c(n) = 100(1.01)^n$

13–24. Find an expression for the general term of each of the following series and add two more terms.

13. $6 + 12 + 18 + 24 + \ldots$
14. $10 + 9 + 8 + 7 + \ldots$
15. $23 + 19 + 15 + 11 + \ldots$
16. $1/2 + 2/3 + 3/4 + 4/5 + \ldots$
17. $5 + 10 + 15 + 20 + \ldots$
18. $4 + 7 + 10 + 13 + \ldots$
19. $1 - 1/4 + 1/9 - 1/16 + \ldots$
20. $0.7 + 0.07 + 0.007 + 0.0007 + \ldots$
21. $(1 \times 2) + (2 \times 3) + (3 \times 4) + (4 \times 5) + \ldots$
22. $32 - 16 + 8 - 4 + \ldots$
23. $5 - 5/3 + 5/9 - 5/27 + \ldots$
24. $\sqrt[3]{1} - \sqrt[3]{4} + \sqrt[3]{9} - \sqrt[3]{16} + \ldots$

25–36. Find the sum of the first four terms of the series in

25. Exercise 13
26. Exercise 14
27. Exercise 15
28. Exercise 16
29. Exercise 17
30. Exercise 18
31. Exercise 19
32. Exercise 20
33. Exercise 21
34. Exercise 22
35. Exercise 23
36. Exercise 24.

37–43. Write each of the following series using sigma notation.

37. $1 + 2 + 3 + 4 + 5 + 6 + 7$
38. $1 + 1/5 + 1/25 + 1/125 + 1/625$

ARITHMETIC PROGRESSIONS

39. $1 + 3 + 5 + 7 + 9 + 11 + 13 + 15 + 17$
40. $1 + 4 + 9 + 16 + 25 + \ldots$
41. $-5 + 0.5 - 0.005 + 0.0005 - \ldots$
42. $7 - 7^2 + 7^3 - 7^4 + \ldots$
43. $1 + 1/2 + 3/4 + 7/8 + \ldots$

44–50. Express each of the following in expanded notation.

44. $\sum_{k=1}^{6} (7k - 2)$

45. $\sum_{i=2}^{7} (i^2 + 1)$

46. $\sum_{k=5}^{10} (k^2 + k)$

47. $\sum_{k=1}^{\infty} \frac{1}{2k - 1}$

48. $\sum_{k=2}^{\infty} 2400(10)^{-2k}$

49. $\sum_{k=1}^{\infty} (-1)^k k^{-1/2}$

50. $\sum_{k=0}^{\infty} (-1)^k \frac{1}{(k + 1)(k + 2)}$

12.2 ARITHMETIC PROGRESSIONS

Definition An **arithmetic progression** is a sequence in which the difference between any two consecutive terms is a constant, d, called the **common difference**.

If the first term of an arithmetic progression is designated by the letter a, then the definition states that the arithmetic progression can be written as follows:

$$a, a + d, a + 2d, a + 3d, a + 4d, \ldots .$$

The nth term, or general term, $c_n = a + (n - 1)d.$[1]

Example 1 Find the 35th term of the arithmetic progression, 5, 9, 13, ….

Solution $a = 5$, $d = 9 - 5 = 4$, and $n = 35$.
The 35th term $= c(35) = 5 + (35 - 1)4 = 5 + 34(4) = 141$.

Definition The **arithmetic means** between two given numbers are the terms of an arithmetic progression excluding the two given numbers which are assigned as the first and the last of the sequence.

---
[1] This statement may be verified by using mathematical induction.

Example 2 Insert three arithmetic means between 4 and 20.

Solution The arithmetic progression is $4, c_2, c_3, c_4, 20$.
$$c_5 = 4 + (5-1)d = 20$$
$$4d = 16 \quad \text{and} \quad d = 4.$$

Thus, $c_2 = 8$, $c_3 = 12$, and $c_4 = 16$. The three arithmetic means are 8, 12, 16.

If one arithmetic mean is inserted between two numbers, then it is called the **average** or the **arithmetic mean** of the two numbers.

Definition The **sum, S_n, of an arithmetic progression** is the sum of the terms of the series associated with the arithmetic sequence. In symbols,
$$S_n = a + (a+d) + (a+2d) + \ldots + (a + [n-1]d).$$

If l designates the last or nth term of the series, then S_n can also be expressed as
$$S_n = l + (l-d) + (l-2d) + \ldots + (l - [n-1]d).$$

By adding the corresponding terms of these two equations for S_n, one obtains
$$2S_n = (a+l) + (a+l) + (a+l) + \ldots + (a+l).$$

Noting that $a + l$ occurs n times,
$$2S_n = n(a+l)$$
or
$$S_n = \frac{n(a+l)}{2}$$
$$= \frac{n}{2}(2a + [n-1]d).$$

Example 3 Find the sum of the first 12 terms of the arithmetic progression $1, 3, 5, 7, \ldots$.

Solution $a = 1, d = 2, n = 12$
$$l = a + (n-1)d = 1 + (12-1)2 = 1 + (11)2 = 23$$
$$S_n = \frac{12(1 + 23)}{2} = 6(24) = 144$$

Example 4 Find the sum of the first 100 positive integers.

ARITHMETIC PROGRESSIONS

Solution $1 + 2 + 3 + \ldots + 100$ is an arithmetic series, with $a = 1$, $d = 1$, $n = 100$, and $l = 100$.

$$S_n = \frac{n(a+l)}{2} = \frac{100(1+100)}{2} = 50(101) = 5050$$

EXERCISES 12.2

1–10. Find the indicated term of each of the following arithmetic progressions.

1. 12th term of $9, 3, -3, \ldots$
2. 40th term of $1, 3, 5, \ldots$
3. 17th term of $10 + \sqrt{2}, 10, 10 - \sqrt{2}, \ldots$
4. 100th term of $\frac{1}{6}, \frac{1}{2}, \frac{5}{6}, \ldots$
5. 26th term of $0.5, 0.75, 1, \ldots$
6. 20th term of $3, 8, 13, \ldots$
7. 38th term of $2, 2.5, 3, \ldots$
8. 51st term of $87, 81, 75, \ldots$
9. 47th term of $8, 7\frac{1}{3}, 6\frac{2}{3}, \ldots$
10. 16th term of $2\sqrt{5}, 4\sqrt{5}, 6\sqrt{5}, \ldots$

11–20. Insert the indicated number of arithmetic means between the two given numbers.

11. Three between 80 and 90
12. One between 60 and 75
13. Five between 24 and -6
14. Four between 1 and 2
15. One between 78 and 65
16. Seven between 1 and 10
17. Three between 10 and -10
18. Two between 45 and 50
19. Eight between 3 and 6
20. Six between 0 and 4

21–30. Find the sum of each of the following arithmetic series.

21. $2 + 4 + 6 + \ldots + 100$
22. $1 + 2 + 3 + \ldots + 200$
23. $12 + 8 + 4 + \ldots + (-20)$
24. $1 + 3 + 5 + \ldots + (2n - 1)$
25. $3 + 6 + 9 + \ldots + 3n$
26. $30 + 29.5 + 29 + \ldots + 5$
27. $\sum_{k=1}^{50} \frac{k+1}{2}$
28. $\sum_{i=1}^{n} (2 - 3i)$
29. $\sum_{k=1}^{10} (5k - 3)$
30. $\sum_{i=0}^{n} (i + 1)$

31. How many positive integers between 5 and 100 are divisible by 7? (*Hint:* $a = 7$, $d = 7$. Find l and n.)

32. A man is offered two jobs: one with a beginning salary of $6000 per year and a $1200 raise at the end of each year, the other with a starting salary of $3000 for the first six months and a raise of $500 every six months thereafter. What is the salary for each job for the sixth year? Which job pays the most and why?

33. A debt of $2000 is paid by paying $100 at the end of each month plus $\frac{1}{2}$ percent interest on the amount unpaid at the end of that month. What is the total payment?

34. The fare charged by a certain taxi cab company is 30¢ for the first $\frac{1}{4}$ mile and 15¢ for each $\frac{1}{4}$ mile thereafter. What is the fare from the center of a city to an airport 10 miles away?

35. If a body falls from rest in a vacuum, it falls 16 feet the first second and 32 feet each second thereafter. How far does the body fall in 20 seconds? How far in k seconds?

36. A man contributes to his savings account by depositing $100 the first month and by increasing his deposit by $5 each month thereafter. How much has he saved at the end of five years?

37. A *harmonic progression* is a sequence of numbers whose reciprocals form an arithmetic progression. For example,

$$\frac{1}{a}, \frac{1}{a+d}, \frac{1}{a+2d}, \ldots, \frac{1}{a+(n-1)d}.$$

   (a) Find the seventh term of the harmonic progression $\frac{1}{2}, \frac{1}{7}, \frac{1}{12}, \ldots$.
   (b) Find a number h so that a, h, b form a harmonic progression. (In this case, h is called the *harmonic mean* of a and b.)
   (c) Find the harmonic mean of 20 and 30.
   (d) Find the sum of the harmonic series, $1 + \frac{1}{2} + \frac{1}{3} + \frac{1}{4} + \frac{1}{5}$.

38. Using mathematical induction, prove that for an arithmetic progression of n terms,
   (a) $c_n = a + (n-1)d$,
   (b) $S_n = \dfrac{n}{2}[2a + (n-1)d]$.

12.3 GEOMETRIC PROGRESSIONS

12.3.1 Finite Geometric Progressions

Definition A **geometric progression** is a sequence in which the ratio between any two consecutive terms is a constant, r, called the **common ratio**.

GEOMETRIC PROGRESSIONS

If the first term of a geometric progression is designated by the letter a, then the definition states that the geometric progression can be expressed as follows:

$$a, ar, ar^2, ar^3, ar^4, \ldots.$$

Thus, the nth term is

$$c(n) = ar^{n-1}.$$

Example 1 Find the sixth term of the geometric progression 3, 15, 75, ...

Solution $c(n) = ar^{n-1}$, $\quad a = 3, \quad r = \dfrac{15}{3} = 5, \quad$ and $\quad n = 6$.

$$c(6) = 3(5)^{6-1} = 3(5)^5 = 3(3125) = 9375$$

Definition The **geometric means** between two given numbers are the terms of a geometric progression excluding the two given numbers which are assigned as the first and last of the sequence.

Example 2 Insert two geometric means between 1 and 64.

Solution The geometric progression is 1, c_2, c_3, 64.

$$a = 1, \quad c_4 = ar^{4-1} = 1(r)^3 = 64$$

Thus $r = 4$.

$$c_2 = 4 \quad \text{and} \quad c_3 = 4 \cdot 4 = 16$$

The two geometric means are 4 and 16.

Example 3 Insert three real geometric means between 3 and 48.

Solution The geometric progression is 3, c_2, c_3, c_4, 48.

$$a = 3 \quad \text{and} \quad c_5 = 3r^4 = 48, \quad \text{or} \quad r^4 = 16$$

Thus,

$$r^2 = \pm 4 \quad \text{and} \quad r = \pm 2, \quad \text{or} \quad r = \pm 2i.$$

For the means to be real $r = 2$ or $r = -2$. There are two possibilities: The real means are 6, 12, 24 or -6, 12, -24.

If one geometric mean is inserted between two numbers, it is called the **mean proportional** or the **geometric mean** of the two numbers.

Definition The **sum, S_n, of a geometric progression** is the sum of the terms of the series associated with the geometric sequence. In symbols,

$$S_n = a + ar + ar^2 + ar^3 + \ldots + ar^{n-1}.$$

If each term of the indicated sum is multiplied by $-r$, then

$$-rS_n = -ar - ar^2 - ar^3 - \ldots - ar^{n-1} - ar^n.$$

Adding these two equations,

$$S_n - rS_n = a - ar^n$$
$$S_n(1 - r) = a(1 - r^n).$$

Now, if $r \neq 1$, then

$$S_n = \frac{a(1 - r^n)}{1 - r} \quad \text{and} \quad S_n = \frac{a(r^n - 1)}{r - 1}.$$

Example 4 Find the sum of the geometric progression 5, 10, ..., 320.

Solution $a = 5$ and $r = \dfrac{10}{5} = 2$

$$c_n = ar^{n-1} = 5(2)^{n-1} = 320$$
$$2^{n-1} = 64 = 2^6$$
$$n - 1 = 6 \quad \text{and} \quad n = 7$$

$$S_7 = \frac{a(r^7 - 1)}{r - 1} = \frac{5(2^7 - 1)}{2 - 1} = 5(128 - 1) = 5(127) = 635$$

12.3.2 Infinite Geometric Progressions

The sum of a finite geometric progression,

$$S_n = \frac{a - ar^n}{1 - r}$$

can also be written as

$$S_n = \frac{a}{1 - r}(1 - r^n).$$

If $|r| < 1$ (or $-1 < r < 1$), then r^n becomes smaller and smaller as n becomes larger and larger. For example, let $r = 1/10 = 0.1$. Then

$$r^2 = (0.1)^2 = 0.01$$
$$r^3 = (0.1)^3 = 0.001$$
$$r^4 = (0.1)^4 = 0.0001$$
$$\vdots$$
$$r^9 = (0.1)^9 = 0.000000001.$$

GEOMETRIC PROGRESSIONS

By taking n large enough, r^n can be made as close to 0 as one wants. Then $1 - r^n$ will be as close to 1 as one wants and, as a result, S_n can be made as close to $a/(1 - r)$ as one wants. Therefore, S, the sum of an infinite geometric progression with $|r| < 1$ is defined as this value, $a/(1 - r)$.

Definition If $|r| < 1$ and $S = a + ar + ar^2 + \ldots$, then

$$S = \frac{a}{1 - r}.$$

A more precise way of stating this result symbolically is as follows:

$$\lim_{n \to \infty} S_n = \frac{a}{1 - r}$$

which is read "the limit of S_n as n increases without bound is $a/(1 - r)$."

Example 5 Find the sum of the infinite geometric series: $12, 3, \frac{3}{4}, \ldots$.

Solution $a = 12, r = \frac{1}{4}$

$$S = \frac{a}{1 - r} = \frac{12}{1 - \frac{1}{4}} = \frac{12(4)}{4 - 1} = \frac{48}{3} = 16$$

Example 6 Express the repeating decimal fraction $2.363636\ldots$ as a common fraction.

Solution $2.363636\ldots = 2 + 0.36 + 0.0036 + 0.000036 + \ldots$
$0.36 + 0.0036 + 0.000036 + \ldots$ is an infinite geometric series with $a = 0.36$ and $r = 0.01$

$$S = \frac{a}{1 - r} = \frac{0.36}{1 - 0.01} = \frac{36}{100 - 1} = \frac{36}{99} = \frac{4}{11}$$

Thus

$$2.363636\ldots = 2 + \frac{4}{11} = 2\frac{4}{11} \quad \text{or} \quad \frac{26}{11}.$$

EXERCISES 12.3

1-8. Find the indicated term of each of the following geometric progressions.

1. Eighth term of $18, 12, 8, \ldots$
2. Fifth term of $1, 7, 49, \ldots$
3. Ninth term of $-16, 20, -25, \ldots$

4. Fifteenth term of 9, 9/10, 9/100, ...
5. Sixth term of 4, $2\sqrt{2}$, 2, ...
6. Seventh term of $\sqrt{2}$, $-\sqrt{6}$, $3\sqrt{2}$, ...
7. Tenth term of -4, 2, -1, ...
8. Eighth term of $2\sqrt[3]{2}$, 4, $4\sqrt[3]{4}$, ...

9–16. Insert the indicated number of real geometric means between the two given numbers.

9. Two between 54 and 16
10. Three between 2 and 162
11. One between 4 and 9
12. Two between 5 and 10
13. One between 6 and 30
14. Two between 56 and 189
15. Three between 1 and 10
16. Two between 3 and 34

17–24. Find the sum of each of the following geometric series.

17. $25 + 10 + \ldots + 0.256$
18. $1 + 7 + \ldots + 7^5$
19. $2 - 12 + \ldots + 2592$
20. $10 + 5 + \ldots + 10(0.5)^6$
21. $\sum_{k=1}^{7} 4(-3/2)^k$
22. $\sum_{i=1}^{8} 2(3^i)$
23. $\sum_{i=1}^{6} 5(1/3)^i$
24. $\sum_{k=2}^{7} 3(-2)^k$

25–32. If the sum exists, find the sum of each of the following infinite geometric series. If the sum does not exist, state why.

25. $10 + 2 + 0.4 + \ldots$
26. $1/50 + 1/40 + 1/32 + \ldots$
27. $1/75 - 1/90 + 1/108 - \ldots$
28. $12 - 2 + \ldots$
29. $1/54 - 1/36 + 1/24 - \ldots$
30. $(3)^{-1/2} + (6)^{-1/2} + \ldots$
31. $\sum_{k=1}^{\infty} 100(\tfrac{3}{4})^k$.
32. $\sum_{i=1}^{\infty} 32(-\tfrac{5}{8})^i$.

33. Convert 0.444 ... to a common fraction.
34. Convert 0.5454 ... to a common fraction.
35. Convert 25.132132 ... to a common fraction.
36. Convert 0.2151515 ... to a common fraction.
37. Each length of the path of the bob of a swinging pendulum is 95 percent of the preceding arc length. If its initial arc length is 18 inches, find the total distance the bob travels before the pendulum comes to rest.
38. On each rebound, a certain ball bounces back $\tfrac{5}{8}$ of its preceding height. If the ball is dropped from a height of 4 feet, find the total distance it travels before coming to rest.

THE BINOMIAL THEOREM

39. How many ancestors has an individual had in the twelve generations preceding him if it is assumed that there were no intermarriages?

40. When the oil used in operating certain machinery is refined so that it can be used again, 25 percent is lost each time it is refined. If there is originally 200 gallons of refined oil, which is refined each time it becomes dirty, find approximately the total amount used in operating the machinery before all of the oil is lost.

41. A piece of equipment costing $12,500 depreciates by 20 percent of its value each year. At the end of each year the amount the equipment has depreciated is placed in a fund. In how many years will the fund contain $8404?

42. If $100 is deposited at the end of each month in a savings fund and if $\frac{1}{2}$ percent interest is paid on the money in the fund each month, how much money is in the fund at the end of one year?

43. If an airpump removes 60 percent of the air in a container with each stroke, find the percentage of air left in the container after four strokes. How many strokes would it require to reduce the air to less than $\frac{1}{2}$ percent?

44. In a certain culture, a bacterium divides into two bacteria every 30 minutes. If there were 50 bacteria in the culture at the beginning, how many bacteria are there at the end of 3 hours?

45. If m, g, and h are the arithmetic, geometric, and harmonic means, respectively, of two numbers a and b, prove that $g^2 = mh$.

46. Using mathematical induction, prove that for a geometric progression of n terms,
    (a) $c(n) = ar^{n-1}$
    (b) $S_n = \dfrac{a(1-r^n)}{1-r}$.

47. If $0 < a < b$, show that the arithmetic mean of a and b is greater than the geometric mean of a and b.

48. If $0 < a < b$, then determine the relationship between
    (a) their arithmetic mean and harmonic mean,
    (b) their geometric mean and harmonic mean.

12.4 THE BINOMIAL THEOREM

The expansions of the following powers of the binomial $a + b$ may be obtained by performing the indicated multiplications.

$$(a + b)^1 = a + b$$
$$(a + b)^2 = a^2 + 2ab + b^2$$

$$(a+b)^3 = a^3 + 3a^2b + 3ab^2 + b^3$$
$$(a+b)^4 = a^4 + 4a^3b + 6a^2b^2 + 4ab^3 + b^4$$
$$(a+b)^5 = a^5 + 5a^4b + 10a^3b^2 + 10a^2b^3 + 5ab^4 + b^5$$

By examining these special cases, the following properties may be observed for $n = 1, 2, 3, 4$, or 5.

(1) Each expansion has $n + 1$ terms.
(2) The first term is a^n and the last term is b^n.
(3) The coefficient of the second term and the next to the last term is n.
(4) The exponent of a is one less than that of the preceding term.
(5) The exponent of b is one more than that of the preceding term.
(6) The sum of the exponents of a and b in each term is n.
(7) The coefficient of any term after the first can be obtained by multiplying the coefficient of the preceding term by its exponent of a and dividing by the number of this preceding term.

Assuming that these properties are valid for any natural number n, an expansion can be written for $(a + b)^n$. This statement is called the **binomial theorem**.

The Binomial Theorem Let n be any natural number. Then

$$(a+b)^n = a^n + na^{n-1}b + \frac{n(n-1)}{1(2)}a^{n-2}b^2 + \frac{n(n-1)(n-2)}{1(2)(3)}a^{n-3}b^3 + \cdots$$
$$+ \frac{n(n-1)(n-2)\cdots(n-r+1)}{1(2)(3)\cdots(r)}a^{n-r}b^r + \cdots + b^n.$$

Proof (by mathematical induction on n) Let S be the set of natural numbers for which the statement is true. We have already observed that 1, 2, 3, 4, and 5 are in S.

PART (1) If $n = 1$, then $(a + b)^1 = a + b$. Thus $1 \in S$.
PART (2) Assume $k \in S$. Then

$$(a+b)^k = a^k + ka^{k-1}b + \cdots + \frac{k(k-1)\cdots(k-r+2)}{1(2)\cdots(r-1)}a^{k-r+1}b^{r-1}$$
$$+ \frac{k(k-1)\cdots(k-r+2)(k-r+1)}{1(2)\cdots(r-1)r}a^{k-r}b^r + \cdots + b^k.$$

Multiplying both sides of this equation by $a + b$, and writing the terms of the product of $a(a + b)^k$ above the corresponding terms of the product of $b(a + b)^k$, then

THE BINOMIAL THEOREM

$$(a+b)^{k+1} = a(a+b)^k + b(a+b)^k$$

$$= a^{k+1} + ka^kb + \ldots + \frac{k(k-1)\ldots(k-r+1)}{1(2)\ldots(r)} a^{k-r+1}b^r + \ldots$$

$$+ a^kb + \ldots + \frac{k(k-1)\ldots(k-r+2)}{1(2)\ldots(r-1)} a^{k-r+1}b^r + \ldots + b^{k+1}$$

$$(a+b)^{k+1} = a^{k+1} + (k+1)a^kb + \ldots$$

$$+ \frac{(k+1)k(k-1)\ldots(k+1-r+1)}{1(2)(3)\ldots(r)} a^{k+1-r}b^r + \ldots + b^{k+1}.$$

Since this expansion is the same as that stated in the theorem for $n = k + 1$, it follows that $k + 1 \in S$ whenever $k \in S$. Thus $S = N$, or the binomial theorem is valid for every natural number n.

The addition of the terms involving b^r in the proof of the binomial theorem is verified as follows:

$$\frac{k(k-1)\ldots(k-r+2)}{1(2)\ldots(r-1)} = \frac{k(k-1)\ldots(k-r+2)r}{1(2)\ldots(r-1)r}$$

and

$$\frac{k(k-1)\ldots(k-r+2)(k-r+1) + k(k-1)\ldots(k-r+2)r}{1(2)\ldots(r)}$$

$$= \frac{k(k-1)\ldots(k-r+2)(k-r+1+r)}{1(2)\ldots(r)}$$

$$= \frac{(k+1)k\ldots(k-r+2)}{1(2)\ldots(r)}$$

$$= \frac{(k+1)k\ldots(k+1-r+1)}{1(2)\ldots(r)}.$$

The term that involves b^r is the $(r+1)$st term in the expansion of $(a+b)^n$. *The $(r+1)$st term of*

$$(a+b)^n = \frac{n(n-1)\ldots(n-r+1)}{1(2)(3)\ldots(r)} a^{n-r}b^r.$$

Example 1 Expand and simplify $(x - 2)^6$.

Solution $a = x$, $b = -2$, and $n = 6$. Substituting in $(a+b)^6$,

$$(x-2)^6 = x^6 + 6x^5(-2) + \frac{6(5)}{2} x^4(-2)^2 + \frac{6(5)(4)}{1(2)(3)} x^3(-2)^3$$

$$+ \frac{6(5)(4)(3)}{1(2)(3)(4)} x^2(-2)^4 + \frac{6(5)(4)(3)(2)}{1(2)(3)(4)(5)} x(-2)^5 + (-2)^6.$$

Simplifying,

$$(x - 2)^6 = x^6 - 12x^5 + 60x^4 - 160x^3 + 240x^2 - 192x + 64.$$

Example 2 Find the ninth term of $(5y + 0.2)^{15}$.

Solution The $(r + 1)$st term $= \dfrac{n(n - 1) \ldots (n - r + 1)}{1(2)(3) \ldots (r)} a^{n-r}b^r$. Since $r + 1 = 9$, then $r = 8$. Also $a = 5y$, $b = 0.2$, $n = 15$, and

$$n - r + 1 = 15 - 8 + 1 = 8.$$

Substituting,

$$\text{The 9th term} = \frac{15(14)(13)(12)(11)(10)(9)(8)}{1(2)(3)(4)(5)(6)(7)(8)} (5y)^{15-8}(0.2)^8$$

$$= 1287y^7.$$

Pascal's Triangle

By including $(a + b)^0 = 1$, the coefficients of the terms of the binomial expansion form an interesting pattern, known as *Pascal's triangle*.

$(a + b)^0$ 1
$(a + b)^1$ 1   1
$(a + b)^2$ 1   2   1
$(a + b)^3$ 1   3   3   1
$(a + b)^4$ 1   4   6   4   1
$(a + b)^5$ 1   5   10   10   5   1
$(a + b)^6$ 1   6   15   20   15   6   1

Each of the numbers different from 1 may be obtained by adding the number on its left to the one on its right in the row immediately above. Thus, $10 = 4 + 6$ and $15 = 5 + 10$ and so on.

The French mathematician Blaise Pascal (1623–1662) discovered many properties of this triangular array which appear in his *Traité du Triangle Arithmétique* written in 1653. The binomial theorem for positive integral exponents is found in this work, as well as one of the first acceptable uses of the method of mathematical induction.

Although the "arithmetical triangle" is named after Pascal because of his work on it, Pascal did not discover this array of numbers. It appears in one of the works of the great Chinese algebraist Chu Shi-kié written around 1303. The array appeared in European publications more than 100 years before Pascal's time.

The great English mathematician Sir Isaac Newton (1642–1727) generalized the binomial theorem for rational values of the exponent. His work appears in letters he wrote in 1676.

GENERALIZED BINOMIAL EXPANSION

EXERCISES 12.4

1–8. Expand by using the binomial theorem and simplify.

1. $(x + 2)^7$
2. $(y - 1)^8$
3. $(x^2 + 2)^5$
4. $(t - 1)^9$
5. $(a^2 - b^2)^6$
6. $\left(x + \dfrac{y}{2}\right)^5$
7. $(2t^2 - 3)^6$
8. $(2y + 1)^8$

9–16. Express, in simplified form, the specified term in the expansion of each of the following.

9. Fifth, $(x - 3)^{10}$
10. Fourth, $\left(\dfrac{x}{2} + 1\right)^{16}$
11. Sixth, $(1 - 0.02y)^8$
12. Fourth, $(y + 4)^{12}$
13. Sixth, $(x^2 - \tfrac{1}{2})^{20}$
14. Seventh, $\left(x + \dfrac{1}{x}\right)^{12}$
15. Fourth, $\left(r^3 + \dfrac{1}{s^3}\right)^9$
16. Eighth, $\left(\dfrac{2}{x} - \dfrac{y}{2}\right)^{15}$

17–24. Write the first four terms of the expansion of each of the following:

17. $(x + y)^{50}$
18. $(t - \sqrt{3})^{35}$
19. $(x^2 - \tfrac{1}{2})^{42}$
20. $(2a + 5b)^6$
21. $(x - y)^{34}$
22. $(t + 1)^{100}$
23. $(y^2 - \sqrt{2})^{15}$
24. $\left(\dfrac{2x + 5}{10}\right)^5$

25. Approximate $(1.02)^{12}$ by finding the sum of the first four terms of the expansion of $(1 + 0.02)^{12}$.
26. Approximate $(0.97)^{10}$ by using the first four terms of $(1 - 0.03)^{10}$.
27. Approximate $(0.98)^8$ correct to the nearest hundredth.
28. Approximate $(1.01)^{10}$ correct to the nearest thousandth.
29. Evaluate $(1 - i)^8$ where $i^2 = -1$.
30. Evaluate $\left(\dfrac{1}{2} + \dfrac{\sqrt{3}}{2}i\right)^9$, where $i^2 = -1$.

12.5 GENERALIZED BINOMIAL EXPANSION

If $|x| < 1$ and if k is any rational number, then it may be shown that it is valid to equate $(1 + x)^k$ with an infinite series; that is,

$$(1 + x)^k = 1 + kx + \frac{k(k-1)}{2} x^2 + \ldots + \frac{k(k-1)\ldots(k-n+1)}{1 \cdot 2 \cdot 3 \ldots n} x^n + \ldots .$$

It should be noted that the coefficient of x^n in this infinite series is identical in form to that of x^n in the binomial expansion of $(1 + x)^k$ when k is a positive integer. For this reason, the infinite series above is called the generalized binomial expansion.

Example 1 Express $(1 + x)^{1/2}$ as an infinite series.

Solution Using $k = \frac{1}{2}$ in the expansion for $(1 + x)^k$,

$$(1 + x)^{1/2} = 1 + \frac{1}{2}x + \frac{\frac{1}{2}(\frac{1}{2} - 1)}{2}x^2 + \frac{\frac{1}{2}(\frac{1}{2} - 1)(\frac{1}{2} - 2)}{1 \cdot 2 \cdot 3}x^3$$

$$+ \frac{\frac{1}{2}(\frac{1}{2} - 1)(\frac{1}{2} - 2)(\frac{1}{2} - 3)}{1 \cdot 2 \cdot 3 \cdot 4}x^4 + \cdots.$$

Simplifying,

$$(1 + x)^{1/2} = 1 + \frac{1}{2}x - \frac{1}{8}x^2 + \frac{1}{16}x^3 - \frac{5}{128}x^4 + \cdots.$$

Example 2 Approximate $\sqrt{1.1} = (1.1)^{1/2}$ by an appropriate binomial expansion using the first four terms.

Solution From Example 1, the sum of the first four terms of $(1 + x)^{1/2}$ is

$$1 + \frac{1}{2}x - \frac{1}{8}x^2 + \frac{1}{16}x^3.$$

Replacing x by 0.1, $(1 + 0.1)^{1/2} \approx 1 + \frac{1}{20} - \frac{1}{800} + \frac{1}{16,000}$,

Simplifying,

$$\sqrt{1.1} \approx 1 + 0.05 - 0.00125 + 0.0000625$$
$$\sqrt{1.1} \approx 1.0488125.$$

A table of square roots gives $\sqrt{1.1} = 1.048809$. Thus the answer to Example 2 is correct to the fifth decimal place.

As Example 2 illustrates, the sum of the first n terms of the generalized binomial expansion can be used to approximate $(1 + x)^k$, if $|x| < 1$, and this sum will differ from the correct value of $(1 + x)^k$ by a small error if n is taken large enough. For some series, only a few terms need to be evaluated for a useful approximation.

Whenever the terms of the series alternate in sign, the error made by using n terms is between 0 and the value of the $n + 1$st term. Referring to Example 2, the fifth term has value -0.00000390625. Thus the error is in the sixth decimal place. This is verified by comparing the result obtained with the tabular value stated; that is, $1.048809 - 1.0488125 = -0.0000035$.

GENERALIZED BINOMIAL EXPANSION

Example 3 Write the first five terms of the binomial expansion of $1/(2-x)^3$.

Solution $\dfrac{1}{(2-x)^3} = (2-x)^{-3} = 2^{-3}\left(1-\dfrac{x}{2}\right)^{-3} = \dfrac{1}{8}\left(1-\dfrac{x}{2}\right)^{-3}$

Replacing x by t and k by -3 in

$$(1+x)^k \approx 1 + kx + \frac{k(k-1)}{2}x^2 + \frac{k(k-1)(k-2)}{2 \cdot 3}x^3$$

$$+ \frac{k(k-1)(k-2)(k-3)}{2 \cdot 3 \cdot 4}x^4$$

$$(1+t)^{-3} \approx 1 - 3t + \frac{-3(-4)}{2}t^2 + \frac{-3(-4)(-5)}{2 \cdot 3}t^3 + \frac{-3(-4)(-5)(-6)}{2 \cdot 3 \cdot 4}t^4$$

$$(1+t)^{-3} \approx 1 - 3t + 6t^2 - 10t^3 + 15t^4$$

Replacing t by $(-x/2)$ and multiplying by $\tfrac{1}{8}$,

$$\frac{1}{8}\left(1-\frac{x}{2}\right)^{-3} \approx \frac{1}{8}\left[1 - 3\left(-\frac{x}{2}\right) + 6\left(-\frac{x}{2}\right)^2 - 10\left(-\frac{x}{2}\right)^3 + 15\left(-\frac{x}{2}\right)^4\right]$$

$$(2-x)^{-3} \approx \frac{1}{8} + \frac{3}{16}x + \frac{3}{16}x^2 + \frac{5}{32}x^3 + \frac{15}{128}x^4.$$

Example 4 Approximate $1/(1.95)^3$, using three terms of the binomial expansion. Round off each term to four decimal places and round off the final result to three decimal places.

Solution $\dfrac{1}{(1.95)^3} = (2 - 0.05)^{-3}$

Using the result of Example 3 and replacing x by 0.05,

$$(2-0.05)^{-3} \approx \frac{1}{8} + \frac{3}{16}\left(\frac{5}{100}\right) + \frac{3}{16}\left(\frac{25}{10,000}\right)$$

$$\approx 0.1250 + 0.0094 + 0.0005$$

$$\approx 0.1349 \approx 0.135.$$

EXERCISES 12.5

1–10. Write the first five terms of the generalized binomial expansion of each of the following. Assume $|x| < 1$.

1. $(1+x)^{1/3}$
2. $(1+x)^{1/4}$
3. $(1-x)^{1/2}$
4. $(1-x^2)^{3/2}$
5. $(1+x^2)^{-2}$
6. $(1+x^3)^{-4}$

7. $(2 + x)^{-1/2}$
8. $(3 + x)^{-1}$
9. $\left(x - \dfrac{1}{x}\right)^{1/4}$, where $x \neq 0$
10. $\left(x - \dfrac{2}{x}\right)^{1/2}$, where $x \neq 0$

11–20. Approximate each of the following to four significant figures.

11. $\sqrt[3]{1.02}$
12. $\sqrt[5]{1.5}$
13. $\sqrt[5]{0.99}$
14. $\sqrt[4]{0.98}$
15. $\dfrac{1}{\sqrt{15}}$
16. $\dfrac{1}{\sqrt{26}}$
17. $\sqrt[5]{34}$
18. $\sqrt[5]{30}$
19. $\dfrac{1}{(1.02)^{10}}$
20. $\dfrac{1}{(0.95)^{8}}$

CHAPTER SUMMARY

A *finite sequence of n terms* is a function c whose domain is the ordered set of natural numbers, 1, 2, 3, ..., n, and whose range is the ordered set, $c(1)$, $c(2)$, ..., $c(n)$ also written as $c_1, c_2, ..., c_n$.

An *infinite sequence* is a function whose domain is the ordered set of all natural numbers and whose range is $c(1), c(2), ..., c(n), ...$ also written as $c_1, c_2, ..., c_n, ...$.

A *series* is the indicated sum of the terms of a sequence.

$$\sum_{k=1}^{n} c(k) = c(1) + c(2) + \ldots + c(n).$$

An *arithmetic progression* is a sequence in which the difference between any two consecutive terms is a constant, d, called the **common difference**.

The *arithmetic means* between two given numbers are the terms of an arithmetic progression excluding the two given numbers which are assigned as the first and the last of the sequence.

The *nth or last term* of an arithmetic progression: $l = a + (n - 1)d$.

The *sum* of an arithmetic progression:

$$S = \dfrac{n(a + l)}{2}.$$

A *geometric progression* is a sequence in which the ratio between any two consecutive terms is a constant, r, called the **common ratio**.

The *geometric means* between two given numbers are the terms of a geometric progression excluding the two given numbers which are assigned as the first and last of the sequence.

The *nth or last term* of a geometric progression: $l = ar^{n-1}$.

The *sum* of a geometric progression:

$$S_n = \frac{a(1-r^n)}{1-r}.$$

The *sum of an infinite geometric progression*: if $|r| < 1$, then

$$S = \frac{a}{1-r}$$

The Binomial Theorem If k is a natural number, then

$$(a+b)^k = a^k + ka^{k-1}b + \ldots + \frac{k(k-1)\ldots(k-n+1)}{1(2)(3)\ldots(n)} a^{k-n}b^n + \ldots + b^k.$$

The Generalized Binomial Expansion If k is a rational number and if $|x| < 1$, then

$$(1+x)^k = 1 + kx + \frac{k(k-1)}{2} x^2 + \ldots + \frac{k(k-1)\ldots(k-n+1)}{1(2)(3)\ldots(n)} x^n + \ldots.$$

If k is negative or fractional, then the expansion is an infinite series.

REVIEW EXERCISES

1–10. Each of the following series is either an arithmetic progression (*A*), a geometric progression (*G*), or a binomial expansion (*B*). Determine the type of series and find the next two terms of each.

1. $\frac{2}{5} + 1 + \frac{5}{2} + \ldots$

2. $\frac{x+1}{x} + \frac{x+2}{x} + \frac{x+3}{x} + \ldots$

3. $3 - 6 - 15 - \ldots$

4. $3 - 6 + 12 - \ldots$

5. $1 - \frac{7}{2} + \frac{21}{4} - \ldots$

6. $\frac{-1}{1+\sqrt{2}} + 1 + \frac{1}{1-\sqrt{2}} + \ldots$

7. $\log 2 + \log 8 + \log 32 + \ldots$

8. $\log 9 + \log 3 + \log \sqrt{3} + \ldots$

9. $(1-i) + 2 + (2+2i) + \ldots$

10. $4\sqrt{2} + 60 + 180\sqrt{2} + \ldots$

11. Solve for x: $\frac{7}{8} = 1 + x + x^2 + x^3 + \ldots$

12. Find the fourth term of $\left(5\sqrt{2} - \frac{x}{5}\right)^9$.

13. Express 5.6363... as a simplified common fraction.
14. Approximate $(1.005)^{12}$ correct to four decimal places.
15. Prove by using mathematical induction:

$$\frac{1}{2(5)} + \frac{1}{5(8)} + \frac{1}{8(11)} + \cdots + \frac{1}{(3n-1)(3n+2)} = \frac{n}{6n+4}$$

16. Prove that the geometric mean of two positive integers is never larger than their arithmetic mean.
17. Find the cost of digging a well 200 yd deep if the charges are \$4 for the first yard, and each successive yard costs 50 cents more than the previous one.
18. An instructor receives a beginning salary of \$6000 a year and a yearly 4 percent raise. What is his salary for his 10th year of teaching?

APPENDIX

APPENDIX

TABLE I POWERS AND ROOTS

No.	Squares	Cubes	Square Roots	Cube Roots	No.	Squares	Cubes	Square Roots	Cube Roots
1	1	1	1.000	1.000	51	2 601	132,651	7.141	3.708
2	4	8	1.414	1.259	52	2 704	140,608	7.211	3.732
3	9	27	1.732	1.442	53	2 809	148,877	7.280	3.756
4	16	64	2.000	1.587	54	2 916	157,464	7.348	3.779
5	25	125	2.236	1.709	55	3 025	166,375	7.416	3.802
6	36	216	2.449	1.817	56	3 136	175,616	7.483	3.825
7	49	343	2.645	1.912	57	3 249	185,193	7.549	3.848
8	64	512	2.828	2.000	58	3 364	195,112	7.615	3.870
9	81	729	3.000	2.080	59	3 481	205,379	7.681	3.892
10	100	1 000	3.162	2.154	60	3 600	216,000	7.745	3.914
11	121	1 331	3.316	2.223	61	3 721	226,981	7.810	3.936
12	144	1 728	3.464	2.289	62	3 844	238,328	7.874	3.957
13	169	2 197	3.605	2.351	63	3 969	250,047	7.937	3.979
14	196	2 744	3.741	2.410	64	4 096	262,144	8.000	4.000
15	225	3 375	3.872	2.466	65	4 225	274,625	8.062	4.020
16	256	4 096	4.000	2.519	66	4 356	287,496	8.124	4.041
17	289	4 913	4.123	2.571	67	4 489	300,763	8.185	4.061
18	324	5 832	4.242	2.620	68	4 624	314,432	8.246	4.081
19	361	6 859	4.358	2.668	69	4 761	328,509	8.306	4.101
20	400	8 000	4.472	2.714	70	4 900	343,000	8.366	4.121
21	441	9 261	4.582	2.758	71	5 041	357,911	8.426	4.140
22	484	10,648	4.690	2.802	72	5 184	373,248	8.485	4.160
23	529	12,167	4.795	2.843	73	5 329	389,017	8.544	4.179
24	576	13,824	4.898	2.884	74	5 476	405,224	8.602	4.198
25	625	15,625	5.000	2.924	75	5 625	421,875	8.660	4.217
26	676	17,576	5.099	2.962	76	5 776	438,976	8.717	4.235
27	729	19,683	5.196	3.000	77	5 929	456,533	8.774	4.254
28	784	21,952	5.291	3.036	78	6 084	474,552	8.831	4.272
29	841	24,389	5.385	3.072	79	6 241	493,039	8.888	4.290
30	900	27,000	5.477	3.107	80	6 400	512,000	8.944	4.308
31	961	29,791	5.567	3.141	81	6 561	531,441	9.000	4.326
32	1 024	32,768	5.656	3.174	82	6 724	551,368	9.055	4.344
33	1 089	35,937	5.744	3.207	83	6 889	571,787	9.110	4.362
34	1 156	39,304	5.830	3.239	84	7 056	592,704	9.165	4.379
35	1 225	42,875	5.916	3.271	85	7 225	614,125	9.219	4.396
36	1 296	46,656	6.000	3.301	86	7 396	636,056	9.273	4.414
37	1 369	50,653	6.082	3.332	87	7 569	658,503	9.327	4.431
38	1 444	54,872	6.164	3.361	88	7 744	681,472	9.380	4.447
39	1 521	59,319	6.244	3.391	89	7 921	704,969	9.433	4.464
40	1 600	64,000	6.324	3.419	90	8 100	729,000	9.486	4.481
41	1 681	68,921	6.403	3.448	91	8 281	753,571	9.539	4.497
42	1 764	74,088	6.480	3.476	92	8 464	778,688	9.591	4.514
43	1 849	79,507	6.557	3.503	93	8 649	804,357	9.643	4.530
44	1 936	85,184	6.633	3.530	94	8 836	830,584	9.695	4.546
45	2 025	91,125	6.708	3.556	95	9 025	857,375	9.746	4.562
46	2 116	97,336	6.782	3.583	96	9 216	884,736	9.797	4.578
47	2 209	103,823	6.855	3.608	97	9 409	912,673	9.848	4.594
48	2 304	110,592	6.928	3.634	98	9 604	941,192	9.899	4.610
49	2 401	117,649	7.000	3.659	99	9 801	970,299	9.949	4.626
50	2 500	125,000	7.071	3.684	100	10,000	1,000,000	10.000	4.641

TABLE II COMMON LOGARITHMS

n	0	1	2	3	4	5	6	7	8	9
1.0	.0000	.0043	.0086	.0128	.0170	.0212	.0253	.0294	.0334	.0374
1.1	.0414	.0453	.0492	.0531	.0569	.0607	.0645	.0682	.0719	.0755
1.2	.0792	.0828	.0864	.0899	.0934	.0969	.1004	.1038	.1072	.1106
1.3	.1139	.1173	.1206	.1239	.1271	.1303	.1335	.1367	.1399	.1430
1.4	.1461	.1492	.1523	.1553	.1584	.1614	.1644	.1673	.1703	.1732
1.5	.1761	.1790	.1818	.1847	.1875	.1903	.1931	.1959	.1987	.2014
1.6	.2041	.2068	.2095	.2122	.2148	.2175	.2201	.2227	.2253	.2279
1.7	.2304	.2330	.2355	.2380	.2405	.2430	.2455	.2480	.2504	.2529
1.8	.2553	.2577	.2601	.2625	.2648	.2672	.2695	.2718	.2742	.2765
1.9	.2788	.2810	.2833	.2856	.2878	.2900	.2923	.2945	.2967	.2989
2.0	.3010	.3032	.3054	.3075	.3096	.3118	.3139	.3160	.3181	.3201
2.1	.3222	.3243	.3263	.3284	.3304	.3324	.3345	.3365	.3385	.3404
2.2	.3424	.3444	.3464	.3483	.3502	.3522	.3541	.3560	.3579	.3598
2.3	.3617	.3636	.3655	.3674	.3692	.3711	.3729	.3747	.3766	.3784
2.4	.3802	.3820	.3838	.3856	.3874	.3892	.3909	.3927	.3945	.3962
2.5	.3979	.3997	.4014	.4031	.4048	.4065	.4082	.4099	.4116	.4133
2.6	.4150	.4166	.4183	.4200	.4216	.4232	.4249	.4265	.4281	.4298
2.7	.4314	.4330	.4346	.4362	.4378	.4393	.4409	.4425	.4440	.4456
2.8	.4472	.4487	.4502	.4518	.4533	.4548	.4564	.4579	.4594	.4609
2.9	.4624	.4639	.4654	.4669	.4683	.4698	.4713	.4728	.4742	.4757
3.0	.4771	.4786	.4800	.4814	.4829	.4843	.4857	.4871	.4886	.4900
3.1	.4914	.4928	.4942	.4955	.4969	.4983	.4997	.5011	.5024	.5038
3.2	.5051	.5065	.5079	.5092	.5105	.5119	.5132	.5145	.5159	.5172
3.3	.5185	.5198	.5211	.5224	.5237	.5250	.5263	.5276	.5289	.5302
3.4	.5315	.5328	.5340	.5353	.5366	.5378	.5391	.5403	.5416	.5428
3.5	.5441	.5453	.5465	.5478	.5490	.5502	.5514	.5527	.5539	.5551
3.6	.5563	.5575	.5587	.5599	.5611	.5623	.5635	.5647	.5658	.5670
3.7	.5682	.5694	.5705	.5717	.5729	.5740	.5752	.5763	.5775	.5786
3.8	.5798	.5809	.5821	.5832	.5843	.5855	.5866	.5877	.5888	.5899
3.9	.5911	.5922	.5933	.5944	.5955	.5966	.5977	.5988	.5999	.6010
4.0	.6021	.6031	.6042	.6053	.6064	.6075	.6085	.6096	.6107	.6117
4.1	.6128	.6138	.6149	.6160	.6170	.6180	.6191	.6201	.6212	.6222
4.2	.6232	.6243	.6253	.6263	.6274	.6284	.6294	.6304	.6314	.6325
4.3	.6335	.6345	.6355	.6365	.6375	.6385	.6395	.6405	.6415	.6425
4.4	.6435	.6444	.6454	.6464	.6474	.6484	.6493	.6503	.6513	.6522
4.5	.6532	.6542	.6551	.6561	.6571	.6580	.6590	.6599	.6609	.6618
4.6	.6628	.6637	.6646	.6656	.6665	.6675	.6684	.6693	.6702	.6712
4.7	.6721	.6730	.6739	.6749	.6758	.6767	.6776	.6785	.6794	.6803
4.8	.6812	.6821	.6830	.6839	.6848	.6857	.6866	.6875	.6884	.6893
4.9	.6902	.6911	.6920	.6928	.6937	.6946	.6955	.6964	.6972	.6981
5.0	.6990	.6998	.7007	.7016	.7024	.7033	.7042	.7050	.7059	.7067
5.1	.7076	.7084	.7093	.7101	.7110	.7118	.7126	.7135	.7143	.7152
5.2	.7160	.7168	.7177	.7185	.7193	.7202	.7210	.7218	.7226	.7235
5.3	.7243	.7251	.7259	.7267	.7275	.7284	.7292	.7300	.7308	.7316
5.4	.7324	.7332	.7340	.7348	.7356	.7364	.7372	.7380	.7388	.7396

TABLE II (continued)

n	0	1	2	3	4	5	6	7	8	9
5.5	.7404	.7412	.7419	.7427	.7435	.7443	.7451	.7459	.7466	.7474
5.6	.7482	.7490	.7497	.7505	.7513	.7520	.7528	.7536	.7543	.7551
5.7	.7559	.7566	.7574	.7582	.7589	.7597	.7604	.7612	.7619	.7627
5.8	.7634	.7642	.7649	.7657	.7664	.7672	.7679	.7686	.7694	.7701
5.9	.7709	.7716	.7723	.7731	.7738	.7745	.7752	.7760	.7767	.7774
6.0	.7782	.7789	.7796	.7803	.7810	.7818	.7825	.7832	.7839	.7846
6.1	.7853	.7860	.7868	.7875	.7882	.7889	.7896	.7903	.7910	.7917
6.2	.7924	.7931	.7938	.7945	.7952	.7959	.7966	.7973	.7980	.7987
6.3	.7993	.8000	.8007	.8014	.8021	.8028	.8035	.8041	.8048	.8055
6.4	.8062	.8069	.8075	.8082	.8089	.8096	.8102	.8109	.8116	.8122
6.5	.8129	.8136	.8142	.8149	.8156	.8162	.8169	.8176	.8182	.8189
6.6	.8195	.8202	.8209	.8215	.8222	.8228	.8235	.8241	.8248	.8254
6.7	.8261	.8267	.8274	.8280	.8287	.8293	.8299	.8306	.8312	.8319
6.8	.8325	.8331	.8338	.8344	.8351	.8357	.8363	.8370	.8376	.8382
6.9	.8388	.8395	.8401	.8407	.8414	.8420	.8426	.8432	.8439	.8445
7.0	.8451	.8457	.8463	.8470	.8476	.8482	.8488	.8494	.8500	.8506
7.1	.8513	.8519	.8525	.8531	.8537	.8543	.8549	.8555	.8561	.8567
7.2	.8573	.8579	.8585	.8591	.8597	.8603	.8609	.8615	.8621	.8627
7.3	.8633	.8639	.8645	.8651	.8657	.8663	.8669	.8675	.8681	.8686
7.4	.8692	.8698	.8704	.8710	.8716	.8722	.8727	.8733	.8739	.8745
7.5	.8751	.8756	.8762	.8768	.8774	.8779	.8785	.8791	.8797	.8802
7.6	.8808	.8814	.8820	.8825	.8831	.8837	.8842	.8848	.8854	.8859
7.7	.8865	.8871	.8876	.8882	.8887	.8893	.8899	.8904	.8910	.8915
7.8	.8921	.8927	.8932	.8938	.8943	.8949	.8954	.8960	.8965	.8971
7.9	.8976	.8982	.8987	.8993	.8998	.9004	.9009	.9015	.9020	.9025
8.0	.9031	.9036	.9042	.9047	.9053	.9058	.9063	.9069	.9074	.9079
8.1	.9085	.9090	.9096	.9101	.9106	.9112	.9117	.9122	.9128	.9133
8.2	.9138	.9143	.9149	.9154	.9159	.9165	.9170	.9175	.9180	.9186
8.3	.9191	.9196	.9201	.9206	.9212	.9217	.9222	.9227	.9232	.9238
8.4	.9243	.9248	.9253	.9258	.9263	.9269	.9274	.9279	.9284	.9289
8.5	.9294	.9299	.9304	.9309	.9315	.9320	.9325	.9330	.9335	.9340
8.6	.9345	.9350	.9355	.9360	.9365	.9370	.9375	.9380	.9385	.9390
8.7	.9395	.9400	.9405	.9410	.9415	.9420	.9425	.9430	.9435	.9440
8.8	.9445	.9450	.9455	.9460	.9465	.9469	.9474	.9479	.9484	.9489
8.9	.9494	.9499	.9504	.9509	.9513	.9518	.9523	.9528	.9533	.9538
9.0	.9542	.9547	.9552	.9557	.9562	.9566	.9571	.9576	.9581	.9586
9.1	.9590	.9595	.9600	.9605	.9609	.9614	.9619	.9624	.9628	.9633
9.2	.9638	.9643	.9647	.9652	.9657	.9661	.9666	.9671	.9675	.9680
9.3	.9685	.9689	.9694	.9699	.9703	.9708	.9713	.9717	.9722	.9727
9.4	.9731	.9736	.9741	.9745	.9750	.9754	.9759	.9763	.9768	.9773
9.5	.9777	.9782	.9786	.9791	.9795	.9800	.9805	.9809	.9814	.9818
9.6	.9823	.9827	.9832	.9836	.9841	.9845	.9850	.9854	.9859	.9863
9.7	.9868	.9872	.9877	.9881	.9886	.9890	.9894	.9899	.9903	.9908
9.8	.9912	.9917	.9921	.9926	.9930	.9934	.9939	.9943	.9948	.9952
9.9	.9956	.9961	.9965	.9969	.9974	.9978	.9983	.9987	.9991	.9996

TABLE III   TRIGONOMETRIC FUNCTIONS

Degrees	Radians	Sin	Csc	Tan	Cot	Sec	Cos		
0° 0′	.0000	.0000	—	.0000	—	1.000	1.0000	1.5708	90° 0′
10′	029	029	343.8	029	343.8	000	000	679	50′
20′	058	058	171.9	058	171.9	000	000	650	40′
30′	.0087	.0087	114.6	.0087	114.6	1.000	1.0000	1.5621	30′
40′	116	116	85.95	116	85.94	000	.9999	592	20′
50′	145	145	68.76	145	68.75	000	999	563	10′
1° 0′	.0175	.0175	57.30	.0175	57.29	1.000	.9998	1.5533	89° 0′
10′	204	204	49.11	204	49.10	000	998	504	50′
20′	233	233	42.98	233	42.96	000	997	475	40′
30′	.0262	.0262	38.20	.0262	38.19	1.000	.9997	1.5446	30′
40′	291	291	34.38	291	34.37	000	996	417	20′
50′	320	320	31.26	320	31.24	001	995	388	10′
2° 0′	.0349	.0349	28.65	.0349	28.64	1.001	.9994	1.5359	88° 0′
10′	378	378	26.45	378	26.43	001	993	330	50′
20′	407	407	24.56	407	24.54	001	992	301	40′
30′	.0436	.0436	22.93	.0437	22.90	1.001	.9990	1.5272	30′
40′	465	465	21.49	466	21.47	001	989	243	20′
50′	495	494	20.23	495	20.21	001	988	213	10′
3° 0′	.0524	.0523	19.11	.0524	19.08	1.001	.9986	1.5184	87° 0′
10′	553	552	18.10	553	18.07	002	985	155	50′
20′	582	581	17.20	582	17.17	002	983	126	40′
30′	.0611	.0610	16.38	.0612	16.35	1.002	.9981	1.5097	30′
40′	640	640	15.64	641	15.60	002	980	068	20′
50′	669	669	14.96	670	14.92	002	978	039	10′
4° 0′	.0698	.0698	14.34	.0699	14.30	1.002	.9976	1.5010	86° 0′
10′	727	727	13.76	729	13.73	003	974	981	50′
20′	756	756	13.23	758	13.20	003	971	952	40′
30′	.0785	.0785	12.75	.0787	12.71	1.003	.9969	1.4923	30′
40′	814	814	12.29	816	12.25	003	967	893	20′
50′	844	843	11.87	846	11.83	004	964	864	10′
5° 0′	.0873	.0872	11.47	.0875	11.43	1.004	.9962	1.4835	85° 0′
10′	902	901	11.10	904	11.06	004	959	806	50′
20′	931	929	10.76	934	10.71	004	957	777	40′
30′	.0960	.0958	10.43	.0963	10.39	1.005	.9954	1.4748	30′
40′	989	987	10.13	992	10.08	005	951	719	20′
50′	.1018	.1016	9.839	.1022	9.788	005	948	690	10′
6° 0′	.1047	.1045	9.567	.1051	9.514	1.006	.9945	1.4661	84° 0′
10′	076	074	9.309	080	9.255	006	942	632	50′
20′	105	103	9.065	110	9.010	006	939	603	40′
30′	.1134	.1132	8.834	.1139	8.777	1.006	.9936	1.4573	30′
40′	164	161	8.614	169	8.556	007	932	544	20′
50′	193	190	8.405	198	8.345	007	929	515	10′
7° 0′	.1222	.1219	8.206	.1228	8.144	1.008	.9925	1.4486	83° 0′
10′	251	248	8.016	257	7.953	008	922	457	50′
20′	280	276	7.834	287	7.770	008	918	428	40′
30′	.1309	.1305	7.661	.1317	7.596	1.009	.9914	1.4399	30′
40′	338	334	7.496	346	7.429	009	911	370	20′
50′	367	363	7.337	376	7.269	009	907	341	10′
8° 0′	.1396	.1392	7.185	.1405	7.115	1.010	.9903	1.4312	82° 0′
10′	425	421	7.040	435	6.968	010	899	283	50′
20′	454	449	6.900	465	6.827	011	894	254	40′
30′	.1484	.1478	6.765	.1495	6.691	1.011	.9890	1.4224	30′
40′	513	507	6.636	524	6.561	012	886	195	20′
50′	542	536	6.512	554	6.435	012	881	166	10′
9° 0′	.1571	.1564	6.392	.1584	6.314	1.012	.9877	1.4137	81° 0′
		Cos	Sec	Cot	Tan	Csc	Sin	Radians	Degrees

TABLE III *(continued)*

Degrees	Radians	Sin	Csc	Tan	Cot	Sec	Cos		
9° 0'	.1571	.1564	6.392	.1584	6.314	1.012	.9877	1.4137	81° 0'
10'	600	593	277	614	197	013	872	108	50'
20'	629	622	166	644	084	013	868	079	40'
30'	.1658	.1650	6.059	.1673	5.976	1.014	.9863	1.4050	30'
40'	687	679	5.955	703	871	014	858	1.4021	20'
50'	716	708	855	733	769	015	853	992	10'
10° 0'	.1745	.1736	5.759	.1763	5.671	1.015	.9848	1.3963	80° 0'
10'	774	765	665	793	576	016	843	934	50'
20'	804	794	575	823	485	016	838	904	40'
30'	.1833	.1822	5.487	.1853	5.396	1.017	.9833	1.3875	30'
40'	862	851	403	883	309	018	827	846	20'
50'	891	880	320	914	226	018	822	817	10'
11° 0'	.1920	.1908	5.241	.1944	5.145	1.019	.9816	1.3788	79° 0'
10'	949	937	164	974	066	019	811	759	50'
20'	978	965	089	.2004	4.989	020	805	730	40'
30'	.2007	.1994	5.016	.2035	4.915	1.020	.9799	1.3701	30'
40'	036	.2022	4.945	065	843	021	793	672	20'
50'	065	051	876	095	773	022	787	643	10'
12° 0'	.2094	.2079	4.810	.2126	4.705	1.022	.9781	1.3614	78° 0'
10'	123	108	745	156	638	023	775	584	50'
20'	153	136	682	186	574	024	769	555	40'
30'	.2182	.2164	4.620	.2217	4.511	1.024	.9763	1.3526	30'
40'	211	193	560	247	449	025	757	497	20'
50'	240	221	502	278	390	026	750	468	10'
13° 0'	.2269	.2250	4.445	.2309	4.331	1.026	.9744	1.3439	77° 0'
10'	298	278	390	339	275	027	737	410	50'
20'	327	306	336	370	219	028	730	381	40'
30'	.2356	.2334	4.284	.2401	4.165	1.028	.9724	1.3352	30'
40'	385	363	232	432	113	029	717	323	20'
50'	414	391	182	462	061	030	710	294	10'
14° 0'	.2443	.2419	4.134	.2493	4.011	1.031	.9703	1.3265	76° 0'
10'	473	447	086	524	3.962	031	696	235	50'
20'	502	476	039	555	914	032	689	206	40'
30'	.2531	.2504	3.994	.2586	3.867	1.033	.9681	1.3177	30'
40'	560	532	950	617	821	034	674	148	20'
50'	589	560	906	648	776	034	667	119	10'
15° 0'	.2618	.2588	3.864	.2679	3.732	1.035	.9659	1.3090	75° 0'
10'	647	616	822	711	689	036	652	061	50'
20'	676	644	782	742	647	037	644	032	40'
30'	.2705	.2672	3.742	.2773	3.606	1.038	.9636	1.3003	30'
40'	734	700	703	805	566	039	628	974	20'
50'	763	728	665	836	526	039	621	945	10'
16° 0'	.2793	.2756	3.628	.2867	3.487	1.040	.9613	1.2915	74° 0'
10'	822	784	592	899	450	041	605	886	50'
20'	851	812	556	931	412	042	596	857	40'
30'	.2880	.2840	3.521	.2962	3.376	1.043	.9588	1.2828	30'
40'	909	868	487	994	340	044	580	799	20'
50'	938	896	453	.3026	305	045	572	770	10'
17° 0'	.2967	.2924	3.420	.3057	3.271	1.046	.9563	1.2741	73° 0'
10'	996	952	388	089	237	047	555	712	50'
20'	.3025	979	357	121	204	048	546	683	40'
30'	.3054	.3007	3.326	.3153	3.172	1.048	.9537	1.2654	30'
40'	083	035	295	185	140	049	528	625	20'
50'	113	062	265	217	108	050	520	595	10'
18° 0'	.3142	.3090	3.236	.3249	3.078	1.051	.9511	1.2566	72° 0'
		Cos	Sec	Cot	Tan	Csc	Sin	Radians	Degrees

513

TABLE III *(continued)*

Degrees	Radians	Sin	Csc	Tan	Cot	Sec	Cos		
18° 0'	.3142	.3090	3.236	.3249	3.078	1.051	.9511	1.2566	72° 0'
10'	171	118	207	281	047	052	502	537	50'
20'	200	145	179	314	018	053	492	508	40'
30'	.3229	.3173	3.152	.3346	2.989	1.054	.9483	1.2479	30'
40'	258	201	124	378	960	056	474	450	20'
50'	287	228	098	411	932	057	465	421	10'
19° 0'	.3316	3256	3.072	.3443	2.904	1.058	.9455	1.2392	71° 0'
10'	345	283	046	476	877	059	446	363	50'
20'	374	311	021	508	850	060	436	334	40'
30'	.3403	3338	2.996	.3541	2.824	1.061	.9426	1.2305	30'
40'	432	365	971	574	798	062	417	275	20'
50'	462	393	947	607	773	063	407	246	10'
20° 0'	.3491	.3420	2.924	.3640	2.747	1.064	9397	1.2217	70° 0'
10'	520	448	901	673	723	065	387	188	50'
20'	549	475	878	706	699	066	377	159	40'
30'	.3578	.3502	2.855	.3739	2.675	1.068	.9367	1.2130	30'
40'	607	529	833	772	651	069	356	101	20'
50'	636	557	812	805	628	070	346	072	10'
21° 0'	.3665	.3584	2.790	.3839	2.605	1.071	.9336	1.2043	69° 0'
10'	694	611	769	872	583	072	325	1.2014	50'
20'	723	638	749	906	560	074	315	985	40'
30'	.3752	.3665	2.729	.3939	2.539	1.075	.9304	1.1956	30'
40'	782	692	709	973	517	076	293	926	20'
50'	811	719	689	.4006	496	077	283	897	10'
22° 0'	.3840	.3746	2.669	.4040	2.475	1.079	.9272	1.1868	68° 0'
10'	869	773	650	074	455	080	261	839	50'
20'	898	800	632	108	434	081	250	810	40'
30'	.3927	.3827	2.613	.4142	2.414	1.082	.9239	1.1781	30'
40'	956	854	595	176	394	084	228	752	20'
50'	985	881	577	210	375	085	216	723	10'
23° 0'	.4014	.3907	2.559	.4245	2.356	1.086	.9205	1.1694	67° 0'
10'	043	934	542	279	337	088	194	665	50'
20'	072	961	525	314	318	089	182	636	40'
30'	.4102	.3987	2.508	.4348	2.300	1.090	.9171	1.1606	30'
40'	131	.4014	491	383	282	092	159	577	20'
50'	160	041	475	417	264	093	147	548	10'
24° 0'	.4189	.4067	2.459	.4452	2.246	1.095	.9135	1.1519	66° 0'
10'	218	094	443	487	229	096	124	490	50'
20'	247	120	427	522	211	097	112	461	40'
30'	.4276	.4147	2.411	.4557	2.194	1.099	.9100	1.1432	30'
40'	305	173	396	592	177	100	088	403	20'
50'	334	200	381	628	161	102	075	374	10'
25° 0'	.4363	.4226	2.366	.4663	2.145	1.103	.9063	1.1345	65° 0'
10'	392	253	352	699	128	105	051	316	50'
20'	422	279	337	734	112	106	038	286	40'
30'	.4451	.4305	2.323	.4770	2.097	1.108	.9026	1.1257	30'
40'	480	331	309	806	081	109	013	228	20'
50'	509	358	295	841	066	111	001	199	10'
26° 0'	.4538	.4384	2.281	.4877	2.050	1.113	.8988	1.1170	64° 0'
10'	567	410	268	913	035	114	975	141	50'
20'	596	436	254	950	020	116	962	112	40'
30'	.4625	.4462	2.241	.4986	2.006	1.117	.8949	1.1083	30'
40'	654	488	228	.5022	1.991	119	936	054	20'
50'	683	514	215	059	977	121	923	1.1025	10'
27° 0'	.4712	.4540	2.203	.5095	1.963	1.122	.8910	1.0996	63° 0'
		Cos	Sec	Cot	Tan	Csc	Sin	Radians	Degrees

TABLE III (continued)

Degrees	Radians	Sin	Csc	Tan	Cot	Sec	Cos		Degrees
27° 0′	.4712	.4540	2.203	.5095	1.963	1.122	.8910	1.0996	63° 0′
10′	741	566	190	132	949	124	897	966	50′
20′	771	592	178	169	935	126	884	937	40′
30′	.4800	.4617	2.166	.5206	1.921	1.127	.8870	1.0908	30′
40′	829	643	154	243	907	129	857	879	20′
50′	858	669	142	280	894	131	843	850	10′
28° 0′	.4887	.4695	2.130	.5317	1.881	1.133	.8829	1.0821	62° 0′
10′	916	720	118	354	868	134	816	792	50′
20′	945	746	107	392	855	136	802	763	40′
30′	.4974	.4772	2.096	.5430	1.842	1.138	.8788	1.0734	30′
40′	.5003	797	085	467	829	140	774	705	20′
50′	032	823	074	505	816	142	760	676	10′
29° 0′	.5061	.4848	2.063	.5543	1.804	1.143	.8746	1.0647	61° 0′
10′	091	874	052	581	792	145	732	617	50′
20′	120	899	041	619	780	147	718	588	40′
30′	.5149	.4924	2.031	.5658	1.767	1.149	.8704	1.0559	30′
40′	178	950	020	696	756	151	689	530	20′
50′	207	975	010	735	744	153	675	501	10′
30° 0′	.5236	.5000	2.000	.5774	1.732	1.155	.8660	1.0472	60° 0′
10′	265	025	1.990	812	720	157	646	443	50′
20′	294	050	980	851	709	159	631	414	40′
30′	.5323	.5075	1.970	.5890	1.698	1.161	.8616	1.0385	30′
40′	352	100	961	930	686	163	601	356	20′
50′	381	125	951	969	675	165	587	327	10′
31° 0′	.5411	.5150	1.942	.6009	1.664	1.167	.8572	1.0297	59° 0″
10′	440	175	932	048	653	169	557	268	50′
20′	469	200	923	088	643	171	542	239	40′
30′	.5498	.5225	1.914	.6128	1.632	1.173	.8526	1.0210	30′
40′	527	250	905	168	621	175	511	181	20′
50′	556	275	896	208	611	177	496	152	10′
32° 0′	.5585	.5299	1.887	.6249	1.600	1.179	.8480	1.0123	58° 0′
10′	614	324	878	289	590	181	465	094	50′
20′	643	348	870	330	580	184	450	065	40′
30′	.5672	.5373	1.861	.6371	1.570	1.186	.8434	1.0036	30′
40′	701	398	853	412	560	188	418	1.0007	20′
50′	730	422	844	453	550	190	403	977	10′
33° 0′	.5760	.5446	1.836	.6494	1.540	1.192	.8387	.9948	57° 0′
10′	789	471	828	536	530	195	371	919	50′
20′	818	495	820	577	520	197	355	890	40′
30′	.5847	.5519	1.812	.6619	1.511	1.199	.8339	.9861	30′
40′	876	544	804	661	501	202	323	832	20′
50′	905	568	796	703	1.492	204	307	803	10′
34° 0′	.5934	.5592	1.788	.6745	1.483	1.206	.8290	.9774	56° 0′
10′	963	616	781	787	473	209	274	745	50′
20′	992	640	773	830	464	211	258	716	40′
30′	.6021	.5664	1.766	.6873	1.455	1.213	.8241	.9687	30′
40′	050	688	758	916	446	216	225	657	20′
50′	080	712	751	959	437	218	208	628	10′
35° 0′	.6109	.5736	1.743	.7002	1.428	1.221	.8192	.9599	55° 0′
10′	138	760	736	046	419	223	175	570	50′
20′	167	783	729	089	411	226	158	541	40′
30′	.6196	.5807	1.722	.7133	1.402	1.228	.8141	.9512	30′
40′	225	831	715	177	393	231	124	483	20′
50′	254	854	708	221	385	233	107	454	10′
36° 0′	.6283	.5878	1.701	.7265	1.376	1.236	.8090	.9425	54° 0′
		Cos	Sec	Cot	Tan	Csc	Sin	Radians	Degrees

TABLE III *(continued)*

Degrees	Radians	Sin	Csc	Tan	Cot	Sec	Cos			
36° 0'	.6283	.5878	1.701	.7265	1.376	1.236	.8090	.9425	54°	0'
10'	312	901	695	310	368	239	073	396		50'
20'	341	925	688	355	360	241	056	367		40'
30'	.6370	.5948	1.681	.7400	1.351	1.244	.8039	.9338		30'
40'	400	972	675	445	343	247	021	308		20'
50'	429	995	668	490	335	249	004	279		10'
37° 0'	.6458	.6018	1.662	.7536	1.327	1.252	.7986	.9250	53°	0'
10'	487	041	655	581	319	255	969	221		50'
20'	516	065	649	627	311	258	951	192		40'
30'	.6545	.6088	1.643	.7673	1.303	1.260	.7934	.9163		30'
40'	574	111	636	720	295	263	916	134		20'
50'	603	134	630	766	288	266	898	105		10'
38° 0'	.6632	.6157	1.624	.7813	1.280	1.269	.7880	.9076	52°	0'
10'	661	180	618	860	272	272	862	047		50'
20'	690	202	612	907	265	275	844	.9018		40'
30'	.6720	.6225	1.606	.7954	1.257	1.278	.7826	.8988		30'
40'	749	248	601	.8002	250	281	808	959		20'
50'	778	271	595	050	242	284	790	930		10'
39° 0'	.6807	.6293	1.589	.8098	1.235	1.287	.7771	.8901	51°	0'
10'	836	316	583	146	228	290	753	872		50'
20'	865	338	578	195	220	293	735	843		40'
30'	.6894	.6361	1.572	.8243	1.213	1.296	.7716	.8814		30'
40'	923	383	567	292	206	299	698	785		20'
50'	952	406	561	342	199	302	679	756		10'
40° 0'	.6981	.6428	1.556	.8391	1.192	1.305	.7660	.8727	50°	0'
10'	.7010	450	550	441	185	309	642	698		50'
20'	039	472	545	491	178	312	623	668		40'
30'	.7069	.6494	1.540	.8541	1.171	1.315	.7604	.8639		30'
40'	098	517	535	591	164	318	585	610		20'
50'	127	539	529	642	157	322	566	581		10'
41° 0'	.7156	.6561	1.524	.8693	1.150	1.325	.7547	.8552	49°	0'
10'	185	583	519	744	144	328	528	523		50'
20'	214	604	514	796	137	332	509	494		40'
30'	.7243	.6626	1.509	.8847	1.130	1.335	.7490	.8465		30'
40'	272	648	504	899	124	339	470	436		20'
50'	301	670	499	952	117	342	451	407		10'
42° 0'	.7330	.6691	1.494	.9004	1.111	1.346	.7431	.8378	48°	0'
10'	359	713	490	057	104	349	412	348		50'
20'	389	734	485	110	098	353	392	319		40'
30'	.7418	.6756	1.480	.9163	1.091	1.356	.7373	.8290		30
40'	447	777	476	217	085	360	353	261		20'
50'	476	799	471	271	079	364	333	232		10'
43° 0'	.7505	.6820	1.466	.9325	1.072	1.367	.7314	.8203	47°	0'
10'	534	841	462	380	066	371	294	174		50'
20'	563	862	457	435	060	375	274	145		40'
30'	.7592	.6884	1.453	.9490	1.054	1.379	.7254	.8116		30'
40'	621	905	448	545	048	382	234	087		20'
50'	650	926	444	601	042	386	214	058		10'
44° 0'	.7679	.6947	1.440	.9657	1.036	1.390	.7193	.8029	46°	0'
10'	709	967	435	713	030	394	173	999		50'
20'	738	988	431	770	024	398	153	970		40'
30'	.7767	.7009	1.427	.9827	1.018	1.402	.7133	.7941		30'
40'	796	030	423	884	012	406	112	912		20'
50'	825	050	418	942	006	410	092	883		10'
45° 0'	.7854	.7071	1.414	1.000	1.000	1.414	.7071	.7854	45°	0'
		Cos	Sec	Cot	Tan	Csc	Sin	Radians	Degrees	

516

PROOF OF THE PARTIAL FRACTION THEOREM

The proof of the general theorem on the resolution of a quotient into partial fractions requires some preliminary theorems or lemmas. A lemma is a theorem that is used to prove another more important theorem.

Lemma 1 If $A(x)$ and $B(x)$ are polynomials over Q with no common polynomial factor of degree ≥ 1, then there exist polynomials $P(x)$ and $Q(x)$ so that

$$\frac{N(x)}{A(x)B(x)} = \frac{P(x)}{A(x)} + \frac{Q(x)}{B(x)}.$$

Proof The proof presented is a construction proof; that is, the proof indicates how the polynomials $P(x)$ and $Q(x)$ are found. For all x, except those such that $A(x)B(x) = 0$,

$$\frac{N(x)}{A(x)B(x)} = \frac{P(x)}{A(x)} + \frac{Q(x)}{B(x)} = \frac{P(x)B(x) + Q(x)A(x)}{A(x)B(x)}.$$

For all x, $P(x)B(x) + Q(x)A(x) = N(x)$ and thus their corresponding coefficients are equal.

$(p_0 + p_1 x + p_2 x^2 + \ldots)(b_0 + b_1 x + b_2 x^2 + \ldots)$
$+ (q_0 + q_1 x + q_2 x^2 + \ldots)(a_0 + a_1 x + a_2 x^2 + \ldots) = n_0 + n_1 x + n_2 x^2 + \ldots.$

Rearranging terms in ascending powers of x,

$(p_0 b_0 + q_0 a_0) + (p_0 b_1 + p_1 b_0 + q_0 a_1 + q_1 a_0)x$
$+ (p_0 b_2 + p_1 b_1 + p_2 b_0 + q_0 a_2 + q_1 a_1 + q_2 a_0)x^2 + \ldots = n_0 + n_1 x + n_2 x^2 + \ldots.$

Equating coefficients,

(1) $p_0 b_0 + q_0 a_0 = n_0,$
(2) $p_0 b_1 + p_1 b_0 + q_0 a_1 + q_1 a_0 = n_1,$
(3) $p_0 b_2 + p_1 b_1 + p_2 b_0 + q_0 a_2 + q_1 a_1 + q_2 a_0 = n_2$

and so on.

If $a_0 \neq b_0$, then p_0 and q_0 may be taken as follows:

$$p_0 = \frac{-n_0}{a_0 - b_0} \quad \text{and} \quad q_0 = \frac{n_0}{a_0 - b_0}$$

as checking in Equation (1) will verify.

If $a_0 = b_0$, then select $p_0 = n_0/2a_0 = q_0$. ($a_0 \neq 0$, otherwise $A(x)$ and $B(x)$ have the common factor x.)

Proceeding to Equation (2), $p_1 b_0 + q_1 a_0 = n_1 - p_0 b_1 - q_0 a_1 = c$.

If $a_0 \neq b_0$, then select

$$p_1 = \frac{-c}{a_0 - b_0} \quad \text{and} \quad q_1 = \frac{c}{a_0 - b_0}.$$

If $a_0 = b_0$, then select

$$p_1 = q_1 = \frac{c}{2a_0}.$$

From Equation (3), $p_2 b_0 + q_2 a_0 = n_2 - p_0 b_2 - p_1 b_1 - q_0 a_2 - q_1 a_1 = d$ and p_2 and q_2 may be selected as before.

Continuing in this way, all the coefficients of $P(x)$ and $Q(x)$ are determined and thus the polynomials $P(x)$ and $Q(x)$ are found so that

$$N(x) = P(x)B(x) + Q(x)A(x)$$

and

$$\frac{N(x)}{A(x)B(x)} = \frac{P(x)}{A(x)} + \frac{Q(x)}{B(x)}.$$

Lemma 2 If n is a natural number and $N(x)$ and $D(x)$ are polynomials over Q, then there exist polynomials over Q, $P_0(x), P_1(x), \ldots, P_n(x)$ so that

$$\frac{N(x)}{[D(x)]^n} = P_0(x) + \frac{P_1(x)}{D(x)} + \frac{P_2(x)}{[D(x)]^2} + \cdots + \frac{P_n(x)}{[D(x)]^n},$$

where the degree of each of $P_1(x), P_2(x), \ldots, P_n(x)$ is less than the degree of $D(x)$.

Proof By the division algorithm, there exist polynomials

$$Q_0(x), P_n(x), Q_1(x), \quad \text{and} \quad P_{n-1}(x)$$

so that

$$N(x) = D(x)Q_0(x) + P_n(x)$$

with degree $P_n(x) <$ degree $D(x)$ and

$$Q_0(x) = D(x)Q_1(x) + P_{n-1}(x)$$

with degree $P_{n-1}(x) <$ degree $D(x)$.

By substituting the expression for Q_0 into the expression for $N(x)$,

$$N(x) = [D(x)]^2 Q_1(x) + D(x)P_{n-1}(x) + P_n(x).$$

Using the division algorithm again,

$$Q_1(x) = D(x)Q_2(x) + P_{n-2}(x)$$

with degree $P_{n-2} <$ degree $D(x)$ and consequently,

$$N(x) = [D(x)]^3 Q_2(x) + [D(x)]^2 P_{n-2}(x) + D(x)P_{n-1}(x) + P_n(x).$$

PROOF OF THE PARTIAL FRACTION THEOREM

By continuing this process,
$$N(x) = [D(x)]^n Q_{n-1}(x) + [D(x)]^{n-1} P_1(x) + \ldots + D(x) P_{n-1}(x) + P_n(x)$$
and
$$\frac{N(x)}{[D(x)]^n} = P_0(x) + \frac{P_1(x)}{D(x)} + \frac{P_2(x)}{[D(x)]^2} + \ldots + \frac{P_{n-1}(x)}{[D(x)]^{n-1}} + \frac{P_n(x)}{[D(x)]^n},$$
where $Q_{n-1}(x) = P_0(x)$.

Corollary If the degree of $N(x) <$ the degree of $[D(x)]^n$, then
$$Q_{n-1}(x) = P_0(x) = 0.$$
The proof is left for the student.

Theorem (The Partial Fractions Theorem) If $N(x)$ and $D(x)$ are polynomials with the degree of $N(x)$ less than the degree of $D(x)$, then $N(x)/D(x)$ can be resolved into a sum of partial fractions as follows:

(1) If $D(x)$ has a linear factor $ax + b$, then one term of the sum is the partial fraction $p/(ax + b)$ where p is a constant.

(2) If $D(x)$ has n repeated linear factors, $(ax + b)^n$, then n terms of the sum are the n partial fractions as follows:
$$\frac{p_1}{ax + b} + \frac{p_2}{(ax + b)^2} + \ldots + \frac{p_n}{(ax + b)^n} \text{ with } p_1, p_2, \ldots, p_n$$
constants.

(3) If $D(x)$ has a quadratic factor, $ax^2 + bx + c$, then one term of the sum is the partial fraction $(px + q)/(ax^2 + bx + c)$ where p and q are constants.

(4) If $D(x)$ has n repeated quadratic factors, $(ax^2 + bx + c)^n$, n terms of the sum are the n partial fractions as follows:
$$\frac{p_1 x + q_1}{ax^2 + bx + c} + \frac{p_2 x + q_2}{(ax^2 + bx + c)^2} + \ldots + \frac{p_n x + q_n}{(ax^2 + bx + c)^n},$$
where the p_i's and q_i's are constants.

Proof Let $D(x) = (a_1 x + b_1)^{n_1} \ldots (a_j x + b_j)^{n_j} (c_1 x^2 + d_1 x + e_1)^{m_1} \ldots (c_k x^2 + d_k x + e_k)^{m_k}$, where no two of the indicated factors have a common linear factor or a common quadratic factor.

By Lemma 1,
$$\frac{N(x)}{D(x)} = \frac{N(x)}{(a_1 x + b_1)^{n_1} B(x)} = \frac{P_1(x)}{(a_1 x + b_1)^{n_1}} + \frac{Q_1(x)}{B_1(x)}.$$
$$\frac{Q_1(x)}{B_1(x)} = \frac{P_2(x)}{(a_2 x + b_2)^{n_2}} + \frac{Q_2(x)}{B_2(x)}$$
and so on.

Finally,
$$\frac{N(x)}{D(x)} = \frac{P_1(x)}{(a_1x + b_1)^{n_1}} + \frac{P_2(x)}{(a_2x + b_2)^{n_2}} + \ldots + \frac{P_{j+k}(x)}{(c_k x^2 + d_k x + e_k)^{m_k}}.$$

By applying Lemma 2 to each of the terms in this sum, it follows that $N(x)/D(x)$ can be resolved into a sum of partial fractions as stated in the theorem.

ANSWERS

to Selected Odd Exercises
and to Review Exercises

CHAPTER 1 (page 9)

Exercises 1.1
1. $\{0, 2, 4, 6, 8\}$
3. $\{0, 1, 2, 3, 4\}$
5. $\{0, 3, 6, 9\}$
7. $\{(0, 0), (0, 1), (1, 0), (1, 1), (2, 0), (2, 1), (3, 0), (4, 0), (5, 0), (6, 0)\}$
9. $\{(0, 5), (1, 3), (2, 1)\}$
11. $\{x \mid x < 6\}$ or $\{x \mid x \leq 5\}$
13. $\{x \mid x = 3n \text{ where } n \in N\}$
15. $\{x \mid x = n^2 \text{ where } n \in N \text{ and } n \leq 6\}$
17. $\{(x, y) \mid y = 2x\}$
19. $\{x \mid x \neq 1 \text{ and } (x = ab \text{ if and only if } a = 1 \text{ or } b = 1)\}$
21. $\{-2, -1, 0, 1, 2\}$
23. $\{-2, -1\}$
25. $\{-3, 2, 3\}$
27. $\{-3, 3\}$
29. $\{-3, 3\}$
31. $\{2\}$
33. $\{-3, -2, -1, 0, 1, 3\}$
35. false, true, false

37. false, true, false
39. true, false, true
41. $\{x \mid -3 < x < 2\}$
43. $\{x \mid x < 2\}$
45. $\{x \mid x \le 5\}$
47. $\{x \mid x \le -3\}$
49. $\{x \mid x < -1 \text{ or } x > 5\}$
51. $\emptyset$
53. $\{x \mid x < 2\}$
55. $\{x \mid x \ge 2\}$
57. $\{x \mid -1 \le x \le 1\}$
59. $\{x \mid -2 \le x \le 1\}$
61. $\{-2, -1, 0, 1\}$
63. infinite, (any ordered pair of the form $(x, 7 - 2x)$)
65. infinite; $\{6, 12, 18, \ldots\}$
67. $\{3, -\frac{1}{2}\}$
69. $\{-2\}$
71. $\{(2, 3)\}$
73. $\{(2, 4), (-2, -4)\}$
75. infinite; (any ordered pair of the form $(x, 0)$ or $(0, y)$)
77. $\{5\}$
79. $\{(3, 4), (-3, 4), (3, -4), (-3, -4)\}$
81. infinite; (x may be any real number greater than 3)
83. $\emptyset$
85. infinite; $\{x \mid x \le 2 \text{ or } x \ge 3\}$
87. infinite; $\{x \mid -1 \le x \le 5\}$
89. (c) the solution set of $p(x) \ge 0$ is the complement of the solution set of $p(x) < 0$

Exercises 1.2 (page 15)

1. TF
3. open
5. TF
7. TF
9. neither
11. open
13. TF
15. open
17. true, true
19. false, true
21. true, true
23. false, false
25. true, true
27. false, true
29. true, true
31. For all x and for all y, for all x and for some y, for some x and for all y, for some x and for some y.
33. For all x there is some y so that, $(y = 0)$; for some x and for some y.
35. For each x there is some y so that, $(y = x)$; there is some x so that for all y, $(x = y)$; for some x and for some y.
37. For all x there is some y so that, for some x and for some y.
39. For some x and for some y, $(x = y = 0)$.

EXERCISES 1.4

41. (a) $\{x \mid x > 5\}$
    (b) $\{6, 7, 8, \ldots\}$
43. (a) $\{(x, y) \mid 3x + y = 12\}$
    (b) $\{(1, 9), (2, 6), (3, 3)\}$
45. (a) $\{x \mid x + 3 = 0 \text{ or } 3x - 1 = 0\}$
    (b) $\{-3, \frac{1}{3}\}$

47. (a) $\{x \mid x = 0\}$
    (b) $\{0\}$
49. (a) $\{x \mid x = \sqrt{9}\}$
    (b) $\{3\}$

Exercises 1.3 (page 21)

1. true
3. false
5. true
7. false
9. true
11. false
13. true
15. true
17. false
19. true

21. true
23. true
25. false
27. $\{x \mid x \leq 0\}$
29. $\{0, 1\}$
31. $\{5\}$
33. $\{x \mid 4 < x < 7\}$
35. $\{(4, 16), (-1, 1)\}$
37. U

39. $\{(2, 4), (-2, -4), (3\sqrt{2}, \sqrt{2}), (-3\sqrt{2}, -\sqrt{2})\}$
41. R
43. $R - \{-3\}$
45. $R - \{5\}$
47. $\emptyset$
49. $R - \{0, 5\}$
51. (a) $\{-4\}, \{5, -4\}$
    (b) not equivalent

53. (a) $\{5, -3\}, \{5, -3\}$
    (b) equivalent
55. (a) $\{3\}, \{3, -2\}$
    (b) not equivalent
57. (a) $\{2, 4\}, \{2, 4\}$
    (b) equivalent

59. (a) $\{(x, y \mid x = 2y + 3\}, \{(x, y) \mid x = 2y + 3\}$
    (b) equivalent
61. (a) $\{(3, \sqrt{7}), (3, -\sqrt{7}), (-3, \sqrt{7}), (-3, \sqrt{7})\}, \{(3, \sqrt{7}), (3, -\sqrt{7}), (-3, \sqrt{7}), (-3, -\sqrt{7})\}$
    (b) equivalent

Exercises 1.4 (page 28)

1. (a) true, true, true, true
   (b) tautology
3. (a) true, false, true, false
   (b) not a tautology
5. (a) true, true, true, true
   (b) tautology

7. (a) true, true, false, true
   (b) not a tautology
9. (a) true, true, true, true
   (b) tautology

11. Converse: If $x = 3$, then $x + 2 = 5$.
    Inverse: If $x + 2 \neq 5$, then $x \neq 3$.
    Contrapositive: If $x \neq 3$, then $x + 2 \neq 5$.

13. Converse: If $x + 2 = x^2$, then $\sqrt{x+2} = x$.
    Inverse: If $\sqrt{x+2} \neq x$, then $x + 2 \neq x^2$.
    Contrapositive: If $x + 2 \neq x^2$, then $\sqrt{x+2} \neq x$.
    Counterexample: Statement and contrapositive are true for all x while the converse and inverse are false for $x = -1$.

15. Converse: If $y = 0$, then $xy = 0$ and $x \neq 0$.
    Inverse: If $xy \neq 0$ or $x = 0$, then $y \neq 0$.
    Counterpositive: If $y \neq 0$, then $xy \neq 0$ or $x = 0$.
    Counterexample: Statement and contrapositive are true for all x and for all y while the converse and the inverse are false for $(x, y) = (0, 0)$.

17. Negation: $ab = 0$ and $a \neq 0$ and $b \neq 0$.
    Contrapositive: If $a \neq 0$ and $b \neq 0$, then $ab \neq 0$.

19. Negation: $(a > 0$ and $b > 0)$ and $(a + b \not> 0$ and $ab \not> 0$).
    Contrapositive: If $a + b \not> 0$ or $ab \not> 0$, then $a \not> 0$ or $b \not> 0$.

21. Negation: $0 < b < 1$ and $n > 0$ and $b^n \not< b$.
    Contrapositive: If $b^n \geq b$, then $b \leq 0$ or $b \geq 1$ or $n \leq 0$.

23. There is some natural number that is not positive.

25. Something is perfect.   29. All problems can be solved.

27. Some squares are not rectangles.   31. There is no solution for $x^2 > 1$.

33. There is some real number so that $x^2 + 1 = 0$.

35. (a) natural numbers (or integers or rationals or reals) (b) digits

37. (a) real numbers
    (b) digits (or natural numbers, or integers, or rationals)

Review Exercises (page 31)

1. $\{-3, -2, -1, 0, 1, 2\}$ 3. $\{2, 1, 0, -1, -2, -3, \ldots\}$

2. $\{1, 4, 9, 16, 25, \ldots\}$ 4. $\{0, 1, \sqrt{2}, -\sqrt{2}\}$

5. $\{x \mid x = 2n$ where n is an integer$\}$

6. $\{x \mid x = \dfrac{n}{5}$ and $n \leq 5$ and n is a natural number$\}$

7. $\{(x, y) \mid y = x^3$ and $x \geq 6$ where x is a digit$\}$

8. $\{(x, y) \mid y = -x$ and x is a natural number$\}$

9. $\{1, 2, 3, 7, 8, 9\}$ 12. $\{2, 3, 4, 5, 6\}$

10. $\{4, 5, 6\}$ 13. $\{2, 3, 7, 8, 9\}$

11. $\{1, 7, 8, 9\}$

EXERCISES 2.1

14. $\{1, 4, 5, 6\}$
15. true
16. true
17. true
18. false
19. true
20. false
21. yes
22. yes
23. yes
24. no
25. For some x, $x \cdot 1 \neq x$.
26. For some x, $1/x = 0$.
27. For all x, $x + 2 \neq 0$.
28. For some x, $x + 2 \neq 2 + x$.
29. For some x, $x + 1 = x$.
30. For all x and for all y
31. For all x, there is no y.
32. For all x, there is some y.
33. For each x, there is some y.
34. For some x, there is some x (or: for each $x \neq 0$, there is some y).
35. For some x and for some y.
36. $\{1, 2, 3, 4\}$
37. $\{3, 4, 5, 6, 7\}$
38. $\{1, 2, 3, 4, 5\}$
39. $\{1, 2, 3, 4, 5, 6, 7, 8\}$
40. $\emptyset$
41. $\{1, 2, 5, 6, 7, 8, \ldots\}$
42. $\{1, 2, 3\}$
43. $\{1\}$
44. $\{4\}$
45. $\{1, 5, 6, 7, \ldots\}$
46. $\{(1, 4), (2, 3), (3, 2), (4, 1)\}$
47. $\{(1, 2), (2, 3), (3, 4), (4, 5), \ldots\}$
48. $\{(2, 3)\}$
49. $\{3, 5\}$
50. $\{(1, 1), (-1, 1), (2, 2), (-2, 2), (3, 3), (-3, 3), \ldots\}$

CHAPTER 2 (page 40)

Exercises 2.1

1. $\{(a, x), (a, y), (b, x), (b, y), (c, x), (c, y)\}$
3. $\{(a, a), (a, b), (a, c), (b, a), (b, b), (b, c), (c, a), (c, b), (c, c)\}$
5. $\{(1, 1), (2, 1), (3, 1), (4, 1), (1, 2), (2, 2), (3, 2), (4, 2), (1, 3), (2, 3), (3, 3), (4, 3), (1, 4), (2, 4), (3, 4), (4, 4)\}$
7. $\{(1, 1), (2, 2), (3, 3), (4, 4)\}$
9. Exercise 6, transitive; Exercise 7, reflexive, symmetric, transitive; Exercise 8, reflexive, symmetric.
11. Symmetric
13. Transitive
15. Symmetric
17. Reflexive, transitive
19. Reflexive, symmetric
21. (Answer not unique), belongs to the same club as, U = people.

23. (Answer not unique), is less than 5 years younger than, U = people, or $x \leq y + k$, k is a natural number, U = natural numbers.
25. (Answer not unique), is an ancestor of, U = people or $x < y$, U = real numbers *or* follows, pages of a book = U.
27. Reflexive because $x \equiv x$ since x and x have the same remainder when divided by 5.
    Symmetric because y and x have the same remainder when x and y have the same remainder.
    Transitive because if r is the remainder for x and y, then r is the remainder for y and z and thus for x and z.
29. Not a function   33. Not a function
31. Function   35. $f(-4) = -2, f(0) = 2, f(2) = 4$
37. Dm = $\{-4, -2, 0, 2\}$, Rn = $\{-2, 0, 2, 4\}$
39. Yes, since each first component of f^{-1} is paired with exactly one second component.

Exercises 2.2 (page 46)

1. (a) 1 is in S since $(2 \cdot 1 - 1) = 1^2$.
   (b) If $1 + 3 + 5 + \ldots + (2n - 1) = n^2$, then
   $1 + 3 + \ldots + (2n - 1) + (2n + 1) = n^2 + (2n + 1) = (n + 1)^2$.
   Thus $n + 1$ is in S when n is in S.

3. (a) $1^2 = \dfrac{1(1 + 1)(2 + 1)}{6} = 1$

   (b) $1^2 + 2^2 + \ldots + n^2 + (n + 1)^2 = \dfrac{n(n + 1)(2n + 1)}{6} + (n + 1)^2$
   $= \dfrac{(n + 1)(n + 2)(2n + 3)}{6}$

5. Induction on n (assume $x \neq 1$)

   (a) $1^{1-1} = 1^0 = 1$ and $\dfrac{x^1 - 1}{x - 1} = 1$

   (b) $1 + x + \ldots + x^{k-1} + x^k = \dfrac{x^k - 1}{x - 1} + x^k = \dfrac{x^{k+1} - 1}{x - 1}$

7. (a) $\dfrac{1}{1 \cdot 3} = \dfrac{1}{3} = \dfrac{1}{2 + 1}$

   (b) $\dfrac{1}{1 \cdot 3} + \dfrac{1}{3 \cdot 5} + \ldots + \dfrac{1}{(2n - 1)(2n + 1)} + \dfrac{1}{(2n + 1)(2n + 3)}$
   $= \dfrac{n}{2n + 1} + \dfrac{1}{(2n + 1)(2n + 3)} = \dfrac{n + 1}{2n + 3}$

EXERCISES 2.3

9. (a) $2^2 = 4$ and $\frac{2}{3}(1)(1 + 1)(2 + 1) = 4$
    (b) $2^2 + 4^2 + \ldots + (2n)^2 + [2(n + 1)]^2 = \frac{2}{3}n(n + 1)(2n + 1) + [2(n + 1)]^2$
    $$= \frac{2(n + 1)(n + 2)(2n + 3)}{3}$$

11. (a) $1 + 2 + 3 + \ldots + n + (n + 1) = \frac{(2n + 1)^2}{8} + (n + 1)$
    $$= \frac{[2(n + 1) + 1]^2}{8}$$
    (b) Not valid because not true for $n = 1$ since $\frac{(2 \cdot 1 + 1)^2}{8} = \frac{9}{8} \neq 1$.

13. (a) $2^{1-1} = 2^0 = 1$
    (b) Not valid because not true for $x = 3$; $2^{3-1} = 2^2 = 4$ and $4 \neq 3$.

15. (a) Part 1 is valid since $3^1 = 3$ and $6 - 12 + 9 = 3$.
    (b) Part 2 is not valid since not true for $x = 4$.

17. (a) Part 1 is not valid: $1^2 + 4 = 5 \neq 1$
    (b) Part 2 is valid:
    $$1 + 3 + \ldots + (2n - 1) + (2n + 1)$$
    $$= n^2 + 4 + 2n + 1 = (n + 1)^2 + 4$$

19. (a) For $x = 1$, $a^2 - b^2 = (a + b)(a - b)$.
    (b) Using the hint, since $a^2 - b^2$ and $a^{2n} - b^{2n}$ are each divisible by $a + b$, then so is $a^{2n+2} - b^{2n+2}$.

Exercises 2.3 (page 48)

1. Assume there is a y so that for all x, $x + y = x$. Then for $x = 1$, $1 + y = 1$ or $s(y) = 1$. This contradicts P(3) so there is no such y.

3. (a) $z = 1$; $(xy) \cdot 1 = xy = x(y \cdot 1)$; Definition of multiplication.
   (b) Assume $(xy)n = x(yn)$. Then

$(xy)(n + 1) = (xy)n + (xy) \cdot 1$	Definition of multiplication
$= x(yn) + x(y \cdot 1)$	Assumption and Part (a)
$= x(yn + y \cdot 1)$	Distributive theorem
$= x[y(n + 1)]$	Definition of multiplication

5. (a) There is at least one element since 1 is in N and $x \cdot 1 = 1 \cdot x = x$ by the commutative axiom and definition of multiplication.
   (b) Assume $1'$ is in N and $x \cdot 1' = 1' \cdot x = x$ for all x. Then for $x = 1$, $1 \cdot 1' = 1' \cdot 1 = 1$. Since $x \cdot 1 = 1 \cdot x = x$ for all x, $1 \cdot 1' = 1' \cdot 1 = 1'$. For each x and y in N, there is exactly one xy in N. Thus, $1 \cdot 1' = 1$ and $1 \cdot 1' = 1'$ requires that $1 = 1'$. Therefore there is at most one such element and thus exactly one.

7. (a) $z = 1$; if $x + 1 = y + 1$, then $x = y$ by Postulate 4.
   (b) Assume if $x + n = y + n$, then $x = y$. Now

$x + (n + 1) = y + (n + 1)$	Given
$(x + n) + 1 = (y + n) + 1$	Associative theorem, addition
$x + n = y + n$	Postulate 4
$x = y$	Assumption

9. (a) For $y = 1$, if $xz = 1 \cdot z$, then $x = 1$ since there is exactly one element, 1, so that $1 \cdot z = z \cdot 1 = z$.
   (b) Assume if $xz = nz$, then $x = n$ and assume $xz = (n + 1)z$. If $x = 1$, then $1 \cdot z = y \cdot z$ and $1 = y$. If $x \neq 1$, then for some k in N, $x = s(k)$ or $x = k + 1$. Then

$(k + 1)z = (n + 1)z$	Substitution
$kz + z = nz + z$	Distributive theorem
$kz = nz$	Cancellation theorem, addition
$k = n$	Assumption
$k + 1 = n + 1$	Addition theorem
$x = n + 1$	Substitution

Exercises 2.4 (page 50)

1. Reflexive because $a \leq a$ since $a = a$.
   Transitive: If $a \leq b$, then $a = b$ or for $n \geq 1$, $a + n = b$.
   If $b \leq c$, then $b = c$ or for $m \geq 1$, $b + m = c$.
   If $a = b$ or $b = c$, then, by substitution, $a \leq c$.
   If $a + n = b$ and $b + m = c$, then $(a + n) + m = c$ or $a < c$.
   Not symmetric since $2 \leq 5$ but $5 \nleq 2$.

3. Proof by induction on n, the number of elements in S.
   (a) If $n = 1$, then $x = x$ for all x in S.
   (b) Let $S_n = \{x_1, x_2, \ldots, x_n\}$ where x_i is any element in N. Assume S is well ordered so that for some k, $x_k \leq x$ for all x in S. Then for $S_{n+1} = \{x_1, x_2, \ldots, x_n, x_{n+1}\}$, $x_k \leq x_i$ for $i \leq n$.
   By the trichotomy axiom, $x_k \leq x_{n+1}$ or $x_k > x_{n+1}$.
   If $x_k \leq x_{n+1}$, then $x_k \leq x$ for all x in S_{n+1}.
   If $x_k > x_{n+1}$, then $x_{n+1} < x_k \leq x_i$ or $x_{n+1} \leq x$ for all x in S_{n+1}.

5. $\{x \mid x < 8 \text{ or } 5 < x\} = N$.

7. $\{x \mid (x < 8 \text{ or } x < 3) \text{ and } (5 < x)\} = \{x \mid 5 < x < 8)\}$.

9. (a) Yes (h) Yes
   (b) No   (i) Yes
   (c) No   (j) Yes
   (d) Yes (k) Yes
   (e) No   (l) Yes
   (f) Yes (m) No
   (g) Yes (n) Yes

REVIEW EXERCISES

(o) Yes (q) No
(p) Yes (r) Yes

11. Proof is similar to that given for Exercise 3.
13. Suppose $a = b + c$ and $a = b + c'$.
    Then $b + c = b + c'$ and $c = c'$ by the cancellation theorem for addition.

Exercises 2.5 (page 58)
1. $\{1, 2, 3, 4, 5, 6, 7, 8, 9\}$
3. $\{11, 12, 13, \ldots\}$
5. $2 \cdot 2 \cdot 2 \cdot 3 \cdot 3$
7. $5 \cdot 7 \cdot 11 \cdot 11$
9. $17 = 4^2 + 1$, $37 = 6^2 + 1$, $101 = 10^2 + 1$
11. For $n = 2$, $n^2 - 1 = 4 - 1 = 3$ and 3 is prime.
13. For $n = 41$, $41^2 + 41 + 41 = (41)(43)$.
15. 70   23. 672
17. 180   25. 216
19. 3   27. 28
21. 252   29. $n = 5$, $496 = 2^4(2^5 - 1) = 16(31)$

Review Exercises (page 61)
1. $\{(1, 1), (1, 2), (1, 3), (1, 4), (2, 1), (2, 2), (2, 3), (2, 4), (3, 1), (3, 2), (3, 3),$
   $(3, 4), (4, 1), (4, 2), (4, 3), (4, 4)\}$
2. $\{(1, 1), (1, 2), (1, 3), (2, 1), (2, 2), (2, 3), (3, 1), (3, 2), (3, 3), (4, 1), (4, 2),$
   $(4, 3)\}$
3. Yes   4. $\{(x, y) \mid y = x \text{ and } x \in A\}$
5. Yes. For each (x, y) in C, (x, x) is in C (reflexive).
   If (x, y) is in C, then (y, x) is in C (symmetric).
   If (x, y) and (y, z) are in C, then (x, z) is in C (transitive).
6. Not reflexive since p is not a son or daughter of himself.
   Not symmetric since if p is a son or daughter of q, then q is not a son or daughter of p.
   Not transitive since a grandchild is not a son or daughter.
7. (a) $3^{1-1} = 3^0 = 1$ and $\dfrac{3^1 - 1}{2} = 1$

   (b) $1 + 3 + \ldots + 3^{k-1} + 3^k = \dfrac{3^k - 1}{2} + 3^k = \dfrac{3^{k+1} - 1}{2}$

8. (a) $1 \cdot 3^1 = 3$ and $\dfrac{(2 - 1)3^2 + 3}{4} = 3$

   (b) $1 \cdot 3 + \ldots + k \cdot 3^k + (k + 1)3^{k+1} = \dfrac{(2k - 1)3^{k+1} + 3}{4} + (k + 1)3^{k+1}$

   $= \dfrac{(2k + 1)3^{k+2} + 3}{4}$

9. (a) $1 \cdot 2 = 2$ and $\dfrac{1(1+1)(1+2)}{3} = 2$

    (b) $1 \cdot 2 + \ldots + k(k+1) + (k+1)(k+2)$
    $= \tfrac{1}{3}k(k+1)(k+2) + (k+1)(k+2) = \tfrac{1}{3}(k+1)(k+2)(k+3)$

10. (a) $\dfrac{1}{(3-1)(3+2)} = \dfrac{1}{10}$ and $\dfrac{1}{6+4} = \dfrac{1}{10}$

    (b) $\dfrac{1}{2 \cdot 5} + \ldots + \dfrac{1}{(3k-1)(3k+2)} + \dfrac{1}{(3k+2)(3k+5)}$
    $= \dfrac{k}{2(3k+2)} + \dfrac{1}{(3k+2)(3k+5)} = \dfrac{k+1}{6(k+1)+4}$

11. $4 \oplus 3 = 4^3 = 64$
12. Yes. $x^y \in W$ for all x and y not both 0.
13. No. $4^3 \neq 3^4$ since $64 \neq 81$.
14. No. $(x^y)^z \neq x^{(y^z)}$ since $(2^3)^2 = 64$ and $2^{(3^2)} = 2^9 = 512$.
15. Yes. 1 is a right-hand identity since $x^1 = x$ for all x in W. There is no left-hand identity.
16. $\{x \mid 1 \leq x \leq 12\}$
17. $\{x \mid 3 < x < 10\}$
18. $\{x \mid 6 < x \leq 12\}$
19. $\{x \mid 3 < x < 10\}$
20. $\{x \mid 3 < x\}$
21. $\{1, 2, 3, 4\}$
22. $\{6, 7, 8, \ldots\}$
23. $\{1, 5\}$
24. $\{5, 10, 15, 20, 25, \ldots\}$
25. $2 \cdot 3^2 \cdot 11$
26. $5 \cdot 59$
27. $2^5 \cdot 3^2$
28. $2 \cdot 3^4 \cdot 5 \cdot 11$
29. 2
30. LCM $= \dfrac{86(124)}{2} = 5332$
31. (29, 31), (41, 43), (59, 61), (71, 73)
32. 8
33. $f(1) = 5, f(2) = 17, f(3) = 257, f(4) = 65{,}537$
34. Yes, $\sqrt{257} < (17)^2$ and no prime ≤ 17 is a factor.
    $\sqrt{65{,}537} < (251)^2$ and no prime ≤ 251 is a factor.
35. $f(5) = 2^{32} + 1 = (256)^4 + 1 = [641(102) + 154]^2 + 1$
    $= (641)^2(102)^2 + 641(204)(154) + (154)^2 + 1$
    $(154)^2 + 1 = 23{,}717 = 641(37)$. Thus $f(5)$ is divisible by 641.
36.
    $220 = 2 \cdot 110$
    $ 4 \cdot 55$
    $ 5 \cdot 44$
    $ 10 \cdot 22$
    $ 11 \cdot 20$
    $ 1$
    $\overline{33 + 251 = 284}$

    $284 = 2 \cdot 142$
    $ 4 \cdot 71$
    $ 1$
    $\overline{7 + 213 = 220}$

CHAPTER 3

Exercises 3.1 (page 70)

1. Distributive axiom
3. Commutative axiom, multiplication
5. Identity axiom, addition
7. Closure axiom, multiplication
9. Addition cancellation theorem
11. $x + 0 = x$ Identity axiom, addition
    $x + 0 = 0 + x$ Commutative axiom, addition
    $0 + x = x$ Substitution
13. $(x + y)z = z(x + y)$ Commutative axiom, multiplication
    $z(x + y) = zx + zy$ Distributive axiom
    $zx = xz$ and $zy = yz$ Commutative axiom, multiplication
    $(x + y)z = xz + yz$ Substitution
15. $x(y + z + t) = x(y + z) + xt$ Distributive axiom
    $x(y + z + t) = xy + xz + xt$ Distributive axiom and substitution
17. $(x + a)(x + b) = (x + a)x + (x + a)b$ Distributive axiom
    $= (x^2 + ax) + (xb + ab)$ Distributive axiom
    $= (x^2 + ax) + (bx + ab)$ Commutative axiom, multiplication
    $= x^2 + (ax + bx) + ab$ Associative axiom, addition
    $= x^2 + (a + b)x + ab$ Distributive and commutative axioms
19. $x - (y + z) = x + (-(y + z))$ Definition of subtraction
    $= x + ((-y) + (-z))$ Opposite of a sum theorem
    $= (x + (-y)) + (-z)$ Associative axiom, addition
    $= (x - y) - z$ Definition of subtraction
21. $-x(y + z) = (-x)y + (-x)z$ Distributive axiom
    $= -xy + (-xz)$ Opposite of a product theorem
    $= -xy - xz$ Definition of subtraction
23. $(-x)(-y)(-z) = -x(yz)$ Product of two opposites theorem
    $= -xyz$ Opposite of a product theorem
25. $2 + 2 = 2 + (1 + 1) = (2 + 1) + 1 = 3 + 1 = 4$
27. $2 \cdot 2 = 2(1 + 1) = 2 + 2 = 4$
29. $1 < 2 \leftrightarrow 2 > 1 \leftrightarrow 2 - 1 > 0 \leftrightarrow 1 > 0$. Now $1 \neq 0$, because if $1 = 0$, then $x \cdot 0 = 0$ and $x \cdot 0 = x \cdot 1 = x$ for all x. If $-1 > 0$, then $(-1)(-1) = 1 > 0$ by the positive closure axiom. But $-1 > 0$ and $1 > 0$ contradicts the trichotomy axiom. Thus $1 > 0$.
31. $2x = 2$ Addition cancellation theorem
    $2x = 2 \cdot 1$ Identity axiom, multiplication
    $x = 1$ Multiplication cancellation theorem

33. $3x = 4$ by the addition cancellation theorem. Since there is no integer x so that $4 = 3x$, this is unsolvable over I. (*Note*: $x = \frac{4}{3}$.)

35. $3 - 2x = 6$
    $(-2x) + 3 = 3 + 3$ Definition of subtraction and commutative axiom
    $-2x = 3$ Addition cancellation theorem
    There is no integer x so that $3 = (-2)x$; so unsolvable. (*Note*: $x = -\frac{3}{2}$.)

37. For each x and y in I, $x - y = x + (-y)$ by the definition of subtraction. For each y in I, there is exactly one element $(-y)$ in I. Therefore, for each x and y in I, there is exactly one element $x + (-y)$ in I by the closure axiom.

39. Domain $= \{(x, y) \mid x = ky$ where k, x, y are integers and $y \neq 0\}$.

Exercises 3.2 (page 74)

1.

·	−1	1
−1	1	−1
1	−1	1

   Closure: The table shows that $xy \in S$ for all x and y in S.
   Associative: $(xy)z = x(yz)$ for all integers.
   Identity: $e = 1$.
   Inverse: $1' = 1$ and $(-1)' = -1$.

3. (a) No, not associative and no identity.
   (b) No, no identity.
   (c) Yes; closed since all entries are in the set, associative by checking all cases (6), $e = 0$, Inverses; $0' = 0$, $1' = 2$, $2' = 1$.
   (d) No, not all elements have inverses (B and D do not).
   (e) No, not all elements have inverses (0 does not).

5. No, not all elements have inverses.

7. A group; closed since $x \in S$, associative since $(x \circ x) \circ x = x \circ (x \circ x) = x$, $e = x$, $x' = x$.

9. *First,* $S = \{\ldots, -4, -2, 0, 2, 4, \ldots\}$; $+$ is a commutative group since closed; $2a + 2b = 2(a + b)$.

   Associative: $(2a + 2b) + 2c = 2(a + b) + 2c$
                                  $= 2(a + b + c) = 2a + (2b + 2c)$
   Identity: 0 and inverse of $2a = -2a$.
   Commutative: $2a + 2b = 2(a + b) = 2(b + a) = 2b + 2a$

   Second, Closed for · since $2a \cdot 2b = 2(a \cdot 2b)$. Associative for · and distributive since these are properties for all integers.
   Third, there is no multiplicative identity. If so, $2x \cdot e = 2x$ for all integers. Then $e = 1$ and $e \notin S$.

11. $(a, b) + (c, d) = (a + c, b + d) = (c + a, d + b) = (c, d) + (a, b)$
    $(a, b) \cdot (c, d) = (ac, bd) = (ca, db) = (c, d) \cdot (a, b)$

13. Let (x, y) be such that for all (a, b) in S, $(a, b) + (x, y) = (a, b)$.
    Then $(a + x, b + y) = (a, b)$ and $a + x = a$ and $b + y = b$.
    Thus $x = 0$ and $y = 0$; that is, $(x, y) = (0, 0)$.

EXERCISES 3.4

15. (a) $(a, b) \cdot [(c, d) + (e, f)] = (a, b) \cdot (c + e, d + f)$
$= (a(c + e), b(d + f))$
$= (ac + ae, bd + bf)$
$= (ac, bd) + (ae, bf)$
$= (a, b) \cdot (c, d) + (a, b) \cdot (e, f)$
   (b) No. $(1, 1) + [(1, 1) \cdot (1, 1)] = (1, 1) + (1, 1) = (2, 2)$
$[(1, 1) + (1, 1)] \cdot [(1, 1) + (1, 1)] = (2, 2) \cdot (2, 2) = (4, 4)$
17. Yes. $(a, b) \cdot (1, 1) = (a, b)$ for all (a, b) in S.
27. No. Not closed for addition, $1 + 3 = 4 \notin S$.
29. Yes
31. Yes

Exercises 3.3 (page 82)
1. $(2a + 1)(2a - 1)(a + 3)$
3. $(r + 2s - 3)(r + 2s + 2)$
5. $(t + 2)(t - 2)(t^2 - 2t + 4)(t^2 + 2t + 4)$
7. $(x - y + z)^2$
13. $9(2x - 5)$
9. $(2p - 3)(p^2 + 1)$
15. $(m + 1)(m + 3)$
11. $(1 - x)(1 - x)(1 + x)$
17. $(x - 3y + 2z)(x - 3y - 2z)$
19. $(x + a)(x - a)(x^2 - ax + a^2)(x^2 + ax + a^2)$
21. $(a - b)(a - b)(a + b)$
23. $(x^2 + 2xy + 3y^2)(x^2 - 2xy + 3y^2)$
25. If $(x + a)(x + b) = x^2 + 4$, then $a + b = 0$ and $ab = 4$.
Thus $b = -a$, $ab = -a^2$ and there is no integer a so that $-a^2 = 4$.
27. If $(x + a)(x + b) = x^2 + x + 1$, then $a + b = 1$ and $ab = 1$.
Thus, $a = b = 1$ or $a = b = -1$. But $1 + 1 = 2$ and $-1 + (-1) = -2$.
29. $x^4 + 4 = x^4 + 4x^2 + 4 - 4x^2 = (x^2 + 2)^2 - (2x)^2$
$= (x^2 + 2x + 2)(x^2 - 2x + 2)$
31. $x^4 - x^3 + 8x - 8 = (x^4 + 8x) - (x^3 + 8)$
$= x(x^3 + 8) - (x^3 + 8) = (x - 1)(x^3 + 8)$
$= (x - 1)(x + 2)(x^2 - 2x + 4)$
33. $x^1 - a^1 = x - a$; $x^{n+1} - a^{n+1} = x^n(x - a) + a(x^n - a^n)$ and so on
35. $x^2 - a^2 = (x + a)(x - a)$
$x^{2n+2} - a^{2n+2} = x^{2n}(x^2 - a^2) + a^2(x^{2n} - a^{2n})$ and so on

Exercises 3.4 (page 87)
1. $(-x^2 + x + 1)(-2x^4 - 2x^3 - 9x^2 - 10x - 19) + 29x + 20$
3. $(x^2 + 3x + 2)(x^2 + x)$
5. $(2x - 3)(2x^2 + 3x - 2) + 3$
7. $3x - 4 + \dfrac{3}{x + 2}$
9. $-x^3 - x^2 + \dfrac{-1}{x - 1}$
11. $2x^3 + 5x^2 + \dfrac{-6}{x - 4}$
13. $x^4 + x^3 + x^2 + x + 1$
15. $(x - 3)(x^2 + x + 6) + 14$

17. $(x+5)(3x^3+2x^2-10x+80)-416$
19. $(x-4)(3x^3-5x^2+10x+40)$ 23. $R=2, P(1)=2$
21. $R=-12, P(2)=-12$ 25. $R=0, P(1)=0$
27. (a) $(x^2-3)(x^3+x+2)+4x+8$
    (b) $(x^2+2)(x^3+5x^2-x)+(-5x+2)$
    (c) $(x^2+1)(x^2+x-1)+(-2x+2)$
    (d) $(x^2-5)(x^2+5)+(5x+23)$
29. Let degree of $B(x) \le$ degree of $A(x)$. Then
    $A(x) = B(x)Q_1(x) + R_1(x)$ and $R_1 = 0$ or deg $R_1 <$ deg B
    $B(x) = R_1(x)Q_2(x) + R_2(x)$ and $R_2 = 0$ or deg $R_2 <$ deg R_1
    ...
    $R_{n-2}(x) = R_{n-1}(x)Q_n(x) + R_n(x)$ and $R_n = 0$ or deg $R_n <$ deg R_{n-1}
    $R_{n-1}(x) = R_n(x)Q_{n+1}(x)$.
    The GCD of $A(x)$ and $B(x)$ is $R_n(x)$, the first nonzero remainder.

Exercises 3.5 (page 92)
1. Yes, $P(-3)=0$ 7. No, $P(1)=2 \ne 0$
3. No, $P(1)=5 \ne 0$ 9. $P(2)=6, P(-1)=-15$
5. Yes, $P(-\frac{1}{3})=0$
11. $P(1)=8, P(-1)=30, P(3)=162$
13. $P(3)=0$ 27. $\{2,-2\}$
15. $P(-a)=0$ 29. $\{2\}$
17. $R=P(2)=-12$ 31. $\{1,2,3\}$
19. $R=P(1)=2$ 33. 66
21. $R=P(1)=0$ 35. -3
23. $\{-1\}$ 37. -1
25. No solutions

Review Exercises (page 95)
1. Definition of $x+y+z$ 5. Identity axiom, multiplication
2. Commutative axiom, addition   6. Distributive axiom
3. Associative axiom, addition   7. Substitution
4. Distributive axiom
8. $x^2 - 2x = 3$ Distributive axiom
   $x^2 - 2x + (-3) = 3 + (-3)$ Addition theorem
   $x^2 - 2x - 3 = 0$ Inverse axiom, addition and definition of subtraction

   $(x-3)(x+1) = 0$ General trinomial theorem, I
   $x-3=0$ or $x+1=0$ Zero product theorem
   $x=3$ or $x=-1$ Addition theorem

REVIEW EXERCISES

9. $\left(\dfrac{x+5}{3}\right)3 = 7(3) = 21$ Multiplication theorem
$(x+5)(\tfrac{1}{3}\cdot 3) = 21$ Associative axiom, multiplication
$(x+5)(1) = 21$ Inverse axiom, multiplication
$x + 5 = 21$ Identity axiom, multiplication
$(x+5) + (-5) = 16$ Addition theorem
$x + (5 + (-5)) = 16$ Associative axiom, addition
$x + 0 = 16$ Inverse axiom, addition
$x = 16$ Identity axiom, addition

10. $(2x + 3) + (-3) = 6 + (-3) = 3$ Addition theorem
$2x + (3 + (-3)) = 3$ Associative axiom, addition
$2x + 0 = 3$ Inverse axiom, addition
$2x = 3$ Identity axiom, addition
Unsolvable since no integer k so that $2k = 3$.

11. $2a + 2b = 2(a + b)$ by distributive axiom.

12. $(2a + 1) + (2b + 1) = [(2a + 1) + 2b] + 1$ Associative axiom, addition
$ = [2a + (1 + 2b)] + 1$ Associative axiom, addition
$ = [2a + (2b + 1)] + 1$ Commutative axiom, addition
$ = (2a + 2b) + (1 + 1)$ Associative axiom, addition
$ = 2(a + b) + 2(1)$ Distributive axiom
$ = 2(a + b + 1)$ Distributive axiom

13. Addition and multiplication by the closure axioms.
Subtraction since by definition $a - b = a + (-b)$ and closure is valid for addition.
Not division since a/b is not an integer for $a = 1$ and $b = 2$.

14. Odd. $(2a + 1)(2b + 1) = (4ab + 2a + 2b) + 1 = 2(2ab + a + b) + 1$

15. $(a + x)(2x - y)$ 18. $(x + y)(x + y)(x + y)$
16. $(4x - 3y)(3x - 20y)$ 19. $(x + y - 2z)(x - y + 2z)$
17. $(p^2 + 3p + 5)(p^2 - 3p + 5)$ 20. $(x - 4)(x^2 + 4x + 16)$
21. $(y - 1)(y + 1)(xy + 2)$
22. $(x^2 + 4xy + 8y^2)(x^2 - 4xy + 8y^2)$
23. $(2a - b + x + y)(2a - b - x - y)$
24. $(y^2 + 2yz + 3z^2)(y^2 - 2yz + 3z^2)$ 25. $(x - 1)(4x^2 + 2x + 1)$
26. $(x + 2)(2x^4 - 4x^3 + 8x^2 + 2x - 3) + 4$
27. $(x + 1)(x^3 + x^2 - x + 2) - 1$ 28. $(x - 3)(3x^4 - x^3 + x + 2) + 2$
29. $(x - 1)(3x^2 + 5x) - 1$
30. $(x - 1)(x^6 + x^5 + x^4 + x^3 + x^2 + x + 1)$
31. -2 33. 33
32. -35 34. 76

35. $\{-1\}$
36. $\{-6\}$
37. $\{-3\}$
38. $\{1, 2, -3\}$
39. $\{-4, -1, 2, 3\}$
40. $\{1, -2\}$

CHAPTER 4

Exercises 4.1 (page 104)

1. $\left(\dfrac{1}{17} + \dfrac{16}{17}\right) + \dfrac{2}{3} = 1 + \dfrac{2}{3} = \dfrac{5}{3}$; Commutative and associative, addition

3. $45\left(\dfrac{22}{37} + \dfrac{15}{37}\right) = 45$; Distributive

5. $\dfrac{125}{3}\left(\dfrac{12}{5}\right)\left(-\dfrac{19}{23}\right)\left(-\dfrac{23}{38}\right) = 100\left(\dfrac{1}{2}\right) = 50$; Commutative and associative, multiplication

7. $450\left(\dfrac{7}{90}\right) - 450\left(\dfrac{11}{150}\right) = 35 - 33 = 2$; Distributive

9. $\dfrac{2}{3}\left(\dfrac{10}{77} + \dfrac{67}{77}\right) = \dfrac{2}{3}(1) = \dfrac{2}{3}$; Distributive and commutative, associative, multiplication

11. $= \dfrac{3}{8} + \left(\dfrac{7}{16} + \dfrac{5}{8}\right)$ Associative, addition

    $= \dfrac{3}{8} + \left(\dfrac{5}{8} + \dfrac{7}{16}\right)$ Commutative, addition

    $= \left(\dfrac{3}{8} + \dfrac{5}{8}\right) + \dfrac{7}{16}$ Associative, addition

    $= 1 + \dfrac{7}{16} = \dfrac{23}{16}$ Addition of quotient theorem, fundamental theorem

13. $= \left(\dfrac{23}{125} \cdot \dfrac{17}{23}\right)\dfrac{3}{8}$ Associative, multiplication

    $= \dfrac{(23)(17)}{(125)(23)} \cdot \dfrac{3}{8}$ Multiplication of quotients theorem

    $= \dfrac{17(23)}{125(23)} \cdot \dfrac{3}{8}$ Commutative, multiplication

    $= \dfrac{17}{125} \cdot \dfrac{3}{8}$ Fundamental theorem

    $= \dfrac{51}{1000}$ Multiplication of quotients theorem

EXERCISES 4.1

15. $\dfrac{11}{15}$

17. $\dfrac{7a - 19}{(a - 2)(2a - 5)}$

19. $\dfrac{4b - 10}{b - 2}$

21. 3

23. $\dfrac{4y + 1}{4y - 1}$

25. 6

27. $\dfrac{a}{b} \equiv \dfrac{a}{b}$ if and only if $ab = ba$ Valid by commutative axiom, multiplication

29. $\dfrac{a}{b} \equiv \dfrac{c}{d}$ if and only if $ad = bc$ Definition

$\dfrac{c}{d} \equiv \dfrac{e}{f}$ if and only if $cf = de$ Definition

$(ad)(cf) = (bc)(de)$ Multiplication theorem

$(af)cd = (be)cd$ Commutative, associative, multiplication

$af = be$ Multiplication cancellation theorem, $cd \neq 0$

$\dfrac{a}{b} \equiv \dfrac{e}{f}$ Definition

31. $\dfrac{a}{b} = \dfrac{c}{d}$ if and only if $ad = bc$ Definition

if and only if $bc = ad$ Symmetric

if and only if $\dfrac{b}{a} = \dfrac{d}{c}$ Definition

33. $\dfrac{1}{m}$

35. $\left(\dfrac{x}{y}\right) \Big/ \left(\dfrac{z}{w}\right) = \left(\dfrac{x}{y} \cdot \dfrac{w}{z}\right) \Big/ \left(\dfrac{z}{w} \cdot \dfrac{w}{z}\right)$ Fundamental theorem

$= \left(\dfrac{xw}{yz}\right) \Big/ \left(\dfrac{wz}{wz}\right)$ Multiplication quotients theorem, commutative multiplication

$= \left(\dfrac{xw}{yz}\right) \Big/ 1$ Fundamental theorem

$= \dfrac{xw}{yz} \cdot \dfrac{1}{1}$ Definition of division

$= \dfrac{xw}{yz}$ Multiplication of quotients theorem, identity axiom multiplication

37. By definition, $\dfrac{a}{b} + \dfrac{c}{d} = \dfrac{ad+bc}{bd}$ Where a, b, c, d are integers, $bd \neq 0$

$\qquad\qquad\qquad = \dfrac{bc+ad}{bd}$ By commutative axiom, addition of integers

$\qquad\qquad\qquad = \dfrac{cb+da}{db}$ By commutative axiom, multiplication of integers

$\qquad\qquad\qquad = \dfrac{c}{d} + \dfrac{a}{b}$ By definition of addition

39. $\dfrac{1}{a} > \dfrac{1}{b}$

Exercises 4.2 (page 107)

1. *Closure*, $+\quad (a + b\sqrt{2}) + (c + d\sqrt{2}) = (a+c) + (b+d)\sqrt{2}$.
   $\quad\cdot\quad (a + b\sqrt{2}) \cdot (c + d\sqrt{2}) = (ac + 2bd) + (ad + bc)\sqrt{2}$.
   Commutative, $+\quad (a+c) + (b+d)\sqrt{2} = (c+a) + (d+b)\sqrt{2}$.
   $\quad\cdot\quad (ac + 2bd) + (ad + bc)\sqrt{2} = (ca + 2db) + (da + cb)\sqrt{2}$.
   Associative, $+\quad [(a+c) + e] + [(b+d) + f]\sqrt{2}$
   $\qquad\qquad\qquad = [a + (c+e)] + [b + (d+f)]\sqrt{2}$.
   $\quad\cdot\quad (ac + 2bd)e + 2(ad + bc)f = a(ce + 2df) + 2(cf + de)b$
   and
   $\qquad [(ac + 2bd)f + (ad + bc)e]\sqrt{2} = [a(cf + de) + b(ce + 2df)]\sqrt{2}$.
   Distributive $(a + b\sqrt{2})[(c+e) + (d+f)\sqrt{2}]$
   $\qquad\qquad\qquad = [(ac + 2bd) + (ad + bc)\sqrt{2}]$
   $\qquad\qquad\qquad\quad + [(ae + 2bf) + (af + be)\sqrt{2}]$.
   Identity, $+\quad 0 = 0 + 0 \cdot \sqrt{2}$.
   $\quad\cdot\quad 1 = 1 + 0 \cdot \sqrt{2}$.
   Inverse, $+\quad -(a + b\sqrt{2}) = (-a) + (-b)\sqrt{2}$
   $\quad\cdot\quad \dfrac{1}{a + b\sqrt{2}} = \dfrac{a - b\sqrt{2}}{a^2 - 2b^2}\quad$ for $a^2 + b^2 \neq 0$.

3. See Exercise 2, replace 2 by 3.

EXERCISES 4.3

5.

+	0	1	a
0	0	1	a
1	1	a	0
a	a	0	1

·	0	1	a
0	0	0	0
1	0	1	a
a	0	a	1

7. Not a field, no multiplication identity and not all nonzero elements have inverses.

9. Not a field, not closed for multiplication and no multiplication inverses for $a \neq 0$.

11. Closure follows from the definitions since the integers are closed with respect to addition and multiplication.
The commutative, associative, and distributive properties are verified by calculating the required expressions and by using the definition, axioms, and theorems for integers.
The identity for $+ = (0, 1)$; the identity for $\cdot = (1, 1)$.
Inverses: $-(a, b) = (-a, b)$ and $1/(a, b) = (b, a)$.
Yes; replace (a, b) by a/b and note that all the definitions are satisfied since they are consequences of the field axioms.

13. Yes. The closure, commutative, identity, and inverse properties can be seen directly from the tables.
The associative and distributive properties are verified by using the definition, axioms, and theorems for integers.

15. No. The nonzero elements 2, 3, and 4 do not have multiplicative inverses. All other properties are satisfied.

Exercises 4.3 (page 114)

1. $\dfrac{1}{15}(3x^3 + 10x - 30)$

3. $\dfrac{1}{84}(56y^5 + 21y^3 - 168y + 12)$

5. $p = 2, q = -1$

7. $p = -\dfrac{5}{21}, q = \dfrac{10}{21}$

9. $\dfrac{x-4}{x+1}; x \neq 3, -1$

11. $\dfrac{1}{x+3}; x \neq 4, 3, -3$

13. $\dfrac{-x^2}{x^3+8}; x \neq -2$

15. $\dfrac{3x}{9+2x}; x \neq 3, -3, -\dfrac{9}{2}$

17. $\dfrac{9x-63}{20x+100}; x \neq 3, -5$

19. $\dfrac{x+2}{3x^3}; x \neq 0, 2, -1$

21. $\dfrac{-5}{y+5}; y \neq 5, -5$

23. $1; x \neq \pm y, a \neq b$

25. $\dfrac{y+3}{y^2-3y-9}$

27. $x^2 - 4; x \neq 0$

Exercises 4.4 (page 121)

1. $a = \frac{13}{9}, b = \frac{14}{9}$
3. $a = -2, b = 1$
5. $a = \frac{2}{5}, b = \frac{3}{5}$
7. $a = 3, b = -1, c = 2$
9. $a = 0, b = 5, c = 1, d = 0$
11. $\dfrac{-1}{x+1} + \dfrac{2}{x+2}$
13. $\dfrac{4}{2x+1} + \dfrac{-5}{3x+2}$
15. $\dfrac{1}{x+1} + \dfrac{-1}{(x+1)^2}$
17. $\dfrac{-1/2}{x+1} + \dfrac{3/2}{x-1}$
19. $x + \dfrac{2}{x+2} + \dfrac{2}{x-2}$
21. $\dfrac{-6}{x+2} + \dfrac{-6}{(x+2)^2} + \dfrac{6}{x+1}$
23. $\dfrac{-3}{x} + \dfrac{4x+5}{x^2+3}$
25. $\dfrac{x}{x^2+1} + \dfrac{-x+1}{(x^2+1)^2}$
27. $\dfrac{8}{x^2+3} + \dfrac{-4}{x^2-2}$
29. $\dfrac{-1/4}{x-3} + \dfrac{5/4}{x+1}$
31. $\dfrac{3x}{3x^2-2} + \dfrac{-2x}{2x^2+3}$
33. $1 + \dfrac{-1/4}{x+3} + \dfrac{9/4}{x-1}$

Exercises 4.5 (page 124)

1. $\frac{1}{3}$
3. -2
5. $\emptyset$
7. 4
9. 1
11. $\emptyset$
13. All x except $\frac{3}{2}$ and $-\frac{5}{4}$
15. 0
17. -5
19. $-\frac{1}{2}$
21. 3
23. $\frac{1}{5}$
25. -6
27. $\frac{3}{2}$
29. $-1, -\frac{1}{2}, \frac{1}{2}$
31. $1, -\frac{1}{2}, -\frac{5}{7}$

Review Exercises (page 128)

1. $\frac{11}{7}$
2. $\dfrac{7x+15}{36}$
3. $\dfrac{1}{18(x+4)}; x \neq \pm 4, \pm\frac{1}{3}$
4. $\dfrac{10-x}{6x(x-4)}; x \neq 0, 4, -4$
5. $\dfrac{x}{2}; x \neq -2$
6. $\dfrac{1}{x-5}; x \neq -3, -5, 5$
7. $\dfrac{-6x}{x-3}; x \neq 3$
8. $x; x \neq 1, -2, 4, 5, -5$
9. $x; x \neq \pm 1, \pm\frac{2}{3}$
10. $8x + 8; x \neq 1, -3$
11. $\dfrac{4/7}{x+2} + \dfrac{2/7}{3x-1}$

EXERCISES 5.1

12. $\dfrac{4/11}{x-6} + \dfrac{-4/11}{x+5}$

13. $\dfrac{7/5}{x+2} + \dfrac{-4/5}{2x-1}$

14. $\dfrac{1/2}{2x+1} + \dfrac{-1/2}{(2x+1)^2}$

15. $x + \dfrac{11/4}{x+2} + \dfrac{13/4}{x-2}$

16. $\dfrac{3}{x+3} + \dfrac{-3}{x+2} + \dfrac{3}{(x+2)^2}$

17. $\dfrac{-1}{x+2} + \dfrac{2x-1}{x^2-3}$

18. $\dfrac{5/4}{2x} + \dfrac{-\frac{5}{8}x + \frac{3}{2}}{x^2+4}$

19. $\frac{3}{7}$
20. $-\frac{1}{4}$
21. -7
22. $\frac{4}{3}$
23. $-\frac{1}{5}$
24. -3
25. 8
26. -1
27. $0, \frac{1}{2}$
28. $3, -3, -\frac{1}{2}$
29. -8
30. $\emptyset$

CHAPTER 5

Exercises 5.1 (page 138)

1. $161/495$
3. $1196/495$
5. $3211/999$
7. $3.\overline{2}$
9. $5.\overline{63}$
11. $8.\overline{307692}$

13. Assume $\dfrac{p}{q} = 3 + \sqrt{2}$, where p and q are relatively prime natural numbers. Then $\sqrt{2} = \dfrac{p}{q} - 3 = \dfrac{p-3q}{q}$, where $\dfrac{p-3q}{q}$ is rational while $\sqrt{2}$ is not. Contradiction.

15. 11.49
17. 8.84
19. 11; No
21. 1; No
23. $\sqrt{3}$; No
25. 0; No
27. $\frac{1}{2}$; Yes
29. 2; Yes

31. $22/7$ ($\pi \approx 3.1416$, $\sqrt{10} \approx 3.1623$, $22/7 \approx 3.1429$)
33. (a) 10.16 miles, (b) 12.72 miles, (c) 0.08 miles
35. (a) to nearest tenth (1.7), (b) to nearest hundredth (1.73)
37. $a + b = \text{lub of } \left\{\dfrac{a_1 + b_1}{10}, \dfrac{10(a_1 + b_1) + a_2 + b_2}{100}, \ldots\right\}$

$ab = \text{lub of } \left\{\dfrac{a_1 b_1}{100}, \dfrac{100 a_1 b_1 + 10(a_1 b_2 + a_2 b_1) + a_2 b_2}{10{,}000}, \ldots\right\}$

Exercises 5.2 (page 145)

1. If $x < y$ and $y < z$, then $y > x$ and $z > y$ by definition of $<$, of $z > x$ by the transitivity order theorem and $x < z$ by definition.

3. $x < y$ and $z > 0$ Given
   $y > x$ Definition of $<$
   $yz > xz$ Multiplication order theorem (1)
   $xz < yz$ Definition of $<$

5. $\{x \mid x > 1\}$

7. $\{x \mid x > 6\}$

9. $\{x \mid x > -2 \text{ or } x \leq -3\}$

11. $\{x \mid x > 6/11\}$

13. $\{x \mid x < 0\}$

15. $\{x \mid x \leq \frac{2}{5}\}$

17. $\{x \mid -1 < x < \frac{1}{2}\}$

19. $\{x \mid x < 2 \text{ or } x > 2\}$

21. $\{x \mid x > 3\}$

23. $\{x \mid -2 < x < 3\}$

EXERCISES 5.3

25. $\{x \mid x < -2\}$
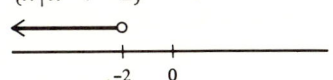

27. $\{x \mid 1 \leq x < 2\}$
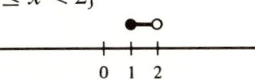

29. $\{x \mid x < 3\}$
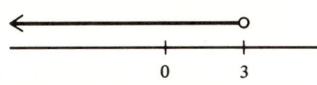

Exercises 5.3 (page 150)

1. $\{x \mid -5 \leq x \leq 1\}$
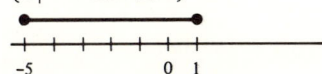

3. $\{x \mid x < -3 \text{ or } x > 4\}$
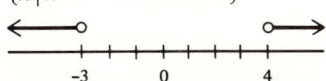

5. $\{1\}$
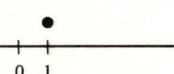

7. $\{x \mid -1 \leq x < 0 \text{ or } x \geq 4\}$
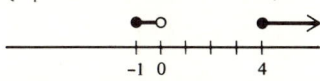

9. $\{x \mid x < -2 \text{ or } -1 < x < 3\}$
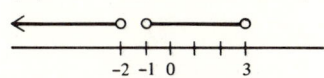

11. $\{x \mid x \leq -2 \text{ or } x \geq 2\}$
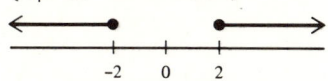

13. $\{x \mid x < 0 \text{ or } x > 2\}$
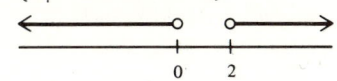

15. $\{x \mid -2 < x < 0 \text{ or } x > 4\}$

17. $\{x \mid -5 < x < -2 \text{ or } x > 0\}$
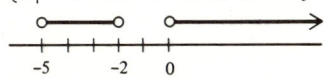

19. $\{x \mid -4 < x < -3 \text{ or } -2 < x < 1 \text{ or } x > 2\}$

21. $\{x \mid \frac{1}{3} < x < \frac{1}{2}\}$
23. $\{x \mid x < -\frac{2}{3} \text{ or } x > \frac{2}{3}\}$
25. $\{x \mid x > -\frac{2}{3}\}$
27. $x < y$ and $z < w$ Given
    $y > x$ and $w > z$ Definition of $<$
    $y + w > x + z$ Theorem 10
    $x + z < y + w$ Definition of $<$
29. $0 < x$ and $x < 1$ Given
    $x > 0$ and $1 > x$ Definition of $>$
    $1 \cdot x > x \cdot x$, $x > x^2$ Multiplication order theorem (1)
    $x^2 < x$ Definition of $>$
31. $x < 4$
33. $x > 1$ 37. $0 < x < 0.25$
35. $x < 120$ 39. $0 < x < \frac{2}{9}$

Exercises 5.4 (page 154)

1. $5, -1$
3. $x \leq -3$ or $x \geq 3$
5. $x < -\frac{1}{2}$ or $x > \frac{7}{2}$
7. $-1 \leq x \leq 1$
9. $-\frac{1}{2} < x < \frac{9}{2}$
11. R (all real numbers)
13. $3, -3$
15. $x \geq 0$
17. $x < -\frac{7}{4}$ or $x > -\frac{1}{4}$
19. R (all real numbers)
21. $-2 < x^2 - 3x + 4 < 14$
23. $2.45 < x < 2.55$
25. $|x - 1| < \varepsilon/5$
27. $0 < \delta \leq \varepsilon/5$
29. $|x - 1| < 1/11$
31. $0 < \delta \leq \varepsilon/6$
33. $0 < \delta \leq 3\varepsilon$
35. $|ab| \geq 0$, $|a| \geq 0$, and $|b| \geq 0$ so $|a| \cdot |b| \geq 0$ and $|ab| = |a| \cdot |b|$
37. $|a + b| = \sqrt{(a + b)^2} = \sqrt{a^2 + 2ab + b^2} \leq \sqrt{a^2 + 2|a||b| + b^2}$
    $\sqrt{(|a| + |b|)^2} = |a| + |b|$
39. $|x^2| \geq 0$, $|x|^2 \geq 0$, $x^2 \geq 0$ so all are equal
41. 1 if $x > 0$ and -1 if $x < 0$

Exercises 5.5 (page 160)

1. x^{m+n-1}
3. x^{m-n}
5. $1 + x - \dfrac{1}{x^2}$

EXERCISES 5.6

7. $\dfrac{y^2}{x^2}$

9. $\dfrac{2}{x^m}$

11. $x^2 + x + 1$
13. $x^{n^2} y^{n^2}$
15. $x - 1$

17. $1 + x + x^2$

19. $\dfrac{x^{2n}}{y^{3n-3}}$

21. 2^{n+1}
23. 3^{n+1}
25. $1/3^{n-1}$

27. (a) If $n = m$, then $x^n/x^m = x^n/x^n = 1$
    (b) If $n > m$, then $n = m + k$ for $k > 0$
    $x^{m+k}/x^m = x^m x^k / x^m = x^k = x^{n-m}$
    (c) If $n < m$, then $m = n + k$ for $k > 0$
    $x^n / x^{n+k} = x^n / x^n x^k = 1/x^k = 1/x^{m-n}$

29. Induction on n.

(1) $\left(\dfrac{x}{y}\right)^1 = \dfrac{x}{y} = \dfrac{x^1}{y^1}$

(2) If $\left(\dfrac{x}{y}\right)^k = \dfrac{x^k}{y^k}$, then $\dfrac{x}{y}\left(\dfrac{x}{y}\right)^k = \dfrac{x}{y}\left(\dfrac{x^k}{y^k}\right)$ and $\left(\dfrac{x}{y}\right)^{k+1} = \dfrac{x^{k+1}}{y^{k+1}}$

31. If $n = 0$, $\left(\dfrac{x}{y}\right)^0 = 1 = \dfrac{x^0}{y^0}$

If $n = -k$, $\left(\dfrac{x}{y}\right)^{-k} = \dfrac{1}{\left(\dfrac{x}{y}\right)^k} = \left(\dfrac{1}{x^k}\right)\dfrac{1}{\dfrac{1}{y^k}} = x^{-k}\dfrac{1}{y^{-k}} = \dfrac{x^n}{y^n}$

33. Induction on n.
    (1) $x^1 > 1$
    (2) If $x^k > 1$ and $x^{k+1} > x$, then $x(x^k) > 1$ (multiplication of inequalities, same sense).

$$x(x^{k+1}) > 1(x) \quad \text{or} \quad x^{(k+1)+1} > x$$

by multiplication order theorem (1)

35. $(x - y)^2 \geq 0$
    $x^2 - 2xy + y^2 \geq 0$
    $x^2 + y^2 \geq 2xy$
37. $>$

39. $<$
41. $<$
43. $<$
45. $<$

Exercises 5.6 (page 165)

1. $2\sqrt{82}$
3. $\sqrt{13}$
5. $13\sqrt{6}$

7. $9\sqrt{2}$
9. 9
11. 5

13. $\sqrt{2}/2$
15. $\sqrt[3]{75}/5$
17. $2\sqrt{x+y}/(x+y)$
19. $3\sqrt[3]{y^2}/y$
21. $(\sqrt{11}-1)/2$
23. $(2\sqrt{3}+3\sqrt{2}-\sqrt{30})/12$
25. $(2\sqrt{30}-3\sqrt{14})/12$
27. $2(\sqrt[3]{x^2}-\sqrt[3]{xy}+\sqrt[3]{y^2})/(x+y)$
29. $(\sqrt[3]{x^2}-\sqrt[3]{x}+1)/(x+1)$
31. $3\sqrt{5}$
33. $\sqrt{6}/3$
35. 0
37. $1/(\sqrt{x}+2)$
39. $1/(\sqrt[3]{x^2}+\sqrt[3]{x}+1)$
41. $1/(8\sqrt[3]{x}-4\sqrt[3]{x^2}+2x)$
43. $x > 25$
45. $\frac{1}{3} < x < 3$
47. $x < -4$ or $x > 4$
49. $-10 < x < 8$
51. $x < -1$ or $x > 3$
53. Since $x^n < y^n$ if and only if $x < y$ and $x^n > y^n$ if and only if $x > y$, thus $x^n = y^n$ if and only if $x = y$.

Exercises 5.7 (page 170)

1. 4
3. 27/125
5. $\frac{1}{9}$
7. 9
9. $2\sqrt[3]{9x}/3$
11. -4
13. $\sqrt[3]{4}$
15. $\sqrt[12]{x}$
17. $(x\sqrt{y}+y\sqrt{x})/x^3y^3$
19. $\sqrt{x}+\sqrt{y}$
21. $(x^2+y^2)/x^4y^4$
23. Given
    Fundamental theorem, fractions
    Definition of $b^{x/y}$
    Product of roots theorem
    Product of integral powers theorem
    Definition of $b^{x/y}$
    Sum of quotients theorem
    Substitution
25. $b^{m/n}/b^{p/q} = b^{mq/nq}/b^{np/nq} = \sqrt[nq]{b^{mq}}/\sqrt[nq]{b^{np}} = \sqrt[nq]{b^{mq}/b^{np}}$
    $= b^{(mq-np)/nq} = b^{m/n-p/q}$
27. $(a/b)^{m/n} = \sqrt[n]{(a/b)^m} = \sqrt[n]{a^m/b^m} = \sqrt[n]{a^m}/\sqrt[n]{b^m} = a^{m/n}/b^{m/n}$
29. $2|x|\sqrt[3]{2x^2}$ or $2x\sqrt[3]{2x^2}$ if $x \geq 0$
31. $2y\sqrt{x^2-y^2}/(x^2-y^2)$ if $x > y$ and $x > -y$
    $-2y\sqrt{x^2-y^2}/(x^2-y^2)$ if $x < y$ and $x < -y$

33. $2xy^2$

35. $\sqrt[8]{320}$

37. $\frac{1}{9} < 3^{-\sqrt{2}} < \frac{1}{3}$

39. $\frac{1}{3} < 3^{-1/\sqrt{2}} < \sqrt{3}/3$

41. $x > 4$

43. $0 < x < \frac{5}{2}$

45. $x > 3$

47. $x < -1$ or $x > 2$

49. $-2 < x < 2$

51. $\sqrt{5}$

53. $\sqrt{2} - 1$

Exercises 5.8 (page 175)

1. 11
3. -4
5. 24
7. -108
9. 8, -2
11. $\emptyset$
13. 2
15. $\emptyset$
17. 49/36
19. 1, 125
21. $-\frac{10}{3} \le x < -2$ or $x > 5$
23. $\frac{6}{5} \le x < 2$ or $x > 3$
25. $x > 0$
27. $x > 7/6$
29. $-2 < x \le 0$
31. $x > \sqrt{7}$ or $x < -\sqrt{7}$
33. $-3 < x < 3$
35. $x > 4$ or $x < -4$

Review Exercises (page 180)

1. 71/330
2. 137/111
3. 23,389/9990
4. 4321/9999
5. lub = 3, glb = -1
6. lub = 4, glb = -4
7. None
8. lub = 1, glb = 0
9. $x \le -3$ or $x > -1$

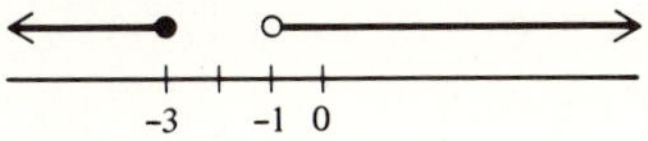

10. $\frac{3}{2} < x < \frac{7}{2}$

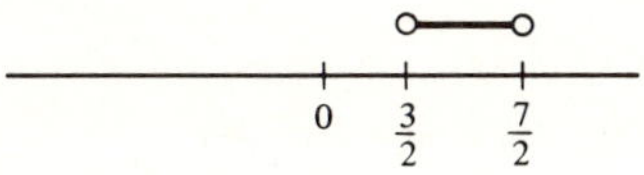

11. $x > \frac{6}{7}$

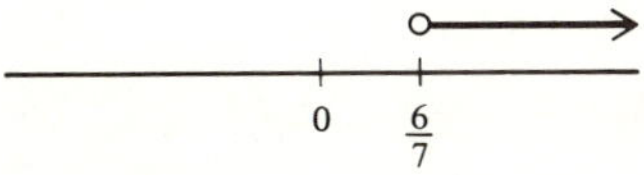

12. $\frac{1}{2} < x < \frac{3}{2}$

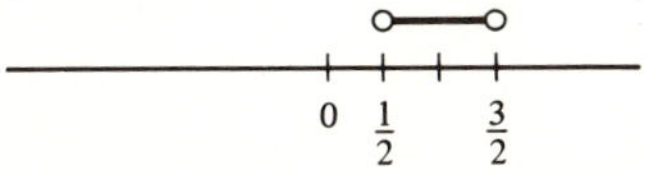

13. $x < 0$

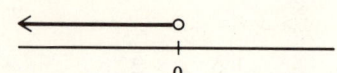

14. $x < 1$

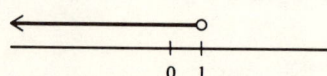

15. $x < 1$ or $x > 3$

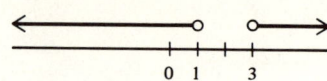

16. $x < -4$ or $-2 < x < 3$

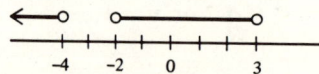

17. $-1 \leq x \leq 0$ or $x \geq 2$

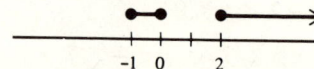

18. $\emptyset$

19. $x < -2$ or $x > 2$

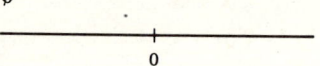

20. $0 \leq x \leq 2$

21. 5, 1
22. $x \geq -4$
23. $\frac{1}{2} < x < \frac{9}{2}$
24. $x < \frac{1}{2}$ or $x > \frac{5}{2}$
25. $x < -\frac{1}{3}$ or $x > 1$
26. $x \leq -\frac{1}{2}$ or $x \geq \frac{5}{2}$
27. All real x
28. $x > 0$
29. $x \geq 0$
30. All real x
31. $11\sqrt{2}$
32. $3(\sqrt{x} - y)/(x - y^2)$
33. $3(\sqrt{x} - \sqrt{y})/(x - y)$
34. $\sqrt[3]{3y}/3y$
35. $(\sqrt{x} + \sqrt{y} - \sqrt{z})(x + y - z - 2\sqrt{xy})/[(x + y - z)^2 - 4xy]$
36. $\sqrt{2}/8$
37. $\frac{1}{16}$
38. -16
39. 64
40. $\frac{1}{8}$
41. 8

EXERCISES 6.1

42. $\frac{16}{3}$
43. $\frac{8}{3}$
44. $x > 3$ or $-\frac{3}{2} \leq x < -1$
45. (a) $x \geq 1$, (b) $x \leq -3$
46. $(x-y)^2 > 0$, $x^2 - 2xy + y^2 > 0$, $x^2 + y^2 > 2xy$, $\dfrac{x^2 + y^2}{xy} > 2$, $\dfrac{x}{y} + \dfrac{y}{x} > 2$

47. $x^2 + y^2 > 2xy$
    $x^2 + z^2 > 2xz$
    $y^2 + z^2 > 2yz$

    $2(x^2 + y^2 + z^2) > 2(xy + yz + xz)$
    $x^2 + y^2 + z^2 > xy + yz + xz$

48. $(x-y)^2 = x^2 - 2xy + y^2 > 0$
    $x^2 + y^2 > 2xy$
    $x^2 + 2xy + y^2 > 4xy$
    $(x+y)^2 > 4xy$
    $x + y > 2\sqrt{xy}$
    $\dfrac{x+y}{2} > \sqrt{xy}$

49. $(x-1)^2 > 0$
    $x^2 - 2x + 1 > 0$
    $x^2 + 1 > 2x$
    $\dfrac{x^2 + 1}{x} > 2$

50. Since $\dfrac{x}{y} > 1$,

    $\left(\dfrac{x}{y}\right)^x > \left(\dfrac{x}{y}\right)^y$

    $\dfrac{x^x}{y^x} > \dfrac{x^y}{y^y}$

    $x^x y^y > x^y y^x$.

CHAPTER 6

Exercises 6.1 (page 191)

1. $\{(x, y) \mid 3y - 4x = 12\}$, all real x, all real y. Yes.
3. $\{(x, y) \mid x = 3\}$, $x = 3$, all real y. No.
5. $\{(x, y) \mid y^2 = \frac{1}{4}(100 - x^2)\}$, $-10 \leq x \leq 10$, $-5 \leq y \leq 5$. No.
7. $\{(x, y) \mid y = x^2 - 3\}$, $x \geq 0$, $y \geq -3$. Yes.
9. $\{(x, y) \mid y = x^3 - 1\}$, all real x, all real y. Yes.

11. (a) $\dfrac{x+6}{2}$

(b) $2\left(\dfrac{x+6}{2}\right) - 6 = x, \quad \dfrac{(2x-6)+6}{2} = x$

(c)

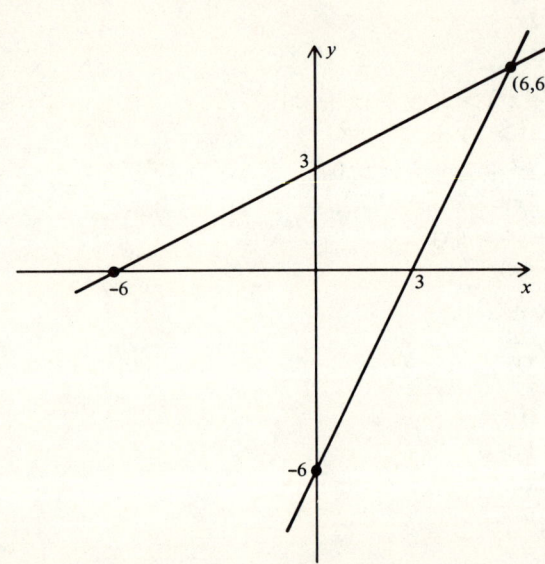

13. (a) $\dfrac{6}{x}$

(b) $\dfrac{6}{6/x} = x$

(c)

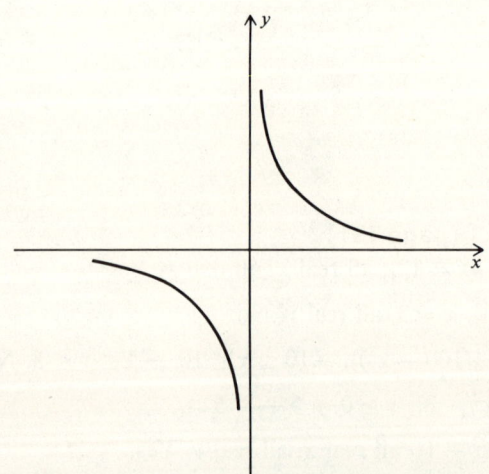

EXERCISES 6.2

15. 2
17. $\frac{8}{3}$
19. $\frac{24}{17}$
21. 1
23. 0
25. -3
27. 0
29. $(\sqrt{3} - 12)/4$
31. 0
33. 2
35. 4
37. $\log_{10} 10{,}000 = 4$
39. $\log_9 3 = \frac{1}{2}$
41. $\log_3 3 = 1$
43. $\log_9 \left(\frac{1}{27}\right) = -\frac{3}{2}$
45. $3^2 = 9$
47. $10^1 = 10$
49. $(36)^{1/2} = 6$
51. $(10)^{-1} = 0.1$

53. 3
55. -4
57. $-\frac{3}{2}$
59. -2.5
61. 6
63. 1
65. 5
67. $\log_b 3 + \log_b 5$
69. $5 \log_b 3$
71. $\frac{1}{2}(\log_b 5 - \log_b 3)$
73. $4 \log_b 3 - 2 \log_b 5$
75. $\frac{3}{2} \log_b 5 - \log_b 7$
77. $\log_b 72$
79. $\log_b 3^5$
81. $\log_b 2025$
83. $\log_b \sqrt[5]{75}$
85. $\log_b \dfrac{9\sqrt[4]{5}}{7^3}$

Exercises 6.2 (page 200)

1. 1.3284
3. $-1 + 0.0899$
5. $-2 + 0.9768$
7. 0.4342
9. 2.2543
11. $-2 + 0.8076$
13. 69.3
15. 8,000,000
17. 38,060
19. 0.08947
21. $\log x = \log 4.92 + \log 0.0658 - \log 786;\ x = 0.000412$
23. $\log x = \log 0.00729 + 8 \log 2.06;\ x = 2.37$
25. $\log x = \log 450 + \frac{1}{3} \log 75.2 - 2 \log 82.4;\ x = 0.280$
27. 0.000305
29. 6540
31. $1 - 0.3010 = 0.6990$
33. $-1 + 0.6020 = -0.3980$
35. 0.3495
37. 3.3010
39. 0.7960
41. 3.808
43. 1.640
45. 0.8544
47. $\frac{8}{3}$
49. $-\dfrac{1}{2} + \dfrac{\log 15}{\log 64} \approx -0.40$
51. -4

53. -1
55. $-\frac{5}{2}$
57. -2
59. $\frac{2}{3}$
61. $-\frac{5}{3}$
63. $\frac{1}{8}$
65. $\dfrac{\log 26}{\log 2} \approx -1.38$
67. $\dfrac{25\sqrt{3}}{6}$
69. 5, 13
71. $\frac{1}{5}$
73. $\sqrt{3}/3$
75. $x = \dfrac{\log 7}{\log 2} \approx 2.81$
77. About seventeen centuries.

Exercises 6.3 (page 215)

1. $(-\sqrt{3}/2, -\frac{1}{2})$, III
3. $(-\sqrt{2}/2, \sqrt{2}/2)$, II
5. $(\frac{1}{2}, \sqrt{3}/2)$, I
7. $(-\frac{1}{2}, -\sqrt{3}/2)$, III
9. (0, 1), between I and II
11. I and IV
13. II
15. III
17. $\pi/2 < s < 3\pi/2$
19. $0 < s < \pi$
21. I, +, +
23. IV, −, +
25. I, +, +
27. III, −, −
29. IV, −, +
31. $\frac{4}{5}$
33. $-\frac{4}{5}$
35. $\frac{4}{5}$
37. $-\frac{4}{5}$
39. $-\frac{4}{5}$
41. $\frac{4}{5}$
43. $\frac{3}{5}$
45. $-\frac{3}{5}$
47. $-\frac{4}{5}$
49. $\frac{4}{5}$
51. $\pi/6$
53. $\pi/6$
55. π
57. $4\pi/3$
59. $3\pi/4$
61. $11\pi/6$
63. $2\pi/3$
65. $3\pi/2$
67. $7\pi/4$
69. $5\pi/4$
71. 0

Exercises 6.4 (page 223)

1. $\sqrt{3}/2$
3. $2\sqrt{3}/3$
5. 1
7. $(\sqrt{6} + \sqrt{2})/4$
9. $2 + \sqrt{3}$
11. 7/25
13. 24/7
15. 25/7
17. 527/625
19. $\sqrt{2}/10$

EXERCISES 6.4

21. $\frac{1}{7}$
23. $5\sqrt{2}/7$
25. $31/17$
27. $(24\sqrt{3}+7)/50$
29. $31/17$
31. 1
33. -1
35. 1
37. 1
39. 0

41. 0
43. $\sqrt{3}/2$
45. $(2+\sqrt{2})/4$
47. $(2-\sqrt{2})/4$
49. $\sqrt{3}$
51. 1
53. $\sqrt{2}/2$
55. $\sqrt{3}/2$
57. $\sqrt{3}/3$
59. $2\sqrt{3}/3$

61. $\tan^2 s - \sin^2 s = \dfrac{\sin^2 s}{\cos^2 s} - \dfrac{\sin^2 s \cos^2 s}{\cos^2 s}$

$\qquad = (1 - \cos^2 s)\dfrac{\sin^2 s}{\cos^2 s} = \sin^2 s \tan^2 s$

63. $\dfrac{2\tan s}{1+\tan^2 s} = \dfrac{2\sin s/\cos s}{1+\sin^2 s/\cos^2 s} = \dfrac{2\sin s \cos s}{\cos^2 s + \sin^2 s} = \sin 2s$

65. $\dfrac{2}{\sin 2s} = \dfrac{2}{2\sin s \cos s} = \dfrac{1}{\sin s \cos s} = \dfrac{\sin^2 s + \cos^2 s}{\sin s \cos s}$

$\qquad = \dfrac{\sin s}{\cos s} + \dfrac{\cos s}{\sin s} = \tan s + \cot s$

67. $\dfrac{\sin 2s \cos s + \cos 2s \sin s}{\sin s} - \dfrac{\cos 2s \cos s - \sin 2s \sin s}{\cos s}$

$\qquad = 2\cos^2 s + \cos 2s - \cos 2s + 2\sin^2 s = 2$

69. $\dfrac{2\tan\frac{s}{2}}{1+\tan^2\frac{s}{2}} = \dfrac{2\sin\frac{s}{2}/\cos\frac{s}{2}}{1+\left(\sin^2\frac{s}{2}\right)/\left(\cos^2\frac{s}{2}\right)} = \dfrac{2\sin\frac{s}{2}\cos\frac{s}{2}}{\cos^2\frac{s}{2}+\sin^2\frac{s}{2}} = \sin s$

71. $\dfrac{\sin 2s}{1-\cos 2s} = \dfrac{2\sin s \cos s}{2\sin^2 s} = \dfrac{\cos s}{\sin s} = \cot s$

73. $(\sin s + \cos s)^2 = \sin^2 s + \cos^2 s + 2\sin s \cos s = 1 + \sin 2s$

75. $\cos 3s = \cos(2s+s)$

$\qquad = \cos 2s \cos s - \sin 2s \sin s$
$\qquad = (\cos^2 s - \sin^2 s)\cos s - 2\sin^2 s \cos s$
$\qquad = (2\cos^2 s - 1)\cos s - 2(1-\cos^2 s)\cos s$
$\qquad = 2\cos^3 s - \cos s - 2\cos s + 2\cos^3 s$
$\qquad = 4\cos^3 s - 3\cos s$

77. -0.8
79. -1.25
81. $\sin\theta = \frac{4}{5}, \cos\theta = \frac{3}{5}$
83. $\sin\theta = \frac{3}{5}, \cos\theta = -\frac{4}{5}$
85. $\sin\theta = -1, \cos\theta = 0$

Exercises 6.5 (page 229)

1. $\pi/6$
3. 0
5. $\pi/3$
7. $-\pi/3$
9. π
11. $\frac{1}{2}$
13. x if $-\pi/2 < x < \pi/2$
15. $12/5$
17. $\pi/6$
19. 0
21. $\pi/2$
23. $\pi/2, 3\pi/2, 7\pi/6, 11\pi/6$
25. $0 \le x < \pi/4, 5\pi/4 < x \le 2\pi$
27. π
29. $0, \pi/2, \pi, 3\pi/2, 2\pi$
31. $24/25, 2\sqrt{5}/5$
33. $-\sqrt{3}, \sqrt{3}/2$
35. Arctan $1 = \pi/4$ and $1 = \tan(2x+y)$, where $\tan x = 1/3$, $\tan y = 1/7$,
$$\tan(2x+y) = \frac{\tan 2x + \tan y}{1 - \tan 2x \tan y} = \frac{3/4 + 1/7}{1 - (3/4)(1/7)} = \frac{21+4}{28-3} = 1, \text{ since}$$
$$\tan 2x = \frac{2(1/3)}{1 - 1/9} = \frac{6}{8} = \frac{3}{4}.$$
37. $\sqrt{2}/2$
39. $(\pi - 1)/2$
41. $-\pi/12$
43. $\pi/2 - 3/5$
45. 1
47. $\pi/4$
49. All real x
51. No, domain of Arcsin x is $-1 \le x \le 1$.

Review Exercises (page 234)

1. 1
2. 0
3. Undefined
4. Undefined
5. -3
6. 2
7. 1.2
8. -0.4
9. 0.2
10. $-\frac{4}{15}$
11. -0.4
12. $2\sqrt{3}/15$
13. -4.6
14. $\sqrt{5}/5$
15. $\log_5 3$
16. $-1 + 0.79$
17. 1.7
18. 40
19. $1/10$
20. -2

REVIEW EXERCISES

21. (a) $\log x = \frac{1}{3}(\log 42.21 + \log 23.50 - \log 5650)$
    (b) 0.5600
22. $(-\sqrt{2}/2, -\sqrt{2}/2)$, III
23. $(-\sqrt{3}/2, \frac{1}{2})$, II
24. $(0, 1)$, between I and II
25. $(\frac{1}{2}, -\sqrt{3}/2)$, IV
26. $\sqrt{3}/2$
27. $\sqrt{3}/3$
28. $\dfrac{\sqrt{2+\sqrt{3}}}{2} = \dfrac{\sqrt{6}+\sqrt{2}}{4}$
29. $\sqrt{3}/2$
30. $2 - \sqrt{3}$

31. $\dfrac{\sin(a+\theta)}{\cos(a+\theta)} = \dfrac{\sin a \cos\theta + \cos a \sin\theta}{\cos a \cos\theta - \sin a \sin\theta} = \dfrac{\tan a + \tan\theta}{1 - \tan a \tan\theta}$

32. $\tan 2\theta = \dfrac{\sin 2\theta}{\cos 2\theta} = \dfrac{2\sin\theta\cos\theta}{\cos^2\theta - \sin^2\theta} = \dfrac{2\tan\theta}{1 - \tan^2\theta}$

33. $\dfrac{2\tan\theta/2}{1+\tan^2\theta/2} = \dfrac{2\tan\theta/2}{1-\tan^2\theta/2} \cdot \dfrac{1-\tan^2\theta/2}{1+\tan^2\theta/2}$

$= \tan\theta \dfrac{\cos^2\theta/2 - \sin^2\theta/2}{\cos^2\theta/2 + \sin^2\theta/2}$

$= \dfrac{\sin\theta}{\cos\theta} \cdot \dfrac{\cos\theta}{1} = \sin\theta$

34. $\dfrac{\tan^2\theta}{1+\sec\theta} + 1 = \dfrac{\tan^2\theta + 1 + \sec\theta}{1+\sec\theta}$

$= \dfrac{\sec^2\theta + \sec\theta}{1+\sec\theta}$

$= \dfrac{\sec\theta(\sec\theta + 1)}{1+\sec\theta} = \sec\theta$

35. $\dfrac{\sin^3\theta - \cos^3\theta}{1+\sin\theta\cos\theta} = \dfrac{(\sin\theta - \cos\theta)(\sin^2\theta + \sin\theta\cos\theta + \cos^2\theta)}{1+\sin\theta\cos\theta}$

$= \sin\theta - \cos\theta$

36. $\dfrac{\sin 2\theta}{1-\cos 2\theta} = \dfrac{2\sin\theta\cos\theta}{1-\cos^2\theta + \sin^2\theta} = \dfrac{2\sin\theta\cos\theta}{2\sin^2\theta} = \dfrac{\cos\theta}{\sin\theta} = \cot\theta$

37. $\tan\theta\tan 2\theta = \dfrac{(\tan\theta)(2\tan\theta)}{1-\tan^2\theta}$

$= \dfrac{2\sin^2\theta + 1 - \sin^2\theta - \cos^2\theta}{\cos^2\theta - \sin^2\theta}$

$= \dfrac{1}{\cos^2\theta - \sin^2\theta} + \dfrac{\sin^2\theta - \cos^2\theta}{\cos^2\theta - \sin^2\theta}$

$= \sec 2\theta - 1$

38. $\dfrac{2}{1-\cos 2\theta} = \dfrac{1}{\dfrac{1-\cos 2\theta}{2}} = \dfrac{1}{\sin^2\theta} = \csc^2\theta$

39. $\pi/3, \pi, 5\pi/3$
40. Arcsin $4/5, \pi$
41. $\pi/6, 5\pi/6, 3\pi/2$
42. $5\pi/12$
43. $\pi/2$
44. π
45. $\pi/4$
46. $\sqrt{3}$
47. 1
48. $\sqrt{2}/10$

49. $\sin\theta = -\sqrt{21}/7 \qquad \sec\theta = \sqrt{7}/2$
    $\cos\theta = 2\sqrt{7}/7 \qquad \csc\theta = -\sqrt{21}/3$
    $\tan\theta = -\sqrt{3}/2 \qquad \cot\theta = -2\sqrt{3}/3$

50. Let $y = $ Arcsin x, then $x = \sin y$ and $-x = \sin(-y)$.
    Thus $-y = $ Arcsin$(-x)$ and Arcsin$(-x) = -$Arcsin x.

CHAPTER 7

Exercises 7.1 (page 244)

1. $5 + 7i$
3. $-2 + 6i$
5. $20 - 10i$
7. $17 - 20i$
9. $\dfrac{2 + 3\sqrt{2}}{3} + \dfrac{(3 - 2\sqrt{2})}{3}i$
11. $278 + 29i$
13. $\dfrac{3}{10} + \dfrac{9}{10}i$
15. $\dfrac{-(6 + 3\sqrt{2} + 2\sqrt{3} + 6)}{6}$
17. $x = 3, y = -4$
19. $x = -1/2, y = 1/4$
23. $\{2 + i, -2 - i\}$

25. $\{\sqrt{3 + \sqrt{10}} + \sqrt{-3 + \sqrt{10}}\,i, -\sqrt{3 + \sqrt{10}} - \sqrt{-3 + \sqrt{10}}\,i\}$

27. Cubing, $z^3 = (2 + 11i) + 3\sqrt[3]{(2 + 11i)^2(2 - 11i)}$
    $\qquad + 3\sqrt[3]{(2 + 11i)(2 - 11i)^2} + 2 - 11i$
    $= 4 + 15z$
    $z^3 - 15z - 4 = 0$
    $(z - 4)(z^2 + 4z + 1) = 0$ or $z = 4$

    (*Note:* A definition of $\sqrt[3]{a + bi}$ is required.)

29. Let $z = a + bi$. Then
    $\sqrt{z} = \sqrt{\dfrac{a + \sqrt{a^2 + b^2}}{2}} + i\sqrt{\dfrac{-a + \sqrt{a^2 + b^2}}{2}} \qquad$ if $b \geq 0$

    $\sqrt{z} = \sqrt{\dfrac{a + \sqrt{a^2 + b^2}}{2}} - i\sqrt{\dfrac{-a + \sqrt{a^2 + b^2}}{2}} \qquad$ if $b < 0$.

EXERCISES 7.2

31. $(x + i)(x - i)$
33. $(x - \sqrt{2} - i\sqrt{2})(x + \sqrt{2} + i\sqrt{2})(x - \sqrt{2} + i\sqrt{2})(x + \sqrt{2} - i\sqrt{2})$
35. $(x + i)^3$
39. $z = \sqrt{3} - i, \bar{z} = \sqrt{3} + i$ or $z = -\sqrt{3} - i, \bar{z} = -\sqrt{3} + i$
41. $z = 4\cos\theta + 3 + (4\sin\theta)i, \bar{z} = 4\cos\theta + 3 - (4\sin\theta)i, 0 \le \theta < 2\pi$
    or $z = (t + 3) \pm \sqrt{16 - t^2}\,i, \bar{z} = (t + 3) \mp \sqrt{16 - t^2}\,i, -4 \le t \le 4$
43. $\emptyset$
51. (2, 1)
53. (2, 7)
55. (0, 5)

Exercises 7.2 (page 250)

1. $\dfrac{3}{2} - \dfrac{3\sqrt{3}}{2}i$

3. $\dfrac{\sqrt{6} + \sqrt{2}}{4} + \dfrac{\sqrt{6} - \sqrt{2}}{4}i$

5. $\dfrac{1}{4} - \dfrac{\sqrt{3}}{4}i$

7. $\dfrac{5\sqrt{3}}{2} + \dfrac{5}{2}i$

9. $\dfrac{\sqrt{6} - \sqrt{2}}{4} - \dfrac{\sqrt{6} + \sqrt{2}}{4}i$

11. $-2\sqrt{2} - 2\sqrt{2}i$

13. $\dfrac{3\sqrt{2}}{2} - \dfrac{3\sqrt{2}}{2}i$

15. $\dfrac{8}{5} + \dfrac{3}{5}i$

17. $5(\cos \pi/2 + i \sin \pi/2)$
19. $5(\cos 3\pi/2 + i \sin 3\pi/2)$
21. $10(\cos 7\pi/6 + i \sin 7\pi/6)$
23. $2\sqrt{2}(\cos 5\pi/4 + i \sin 5\pi/4)$
25. $8(\cos 2\pi/3 + i \sin 2\pi/3)$
27. $2(\cos 3\pi/2 + i \sin 3\pi/2)$
29. $6\sqrt{2}(\cos \pi/4 + i \sin \pi/4)$

31. (a) $\sqrt{3} - i$ (b) $-\sqrt{3} - i$
33. (a) $\dfrac{3}{2}(\sqrt{2} - \sqrt{6}) + \dfrac{3}{2}(\sqrt{6} + \sqrt{2})i$ (b) $\dfrac{\sqrt{2}}{3} - \dfrac{\sqrt{2}}{3}i$
35. (a) $0 + 4i$ (b) $-2 + 0i$
37. (a) $-6 + 0i$ (b) $\tfrac{3}{2}(\cos 40° + i \sin 40°) \approx 1.1490 + 0.9642i$
39. (a) $\sin 2\theta = \sin 36° = \cos(90° - 36°) = \cos 54° = \cos 3\theta$
    (b) $2\sin\theta\cos\theta = \cos 2\theta \cos\theta - \sin 2\theta \sin\theta$
    $= (1 - 2\sin^2\theta)\cos\theta - 2\sin^2\theta \cos\theta$
    $2\sin\theta = 1 - 4\sin^2\theta$
    $4\sin^2\theta + 2\sin\theta - 1 = 0$
    (c) $\sin\theta = \dfrac{-1 + \sqrt{5}}{4}$ by quadratic formula.
    Since $\theta = 18°$, $\sin\theta > 0$ and
    $\sin\theta = \sin 18° = \dfrac{-1 + \sqrt{5}}{4} = \cos(90° - 18°) = \cos 72°.$

(d) $\sin 72° = \sin 4\theta$

$$\sin 4\theta = \cos\theta = \sqrt{1 - \sin^2\theta} = \sqrt{1 - \left(\frac{6 - 2\sqrt{5}}{16}\right)} = \frac{\sqrt{10 + 2\sqrt{5}}}{4}$$

(e) $\sin 36° = \cos 54° = 2\sin\theta\cos\theta = \dfrac{\sqrt{10 - 2\sqrt{5}}}{4}$

$$\sin 54° = \cos 36° = \cos^2\theta - \sin^2\theta = \frac{1 + \sqrt{5}}{4}$$

41. $4(\cos \pi/10 + i \sin \pi/10)$
43. $2\sqrt{2}(\cos \pi/12 + i \sin \pi/12)$
45. $\sqrt{2(\cos\alpha + 1)}\,(\cos\alpha/2 + i\sin\alpha/2)$
47. $\sqrt{2}\,[\cos(\beta + \pi/4) + i\sin(\beta + \pi/4)]$
59. (a) on a vertical line three units to the right of the y-axis;
    (b) on a horizontal line three units above the x-axis.
61. Symmetric to the x-axis.
63. $\sqrt{2}/2 + (\sqrt{2}/2)i,\ \sqrt{2}/2 - (\sqrt{2}/2)i,\ -\sqrt{2}/2 + (\sqrt{2}/2)i,\ -\sqrt{2}/2 - (\sqrt{2}/2)i$
67. $w = r_1(\cos\theta_1 + i\sin\theta_1),\ z = r_2(\cos\theta_2 + i\sin\theta_2)$
    $wz = r_1 r_2[\cos(\theta_1 + \theta_2) + i\sin(\theta_1 + \theta_2)];\ |w| = r_1,\ |z| = r_2$
    $|wz| = r_1 r_2 = |w| \cdot |z|$
69. $z = a + bi,\ \bar{z} = a - bi,\ z\bar{z} = a^2 + b^2,\ |z| = \sqrt{a^2 + b^2} = \sqrt{z\bar{z}}$
71. If $a + bi = c + di$, then $a = c$ and $b = d$ and $\sqrt{a^2 + b^2} = \sqrt{c^2 + d^2}$

Exercises 7.3 (page 257)

1. $64i$
3. -256
5. $-8i$
7. $-128 + 128\sqrt{3}\,i$
9. $-1/4$
11. $1,\ -1/2 + i\sqrt{3}/2,\ -1/2 - i\sqrt{3}/2$
13. $1,\ -1,\ -1/2 + i\sqrt{3}/2,\ -1/2 - i\sqrt{3}/2,\ 1/2 - i\sqrt{3}/2,\ 1/2 + i\sqrt{3}/2$
15. $1,\ -1,\ i,\ -i,\ \sqrt{2}/2 + i\sqrt{2}/2,\ \sqrt{2}/2 - i\sqrt{2}/2,\ -\sqrt{2}/2 + i\sqrt{2}/2,$
    $-\sqrt{2}/2 - i\sqrt{2}/2$
17. $9i,\ -9i$
19. $\sqrt{6}/2 - i\sqrt{2}/2,\ -\sqrt{6}/2 + i\sqrt{2}/2$
21. $-\sqrt[6]{2}\sqrt{2}/2 + i\sqrt[6]{2}\sqrt{2}/2,\ \sqrt[6]{2}(\sqrt{6} + \sqrt{2})/4 + i\sqrt[6]{2}(\sqrt{6} - \sqrt{2})/4,$
    $-\sqrt[6]{2}(\sqrt{6} - \sqrt{2})/4 - i\sqrt[6]{2}(\sqrt{6} + \sqrt{2})/4$
23. $2i,\ -\sqrt{3} - i,\ \sqrt{3} - i$

25. $\cos 2\pi/9 + i \sin 2\pi/9$
    $-\cos \pi/9 + i \sin \pi/9 \; (= \cos 8\pi/9 + i \sin 8\pi/9)$
    $-\sin 2\pi/9 - i \cos 2\pi/9 \; (= \cos 14\pi/9 + i \sin 14\pi/9)$

27. $\sqrt{2-\sqrt{2}} + i\sqrt{2+\sqrt{2}}, \; -\sqrt{2+\sqrt{2}} + i\sqrt{2-\sqrt{2}},$
    $-\sqrt{2-\sqrt{2}} - i\sqrt{2+\sqrt{2}}, \; \sqrt{2+\sqrt{2}} - i\sqrt{2-\sqrt{2}}$

29. 2
    $(\sqrt{5}-1)/2 + i\sqrt{10+2\sqrt{5}}/2$
    $-(\sqrt{5}+1)/2 + i\sqrt{10-2\sqrt{5}}/2$
    $-(\sqrt{5}+1)/2 - i\sqrt{10-2\sqrt{5}}/2$
    $(\sqrt{5}-1)/2 - i\sqrt{10+2\sqrt{5}}/2$

31. $\dfrac{\sqrt{2+\sqrt{3}}}{2} + \dfrac{\sqrt{2-\sqrt{3}}}{2} i$

39. The solutions of $z^7 = 1$ are
    $$z = \cos 2k\pi/7 + i \sin 2k\pi/7 \quad \text{for } k = 0, 1, 2, 3, 4, 5, 6.$$
    $k = 0$ corresponds to the solution of $z = 1$. $\quad x = z + 1/z = 2 \cos 2k\pi/7$.

    $k = 1, \quad x = 2 \cos 2\pi/7$
    $k = 2, \quad x = 2 \cos 4\pi/7$
    $k = 3, \quad x = 2 \cos 6\pi/7$
    $k = 4, \quad x = 2 \cos 8\pi/7 = 2 \cos 6\pi/7$
    $k = 5, \quad x = 2 \cos 10\pi/7 = 2 \cos 4\pi/7$
    $k = 6, \quad x = 2 \cos 12\pi/7 = 2 \cos 2\pi/7$

Review Exercises (page 267)

1. $5 + i$
2. $-1 + 5i$
3. $12 + 5i$
4. $-5 + 12i$
5. i
6. $(3 + \sqrt{2})/3 + i(1 - 3\sqrt{2})/3$
7. $-(8 + 5\sqrt{2})/7$
8. $-(\sqrt{6} + 3\sqrt{3} + 2\sqrt{2} + 6)/7$
9. $-2/3 - i$
10. $29 - 11i$
11. $x = 1, y = 8$
12. $x = -1, y = 2$
13. $1 + i, -1 - i$
14. $3\sqrt{2}/2 - i\sqrt{2}/2, \; -3\sqrt{2}/2 + i\sqrt{2}/2$
15. $\sqrt{26}/2 - i\sqrt{14}/2, \; -\sqrt{26}/2 + i\sqrt{14}/2$
16. $(\tfrac{8}{5}, \tfrac{1}{5})$
17. All real x and y
18. $(0, 6)$
19. $-1 + i\sqrt{3}$
20. $-3\sqrt{3}/2 + (\tfrac{3}{2})i$
21. $\sqrt{6} - \sqrt{2} + (\sqrt{6} + \sqrt{2})i$

22. $1 + i\sqrt{3}$
23. $2(\cos 2k\pi + i \sin 2k\pi)$, k an integer
24. $3(\cos 3\pi/2 + i \sin 3\pi/2)$
25. $3\sqrt{2}(\cos 7\pi/4 + i \sin 7\pi/4)$
26. $8(\cos 2\pi/3 + i \sin 2\pi/3)$
27. (a) $-\sqrt{3} - i$ (b) $\sqrt{3} - i$
28. (a) $\frac{9}{4}(\sqrt{2} - \sqrt{6}) + \frac{9}{4}(\sqrt{2} + \sqrt{6})i$
    (b) $\frac{1}{4}(\sqrt{6} + \sqrt{2}) + \frac{1}{4}(\sqrt{6} - \sqrt{2})i$
29. (a) $i\sqrt{2}$ (b) $-\sqrt{2}/2$
30. (a) $6 + 4i$ (b) $-6/13 + (4/13)i$
35. -64
36. $-81/2 + \frac{81\sqrt{3}}{2}i$
37. $-119 - 120i$
38. $-\frac{7}{625} + \frac{24}{625}i$
39. $-1, 1/2 + i\sqrt{3}/2, 1/2 - i\sqrt{3}/2$
40. $3\sqrt{7}/7 - i\sqrt{35}/7, -3\sqrt{7}/7 + i\sqrt{35}/7$
41. $-2i, \frac{\sqrt{10 - 2\sqrt{5}}}{2} \pm \frac{1 + \sqrt{5}}{2}i, -\frac{\sqrt{10 - 2\sqrt{5}}}{2} \pm \frac{1 + \sqrt{5}}{2}i$
42. $-\sqrt[3]{4}/2 + i\sqrt[3]{4}/2, \sqrt[6]{2}(\sqrt{6} + \sqrt{2})/4 + i\sqrt[6]{2}(\sqrt{6} - \sqrt{2})/4,$
    $-\sqrt[6]{2}(\sqrt{6} - \sqrt{2})/4 - i\sqrt[6]{2}(\sqrt{6} + \sqrt{2})/4$
43. $2, -2, 2i, -2i$
44. $\sqrt{\sqrt{2} - 1} + i\sqrt{\sqrt{2} + 1}, -\sqrt{\sqrt{2} - 1} - i\sqrt{\sqrt{2} + 1}$
52. $\dfrac{-2 + 3i}{3 + 2i} = \dfrac{\sqrt{13}\,[\cos(\pi/2 + \theta) + i \sin(\pi/2 + \theta)]}{\sqrt{13}\,(\cos \theta + i \sin \theta)}$
    $= \cos \pi/2 + i \sin \pi/2$
    $= i$

CHAPTER 8

Exercises 8.1 (page 276)
1. $1, 6$
3. $-4, \frac{1}{3}$
5. $2 + \sqrt{2}, 2 - \sqrt{2}$
7. $-\sqrt{2}/2 + i, -\sqrt{2}/2 - i$
9. $1 \pm \sqrt{5}$
11. $2i, i/2$
13. $(1 \pm i\sqrt{2})/2$
15. $i, -5i/2$
17. $2\sqrt{3}, -2\sqrt{3}$
19. $(2 \pm \sqrt{14})/5$
21. 8
23. $4 \pm 2\sqrt{3}$
25. $9, -1$

EXERCISES 8.1

27. $ax^2 + bx + c = a\left(x + \dfrac{b}{2a}\right)^2 + (4ac - b^2)/4a > 0$ if and only if $a > 0$ and $b^2 - 4ac < 0$.
29. 12, 7, 2
33. -2
31. -33
35. 1
37. Yes. $P(2 + 4i) = 0$.
39. (a) $(x^2 - 2)(x^2 + 2)$
    (b) $(x + \sqrt{2})(x - \sqrt{2})(x^2 + 2)$
    (c) $(x + \sqrt{2})(x - \sqrt{2})(x + i\sqrt{2})(x - i\sqrt{2})$
41. (a) and (b) $(x + 1)(x^2 - x + 1)$
    (c) $(x + 1)\left(x - \dfrac{1 + i\sqrt{3}}{2}\right)\left(x - \dfrac{1 - i\sqrt{3}}{2}\right)$
43. (a) and (b) $(x + 1)(x + 2)(x^2 + 4)$
    (c) $(x + 1)(x + 2)(x + 2i)(x - 2i)$
45. (a) and (b) $(x^2 + 1)^2$
    (c) $(x + i)^2(x - i)^2$
47. $x^3 - 4x^2 + 2x + 4$
49. $x^3 + x + 10$
51. -1, 5, simple roots; 0, double root.
53. 2, $-1 + i\sqrt{2}$, $-1 - i\sqrt{2}$, simple roots.
55. 1, -2, simple roots; -3, double root.
57. -2, triple root.
59. 8
61. $P(k) = k^3 + bk^2 + ck + d = 0$ if and only if $x - k$ is a factor.
    $P(-k) = -k^3 + bk^2 - ck + d = 0$ if and only if $x + k$ is a factor.
    $Q(x) = x^2 + (b + k)x + k^2 + bk + c$
    $Q(-k) = k^2 + c = 0$ if and only if $x + k$ is a factor.
    Thus $bk^2 + d = b(-c) + d = 0$, or $d = bc$.
63. $(x - r_1)(x - r_2)(x - r_3)(x - r_4)$
    $= x^4 - (r_1 + r_2 + r_3 + r_4)x^3 + \ldots + r_1 r_2 r_3 r_4$
    $= x^4 - \dfrac{b}{a}x^3 + \ldots + \dfrac{e}{a},$

thus
$$r_1 + r_2 + r_3 + r_4 = -\dfrac{b}{a}$$

and
$$r_1 r_2 r_3 r_4 = \dfrac{e}{a}.$$

Generalizing for n, $-b/a =$ the sum of the roots, and the constant term $= (-1)^n$ product of the roots.

Exercises 8.2 (page 282)

1. $2, -1$
3. $2, -1, -\frac{3}{2}$
5. $2, -\frac{7}{4}$
7. None
9. None
11. 4
13. None
15. $5, -3, \frac{1}{2}$
17. $1, 3, -6$
19. 2
21. $3 + \sqrt{2}, 3 - \sqrt{2}, \sqrt{2}, -\sqrt{2}$
23. $-1 \pm \sqrt{2}, i, -i$
25. $2 \pm i, -2 \pm \sqrt{2}$
27. $-1 \pm 2i, 1 \pm \sqrt{2}$
29. $2i, -2i, (-1 \pm \sqrt{17})/2$
31. $1 - i\sqrt{5}, 4; c = 14, d = -24$
33. $x^3 - 9x^2 + 33x - 25 = 0$
35. $x^4 + 2x^3 - 5x^2 - 30x - 28 = 0$
37. $x^4 - 4x^3 + 8x + 4 = 0$

Exercises 8.3 (page 289)

1. $-1 \leq x \leq 3$
3. $-2 < x < 4$
5. $-1 < x < 0$
7. $-2 < x < 2$
9. $-2 < x < 4$
11. One positive real root, zero or two negative real roots.
13. One positive real root, zero or two negative real roots.
15. Zero positive real roots, one negative real root, four imaginary roots.
17. No positive real roots, no negative real roots, six imaginary roots.
19. One or three positive real roots, one negative real root.
21. $1, -2$
23. $\frac{1}{2}, -\frac{2}{3}, -\frac{3}{2}$
25. $5, 6$
27. $\frac{3}{2}, -\frac{3}{2}$
29. $-6, -6$
31. $\sqrt{2}, -\sqrt{2}$
33. None
35. $1/10$
37. $v^+ = 1, v^- = 0$
39. $v^+ = 0, v^- = 1$
41. If $p < v^+$, then the degree of $Q(z) = n - p$, where
$$P(z) = a_n(z - z_1) \ldots (z - r_p)Q(z),$$
and $r_1, r_2, \ldots, r_p$ are the positive real roots. Then $v^- \leq n - p$, the degree of $Q(z)$ and $N = n - p \leq v^-$; thus $n = v^-$. Now $p < v^+ \leq n$, the degree of $P(z)$, thus $N + p < n$, or there are fewer than n real roots, which is a contradiction.
43. $v^+ = 2$, so that are zero or two positive real roots. $v^- = 0$, so there are no negative real roots. Let $x = y + 1$; then $x^4 - 15x + 3 = y^4 + 4y^3 + 6y^2 - 11y - 11$ which has one variation in sign, so there exists exactly one real root, r. Then $x = r + 1$ is one real root of $x^4 - 15x + 3 = 0$, and thus it has exactly two real roots.

Exercises 8.4 (page 294)
1. $P(2) = -5, P(3) = 15$
3. $P(1) = -5, P(2) = 7$
5. $P(0) = 6, P(1) = -10$
7. $-2 < r < -1, -1 < r < 0, 0 < r < 1$
9. $-2 < r < -1$
11. 2.65
13. 3.25
15. 1.38

Exercises 8.5 (page 301)
1. $\sqrt[3]{2} - \sqrt[3]{4}, \omega\sqrt[3]{2} - \omega^2\sqrt[3]{4}, \omega^2\sqrt[3]{2} - \omega\sqrt[3]{4}$
3. $-5, 1, 1$
5. $y_1 = \sqrt[3]{i} + \sqrt[3]{i} = 2\sqrt[3]{i} = -2i$
   $y_2 = (\omega + \omega^2)\sqrt[3]{i} = (-1)(-i) = i$
   $y_3 = (\omega^2 + \omega)\sqrt[3]{i} = (-1)(-i) = i$
7. $y_1 = 1 + \sqrt{-2/3} - \sqrt{-2/3} = 1$
   $y_2 = 1 + (\omega - \omega^2)\sqrt{-2/3} = 1 + i\sqrt{3}\,i\sqrt{2/3} = 1 - \sqrt{2}$
   $y_3 = 1 + (\omega^2 - \omega)\sqrt{-2/3} = 1 - i\sqrt{3}\,i\sqrt{2/3} = 1 + \sqrt{2}$
9. $[2\omega^2, -\omega^2, -\omega^2]$ where $1 + \omega = -\omega^2$
11. $y_1 = 7 - 3 = 4$
    $y_2 = 7\omega - 3\omega^2 = -2 + 5i\sqrt{3}$
    $y_3 = 7\omega^2 - 3\omega = -2 - 5i\sqrt{3}$
13. $1, 2, 3$
15. $-6, i\sqrt{3}, -i\sqrt{3}$
17. $y_1 = \sqrt[3]{10 + \sqrt{108}} + \sqrt[3]{10 - \sqrt{108}}$
    $y_2 = \omega\sqrt[3]{10 + \sqrt{108}} + \omega^2\sqrt[3]{10 - \sqrt{108}}$
    $y_3 = \omega^2\sqrt[3]{10 + \sqrt{108}} + \omega\sqrt[3]{10 - \sqrt{108}}$
19. $A = \sqrt[3]{(3 + \sqrt{5})/2}, B = \sqrt[3]{(3 - \sqrt{5})/2}$
    $y_1 = -1 + A + B, y_2 = -1 + \omega A + \omega^2 B, y_3 = -1 + \omega^2 A + \omega B$

Exercises 8.6 (page 304)
1. $-1, \frac{1}{2}, \frac{1}{2}$
3. $\cos 40°, \cos 80°, -\cos 20°$
5. $\sqrt{2}/2, (\sqrt{6} - \sqrt{2})/4, -(\sqrt{6} + \sqrt{2})/4$
7. $\cos 50°, \cos 70°, -\cos 10°$
13. $4\cos 40°, 4\cos 80°, -4\cos 20°$
9. $-2, 1, 1$
15. $5, -\frac{5}{2}, -\frac{5}{2}$
11. $2\cos 40°, 2\cos 80°, -2\cos 20°$

17. This is the irreducible case, three irrational real roots.

19. (a) $z_1 = \dfrac{\sqrt[3]{4}}{2}(-1+i)$

$z_2 = \dfrac{2+\sqrt{3}}{\sqrt[3]{20+12\sqrt{3}}} + \dfrac{1}{\sqrt[3]{20+12\sqrt{3}}}i$

$z_3 = \dfrac{2-\sqrt{3}}{\sqrt[3]{20-12\sqrt{3}}} + \dfrac{1}{\sqrt[3]{20-12\sqrt{3}}}i$

(b) $z_1 = \sqrt[6]{2}(\cos 3\pi/4 + i\sin 3\pi/4) = \dfrac{\sqrt[3]{4}}{2}(-1+i)$

$z_2 = \sqrt[6]{2}(\cos 15° + i\sin 15°) = \sqrt[6]{2}\left[\dfrac{\sqrt{2+\sqrt{3}}}{2} + i\dfrac{\sqrt{2-\sqrt{3}}}{2}\right]$

$z_3 = -\sqrt[6]{2}(\sin 15° + i\cos 15°) = -\sqrt[6]{2}\left[\dfrac{\sqrt{2-\sqrt{3}}}{2} + i\dfrac{\sqrt{2+\sqrt{3}}}{2}\right]$

(c) $\dfrac{\sqrt{1+(2+\sqrt{3})^2}}{\sqrt[3]{20+12\sqrt{3}}} = \dfrac{2\sqrt{2+\sqrt{3}}}{\sqrt[3]{20+12\sqrt{3}}} = \dfrac{\sqrt{6}+\sqrt{2}}{\sqrt[3]{20+12\sqrt{3}}} = \sqrt[6]{2}$

$\sin 15° = \dfrac{1}{\sqrt[6]{2}\sqrt[3]{20+12\sqrt{3}}} = \dfrac{1}{2\sqrt{2+\sqrt{3}}} \cdot \dfrac{\sqrt{2-\sqrt{3}}}{\sqrt{2-\sqrt{3}}} = \dfrac{\sqrt{2-\sqrt{3}}}{2}$

21. (a) $z_1 = \dfrac{\sqrt{5+2\sqrt{5}}}{\sqrt[5]{176+80\sqrt{5}}} + \dfrac{1}{\sqrt[5]{176+80\sqrt{5}}}i$

$z_2 = \dfrac{\sqrt{5-2\sqrt{5}}}{\sqrt[5]{176-80\sqrt{5}}} + \dfrac{1}{\sqrt[5]{176-80\sqrt{5}}}i$

$z_3 = \dfrac{-\sqrt{5+2\sqrt{5}}}{\sqrt[5]{176+80\sqrt{5}}} + \dfrac{1}{\sqrt[5]{176+80\sqrt{5}}}i$

$z_4 = \dfrac{-\sqrt{5-2\sqrt{5}}}{\sqrt[5]{176-80\sqrt{5}}} + \dfrac{1}{\sqrt[5]{176-80\sqrt{5}}}i$

$z_5 = i$

(b) $z_1 = \cos 18° + i\sin 18°$
$z_2 = -\sin 36° - i\cos 36°$
$z_3 = -\cos 18° + i\sin 18°$
$z_4 = \sin 36° - i\cos 36°$
$z_5 = \cos 90° + i\sin 90°$

REVIEW EXERCISES

(c) $z_1 = \dfrac{\sqrt{10 + 2\sqrt{5}}}{4} + \dfrac{\sqrt{5} - 1}{4} i$

$z_2 = \dfrac{-\sqrt{10 - 2\sqrt{5}}}{4} - \dfrac{\sqrt{5} + 1}{4} i$

$z_3 = \dfrac{-\sqrt{10 + 2\sqrt{5}}}{4} + \dfrac{\sqrt{5} - 1}{4} i$

$z_4 = \dfrac{\sqrt{10 - 2\sqrt{5}}}{4} - \dfrac{\sqrt{5} + 1}{4} i$

$z_5 = i$

(d) $\left(\dfrac{1}{\sqrt[5]{176 + 80\sqrt{5}}}\right)^5 = \dfrac{(11 - 5\sqrt{5})}{16(11 + 5\sqrt{5})(11 - 5\sqrt{5})} = \dfrac{5\sqrt{5} - 11}{64}$

$\left(\dfrac{\sqrt{5} - 1}{4}\right)^5 = \left(\dfrac{3 - \sqrt{5}}{8}\right)^2 \left(\dfrac{\sqrt{5} - 1}{4}\right) = \left(\dfrac{7 - 3\sqrt{5}}{32}\right)\left(\dfrac{\sqrt{5} - 1}{4}\right)$

$= \dfrac{5\sqrt{5} - 11}{64}$

Exercises 8.7 (page 308)

1. $\dfrac{-\sqrt{2} \pm \sqrt{2\sqrt{2} - 2}}{2}$, $\dfrac{\sqrt{2} \pm \sqrt{-2\sqrt{2} - 2}}{2}$

3. $3 \pm \sqrt{2}, -1 \pm i\sqrt{2}$
5. $(-3 \pm \sqrt{5})/2, (-1 \pm i\sqrt{11})/2$
7. $(-2 \pm \sqrt{10})/2, \pm i\sqrt{2}/2$

9. $1 \pm i, 1 \pm \sqrt{3}$
11. $-4 \pm \sqrt{5}, 1 \pm \sqrt{2}$

Review Exercises (page 311)

1. $(-3 \pm \sqrt{41})/4$
2. $(-\sqrt{3} \pm i\sqrt{5})/2$
3. $i, 2i$
4. $(-1 \pm \sqrt{10})/3$
5. $(1 \pm \sqrt{29})/2$
6. $\tfrac{1}{3}$
7. $\pm 2\sqrt{10}$
8. $2 \pm 2i\sqrt{2}$
9. -12
10. 138
11. No, $P(2 + 3i) = -45 + 14i$
12. (a) $(x + 2)(2x^2 - x + 6)$
    (b) $(x + 2)(2x^2 - x + 6)$
    (c) $(x + 2)\left(x - \dfrac{1}{4} - \dfrac{\sqrt{47}}{4} i\right)\left(x - \dfrac{1}{4} + \dfrac{\sqrt{47}}{4} i\right)$

13. (a) $(x-2)(x^2-3)$
    (b) and (c) $(x-2)(x+\sqrt{3})(x-\sqrt{3})$
14. (a) $(x^2-2)(x^2+4)$
    (b) $(x+\sqrt{2})(x-\sqrt{2})(x^2+4)$
    (c) $(x+\sqrt{2})(x-\sqrt{2})(x+2i)(x-2i)$
15. $2, -2, -3$
16. 2
17. $3, -2$
18. $4, -2, -\frac{3}{2}$
19. $\frac{1}{2}, \frac{1}{3}$
20. $\frac{1}{2}, \frac{2}{3}, -\frac{3}{4}$
21. $x^4 - 4x^3 + 32x - 13 = 0$
22. $-2 < r < 5$
23. $-1 < r < 3$
24. $-2 \le r \le 2$
25. No negative real roots, one or three positive real roots, two or no imaginary roots.
26. One negative real root, one positive real root, two imaginary roots.
27. One or two positive real roots, zero or two negative real roots, two, four, or six imaginary roots.
28. One real root ($r = 0$) and four imaginary roots.
29. $0 < r < 1, -1 < r < 0, -5 < r < -4$
30. $r = 1, r = 2, r = 3$
31. $0 < r < 1$
32. $-3 < r < -2, -2 < r < -1, 0 < r < 1, r = 2$ (double root)
33. 0.79
34. -1.61
35. $2\cos 20°, -2\cos 40°, -2\cos 80°$
36. $\sqrt[3]{9} + \sqrt[3]{3}, \omega\sqrt[3]{9} + \omega^2\sqrt[3]{3}, \omega^2\sqrt[3]{9} + \omega\sqrt[3]{3}$
37. $-\frac{1}{3}, 1+i, 1-i$
38. $1, 2, -3$
39. $(-\sqrt[3]{1+\sqrt{2}} + \sqrt[3]{1-\sqrt{2}})i, (-\omega\sqrt[3]{1+\sqrt{2}} - \omega^2\sqrt[3]{1-\sqrt{2}})i$
    $(-\omega^2\sqrt[3]{1+\sqrt{2}} - \omega\sqrt[3]{1-\sqrt{2}})i$
40. $\sqrt{2}+\sqrt{6}, \sqrt{2}-\sqrt{6}, -2\sqrt{2}$

CHAPTER 9

Exercises 9.1 (page 319)

1. 13
3. $\sqrt{2}$
5. 2.5
7. $2\sqrt{26}$
9. $\sqrt{(a-1)^2 + (b-2)^2}$
11. $(0, 2)$
13. $(-2, -\frac{9}{2})$
15. $\left(\dfrac{a+c}{2}, \dfrac{b+d}{2}\right)$
17. $OA = OB = OC = \sqrt{61} = r$

EXERCISES 9.3   567

19. $|AB|^2 = 13$, $|BC|^2 = 13$, $|AC|^2 = 26$; $|AB|^2 + |BC|^2 = |AC|^2$
21. Midpoint, $M: (\frac{7}{2}, \frac{3}{2})$; $|MB| = \sqrt{26}/2 = \frac{1}{2}|AC|$
23. 8
25. 8 or 4
27. $(5, -6)$ or $(-10, 3)$

Exercises 9.2 (page 324)

1. -3 9. -3
3. 1   11. $45°$ or $\pi/4$
5. 0   13. $135°$ or $3\pi/4$
7. -1 15. $180° - \text{Tan}^{-1} 2$
17. m of $AB = \frac{1}{3} = m$ of DC
    m of $AD = 1 = m$ of BC
19. m of $AB = -2$; m of $BC = \frac{1}{2}$; m of $CD = -2$; m of $AD = \frac{1}{2}$
    $|AB| = 3\sqrt{5} = |BC|$
21. (a) 28   (b) 2   25. (a) $\frac{3}{4}$ (b) $-\frac{4}{3}$
23. 0   27. -4

Exercises 9.3 (page 328)

1. Let m = slope of given line; then the slope of each of the perpendicular lines is $-1/m$, so they are parallel.
3. $M_1: \left(\frac{b}{2}, \frac{c}{2}\right)$ and $M_2: \left(\frac{2a-b}{2}, \frac{c}{2}\right)$; $|M_1 M_2| = (a - b)$
    $\frac{1}{2}(DA + CB) = \frac{1}{2}(a + a - b - b) = a - b$

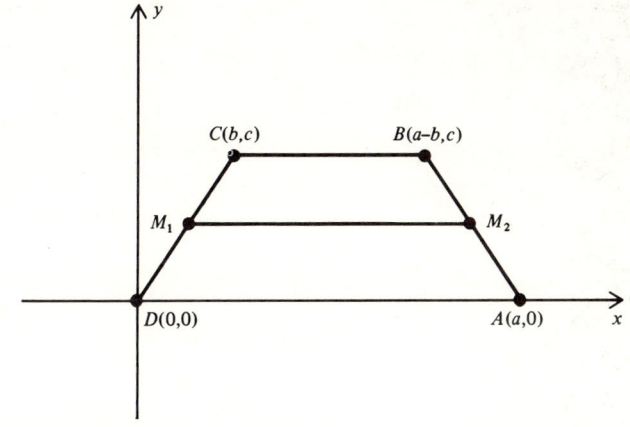

5. Midpoint of DB is $\left(\dfrac{a+b}{2}, \dfrac{h}{2}\right)$, Midpoint of AC is $\left(\dfrac{a+b}{2}, \dfrac{h}{2}\right)$.

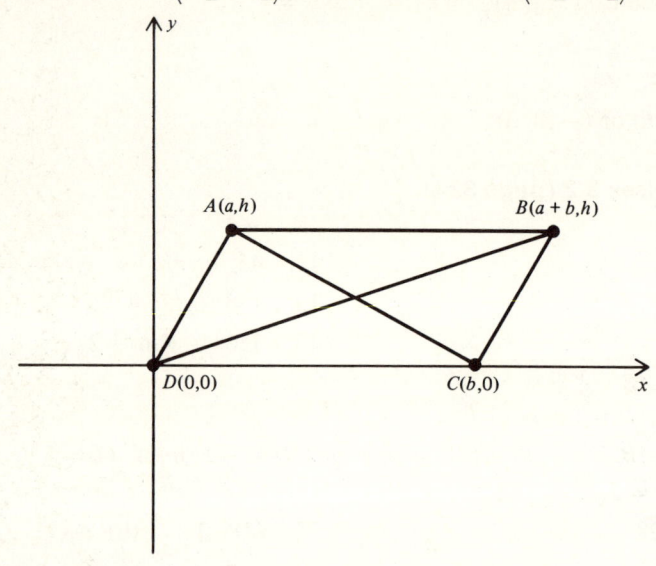

7. $A: \left(\dfrac{b}{2}, \dfrac{c}{2}\right)$, $B: \left(\dfrac{b+d}{2}, \dfrac{c+e}{2}\right)$; $C: \left(\dfrac{a+d}{2}, \dfrac{e}{2}\right)$; $D: \left(\dfrac{a}{2}, 0\right)$

m of $AB = \dfrac{e}{d} = m$ of DC

m of $BC = \dfrac{c}{b-a} = m$ of AD

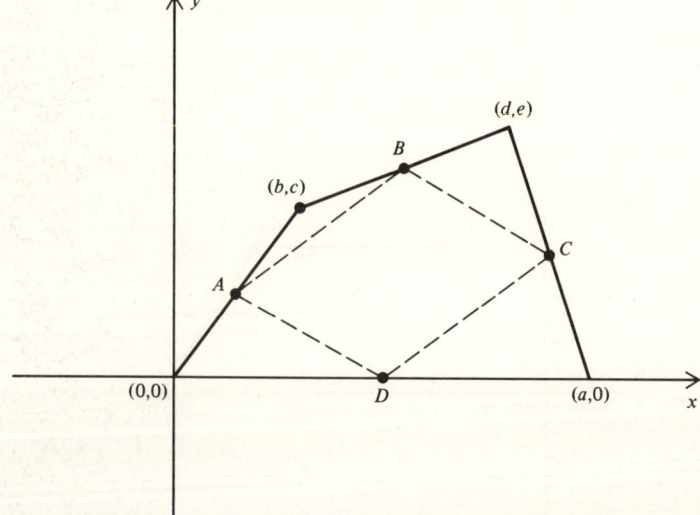

9. If $DB = AC$, then $\sqrt{(a+b)^2 + h^2} = \sqrt{(b-a)^2 + h^2}$ or $a+b = b-a$ or $a = 0$ and $DA \perp DC$.

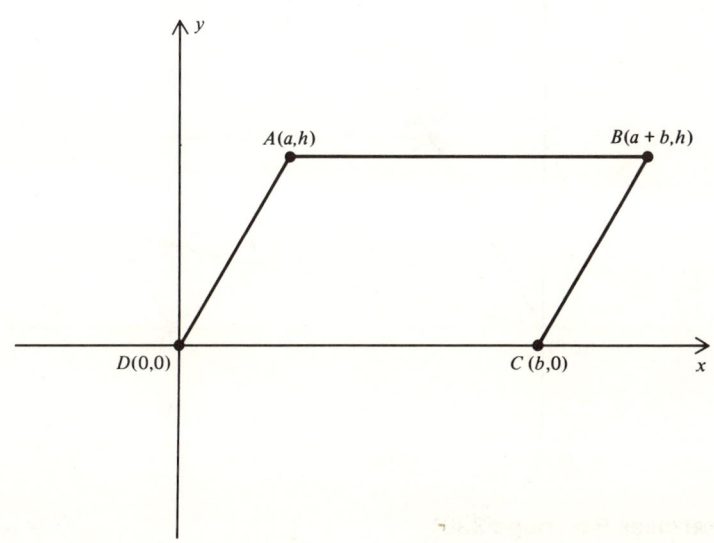

Exercises 9.4 (page 331)

1. $x + y - 5 = 0$
3. $2x - 3y = 0$
5. $x + y - 5 = 0$
7. $bx + ay - ab = 0$
9. $x = 3$
11. $x - 2y + 4 = 0$
13. $3x + 4y - 10 = 0$
15. $7x + 8y + 16 = 0$
17. $3x + 4y - 24 = 0$

19. $4x - 3y = 0$
21. $(-5, -1)$
25. $\theta = \text{Tan}^{-1}(1/3) \approx 18°$
27. $\theta = \text{Tan}^{-1}(-9/2) \approx 103°$
29. $\theta = \text{Tan}^{-1}(2/25) \approx 2°$
31. $30°, 34°, 117°$
33. $105°$
35. $35/2$

37. $AM: y = \dfrac{h}{2a+b}x$. $BD: y = \dfrac{h}{b-a}(x-a)$

$P: \left(\dfrac{2a+b}{3}, \dfrac{h}{3}\right)$, $|PD| = \dfrac{\sqrt{(b-a)^2 + h^2}}{3}$

$|BD| = \sqrt{(b-a)^2 + h^2}$

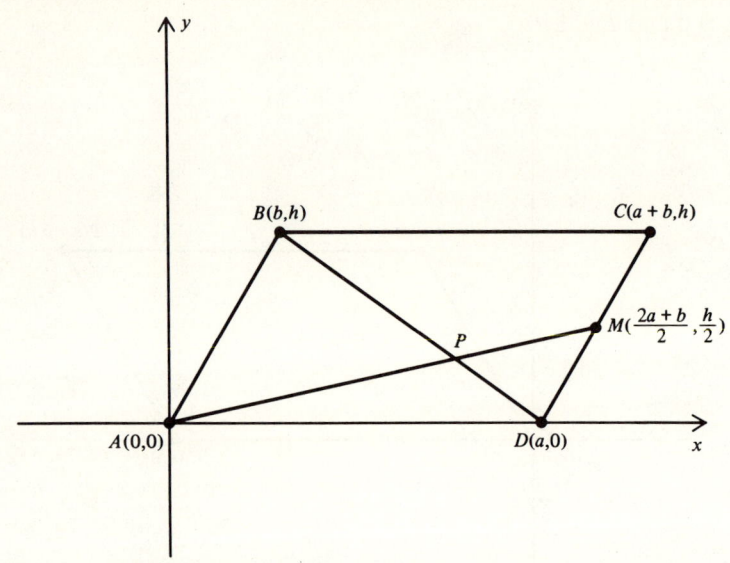

Exercises 9.5 (page 339)

1. x-intercepts $= \pm\frac{3}{2}$, y-intercepts $= \pm 3$;
   domain $-\frac{3}{2} \leq x \leq \frac{3}{2}$; range $-3 \leq y \leq 3$;
   symmetry: origin, x-axis, y-axis.

3. No x-intercepts; y-intercepts $= \pm 3$;
   domain, all real x; range, $|y| \geq 3$;
   symmetry: origin, x-axis, y-axis.

5. x-intercept $= 0$, y-intercept $= 0$;
   domain and range, all real numbers;
   symmetry: origin, x-axis, y-axis.

7. x-intercepts $= \pm 3$, y-intercepts $= 6$ and -2;
   domain, $|x| \leq 2\sqrt{3}$; range $-2 \leq y \leq 6$;
   symmetry: y-axis.

9. x-intercepts $= \pm 4$, no y-intercepts;
   domain, $|x| \geq 4$; range, all real y;
   symmetry: origin, x-axis, y-axis.

11. No x or y intercepts;
    domain, all real x except 0; range, $y > 0$;
    symmetry: y-axis.

13. x-intercepts $= \pm 1$; no y-intercepts;
    domain, $|x| \geq 1$; range $|y| < 1$;
    symmetry: origin, x-axis, y-axis.

EXERCISES 9.6

15. x-intercepts $= -2, 1, 3$; y-intercepts $= \pm\sqrt{6}$;
domain $-2 \leq x \leq 1$ or $x \geq 3$; range, all real y;
symmetry: x-axis.
17. x-intercept $= 0$, y-intercept $= 0$;
domain, all real x except -1 and 2;
range, $y \leq 0$ or $y \geq \frac{8}{9}$;
not symmetric to origin, x-axis, or y-axis.
19. x-intercept $= -2$, y-intercept $= -3$;
domain, all real x except 2; range, all real y except 3;
not symmetric to origin, x-axis, or y-axis.

Exercises 9.6 (page 343)

1. Center $(2, -3)$, $r = 4$
3. Center $(\frac{7}{2}, 1)$, $r = 2$
5. Not a circle but a point $(3, 5)$
7. $(x + 1)^2 + (y - 2)^2 = 9$
9. $(x - 2)^2 + (y + 1)^2 = 13$
11. $(x - 5)^2 + (y - 4)^2 = 1$
13. $(x - 4)^2 + (y - 2)^2 = 9$
    $(x - 1)^2 + (y - 5)^2 = 9$
15. $(x - 4)^2 + (y - 6)^2 = 52$
17. $(-\frac{1}{2}, 0)$, $x = \frac{1}{2}$
19. $(\frac{3}{4}, 0)$, $x = -\frac{3}{4}$
21. $(0, -\frac{5}{2})$, $y = \frac{5}{2}$
23. $(0, 3/16)$, $y = -3/16$
25. $y^2 = 8x$
27. $y^2 = 12x$
29. $y^2 = \frac{12}{7}x$
31. $x^2 = y$
33. $x^2 = 3y$
35. $y^2 = 8x$ or $x^2 = y$
37. $y^2 = 16x$ or $x^2 = -2y$
39.

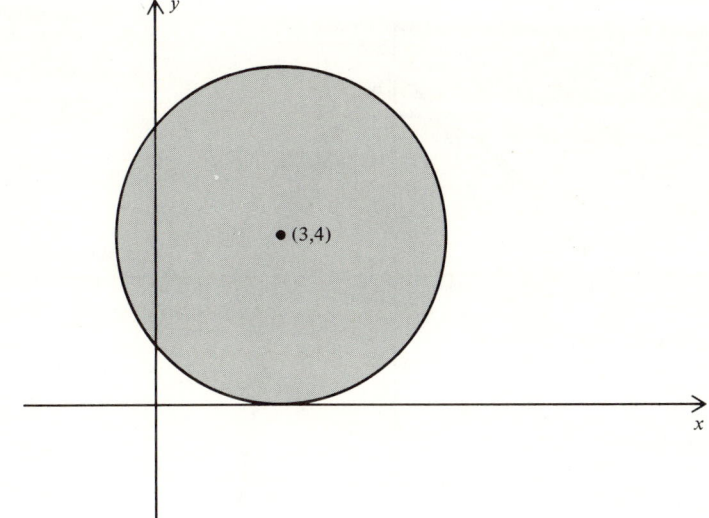

$(x - 3)^2 + (y - 4)^2 \leq 16$

41.

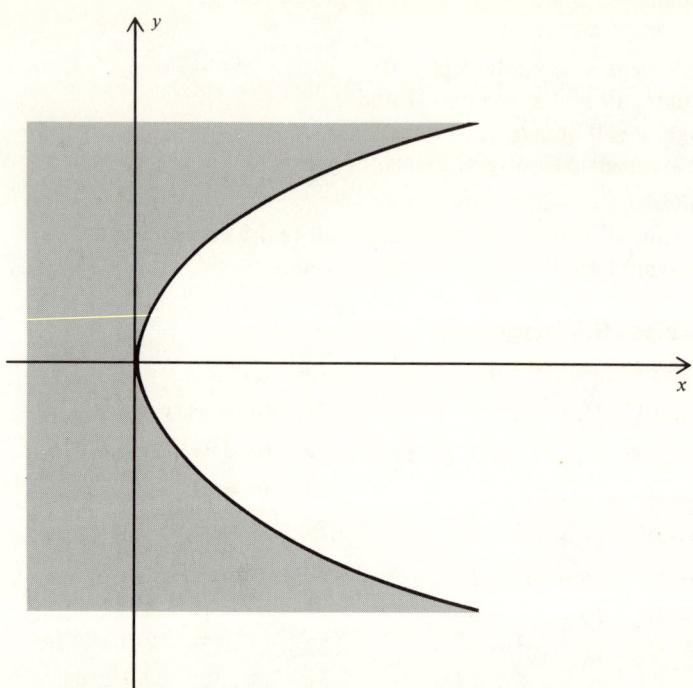

43.

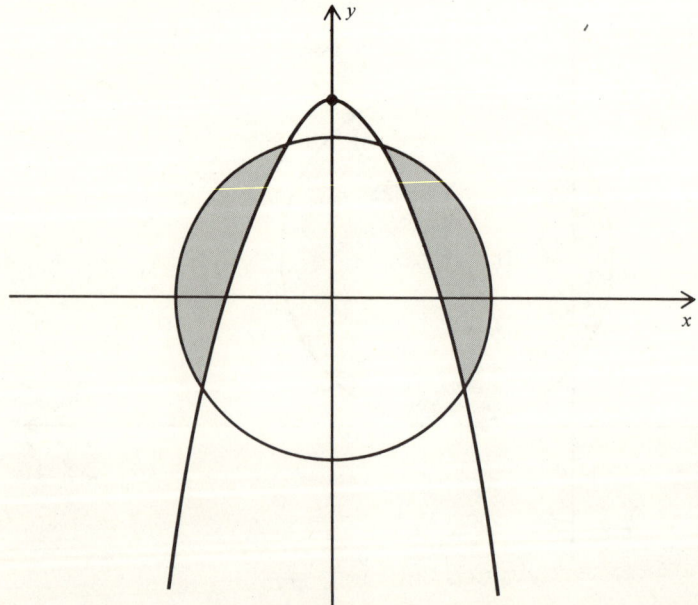

Exercises 9.7 (page 352)

1. Ellipse: x-intercepts $= \pm 10$, y-intercepts $= \pm 2$
   domain, $|x| \leq 10$; range, $|y| \leq 2$
   symmetric to origin, x-axis, y-axis.

3. Ellipse: x-intercepts $= \pm\sqrt{10}$, y-intercepts $= \pm 5$
   domain, $|x| \leq \sqrt{10}$; range, $|y| \leq 5$
   symmetric to origin, x-axis, y-axis.

5. Hyperbola: no x-intercepts, y-intercepts $= \pm 3$
   domain, all real x; range, $|y| \geq 3$
   symmetric to origin, x-axis, y-axis
   asymptotes, $y = \pm x$.

7. Hyperbola: x-intercepts $= \pm 4$, no y-intercepts
   domain, $|x| \geq 4$; range, all real y
   asymptotes, $4y = \pm 3x$
   symmetric to origin, x-axis, y-axis.

9. $\dfrac{x^2}{9} + \dfrac{y^2}{5} = 1$

11. $\dfrac{x^2}{16} + \dfrac{y^2}{1} = 1$ or $\dfrac{x^2}{1} + \dfrac{y^2}{16} = 1$

13. $\dfrac{x^2}{25} + \dfrac{y^2}{16} = 1$ or $\dfrac{x^2}{16} + \dfrac{y^2}{25} = 1$

15. $\dfrac{y^2}{1} - \dfrac{x^2}{3} = 1$

17. $\dfrac{x^2}{36} - \dfrac{y^2}{25} = 1$ or $\dfrac{y^2}{36} - \dfrac{x^2}{25} = 1$

19. $\dfrac{x^2}{5} - \dfrac{y^2}{4} = 1$ or $\dfrac{y^2}{5} - \dfrac{x^2}{4} = 1$

21. $4x^2 + 3y^2 = 48$

23. $16x^2 + 15y^2 - 114y + 207 = 0$

25. $\dfrac{x^2}{4} + \dfrac{y^2}{9} = 1$

27. $3y^2 - x^2 = 12$

29. $7x^2 - 9x^2 - 20y = 32$

31. $9x^2 - 4y^2 = 81$

33. $e = 1$, parabola

35. 4 ft, 10 ft, 28 ft, 58 ft, 100 ft

Exercises 9.8 (page 359)

1. $x'^2 + y'^2 = 16$
3. $x'^2 = 16y'$
5. $\dfrac{x'^2}{81/16} + \dfrac{y'^2}{81/4} = 1$
7. $x'^2 + y'^2 = \tfrac{1}{2}$
9. $\dfrac{x'^2}{10} - \dfrac{y'^2}{8} = 1$
11. $\dfrac{x'^2}{4} - \dfrac{y'^2}{1} = 4$
13. $y'^2 = 4x'^2$
15. $3x'^2 + y'^2 = 0$
17. $y' = x'^3 - 4x'$
25. Any line through $(2, -3)$
27. $x = -3$
29. $x = -2$, $y = \tfrac{5}{2}$
31. Any line through $(\tfrac{1}{2}, \tfrac{1}{2})$
33. $x = 2$, $y = 1$
35. $x = 3$, $y = 2$
37. $x = -1$, $y = 2$
39. Any line through $(1, -5)$

Exercises 9.9 (page 365)

1. $x'y' = -\frac{9}{2}$
3. $7y'^2 - 3x'^2 = 3$
5. $221x'^2 - 117y'^2 + 52 = 0$; hyperbola
7. $31x'^2 + y'^2 = 10$; ellipse
9. $x'^2 - 4y'^2 = 16$; hyperbola
11. $y'^2 = x'$; parabola
13. Hyperbola
15. Parabola
17. Ellipse
19. $x''^2 - y''^2 = 2$; $x'' = x' - \sqrt{26}$, $y'' = y' - \sqrt{26}$; $x = \dfrac{1}{\sqrt{26}}(x' - 5y')$

$y = \dfrac{1}{\sqrt{26}}(5x' + y')$

21. $x''^2 - y''^2 = 0$; $x'' = x' - \sqrt{2}/2$; $x = \dfrac{1}{\sqrt{2}}(x' - y')$

$y'' = y' - 3\sqrt{2}/2 \quad y = \dfrac{1}{\sqrt{2}}(x' + y')$

23. $y''^2 = 4\sqrt{2}\,x''$; $x'' = x' - 3\sqrt{2}/4$; $x = \dfrac{1}{\sqrt{2}}(x' - y')$

$y'' = y' + \sqrt{2} \quad y = \dfrac{1}{\sqrt{2}}(x' + y')$

25. $m' = \dfrac{m + \tan\theta}{1 - m\tan\theta}$

Exercises 9.10 (page 373)

11. $(2\sqrt{2}, \pi/4)$, $(-2\sqrt{2}, 5\pi/4)$
13. $(4, \pi/6)$, $(4, 13\pi/6)$
15. $(2, \pi/6)$, $(-2, -5\pi/6)$
17. $(\sqrt{2}, 5\pi/4)$, $(-\sqrt{2}, \pi/4)$
19. $(2, \pi)$, $(2, -\pi)$
21. $x^2 + y^2 = 2x$
23. $x^2 = -8(y - 2)$
25. $x^2 + y^2 - 2x + 2y = 0$
27. $x = 2$
29. $x + y = 5$
31. $x^2 - y^2 = 4$
33. $r\cos\theta = 4$
35. $r = 6$

EXERCISES 9.10

37. $r = 8 \sin \theta$
39. $r^2 \sin 2\theta = 12$
41. $r^2 \cos 2\theta = 9$
43.

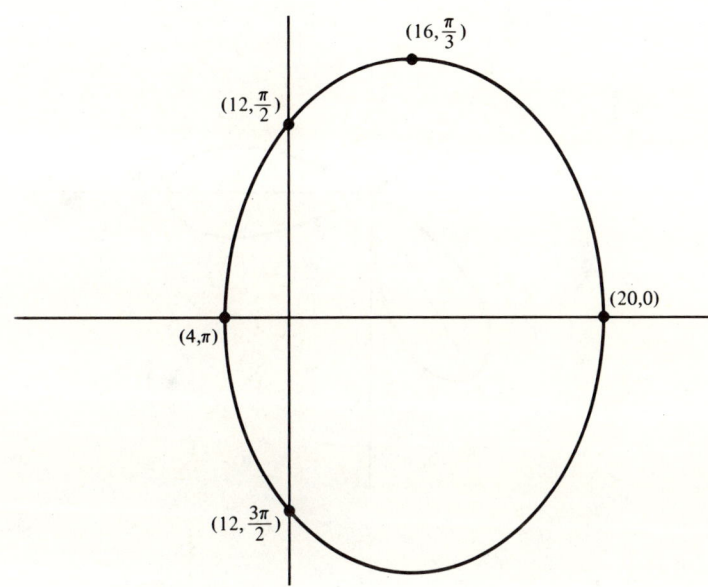

45.

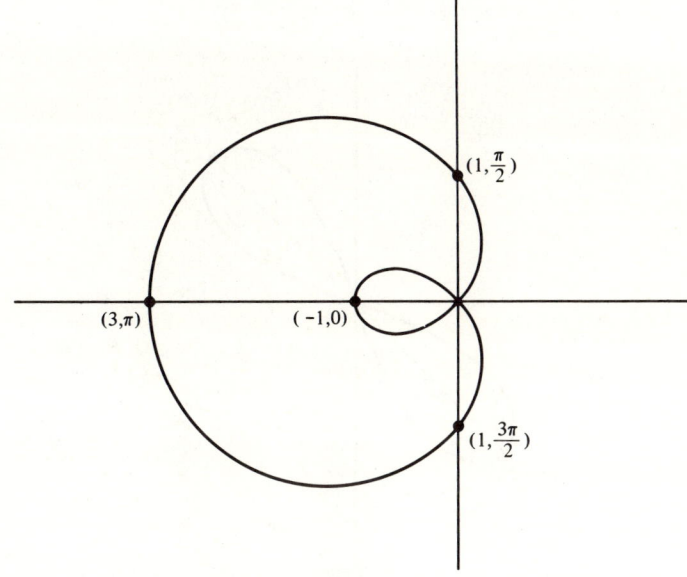

Limacon

47.

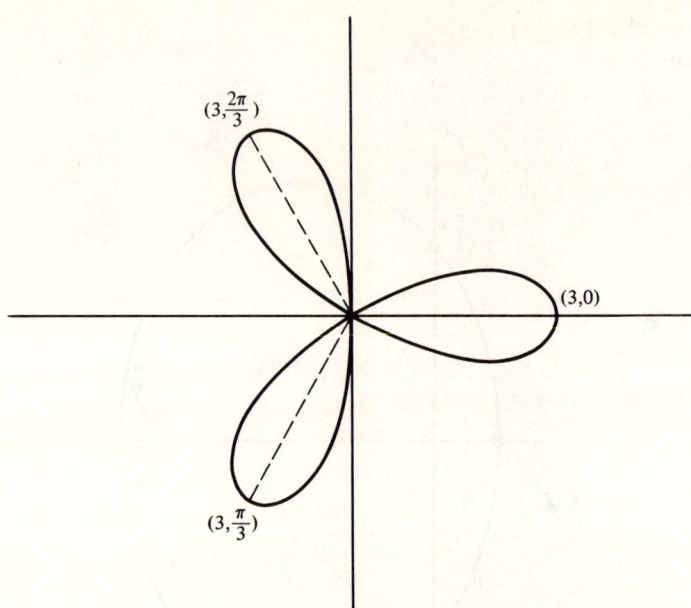

49.

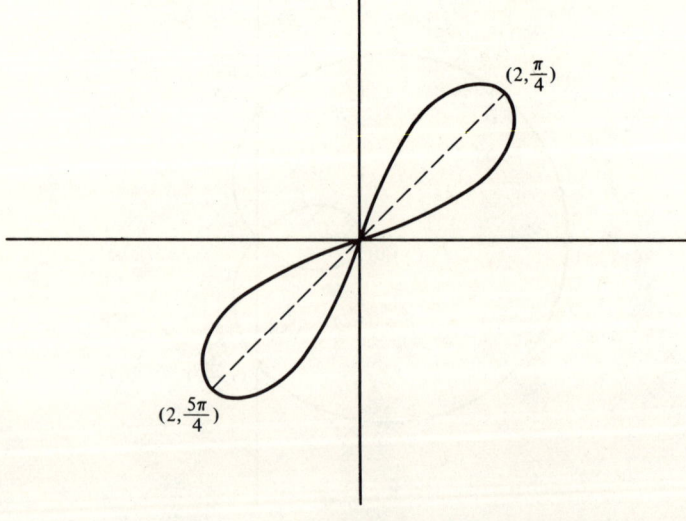

Lemniscate

51.

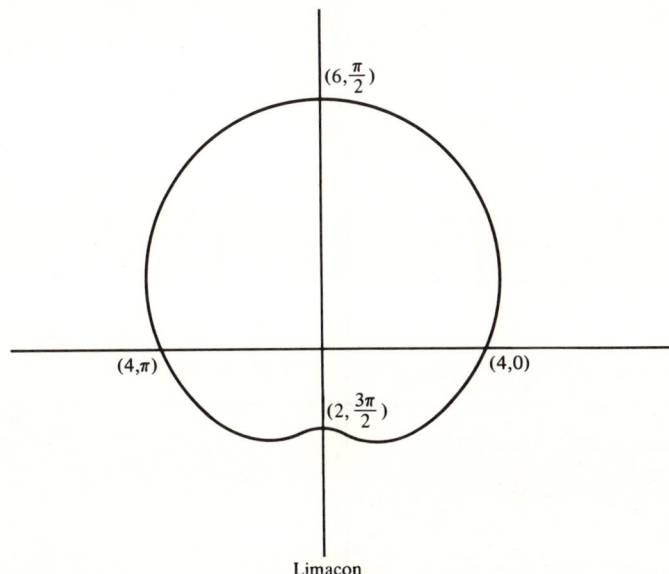

Limacon

53. $r = -4 \cos \theta$ 55. $r = 5$
57. $r = 6 \cos \theta + 8 \sin \theta$, center $(3, 4)$ rectangular, center $(5, \text{Tan}^{-1} 4/3)$ polar
59. $\theta = 3\pi/4$ 63. 1 61. $2\sqrt{3}$

Exercises 9.11 (page 378)
1. $y = 3x + 7$
3. $16x^2 - 9y^2 = 144$
5. $xy = 1$
7. $4(x + 1) = (x - y)^3$
9. $y = 1 - 2x^2$
11. $r = \cos 2\theta$

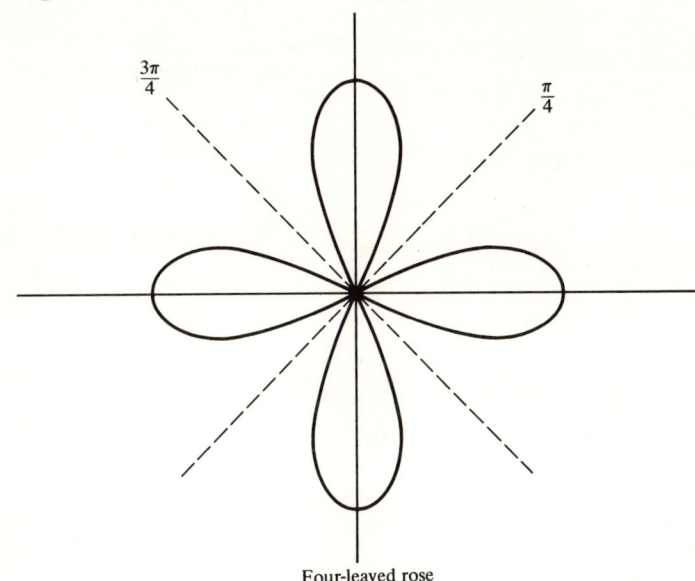

Four-leaved rose

13. $r = \sin\theta$

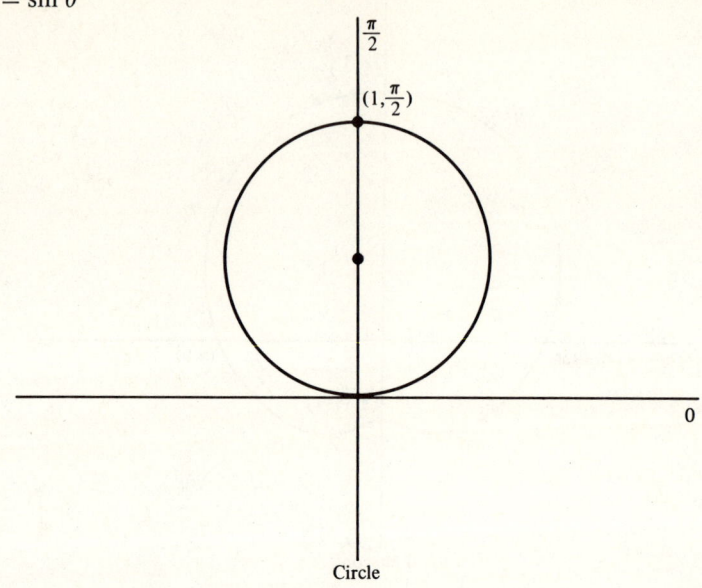

Circle

15. $r = 1 - 2\cos\theta$

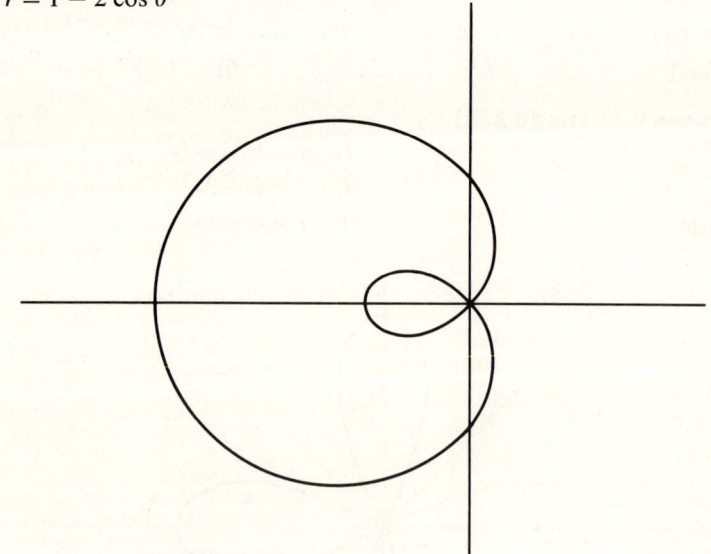

Limacon

17. Part, $x > 0, y > 0$ 19. Part, $-2 \leq x \leq 2$

21. $x^2 + y^2 = 9$, part of circle, $y > 0$.

23. $\dfrac{(x-3)^2}{4} - \dfrac{(y-2)^2}{9} = 1$, all of hyperbola.

25. $y^2 - x^2 = 4$, all of hyperbola.
27. $4(x - 2)^2 + y^2 = 16$, all of ellipse.
29. $y^2 = 4x + 4$, part of parabola, $x > 0$, $y > 0$
31. $x = -3 + 3 \sin t$
    $y = 2 + \sqrt{3} \cos t$
33. $x = \frac{1}{2} \sin t$
    $y = \cos t$
35. $x = t^2$
    $y = t^3$
37. $x = 8 \cos^3 t$
    $y = 8 \sin^3 t$
39. $x = \dfrac{5t^2}{t^5 + 1}$
    $y = \dfrac{5t^3}{t^5 + 1}$
41. $t = x - 4$
    $y - 6 = 4(x - 4)$
    $m = 4$
43. $\dfrac{x - x_1}{x_2 - x_1} = t$, $\dfrac{y - y_1}{y_2 - y_1} = t$ or $x = x_1 + (x_2 - x_1)t$
    $y = y_1 + (y_2 - y_1)t$
45. $x = -2 + 3t$
    $y = 1 - 6t$

Exercises 9.12 (page 386)
1. Sum $= (3, 4)$
3. Sum $= (\pi - 1, \sqrt{3} - 2)$
5. Sum $= (-\sqrt{2}/2, 0)$
7. $(-4, 4)$
9. $(-2, 2)$
11. $(2, 2)$
13. $(4, -3)$
15. $((\sqrt{2} - \sqrt{3})/2, 0)$
17. $8, \theta = \pi/6$
19. $\sqrt{2}, \theta = 3\pi/4$
21. $2.5, \cos\theta = \frac{4}{5}, \sin\theta = -\frac{3}{5}$
23. $7, \theta = \pi$
25. $4, \theta = \pi/12$
27. (a) 4   (b) 4/5   (c) $4/\sqrt{5}$
29. (a) -6 (b) $-\sqrt{2}/10$ (c) $-\sqrt{10}/5$
31. (a) $2\sqrt{3}$ (b) 1/2   (c) $\sqrt{3}$
33. (a) 0   (b) 0   (c) 0
35. (a) 3   (b) $\sqrt{2}/10$ (c) $3\sqrt{2}/4$
37. 5
39. $2\sqrt{5}$
41. $2x + 5y = 19$
43. $x^2 + y^2 = 25$
45. $9x^2 + 25y^2 = 225$
47. $9x^2 - 4y^2 = 36$
49. $y = 9 - x^2$

Review Exercises (page 390)
1. (a) $\sqrt{10}$ (b) -3 (c) $3x + y = 11$ (d) $(\frac{7}{2}, \frac{1}{2})$
2. (a) $\sqrt{17}$ (b) -4 (c) $4x + y + 5 = 0$ (d) $(-\frac{3}{2}, 1)$

3. (a) $\sqrt{481}/12$ (b) 20/9 (c) $20x - 9y = 13$ (d) $(\frac{1}{8}, -\frac{7}{6})$
4. (a) 5 (b) $\frac{4}{3}$ (c) $4x - 3y = 12$ (d) $(\frac{3}{2}, -2)$
5. (a) $2\sqrt{2}$ (b) 1 (c) $x - y = 1$ (d) $(-2, -3)$
6. $P: \left(\frac{d}{2}, \frac{e}{2}\right)$, $Q: \left(\frac{b+d}{2}, \frac{c+e}{2}\right)$, $R: \left(\frac{a+b}{2}, \frac{c}{2}\right)$, $S: \left(\frac{a}{2}, 0\right)$

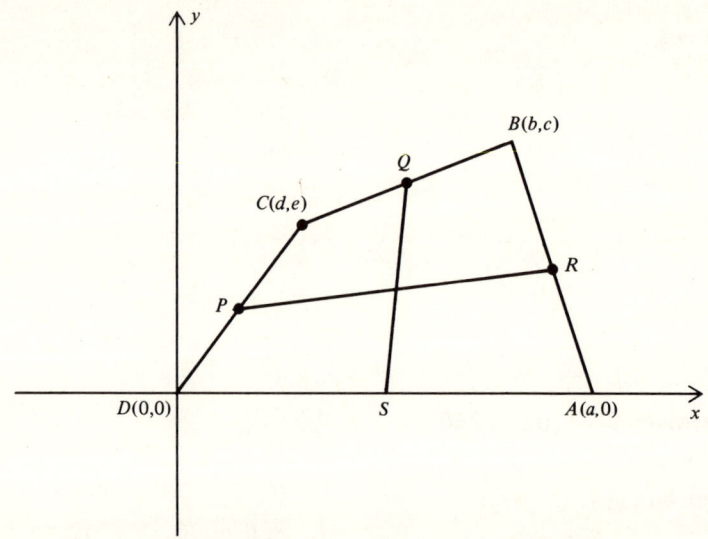

Midpoint of PR is $\left(\frac{a+b+d}{4}, \frac{c+e}{4}\right)$ = midpoint of QS

7. $|DB| = \sqrt{(a+b)^2 + h^2} = |AC|$

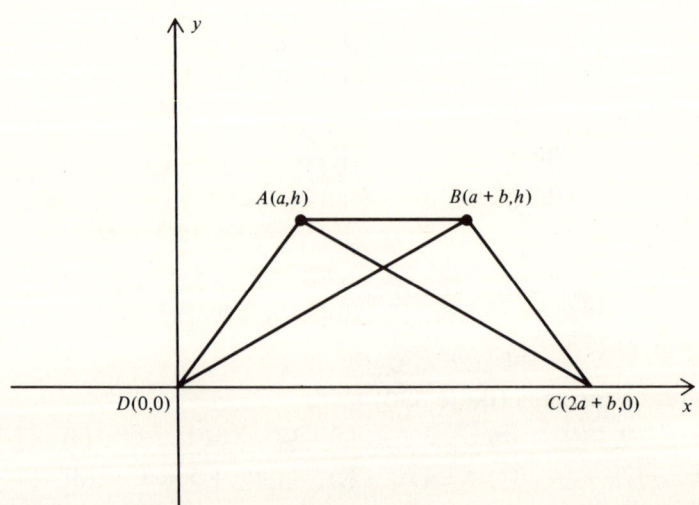

8. $x - 2y + 8 = 0$
9. x-intercept $= 0$, y-intercept $= 0$; domain, $|x| < 2$; range, all real y; symmetric to origin, x-axis, and y-axis.
10. No x-intercepts; y-intercept $= 2$; domain, all real x; range $0 < y \leq 2$; symmetric to y-axis.
11. x-intercept $= 5$; y-intercepts $= \pm 1$; domain, $-5 < x \leq 5$; range, all real y; symmetric to x-axis.
12. x-intercept $= 0$; y-intercepts $= 0, \sqrt{3}, -\sqrt{3}$; domain, all real x; range, all real y; symmetric to origin.
13. $x = 2, y = 3$
14. $y = 4$
15. Circle
16. Hyperbola
17. Ellipse
18. Parabola
19. Ellipse. $5x'^2 + 4y'^2 = 36$
20. Point. $(x - 2)^2 + (y - 6)^2 = 0$; $x'^2 + y'^2 = 0$
21. Parabola. $y'^2 = 6x'$, $\cos \theta = 3/5$, $\sin \theta = 4/5$
22. Ellipse. $3x''^2 + y''^2 = 12$, $\quad x'' = x' + 1, \quad \theta = \pi/4$
    $\quad y'' = y' + 3$
23. Hyperbola. $4x''^2 - y''^2 = 4$, $\quad x'' = x' - 1, \quad \theta = \pi/4$
    $\quad y'' = y' + 2$
24. $(x - 3)^2 = -4(y + 1)$
25. Vertices $(\pm 9/5, 0)$, $(0, \pm 9/13)$, $2a = 18/5$, $2b = 18/13$, foci $(\pm 108/65, 0)$, $e = 12/13$.
26. Vertices $(0, \pm 4)$, $2a = 8$, $2b = 6$ foci $(0, \pm 5)$, $e = \frac{5}{4}$.
27.

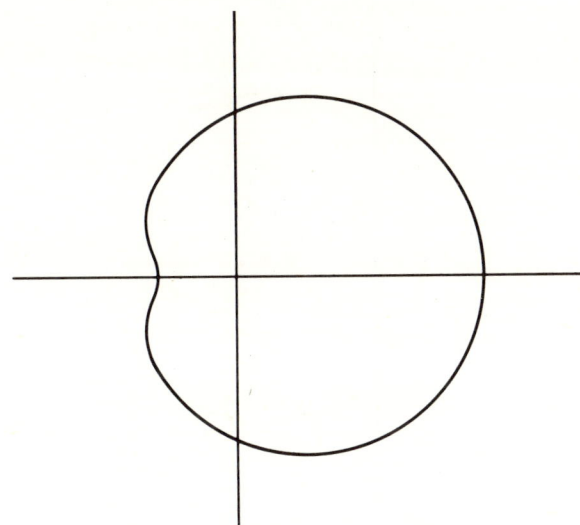

Limacon

28.

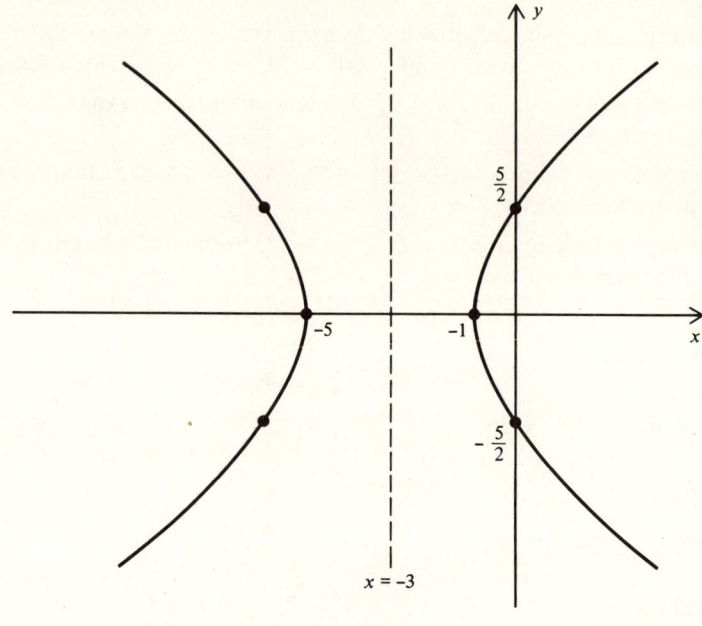

Hyperbola

29.

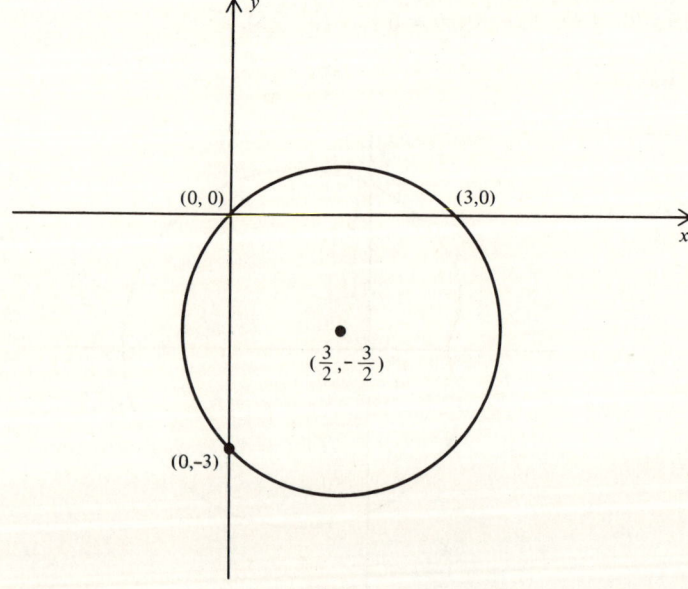

Circle

REVIEW EXERCISES

30.

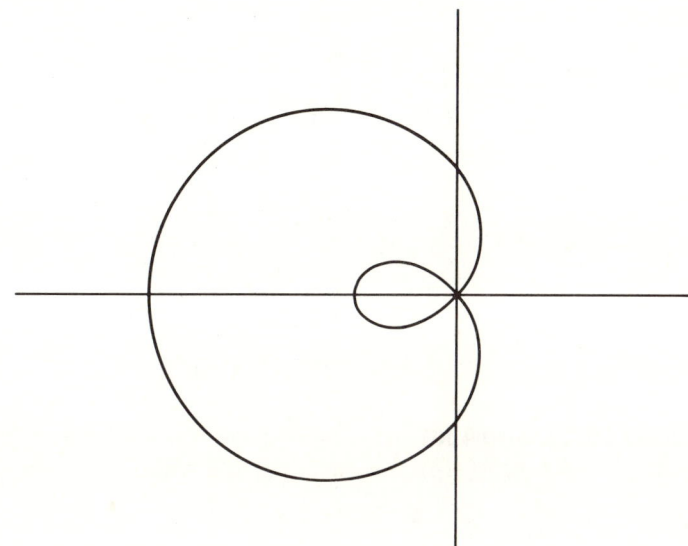

Limacon

31. $r = \dfrac{6}{2\cos\theta + \sin\theta}$

32. $\dfrac{(x+2)^2}{9} + \dfrac{(y-2)^2}{25} = 1$

33. $x + 4y = 11$

34. $x^2 + (y-1)^2 = 1$

35. $y^2 = 2x$

36. $\dfrac{y^2}{9} - \dfrac{x^2}{4} = 1$

37. $x^2 - y^2 = 1$

38. $5\mathbf{i} + 5\mathbf{j}, \mathbf{i} - \mathbf{j}, 12, |\mathbf{a}| = \sqrt{13}, |\mathbf{b}| = \sqrt{13}$

39. $-2\mathbf{i} + 2\mathbf{j}, -4\mathbf{i} - 6\mathbf{j}, -11, |\mathbf{a}| = \sqrt{13}, |\mathbf{b}| = \sqrt{17}$

40. $6\mathbf{i} - 10\mathbf{j}, -4\mathbf{i} + 4\mathbf{j}, 26, |\mathbf{a}| = \sqrt{10}, |\mathbf{b}| = \sqrt{74}$

41. $-\mathbf{i} - \mathbf{j}, -3\mathbf{i} + 3\mathbf{j}, -4, |\mathbf{a}| = \sqrt{5}, |\mathbf{b}| = \sqrt{5}$

42. $3\mathbf{i} - 7\mathbf{j}, 9\mathbf{i} + 3\mathbf{j}, -8, |\mathbf{a}| = 2\sqrt{10}, |\mathbf{b}| = \sqrt{34}$

43. (a) $5\pi/12$ (b) $-2\sqrt{2} + 2\sqrt{6}$

44. (a) $\pi/2$ (b) 0

45. (a) $\pi/6$ (b) 1

46. (a) $\pi/3$ (b) $\sqrt{3}$

47. (a) $\text{Arccos } \tfrac{4}{5}$ (b) $-4\sqrt{5}/5$

CHAPTER 10

Exercises 10.1 (page 397)
1. $\sqrt{17}$
3. 9
5. $\sqrt{19}$
7. $2\sqrt{3}\sqrt{a^2 + b^2 + c^2}$
9. $5\sqrt{2}$
11. $(\frac{1}{2}, 3, 1)$
13. $(1, 0, \frac{1}{2})$
15. $(-\frac{1}{2}, \frac{3}{2}, -\frac{1}{2})$
17. $(0, 0, 0)$
19. $(\frac{3}{2}, -\frac{1}{2}, 0)$
21. $(4, 0, 0)$
23. Right
25. Equilateral
27. $(2, 4, 1)$

Exercises 10.2 (page 406)
1. $(-\frac{2}{9}, \frac{1}{9}, \frac{2}{9})$, $\alpha = \text{Arccos}(-\frac{2}{9})$, $\beta = \text{Arccos}\frac{1}{9}$, $\gamma = \text{Arccos}\frac{2}{9}$
3. $(-\sqrt{2}/2, -\frac{1}{2}, \frac{1}{2})$, $\alpha = 135°$, $\beta = 120°$, $\gamma = 60°$
5. $(0, 0, -1)$, $\alpha = 90°$, $\beta = 90°$, $\gamma = 180°$
7. $(2, 2, -1)$
9. $(-6, 2, 9)$
11. $(2, -4, 2)$
13. $(\frac{2}{3}, \frac{2}{3}, -\frac{1}{3})$ or $(-\frac{2}{3}, -\frac{2}{3}, \frac{1}{3})$
15. $(-\frac{6}{11}, \frac{2}{11}, \frac{9}{11})$ or $(\frac{6}{11}, -\frac{2}{11}, -\frac{9}{11})$
17. $(\sqrt{6}/2, -\sqrt{6}/6, \sqrt{6}/2)$ or $(-\sqrt{6}/2, \sqrt{6}/6, -\sqrt{6}/2)$
19. $x = -1 + 2t, y = 2 + 2t, z = 3 - t$
21. $x = 2 - 6t, y = -4 + 2t, z = -7 + 9t$
23. $x = 3 + 2t, y = -1 - 4t, z = 1 + 2t$
25. $c = \frac{9}{5}$
27. $\pi/3$ or $2\pi/3$
29. $AB = (-9, -12, 6) = -3(BC)$
    $BC = (3, 4, -2)$
    $AC = (-6, -8, 4) = -2(BC)$
31. $\text{Cos}^{-1}(\frac{11}{29})$
33. $\text{Cos}^{-1}(\frac{2}{3})$
35. $60°$
37. $45°$
39. $x = 2 - 2t, y = 1 + 3t, z = -3 + t$
41. $x = 2 + t, y = 1, z = -3 - t$
43. $x = 1 + t, y = -4 + 2t, z = 2 - 4t$
45. $x = 1 + t, y = -4 + 2t, z = 2 + t$

Exercises 10.3 (page 413)
1. 15, $\cos \alpha = \frac{2}{3}$, $\cos \beta = -\frac{2}{15}$, $\cos \gamma = \frac{11}{15}$
3. 9, $\cos \alpha = \frac{1}{9}$, $\cos \beta = -\frac{4}{9}$, $\cos \gamma = -\frac{8}{9}$

EXERCISES 10.4

5. $7, \cos \alpha = \frac{3}{7}, \cos \beta = \frac{6}{7}, \cos \gamma = -\frac{2}{7}$
7. $3, \cos \alpha = -\frac{2}{3}, \cos \beta = -\frac{1}{3}, \cos \gamma = \frac{2}{3}$
9. $\sqrt{29}, \cos \alpha = 4/\sqrt{29}, \cos \beta = 3/\sqrt{29}, \cos \gamma = -2/\sqrt{29}$
11. $4\mathbf{i} + 5\mathbf{j} + 7\mathbf{k}, -1, -\mathbf{i} + 5\mathbf{j} - 3\mathbf{k}$
13. $12\mathbf{i} + 7\mathbf{j} - \mathbf{k}, -7, -3\mathbf{i} + 5\mathbf{j} - \mathbf{k}$
15. $\pi/2$
17. $\pi/4$
19. $(1/\sqrt{66}, -4/\sqrt{66}, -7/\sqrt{66})$
21. $(2/\sqrt{5}, 1/\sqrt{5}, 0)$
23. $-\frac{2}{9}$
25. $\frac{15}{11}$
27. $\sqrt{5}$
29. $\sqrt{5}/2$
31. $4\sqrt{3}\mathbf{i} + 4\sqrt{3}\mathbf{j} + 4\sqrt{3}\mathbf{k}$
33. $|\mathbf{a} \times \mathbf{b}| = |\mathbf{a}||\mathbf{b}|\sin \theta = 0 \rightarrow \sin \theta = 0$ and $\mathbf{a}$ and $\mathbf{b}$ are $||$
35. $-4\mathbf{i} + 2\mathbf{j}$
37. -2
39. $-3\mathbf{i} + 2\mathbf{j} - \mathbf{k}$
41. 2

Exercises 10.4 (page 419)

1. $x + y - z = 1$
3. $3x - 4y - z = 3$
5. $(3/7, -2/7, 6/7)$
7. $(1/\sqrt{3}, -1/\sqrt{3}, 1/\sqrt{3})$
9. $x = 3, y = 2, z = 6$

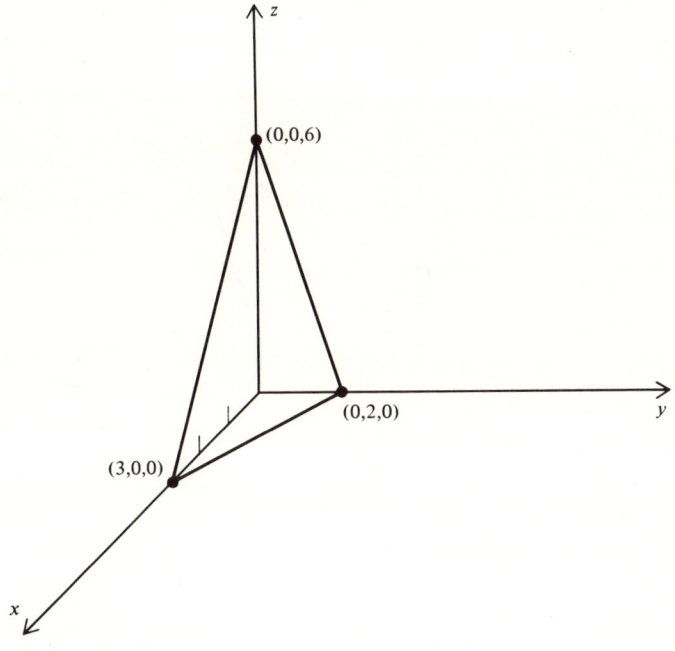

11. $x = 0$, $y = 0$, $z = k$, where k is any real number.

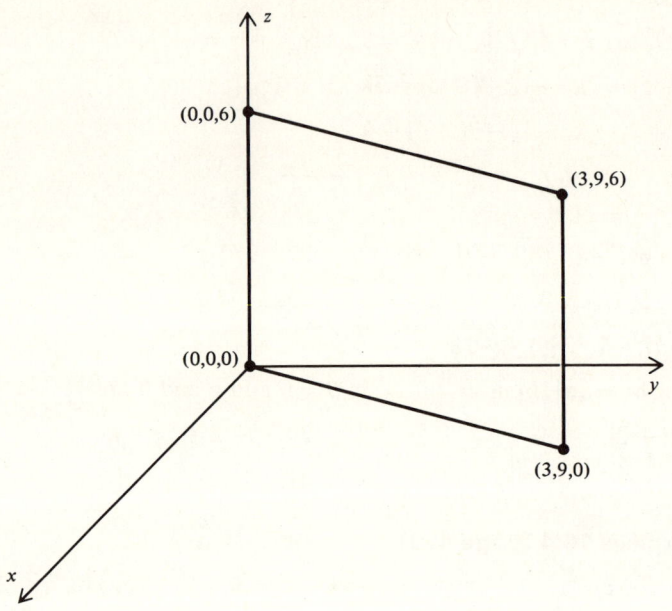

13. $x = 6$, $y = -4$, $z = 3$

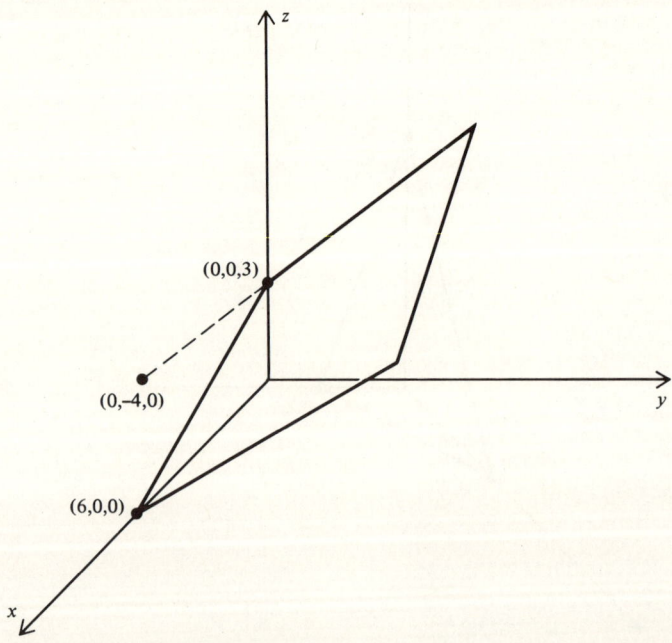

EXERCISES 10.5

15. (a) $x = -3$ (b) $y = -2$ (c) $z = -1$

17. $x - 2y - 2z = 1$ 19. $3x + y - z + 10 = 0$

21. $x = 5, y = t, z = 2t - 2$ (answer not unique).

23. $x = t, y = 2 - 2t, z = 2 + t$ (answer not unique).

25. $x = t, y = 2t, z = 3t$ 27. $x = 2, y = 3, z = t$

29. Any point on the plane has coordinates $(a, b, 0)$.
    The unit vector perpendicular to the plane is $(0, 0, 1)$.
    $$0 \cdot (x - a) + 0 \cdot (y - b) + 1 \cdot (z - 0) = 0 \rightarrow z = 0$$

31. Any point on the plane has coordinates $(0, b, c)$.
    Unit normal to the plane is $(1, 0, 0)$.
    $$1(x - 0) + 0(y - b) + 0(z - c) = 0 \rightarrow x = 0$$

33. Unit normal is $(0, 1, 0)$.
    $$0(x - a) + 1(y - k) + 0(z - c) = 0 \rightarrow y = k$$

Exercises 10.5 (page 425)

1. $(x - 1)^2 + (y - 2)^2 + (z - 1)^2 = 16$

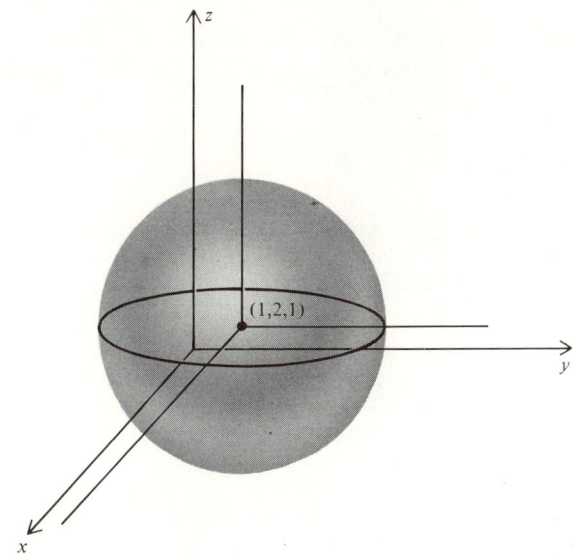

3. $(x+1)^2 + (y+1)^2 + (z-1)^2 = 49$

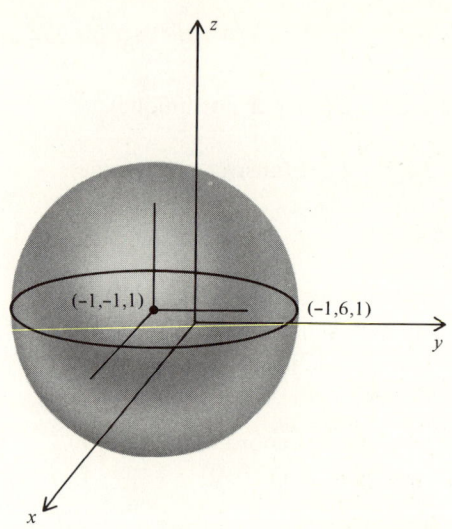

5. $(x+3)^2 + (y-1)^2 + (z-4)^2 = 36$
   $C: (-3, 1, 4)$ and $r = 6$

7. $(x+1)^2 + (y-3)^2 + (z+11)^2 = 9$
   $C: (-1, 3, -11)$ and $r = 3$

9. Right circular cylinder.

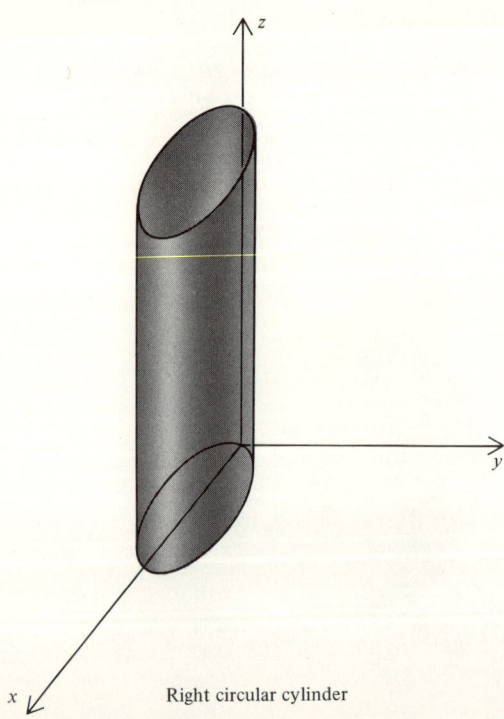

Right circular cylinder

EXERCISES 10.5

11. Parabolic cylinder.

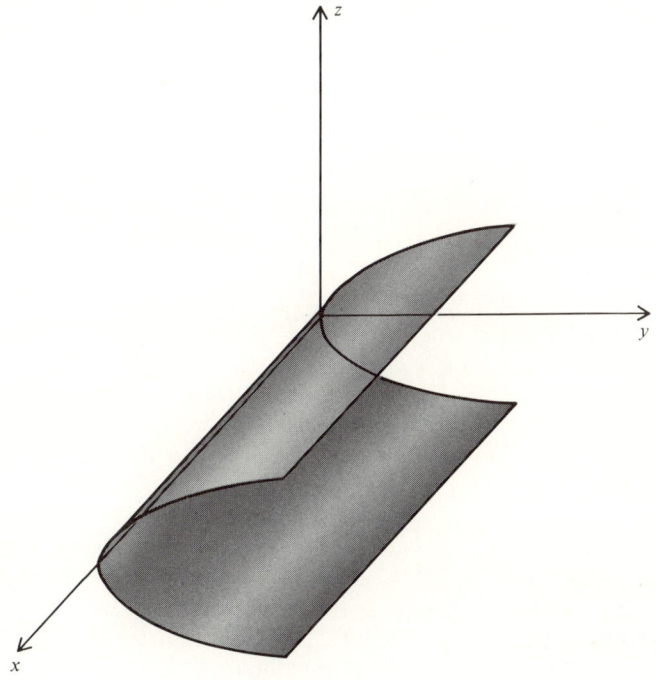

Parabolic cylinder

13. $(x-2)^2 + (y+1)^2 = 4$. Right circular cylinder.

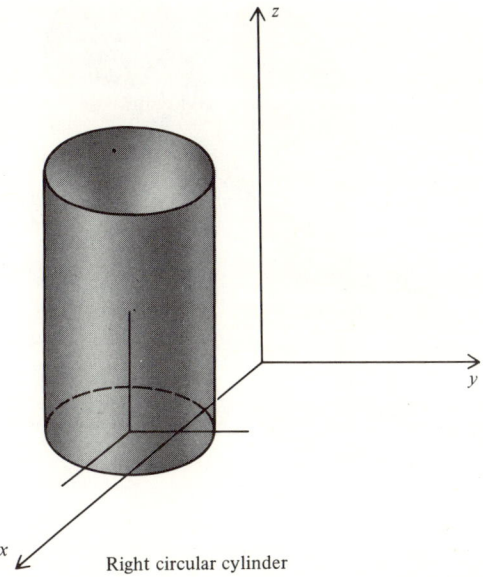

Right circular cylinder

15. Right circular cone.

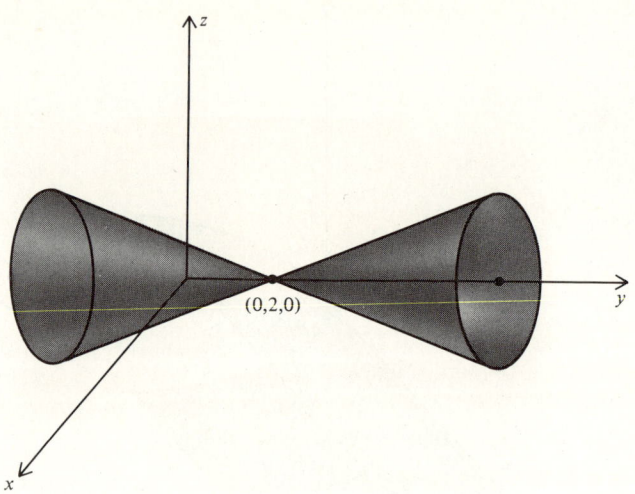

Right circular cone

17. Right circular cone.

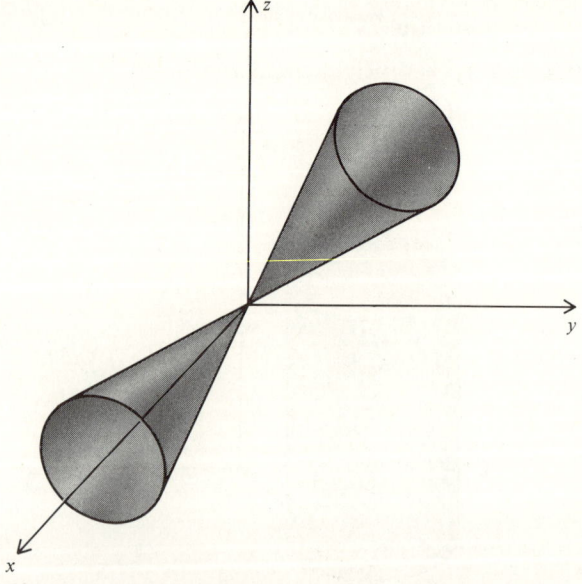

Right circular cone

19. Right elliptic cylinder.

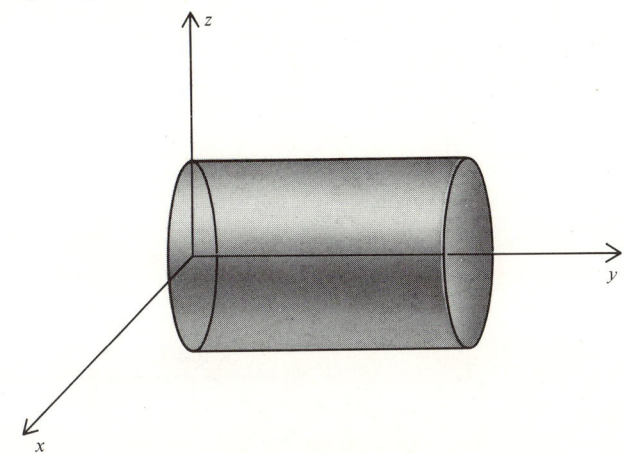

Right elliptic cylinder

Review Exercises (page 427)

1. $\sqrt{21}$, $(\frac{5}{2}, 3, 0)$
2. $\sqrt{66}$, $(\frac{3}{2}, \frac{5}{2}, 4)$
3. $2\sqrt{21}$, $(0, 0, 0)$
4. $\sqrt{a^2 + 2b^2 + 1}$, $(a/2, 0, \frac{1}{2})$
5. $(1/\sqrt{14}, 2/\sqrt{14}, 3/\sqrt{14})$
6. $(-1/\sqrt{14}, -2/\sqrt{14}, -3/\sqrt{14})$
7. $(1/\sqrt{5}, 0, 2/\sqrt{5})$
8. $(\frac{1}{3}, -\frac{2}{3}, \frac{2}{3})$
9. $(-\frac{2}{15}, \frac{11}{15}, -\frac{2}{3})$
10. $\overrightarrow{AB} = (-3, -6, 6) = -3(1, 2, -2)$
    $\overrightarrow{BC} = (5, 10, -10) = 5(1, 2, -2)$
11. $x = 2 + t$, $y = -3 - t$, $z = 2 - \sqrt{2}t$ (answer not unique).
12. (a) parallel (b) perpendicular (c) neither (d) perpendicular (e) neither (coincident).
13. (a) perpendicular (b) parallel (c) neither.
14. $\sqrt{14}$
15. -4
16. $(10, 4, 8)$
17. $(-10, -4, -8)$
18. $(-3, 1, -9)$
19. $(6, -2, -4)$
20. 22
21. $6\sqrt{5}$
22. $\text{Arccos } \frac{2}{7}$
23. $\text{Arccos } \sqrt{21}/14$
24. $2x + y + z = 7$
25. $2x - y - 3z = 5$
26. $(\frac{9}{11}, \frac{2}{11}, \frac{6}{11})$

27. $x = 12$, $y = 4$, $z = 6$

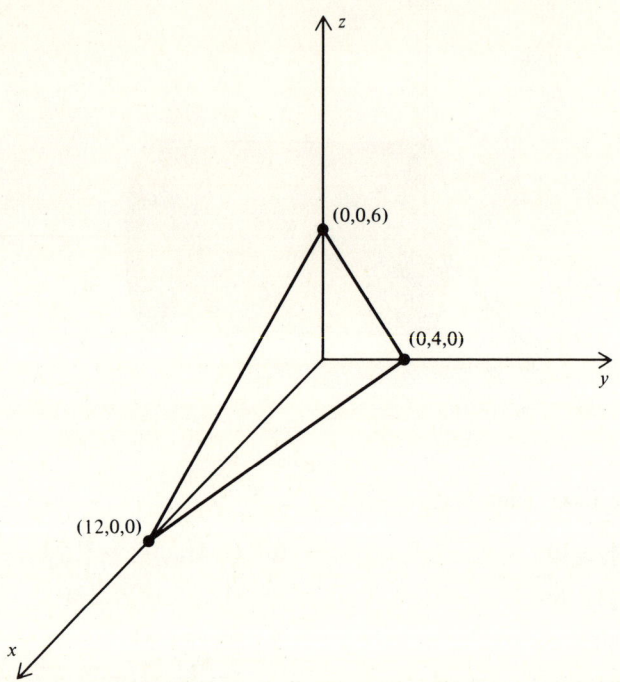

28. $x = -2$, $y = 4$, $z = 2$

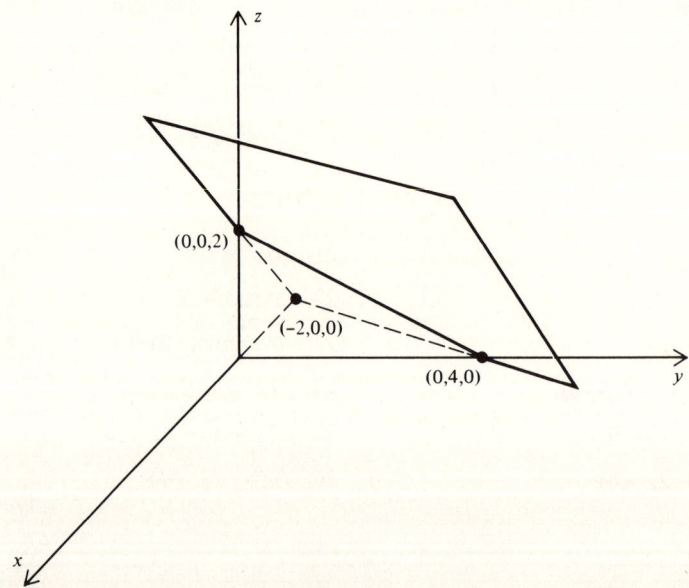

29. No x-intercept, $y = 5$, $z = 2$

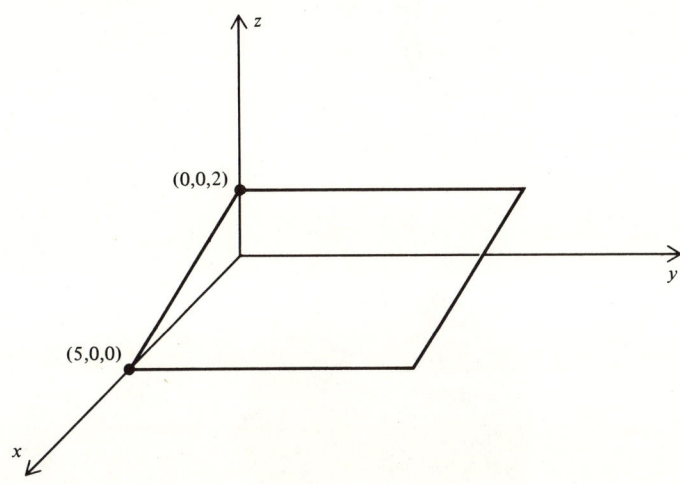

30. $x = 3$, no y- or z-intercepts

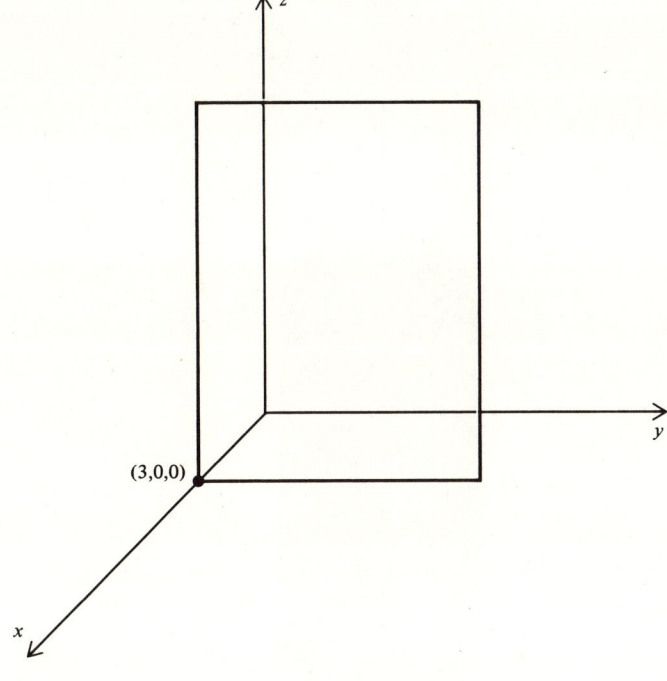

31. $x + 3y - 4z = 5$
32. $x = t$, $y = 7t + 13$, $z = 9t + 17$
33. $x = 2 + t$, $y = 1 - 3t$, $z = 5 - 6t$

34. $\sqrt{3}/9\,x + 5\sqrt{3}/9\,y - \sqrt{3}/9\,z = 1$
35. $(x-1)^2 + (y-3)^2 + (z-2)^2 = 25$
36. $(x-1)^2 + (y+1)^2 + (z-1)^2 = 9$
37. $(x-2)^2 + (y+3)^2 + (z-2)^2 = 49;\ C: (2, -3, 2)$ and $r = 7$
38. Sphere.

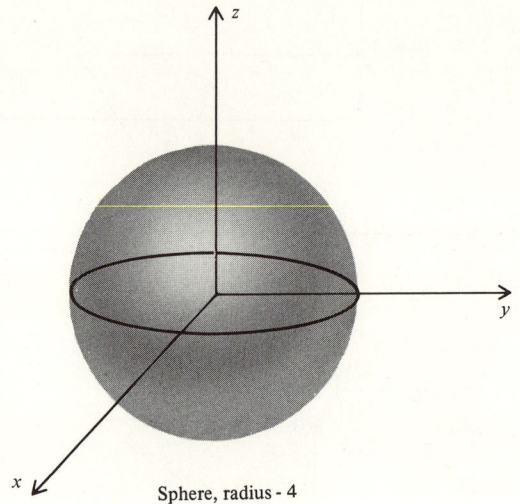

Sphere, radius - 4

39. Right circular cylinder.

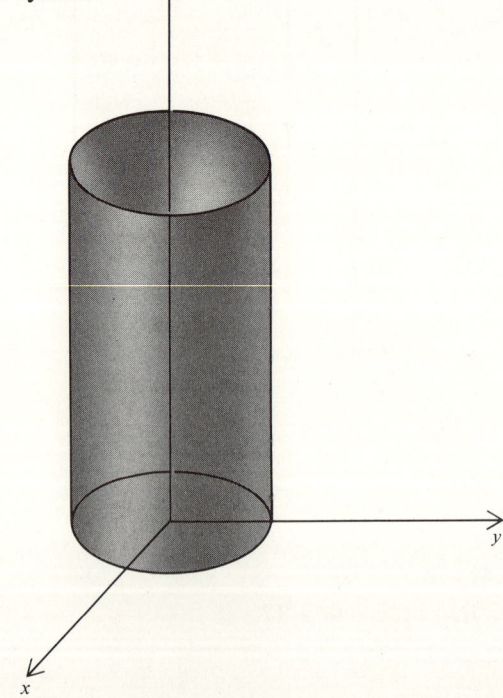

Right circular cylinder

REVIEW EXERCISES

40. Elliptic cylinder.

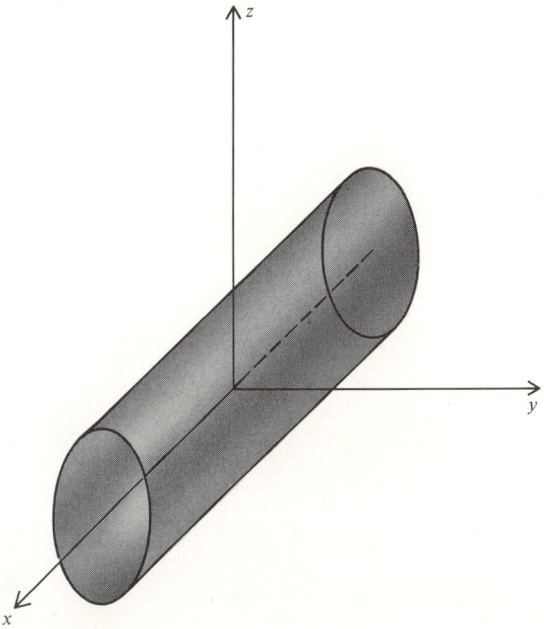

41. Right circular cone.

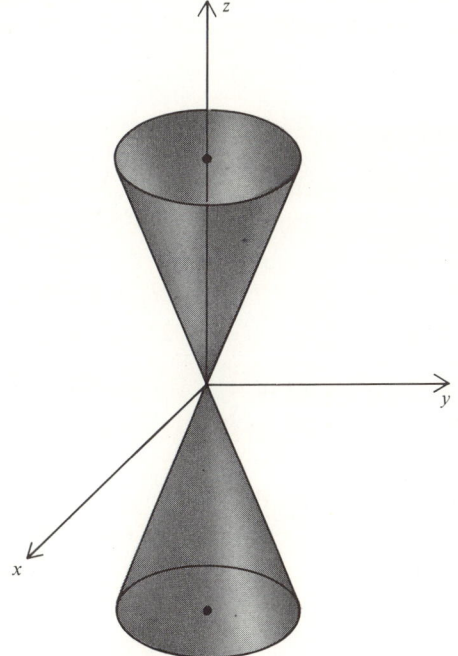

Right circular cone

42.

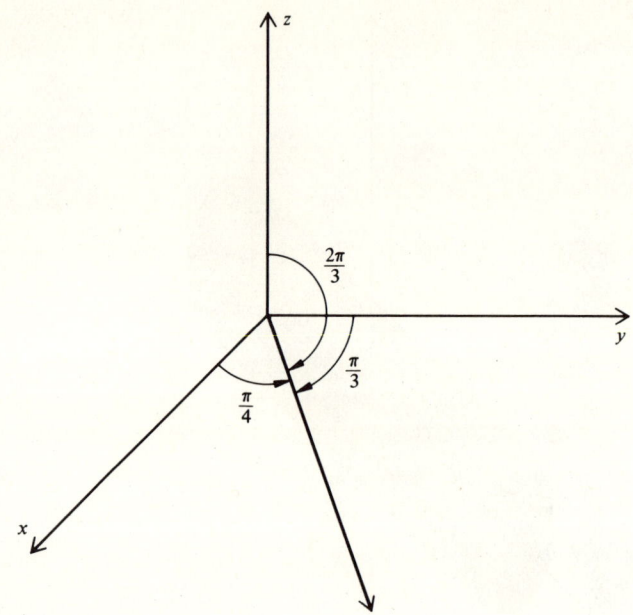

CHAPTER 11

Exercises 11.1 (page 436)
1. $(-1, 1, -2)$
3. $(\frac{1}{6}, \frac{5}{6}, \frac{2}{3})$
5. $(a, -4a + 13, 3a - 8)$
7. $(0, 0, 0)$
9. $\left(a, 8 - 9a, \dfrac{11 - 13a}{2}\right)$
11. $\emptyset$
13. $(1, -2, -1, 3)$
15. $(5, 1, -2, -3)$
17. $\left(x, \dfrac{4x}{5}, \dfrac{7x + 5}{5}\right)$
19. $\left(\dfrac{w + 7}{4}, 3 - 4w, \dfrac{3 - 7w}{4}, w\right)$
21. $(2, -1, 3)$
23. $(2, \frac{5}{2}, \frac{3}{2})$

Exercises 11.2 (page 442)
1. $\begin{bmatrix} 4 & 5 \\ -7 & -1 \end{bmatrix}$
3. Undefined, dimensions are different.
5. $\begin{bmatrix} 5 & 3 & -4 \\ 2 & -3 & 0 \\ 3 & -1 & -1 \end{bmatrix}$
7. $\begin{bmatrix} 6 & 15 & 14 \end{bmatrix}$
9. $\begin{bmatrix} 6 & -9 & 12 \end{bmatrix}$
11. $\begin{bmatrix} -2 & 10 \\ 16 & 11 \end{bmatrix}$

EXERCISES 11.3

13. Undefined, dimensions not compatible.
15. $\begin{bmatrix} a-2b+c+3d & 7 \\ 2a & +5c+4d & -1 \end{bmatrix}$
17. $\begin{bmatrix} a & b & c \\ d & e & f \\ g & h & i \end{bmatrix}$
19. Undefined, different dimensions.
21. $\begin{bmatrix} 6 & x+2 & 4 & 0 \\ y+3 & 0 & -1 & 2 \end{bmatrix}$
23. $w = -2, x = 3, y = 1, z = 5$
25. $x = -3, y = \frac{12}{5}$
27. $x = \frac{9}{2}, y = -2, z = -\frac{15}{4}$
29. $\begin{bmatrix} -1 & 1 \\ -2 & -1 \end{bmatrix}$
31. $\begin{bmatrix} 5 & 1 \\ 1 & 2 \end{bmatrix}$
33. $\begin{bmatrix} -5 & -1 \\ -1 & -2 \end{bmatrix}$
35. $\begin{bmatrix} \frac{2}{9} & -\frac{1}{9} \\ -\frac{1}{9} & \frac{5}{9} \end{bmatrix}$
37. $\begin{bmatrix} 1 & -1 \\ 1 & 0 \end{bmatrix}$
39. Unequal.
41. $AB = \begin{bmatrix} 0 & 0 \\ 0 & 0 \end{bmatrix}$, no.
43. $A + B = \begin{bmatrix} a+1 & b+1 \\ c-1 & d+1 \end{bmatrix} = \begin{bmatrix} 1+a & 1+b \\ -1+c & 1+d \end{bmatrix} = B + A$
45. $(A + B) + C = \begin{bmatrix} a+2 & b+2 \\ c & d \end{bmatrix} = A + (B + C)$

$(AB)C = \begin{bmatrix} 2a & -2b \\ 2c & -2d \end{bmatrix} = A(BC)$

47. $I = \begin{bmatrix} 1 & 0 \\ 0 & 1 \end{bmatrix}$

49. $A^{-1} = \begin{bmatrix} \frac{d}{ad-bc} & \frac{-b}{ad-bc} \\ \frac{-c}{ad-bc} & \frac{a}{ad-bc} \end{bmatrix}$. Yes, $A^{-1}A = AA^{-1}$.

51. $Y = A^{-1}B$ if $ad - bc \neq 0$

$\frac{B}{A} = A^{-1}B$ if $ad - bc \neq 0$

53. $x = -\pi/6, y = 3, z = 32, w = -3$
55. $x = \pm 3, y = \pm 2$

Exercises 11.3 (page 450)

1. $(-4, -8)$
3. $(3, -2)$
5. $(3, 6, 12)$
7. $(2, 3, 4)$
9. $\emptyset$
11. $(\frac{1}{2}, \frac{1}{4}, \frac{3}{4})$
13. $\emptyset$
15. $(3z - 4, 4z - 10, z)$
17. $(4, -4, 5, -5)$
19. $(\frac{7}{2}, -\frac{13}{4}, \frac{3}{4}, -3)$

Exercises 11.4 (page 457)

1. 7
3. -36
5. $\begin{vmatrix} -2 & 7 & 7 \\ 1 & 0 & 0 \\ 2 & -1 & -3 \end{vmatrix} = 14$
7. $\begin{vmatrix} 1 & 2 & 0 \\ 1 & 4 & 0 \\ 1 & -1 & 2 \end{vmatrix} = 4$
9. $\begin{vmatrix} 1 & 2 & 3 \\ 0 & 2 & 6 \\ 0 & 0 & 6 \end{vmatrix}$
11. $x = -3$
13. $x = 8$
15. $(x-y)\begin{vmatrix} 1 & 1 & 2 \\ 0 & x & x \\ x & y & x+y \end{vmatrix} = (x-y)\begin{vmatrix} 1 & 1 & 1 \\ 0 & x & 0 \\ x & y & x \end{vmatrix} = (x-y)(0) = 0$
17. $\begin{vmatrix} 1 & 1 & 1 & 1 \\ 0 & 1 & 2 & 3 \\ 0 & 0 & 2 & 6 \\ 0 & 0 & 0 & 6 \end{vmatrix} = 12$
19. 92
21. $\begin{vmatrix} 1 & 1 & 1 & 1 \\ 0 & a-1 & a^2-1 & a^3-1 \\ 0 & b-1 & b^2-1 & b^3-1 \\ 0 & c-1 & c^2-1 & c^3-1 \end{vmatrix}$

$= (a-1)(b-1)(c-1)\begin{vmatrix} 1 & a+1 & a^2+a+1 \\ 1 & b+1 & b^2+b+1 \\ 1 & c+1 & c^2+c+1 \end{vmatrix}$

$= (a-1)(b-1)(c-1)\begin{vmatrix} 1 & a & a^2 \\ 1 & b & b^2 \\ 1 & c & c^2 \end{vmatrix}$

$= (a-1)(b-1)(c-1)\begin{vmatrix} 1 & a & a^2 \\ 0 & b-a & b^2-a^2 \\ 0 & c-a & c^2-a^2 \end{vmatrix}$

$= (a-1)(b-1)(c-1)(b-a)(c-a)\begin{vmatrix} 1 & b+a \\ 1 & c+a \end{vmatrix}$

$= (a-1)(b-1)(c-1)(b-a)(c-a)(c-b)$

23. $f(x) = k(x-2)(x-3)(x-4)$
    $f(0) = 48 = -24k$, and $k = -2$
    $f(x) = -2(x-2)(x-3)(x-4)$
    $f(1) = 12$

Exercises 11.5 (page 465)

1. $(1, 6, 3)$
3. $(\frac{3}{4}, -\frac{1}{2}, -\frac{2}{3})$
5. $(0, 2, 10)$
7. $(0, 5, 3, -1)$
9. $(-5, 10, 5)$
11. $\emptyset$
13. $\left(\dfrac{9 - 5z}{8}, \dfrac{23z - 11}{16}, z\right)$
15. $(-2, 3, 5, -4)$
17. 15 nickels, 25 dimes, 7 quarters.
19. $I_1 = 8, I_2 = 5, I_3 = 3$
21. 40 of P, 25 of Q, 6 of R
23. 12 minutes by cab, 36 minutes by bus, 2 hours 24 minutes by plane.
25. $\left(8 - 5y, y, \dfrac{2y + 4}{3}, \dfrac{8y + 10}{3}\right)$
    No, for $y = 3$, $x = -17$, impossible.
    Yes, for $y = 1$, $(3, 1, 2, 6)$.
29. $\{(x, 3x, 5x) \mid x \in R\}$
31. $k = 8$

Exercises 11.6 (page 475)

1. $(-2, 3), (3, -2)$
3. $(4.2, 1.4), (-4.2, -1.4)$
5. $(2.8, 2.1), (-2.8, -2.1)$
7. $(2, 0), (-1, 9)$
9. No real solutions.
11. $(3, 2), (3, -2), (-3, 2), (-3, -2)$
13. $(2, -3), (-2, -1)$
15. $(3 + i, -1 - 3i), (3 - i, -1 + 3i)$
17. $(6, -2), (-3, 4)$
19. $(\pm\sqrt[4]{12}, -2 + 2\sqrt{3}), (\pm i\sqrt[4]{12}, -2 - 2\sqrt{3})$
21. $(4, 6), (\frac{7}{3}, 1)$
23. $(-2, -2), (\frac{22}{9}, \frac{2}{9})$
25. $(2, 3), (2, -3), (-\frac{2}{9}, \frac{1}{3}), (-\frac{2}{9}, -\frac{1}{3})$
27. $(7, 4), (136/25, -\frac{19}{5})$
31. $(\pm\sqrt{2}, \pm 1)$
29. $(\pm 2, \pm\sqrt{5})$
33. $(4, 2), (-4, -2), (\sqrt{6}, -2\sqrt{6}), (-\sqrt{6}, 2\sqrt{6})$
35. $(\sqrt{3}, 2\sqrt{3}), (-\sqrt{3}, -2\sqrt{3})$
37. $(24, 5), (-20, -6)$
39. $(2 + i\sqrt{2}, 2 - i\sqrt{2}), (2 - i\sqrt{2}, 2 + i\sqrt{2}), (-3 + i\sqrt{7}, -3 - i\sqrt{7}), (-3 - i\sqrt{7}, -3 + i\sqrt{7})$
41. $(-1 + \sqrt{2}, -1 - \sqrt{2}), (-1 - \sqrt{2}, -1 + \sqrt{2}), (4 + i, 4 - i), (4 - i, 4 + i)$

Exercises 11.7 (page 480)
1. $(-\frac{5}{2}, 1)$ and $(0, 0)$
3. $(0, 0), (2, 1), (-1, -\frac{1}{8})$
5. $(0, 0), (2, 1)$
7. $(0, 1), (0, -1)$
9. $(-1, 2)$

Review Exercises (page 483)
1. $\begin{bmatrix} 6 & -2 & 3 \\ 10 & -3 & -7 \\ 0 & 8 & 10 \end{bmatrix}$

2. $\begin{bmatrix} 16 & 2x+5y+3z & \\ 6 & 3x+y+2z- & t \\ 21 & x+7y & -2t \end{bmatrix}$

3. $\begin{bmatrix} 1 & -3 & 2 & -5 \\ 0 & 1 & 0 & 2 \\ 0 & 0 & 7 & -7 \end{bmatrix}$

4. $\begin{vmatrix} 2 & -4 & 6 \\ 0 & -5 & 10 \\ 0 & 0 & 2 \end{vmatrix}$

5. $(-4, 2, 3)$
6. $(-11/4, 3/2, -33/4)$
7. $\{(10t, -24t, -13t) \mid t \text{ is any real or complex number}\}$
8. $P = 50, Q = 35, R = 40$
9. $\left(\frac{\pm 4\sqrt{170}}{17}, \frac{\pm 2\sqrt{119}}{17}\right)$
10. $(4, 3), (4, -3), (-5, 0)$
11. $(2i, \frac{3}{2}), (-2i, \frac{3}{2}), (2\sqrt{6}, -11/2), (-2\sqrt{6}, -11/2)$
12. $(2, -1), (-2, 1), (17/9, -11/9), (-17/9, 11/9)$
13. 40 mph, 20 hours.
14. 10 in. by 20 in.
15. 6 and $\frac{1}{3}$
16. 8 ft by 15 ft.
17. $k = \pm 10$

CHAPTER 12

Exercises 12.1 (page 488)
1. 1, 3, 5, 7, 9
3. 0.5, 0.05, 0.005, 0.0005, 0.00005
5. 10, 20, 40, 80, 160
7. 1, 2, 3, 4, 5
9. $-\frac{3}{2}, \frac{3}{4}, -\frac{3}{8}, \frac{3}{16}, -\frac{3}{32}$
11. 0.43, 0.0043, 0.000043, 0.00000043, 0.0000000043

EXERCISES 12.3

13. $6n$, $30 + 36$ (answer not unique).
15. $27 - 4n$, $7 + 3$ (answer not unique).
17. $5n$, $25 + 30$ (answer not unique).
19. $\dfrac{(-1)^{n+1}}{n^2}$, $\dfrac{1}{25} - \dfrac{1}{36}$ (answer not unique).
21. $n(n + 1)$, $(5 \times 6) + (6 \times 7)$ (answer not unique).
23. $5(-\tfrac{1}{3})^{n-1}$, $\dfrac{5}{81} - \dfrac{5}{243}$ (answer not unique).
25. 60
27. 68
29. 50
31. 115/144
33. 40
35. 100/27
37. $\sum_{n=1}^{7} n$
39. $\sum_{n=1}^{9} 2n - 1$
41. $\sum_{n=1}^{\infty} 50(-0.1)^n$
43. $1 + \sum_{n=1}^{\infty} \dfrac{2^n - 1}{2^n}$ or $\sum_{n=0}^{\infty} \dfrac{2^n - 1}{2^n}$
45. $5 + 10 + 17 + 26 + 37 + 50$
47. $1 + \tfrac{1}{3} + \tfrac{1}{5} + \tfrac{1}{7} + \ldots$
49. $-1 + 1/\sqrt{2} - 1/\sqrt{3} + 1/2 - 1/\sqrt{5} + \ldots$

Exercises 12.2 (page 491)

1. -57
3. $10 - 15\sqrt{2}$
5. 6.75
7. 20.5
9. $-22\tfrac{2}{3}$
11. 82.5, 85, 87.5
13. 19, 14, 9, 4, -1
15. 71.5
17. 5, 0, -5
19. $3\tfrac{1}{3}, 3\tfrac{2}{3}, 4, 4\tfrac{1}{3}, 4\tfrac{2}{3}, 5, 5\tfrac{1}{3}, 5\tfrac{2}{3}$
21. 2550
23. -36
25. $3n(n + 1)/2$
27. 1325/2
29. 245
31. 14
33. $2105
35. 6400, $16k^2$
37. (a) 1/32 (b) $2ab/(a + b)$ (c) 24 (d) 137/60

Exercises 12.3 (page 495)

1. 256/243
3. -390, 625/4096
5. $\sqrt{2}/2$
7. 1/128
9. 36, 24
11. 6 or -6
13. $6\sqrt{5}$ or $-6\sqrt{5}$
15. $\sqrt[4]{10}, \sqrt[4]{100}, \sqrt[4]{1000}$, or $-\sqrt[4]{10}, \sqrt[4]{100}, -\sqrt[4]{1000}$

17. 41, 496
19. 2222
21. $-1389/32$
23. $1820/729$
25. $25/2$
27. $2/275$
29. Sum does not exist, $r = -\frac{3}{2}$, $|-\frac{3}{2}| > 1$.
31. 300
33. $\frac{4}{9}$
35. $8369/333$
37. 360 in.
39. 8190
41. 5
43. (a) 2.56 percent (b) 6 strokes
45. $m = \dfrac{a+b}{2}, g = \sqrt{ab}, h = \dfrac{2ab}{a+b}, g^2 = ab,$

$mh = \dfrac{a+b}{2} \cdot \dfrac{2ab}{a+b} = ab, g^2 = mh$

47. $(a-b)^2 > 0$
$a^2 - 2ab + b^2 > 0$
$a^2 + b^2 > 2ab$
$a^2 + 2ab + b^2 > 4ab$
$(a+b)^2 > 4ab$
$|a+b| > 2\sqrt{ab}$
$\dfrac{a+b}{2} > \sqrt{ab}$

Exercises 12.4 (page 501)

1. $x^7 + 14x^6 + 84x^5 + 280x^4 + 560x^3 + 672x^2 + 448x + 128$
3. $x^{10} + 10x^8 + 40x^6 + 80x^4 + 80x^2 + 32$
5. $a^{12} - 6a^{10}b^2 + 15a^8b^4 - 20a^6b^6 + 15a^4b^8 - 6a^2b^{10} + b^{12}$
7. $64t^{12} - 576t^{10} + 2160t^8 - 4320t^6 + 4860t^4 - 2916t^2 + 729$
9. $17010x^6$
11. $-1792y^5$
13. $-\dfrac{969}{2}x^{30}$
15. $84r^{18}/s^9$
17. $x^{50} + 50x^{49}y + 1225x^{48}y^2 + 19{,}600x^{47}y^3$
19. $x^{84} - 21x^{82} + \dfrac{861}{4}x^{80} - 1435x^{78}$
21. $x^{34} - 34x^{33}y + 561x^{32}y^2 - 5984x^{31}y^3$
23. $y^{30} - 15\sqrt{2}y^{28} + 210y^{26} - 910\sqrt{2}y^{24}$
25. 1.268
27. 0.85
29. 16

REVIEW EXERCISES

Exercises 12.5 (page 503)
1. $1 + x/3 - x^2/9 + 5x^3/81 - 10x^4/243$
3. $1 - x/2 - x^2/8 - x^3/16 - 5x^4/128$
5. $1 - 2x^2 + 3x^4 - 4x^6 + 5x^8$
7. $\sqrt{2}/2(1 - x/4 + 3x^2/32 - 5x^4/128 + 35x^6/2048)$
9. $x^{1/4}(1 - 1/4x^2 - 3/32x^4 - 7/128x^6 - 77/2048x^8)$
11. 1.007
13. 0.9980
15. 0.2581
17. 2.024
19. 0.8204

Review Exercises (page 505)
1. G, 25/4, 125/8
2. A, $(x + 4)/x$, $(x + 5)/x$
3. A, -24, -33
4. G, -24, 48
5. B, $-35/8$, $35/16$; $(1 - 1/2)^7$
6. G, $1/(3 - 2\sqrt{2})$, $1/(7 - 5\sqrt{2})$
7. A, log 128, log 512
8. G, $\log \sqrt[4]{3}$, $\log \sqrt[8]{3}$
9. G, $4i$, $4i - 4$
10. B, 540, $405\sqrt{2}$; $(\sqrt{2} + 3)^5$
11. $-\frac{1}{7}$
12. $-147{,}000x^3$
13. 62/11
14. 1.0617

15. (a) $\dfrac{1}{2(5)} = \dfrac{1}{(6+4)} = \dfrac{1}{10}$

   (b) $\dfrac{1}{2(5)} + \cdots + \dfrac{1}{(3k-1)(3k+2)} + \dfrac{1}{(3k+2)(3k+5)}$

   $= \dfrac{k}{(6k+4)} + \dfrac{1}{(3k+2)(3k+5)}$

   $= \dfrac{3k^2 + 5k + 2}{2(3k+2)(3k+5)} = \dfrac{k+1}{2(3k+5)} = \dfrac{k+1}{6k+10} = \dfrac{k+1}{6(k+1)+4}$

16. $x + y - 2\sqrt{xy} = (\sqrt{x} - \sqrt{y})^2 \geq 0$
    $x + y \geq 2\sqrt{xy}$ and $\sqrt{xy} \leq (x + y)/2$

17. $10,750
18. $8540

INDEX

A

Abscissa, 317
Absolute value, 151, 247
Additive inverse, 66, 110
Analytic proofs, 325
Angles, 220
Antilogarithm, 194
Arccosine, 226
Arcsine, 226
Arctangent, 226
Argument of complex number, 247
Arithmetic mean, 490
Arithmetic progressions, 489

B

Binomial equation, 254
Binomial theorem, 497
Bounds, upper and lower, 135–136

C

Cartesian coordinate system, 317
Cartesian product, 37, 59
Characteristic, 194

Circle, 339–340
Coefficient, 79, 269
Common logarithms, 193
Complement of set, 7–8
Complex numbers, 238
Completeness axiom, 137
Composite number, 52
Cones, 424
Conjugate imaginary roots, 280
Conjugate surd roots, 281
Conjugates, 243
Constant, 12
Coordinates, 315, 367, 392
Cosecant, 219
Cosine, 205
Cotangent, 219
Counting numbers, 42
Cramer's rule, 461
Cross product, 409
Cubic equation, 294
Cylinders, 422

D

Degree, 221

Degree of polynomial, 79, 269
DeMoivre's theorem, 253
Descartes' rule of signs, 287
Determinants, 451
Direction cosines, 401
Direction numbers, 402
Direction of vector, 381, 408
Discriminant of cubic, 300
Discriminant of quadratic, 271
Distance, 315, 392
Distance formula, 317, 394
Divisible, 52
Division algorithm, 54, 60, 83, 273
Divisor, 52
Domain, 37, 59
Dot product, 385, 408

E

Eccentricity, 353, 354
Ellipse, 345
Empty set, 3
Equal sets, 4
Equality, 38, 59
Equivalence relation, 38, 60
Equivalent sets, 4
Equivalent statements, 21, 122
Exponential equations, 198
Exponential functions, 184
Exponents, 155
Extent, 334

F

Factor, 52, 80
Factor theorem, 89, 274
Factoring, 80
Field, 106
Field axioms, 102, 137
Finite set, 4
Function, 39, 59
Fundamental theorem of algebra, 272
Fundamental theorem of arithmetic, 53, 60

G

GCD, 55
Gauss' reduction method, 448
General form of line, 331
Geometric mean, 493
Geometric progressions, 492
Group, 72–73

H

Hyperbola, 347

I

Identity, addition, 66, 103, 109, 137
  multiplication, 48, 66, 103, 110, 137
Identity function, 186
Inclination, 321
Inequalities, 142
Infinite set, 4
Inner product, 385, 408
Integers, 63
Integral domain, 73
Integral exponents, 155
Intercepts, 333, 334, 416
Intersection of sets, 8
Inverse, additive, 66, 110, 137
  multiplicative, 103, 110, 137
Inverse function, 187
Inverse trigonometric functions, 225
Irrational numbers, 131, 136

L

LCM, 56
Leading coefficient, 79, 269
Linear interpolation, 195
Linear systems, 430
Logarithmic equations, 199
Logarithmic function, 189
Logic, 11
Lower bounds of real roots, 284

INDEX

M

Magnitude of vector, 381, 407
Mantissa, 194
Mathematical induction, 44
Matrices, 438
Matrix, 438
Midpoint formula, 318, 397
Minor of determinant, 452
Modulus of complex number, 247
Multiple, 52
Multiplicative inverse, 103, 110
Multiplicity of root, 275

N

Natural logarithm, 197
Natural numbers, 34
Negative product theorem, 146
Normal equation of plane, 421
Normal to plane, 414

O

One-to-one function, 187
Open statement, 12
Operation, 40, 59
Opposite, 66, 103
Order axioms, 103, 137
Order relation, 48
Order theorems, 140
Ordered pair, 36
Ordinate, 317
Outer product, 409

P

Parabola, 341
Parallel lines, 322, 399
Parallel planes, 417
Parametric equations, 375, 402
Partial fractions, 116
Pascal's triangle, 500
Peano postulates, 42, 60
Perpendicular lines, 323, 404

Perpendicular planes, 417
Planes, 414
Point-slope form of line, 329
Polar coordinates, 367
Polar form of complex number, 246
Polynomial, 79, 269
Position vector, 382, 409
Positive product theorem, 146
Prime number, 52

Q

Quadratic formula, 270
Quadratic systems, 468
Quantifiers, 13
Quartic equation, 306

R

Radian, 221
Radical equations, 171
Radical inequalities, 171
Radicals, 162
Range, 38, 59
Rational equations, 122
Rational exponents, 166
Rational function, 108
Rational roots theorem, 278
Rationals, 98
Real number line, 132
Reals, 131
Reciprocal, 103
Rectangular coordinates, 315, 392
Reference angle, 222
Relation, 37, 38, 59
Remainder theorem, 88, 274
Resultant of equations, 477
Resultant of vectors, 382
Ring, 73
Root, 90, 273
Rotation of axes, 360

S

Secant, 219
Sequences, 485

Series, 486
Sets, 3
Sigma notation, 487
Sine, 205
Slope, 320
Slope-intercept form of line, 330
Spheres, 421
Subset, 3, 4
Symmetry, 336
Synthetic division, 84

T

Tangent, 219
Tautologies, 25
Translation of axes, 356
Trichotomy, 50, 60, 103
Trigonometric functions, 205
Truth sets, 14

U

Union of sets, 7, 8
Universal set, 3
Upper bounds for real roots, 284

V

Variable, 12
Vectors, three-dimensional, 407
  two-dimensional, 380
Venn diagrams, 5
Weierstrass zero theorem, 290

W

Well-ordering, 50

Z

Zero of polynomial, 90